REACTION MECHANISMS AND CONTROL PROPERTIES OF PHOSPHOTRANSFERASES

Internationales Symposium Reinhardsbrunn Mai 1971

AKADEMIE-VERLAG · BERLIN
1973

Honorary President: Prof. Dr. Dr. Drs. h. c. Karl Lohmann
Präsident der Deutschen Gesellschaft
für experimentelle Medizin

Chairmen: Academician Prof. Dr. S. E. Severin
President of the All-Union Biochemical Society of USSR
Prof. Dr. E. Hofmann
Vorsitzender der Biochemischen Gesellschaft der DDR

Scientific Secretary: H.-J. Böhme

Erschienen im Akademie-Verlag GmbH, 108 Berlin, Leipziger Str. 3–4

Lizenznummer: 202 · 100/541/73
Offsetdruck: VEB Druckerei „Thomas Müntzer", 582 Bad Langensalza
Bestellnummer 5996 • ES 18 G 1 · EDV: 761 706 7
Printed in the German Democratic Republic
68,–

WELCOME ADDRESS by E. HOFMANN, Leipzig

Dear and distinguished participants of the first joint symposium of the Allunion-Biochemical Society of USSR and the Biochemical Society of GDR!
It is a very great honour and privilege for me to welcome you in our country, in the German Democratic Republic, especially in this pleasant place here in Thuringia. I greet you on behalf of the committee and the members of the Biochemical Society of GDR.
In the first place I greet cordially our friends and colleagues from USSR, particularly the President of the Allunion-Biochemical Society of USSR, Professor SEVERIN from Moscow. Owing to his great activity in all stages of organization the first joint symposium of the Biochemical Societies of USSR and GDR actually could take place. We welcome all members of the delegation of USSR. We hope indeed that all of you will have enjoyable and interesting days in our German Democratic Republic, a country which is tightly connected with the Soviet-Union in political, scientific and economical relations and in the struggle for peace which is essential for the success of all efforts of people all over the world. We are very satisfied and glad about your participation and we are sure that this metting will improve our friendship and contribute to a still closer collaboration of biochemists between our two countries.
We are happy to greet distinguished and world-famous scientists from others countries too, who followed our invitation to participate and to give lectures about their experimental and theoretical work, in fields which are closely connected with the topics of this meeting.
It is an honour for me to welcome outstanding scientists from the other socialistic countries like Czechoslowakia and Hungary.
We greet cordially our friends and colleagues from the German Federal Republic, most of them are not for the first time in GDR, and all of them belong to the large group of people who are welcome in our country everytime.
Last not least we are deeply impressed that also international wellknown scientists from the other westeuropean countries followed our invitation. We cordially greet

our friends and guests from The Netherlands, Belgium, France, Spain, Denmark and Austria.
We hope that all of you will have in this week interesting and enjoyable days, filled with stimulating discussions. In addition, we are convinced that this lovely place and its surroundings will encourage us to improve our personal contacts during the days and the evenings.
The subject of this symposium covers a large field, the enzymatic transfer of phosphate- and phosphoryl-groups, but also the transfer of other phosphorous-containing residues like pyrophosphate and nucleotides, especially having in mind the polymerases catalyzing the synthesis of nucleic acids. In addition we are able to discuss also the role of pyridoxalphosphate in phosphorylase and in other enzymes. A round-table-conversation will be devoted to the interactions of subunits in proteins and in general to protein-protein-interactions. Especially association-dissociation-reactions of proteins should be discussed.
We are convinced that the wellknown high standard of the scientists giving lectures and the selected audience will be the best guarantee for a successfull symposium, and we hope indeed that we have organized the programme in such a way that all participants will realize this for themselves.
Our knowledge about metabolic transfer-reactions started with the splendid discoveries in MEYERHOFs laboratory, performed by Professor KARL LOHMANN, which are concerned with phosphate transfer reactions. It was KARL LOHMANN, the honorary president of our meeting, and discoverer of ATP who first described the reversible phosphate transfer from ATP to creatine in muscle catalyzed by creatine phosphokinase, a reaction today still called as LOHMANN reaction.
The famous and pioneering contributions of biochemists from Soviet-Union in the field of transfer reactions are well known to the international scientific community. BRAUNSTEIN and KRITSMAN discovered 1937 the transfer of amino groups between amino and ketoacids. ENGELHARDT founded the concept of respiratory chain phosphorylation working with avian erythrocytes and BELITZER and TSIBAKOWA discovered independently from KALCKAR 1939 the formation of ATP in the respiratory chain. Another most important disclosure from this time is the discovery of the ATPase nature of myosin and the development of Mechano-biochemistry by ENGELHARDT. Especially in this field the combined endeavours of sovietic, hungarian and german workers in the classical era of biochemistry should be mentioned. So we see, that the subject of this symposium is not an accident, but it has its roots in the brilliant history of biochemistry in our countries.

Now, ladies and gentlemen, permit me to tell you something about this place and this region.

The history of Thuringia as one of the central parts of Germany is very eventful. It consisted until the revolution in 1918 of not less than nearly ten minor states, especially small counties and duchies. Reinhardsbrunn castle belonged to the dukedom of Saxonia-Coburg-Gotha. For the dukes it served as Summer residence and hunting castle.

In the early middleage here in Reinhardsbrunn existed a Benedictine monastery of the house of LUDOWINGER. It has been destroyed completely during the 30 year war.

The thuringian country, especially Erfurt, the Wartburg, Weimar and Jena played an important role in German literature and culture.

Erfurt had an old university founded 1392, that is 17 years before the Leipzig university has been founded, however the Erfurt university has been closed in 1816 under the influence and the force of the foreign rule of Napoleon. In the Erfurt university MARTIN LUTHER studied. In the 16^{1h} century this university was one of the most important central places of German Humanism. Today Erfurt is the capital of this district, it has a Medical Academy and a Pedagogical High School. Gothic cathedrals and small old houses, narrow streets and pithoresque sceneries in the city remember the visitor to the old history of this town.

Not far from here is the Wartburg, wellknown from "Sängerkrieg" ("struggle of minstrels") and from the famous opera "Tannhäuser" by RICHARD WAGNER. This castle served as hiding-place of MARTIN LUTHER, the great german reformator and humanist. In this castle MARTIN LUTHER translated the Bible in the 1520^{ieth} and layed the most important fundaments for the development of the german literary language.

In the other direction from here we come to Weimar, the town of german classical literature and culture, where GOETHE and SCHILLER, as well as WIELAND and HERDER lived. Later on, world famous composers worked in Weimar, like FRANZ LISZT and RICHARD STRAUSS. Until today Weimar respires the air of this famous tradition. On Wednesday, you can visit the abundant and splendid houses of JOHANN WOLFGANG von GOETHE and the more humble home of FRIEDRICH SCHILLER.

In preparation of this introductory lecture I read again the inaugural lecture of FRIEDRICH SCHILLER as unsalaried professor of history at the university of Jena, now called Friedrich-Schiller-University: "Was heißt und zu welchem

Ende studiert man Universalgeschichte?" (Translated: "What means and to which intention people study universal history?"). In this lecutre, which did not lose its significance also for our time of its high technical standard and its rapid growth of knowledge, SCHILLER divided scholars into two groups, the philosophical, intellectual ones and the socalled, by SCHILLER called "Brotgelehrter", translated perhaps as "scholar only living for his earning".
He pointed out that for the last type of scholars each progress in science causes annoyance and despair, because his wellformed and in his view socalled scientific system goes to rack and ruin als knowledge proceeds. He is too lazy and any effort would be terrible for him to complete his knowledge and to support progress. Opposite to him is the intellectual and philosophical minded scholar, the true scientist, who is satisfied and happy about each progress in science and society and who supports all movements towards a deeper insight into nature, history, society and the regularities of their development. No doubt, to which group of people FRIEDRICH SCHILLER, one of the greatest dramatists of all times and freeman of the French revolution belonged!
GOETHE, complete different in his individuality and character in comparison to his friend FRIEDRICH SCHILLER composed most of his beautiful poems in the lovely country of the Thuringian forests like "Über allen Wipfeln ist Ruh", "Erlkönig", "Ilmenau" and others.
In contrast to this classical and cultural tradition of this district, we have to remember ourselves on another aspect of history. Not far from Weimar, on the Ettersberg, we found now a National Memorial reminding us of the blackest chapter of german history. On this mountain, 10 km from the town of GOETHE and SCHILLER, during the time of Nazism one of the most awful concentration camps, that of Buchenwald, was erected. In this camp nearly 60 000 prisoners were assassinated, among them the leaders of the german working class, the communist ERNST THÄLMANN and the social democrat RUDOLF BREITSCHEID. Both, Weimar and Buchenwald, are National Memorial Places for our people and also for foreign visitors remembering both on great cultural and scientific efforts and also on the most evil antihumanism in history.
In GDR social liefe and prosperity, development of science, public health service and industry have made great achievements, predominantly in the last 10 years.

Now, ladies and gentlemen, I am going to open this symposium, hoping that all of us will have high scientific profit. Now, I will ask Professor SEVERIN to act as chairman in this first section in the morning.

CONTENTS

2. ADENOSINETRIPHOSPHATASES

3. MECHANISM OF ACTION AND CONTROL PROPERTIES OF SELECTED ENZYME SYSTEMS IN MAIN ROUTES OF METABOLISM

General Problems

Selected Enzymes

I. ENZYMATIC HYDROLYSIS OF INORGANIC PHOSPHATES

Laboratory of Bioorganic Chemistry, Lomonosov State University, Moscow

S.M. AVAEVA, T.I. NAZAROVA

HYDROLYSIS OF THE SUBSTRATES BY INORGANIC PYROPHOSPHATASE FROM YEAST

A considerable body of evidences has been accumulated by enzymology which show that general conclusions and prognostications are possible as far as certain groups of enzymes are concerned.

It would be valuable to discuss properties of the large group of enzymes responsible for various pyrophosphates metabolism just from this point of view. According to the modern classification these enzymes belong to three classes comprising about 200 enzymes: hydrolases, transferases and ligases. The enzymes of the first class which hydrolyse polyphosphate bonds include numerous ATP-ases and pyrophosphatases. Transferases utilize ATP or some other NTP's as one of the substrates, the cleavage of polyphosphate bond being accompanied by transfer of the phosphoric or pyrophosphoric residue to an acceptor. A great number of enzymes belongs to the ligase class. They participate in the synthesis of the C-O, C-S, C-N and C-C bonds and their action is conjugated with degradation of ATP.

All the above enzymes are functionally different, but the fact of their utilizing the same compound as one of the substrates, certainly, makes them similar in some ways. Indeed, all the enzymes mentioned require high metal ion concentrations for their activity to be revealed. Uncomplexed polyphosphate, appearing in the reaction mixture at low concentrations of metal ions, inhibits the enzyme. When concentration of metal exceeds that of polyphosphate the rate of enzymatic reaction also may decrease. These data could be interpreted to mean that the formation of the active complex of these enzymes proceeds via a similar mechanism.

It is also essential that whatever the mechanism of the ATP-enzyme interactions (ping-pong, ordered or random sequential mechanisms), - in any case the enzyme should have a locus for ATP binding and a similar structure of certain part of the active centre. There are some aspects pertaining to the problem of possible similarity of the active sites of the enzymes digesting the same substrate. When one is dealing with the chemical nature of amino acids forming a protein molecule the possibility of synthesis of an innumerable number of proteins comes into mind. However, the number of amino acids, which are present in the active centre and take part in the chemical conversions of the substrate, is relatively small. If one takes into account the pH optimum of the enzyme and the pK's of the functional groups of the protein, this set of amino acids is further reduced. Thus, it may be concluded that the enzymatic conversion of polyphosphates involve just a small number of functional groups of proteins the groups being the same in many enzymes. Present day enzymology possesses rather scanty knowledge about the primary structure and the mechanism of action of these enzymes. But since these enzymes are believed to have more or less similar mechanism of action, the understanding of the mechanism of any enzyme of the group may be useful for elucidating the mechanisms of the others. The present paper describes the results of our study of yeast inorganic pyrophosphatase. Pyrophosphatase activity in yeast was discovered long ago, and by 1952 KUNITZ (1) had described the methods of isolation, purification and crystallization of the enzyme. Inorganic pyrophosphatase from yeast is a protein of the albumin type with a molecular weight about 60, 000 (2). Two main questions - the quaternary structure and the mechanism of action of this enzyme will be discussed below.

Manifestation of enzymatic activity is undoubtedly associated with the quaternary structure of the molecule. Our first experiments with sodium dodecyl sulphate resulted in the suggestion that the molecule of yeast inorganic pyrophosphatase dissociates into subunits (3). Most important data on the enzyme quaternary structure were obtained when the protein was treated with maleic anhydride. The protein molecule undergoes noticeable changes even when there are only 9-10 maleic acid residues per molecule of the protein (4). Determination of molecular weight of such "partially" acylated protein and that of completely acylated one showed that the molecule of inorganic pyrophosphatase from yeast consists of four subunits (Table 1).

Dissociation into dimers takes place after acylation of some part of ε-amino groups, while dissociation into monomers requires acylation of all lysine residues.

The next step was characterization of the subunits. Even the data on the consecutive dissociation of the protein by the action of maleic anhydride gave grounds to believe

Table 1

PHYSICO-CHEMICAL CONSTANTS OF YEAST PYROPHOSPHATASE AND ITS SUBUNITS

Constants Proteins	Native (tetramer)	Partially acylated (dimer)	Totally acylated (monomer)
S_{20}	4.2(c ⟶ o)	1.8(c ⟶ o)	1(c=0.5%)
$D \times 10^7\ cm^2 \cdot sec^{-1}$	6.68[1]	4.6	7.4
Molecular weight:			
Trotman's method	56 000	Mz 36 000	
		Mw 30 000	
Svedberg's method	60 000[1)]	36 000	13 000
Sedimentation equilibrium method			14 000
disc electrophoresis method		30 000	17 000

1) H. SCHACHMAN (2)

that the subunit s were not identical. The suggestion of $\alpha_2\beta_2$ structure was favoured by the study of the action of carboxypeptidase A on pyrophosphatase (5). Four amino acids - leucine, phenylalanine, tyrosine and alanine have been detected after one hour hydrolysis. Other amino acids appeared when the incubation time was increased.

However, only two moles of every C-terminal amino acids were found per mole of protein.

The $\alpha_2\beta_2$ structure has received a great impetus from the results of the Edman degradation experiments in an automatic Beckman sequencer. It was revealed that after every five steps two amino acids cleave off in equal amounts. Prior to the study of mechanism of the enzyme action we tried to answer the following questions: what are the pathways of the active enzyme-substrate complex formation. Whether the enzyme-substrate interaction involves a chemical modification of the enzyme, i.e. formation of the covalent intermediates. What functional groups of the enzyme are responsible for these reactions.

In the case of yeast inorganic pyrophosphatase the active complex consists of the enzyme, pyrophosphate and metal ion. The most general model describing the first steps of enzymatic reaction is the following

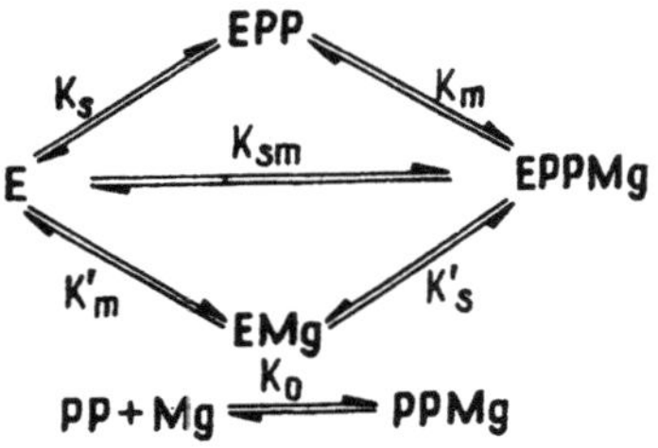

Examination of the velocity dependence upon free Mg^{+2} or PP concentrations while one of them was held constant resulted in the determination of the dissociation constants (Table 2).

Table 2

DISSOCIATION CONSTANTS OF THE ENZYME-PYROPHOSPHATE-METAL COMPLEXES (pH 7, 5 mM Tris-HCl buffer, 25°C)

Reaction	Symbol	Value x 10^4M
EPP ⇌ E + PP	K_s	0, 085 ± 0, 008
EPPMg ⇌ EPP + Mg	K'_m	1, 5 ± 0, 15
EPPMg ⇌ E + PPMg	K_{sm}	0, 025 ± 0, 005
EMg ⇌ E + Mg	K_m	20
PPMg ⇌ PP + Mg	K_o	5

The data presented enabled to discuss the organization of enzyme-substrate complex. The enzyme-metal complex is very unstable, that is why the formation of the active complex by the way shown in the lower part of the general scheme is hardly probable. Uncomplexed pyrophosphate may bind to the enzyme rather strongly. It accounts for the fact of inhibition of the enzyme by pyrophosphate. As pyrophosphate may bind to the enzyme without metal the latter cannot play the role the bridge between the enzyme and pyrophosphate. The comparison between K_s and K_{sm} shows that magnesium pyrophosphate binds to the enzyme more strongly than pyrophosphate. On the other hand the metal is bound more tightly in the enzyme-pyrophosphate complex than in the pyrophosphate one. These facts should be evidently interpreted as favouring the structure given below. Such structure involves the attachment of the metal both to the enzyme and pyrophosphate.

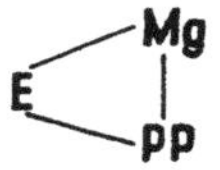

Ability of pyrophosphatase to be modified by substrates was elucidated in experiments with a number of labelled compounds:

The first compounds were derivatives of serylpyrophosphate (PPSer) with the ^{32}P atom in the β-position or with the ^{14}C atom in the serine moiety (6). Both phosphorous atoms were labelled in inorganic pyrophosphate. Reaction with labelled inorganic phosphate.was also studied. All these compounds were incubated with the enzyme at 5°C. After 10 min. of incubation the reaction was interrupted and the protein was separated from the low molecular weight radioactive compounds by passing the mixture through a Sephadex G-75 column (Fig. 1).

^{32}PPSer
PPSer (^{14}C)
^{32}P ^{32}P
^{32}P

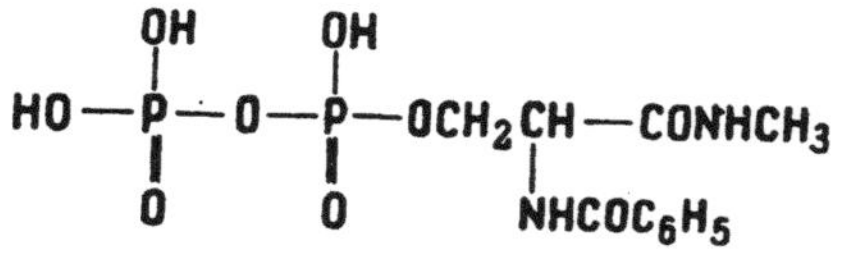

PPSer

These experiments revealed some facts which are essential for the understanding of the mechanism of the enzymatic action. Firstly, regardless of the substrate part which was labelled, both with the ^{32}P- and ^{14}C-substrates labelled protein could be obtained. Secondly, the label incorporates in protein both in the presence and absence of the metal. It points to the fact that at the first step (a) of the enzyme-pyrophosphate interaction the whole substrate molecule is covalently bound. At this step no metal ion is required (7).

$$(a)\quad E + ROPP \rightleftharpoons E-PPOR$$

$$(b)\quad E-PPOR \xrightarrow{Me^{+2}} E-P + ROP$$

$$(c)\quad E-P \rightleftharpoons E + P$$

In the presence of metal ions the next step (b), i. e. hydrolysis of the pyrophosphate bond may proceed. Metal ions do cause a change in the quality of the label in the protein (Table 3).

Table 3

LABEL RELEASE FROM THE PHOSPHORYLATED PROTEIN IN PRESENCE OF METAL IONS

$$E + PPSer(^{14}C) \rightleftharpoons E - PPSer(^{14}C) \xrightarrow{Zn^{+2}} EP + PSer(^{14}C)$$

$$E + {}^{32}P^{32}P \rightleftharpoons E - {}^{32}P^{32}P \xrightarrow{Mg^{+2}} E - {}^{32}P + {}^{32}P$$

Substrates	Specific activity cpm	Me^{+2}	Protein cpm
$PPSer(^{14}C)$	4 000	-	3 840
$PPSer(^{14}C)$	4 000	Zn	-
${}^{32}P^{32}P$	3 125	-	3 095
${}^{32}P^{32}P$	3 125	Mg	1 500

When the ^{14}C-labelled substrate was used, no label was found in the protein at this step. In the case of pyrophosphate the addition of magnesium ions to the labelled protein decreases the amount of the label in it twofold.

At this step of the reaction the phosphorylated enzyme is formed which is then hydrolyzed to yield free enzyme and inorganic phosphate (step c). This reaction has been proved to be reversible and the most convenient one for preparation of phosphorylated enzyme. It has been proved that the interaction of the enzyme with inorganic phosphate results in phosphorylation of the enzyme active site. Thus, the enzyme-pyrophosphate interaction proceeds via formation of several covalent intermediates. The pyrophosphorylated enzyme is formed at the first step and phosphorylated one - at the next. It should be noted that formation of phosphorylated enzymes has been shown to occur in many cases. But much more surprizing is the fact that whole substrate molecule must bind to enzyme prior to its hydrolysis. Formation of a similar substrate enzyme compounds was also reported last year for nucleoside-triphosphate-adenosinemonophosphate transphosphorylase (8) and myosine (9).

The data obtained gave rise to the next problem - which amino acid is phospory-lated in the enzyme?
The effect of pH on the release of phosphoric acid from the phosphorylated protein was examined after 2 hour incubation at 37°C (Fig. 2).
It is clear that the phosphorous bond stability varies with pH, the hydrolysis being most intensive in acidic medium. These data are not sufficient to rich the conclusion about the nature of the linkage between phosphate and protein. It is a matter of common knowledge that the stability of phosphate residue in proteins is determined by interactions with other functional groups of the protein. It is important that different methods of degradation of the phosphorylated protein yield various phosphorylated fragments (10, 11). It must be borne in mind that phosphorylation of the following residues in proteins - serine and cysteine, basic and carboxylic groups may occur. In the case of inorganic pyrophosphatase there are much fewer possibilities, since serine and cysteine should be excluded. The conclusion is based on the experiments with diisopropylfluorophosphate (12).
This well-known inhibitor of serine enzymes suppresses the activity of pyrophosphatase only when used at extremely high concentrations. But at such concentrations the activity of the enzyme also decreases in the presence of alcohols, acetone and ether. It should be noted that diisopropylfluorophosphate behaves as a competitive and reversible inhibitor. As to the participation of cysteine in the active centre of the molecule, we failed to confirm the evidence of the presence of the SH-group in the active centre (13). The activity of the enzyme remains constant in the presence of p-chloromercuribenzoate, 5, 5' -dithiobis (2-nitrobenzoic acid) and alkylating agents.
Consequently, as we could expect either the amino or carboxyl group to be phosphorylated in pyrophosphatase we were first to try the phosphorylated protein for phosphohistidine and acylphosphate.
The routine method separation of phosphohistidine from protein is alkaline hydrolysis. This procedure has been employed to prove the presence of phosphohistidine in a number of proteins. We hydrolyzed the phosphorylated pyrophosphatase with 3N sodium hydroxide for 4 hours and separated the mixture on a Dowex column (Fig. 3).
Phosphoric acid proved to be the only radioactive substance. These experiments have proved the absence of the phosphohistidine in phosphorylated enzyme. But alkaline treatment of phosphorylated pyrophosphatase was to be studied more closely. Phosphorylated enzyme was preincubated in subalkaline medium

(pH 10, 24 hours, 20°), than hydrolyzed with 3N sodium hydroxide. In this case hydrolyzate contained phosphohistidine (Fig. 4). Phosphohistidine can be also found after preincubation of phosphorylated protein with pronase (pH 7, 4 hours, 37°C) or 1N sodium hydroxide at room temperature (12 hours).
So it was shown that phosphohistidine could be isolated but it was a by-product and was formed as a result of the phosphoric acid residue migration to histidine. This migration takes place owing to the convertion of ammonium groups of protein into active nucleophilic agents in subalkaline medium.
It should be noted that phosphohistidine is known to be a part of phosphorylated enzymes, but it seems in the light of our findings that at least some of the published data should be reviewed.
Important results were got in experiments with hydroxylamine. As rule interaction of phosphorylated protein with hydroxylamine at pH 7 is used for revealing acylphosphate group. We have shown that phosphohistidine also reacts with hydroxylamine, the rate constant of phosphate release being 0, 16 $min^{-1}M^{-1}$. So both acylphosphate and phosphohistidine residues in protein can react with hydroxylamine.

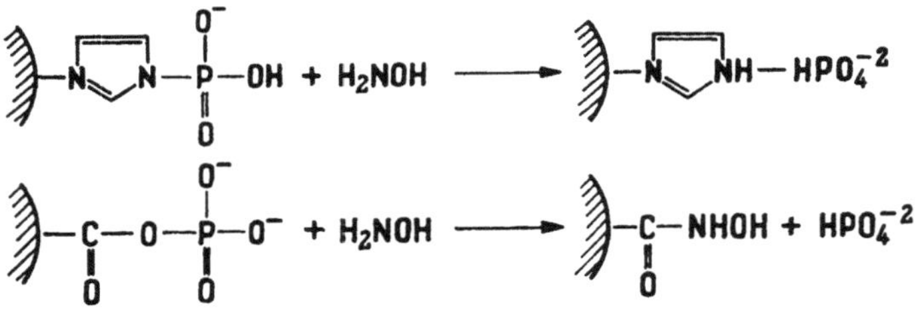

In both reactions phosphoric acid is liberated but the behaviour of the enzyme is quite different in the two cases. Inhibition of the enzyme must be observed only after reaction of acylphosphate group. Really the activity of pyrophosphatase after 15 min incubation with pyrophosphate and 0.4M NH_2OH at pH 7 decreased twofold. The fact can be explained by the existence of the acylphosphate bond in the phosphorylated protein.
But as we had only indirect data about acylphosphate group in the protein it was necessary to find a chemical reaction resulting in a derivative of the carboxyl group which is stable under the conditions of protein peptide bonds hydrolysis. We turned to the well-known reaction of cyanide with carbonic acid chlorides yielding a ketonitril.

$$RCOCl + CN^- \longrightarrow RCOCN + Cl^-$$

If acylphosphate behaves like acidchloride, in the case of protein the product would be a protein ketonitril and after hydrolysis - ketoamino acid nitril and then ketoaminodicarbonic acid.

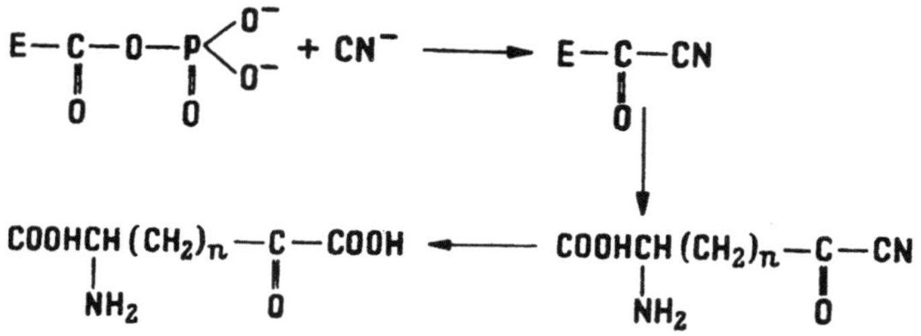

We studied the reaction of inorganic pyrophosphatase with potassium cyanide. These experiments revealed three facts of basic importance.
First, cyanide was proved to inhibit pyrophosphatase (Fig. 5).
The reaction mixture contained the enzyme, pyrophosphate and cyanide. The enzymatic activity was determined after dilution and addition of magnesium ions. The enzyme completely loses its activity after three hours of incubation with cyanide.
Second, in accordance with the chemical scheme, cyanide should push phosphoric acid from phosphorylated protein (Fig. 6).
In the diagramme one can see the experimental data on the composition of the reaction mixture a) after the reaction of the enzyme with labelled phosphoric acid for 10 min., b) after the reaction of the enzyme with phosphate and then with cyanide. Great discrepancy between the two curves means that the cyanide pushes the phosphoric acid residue from the active site of the enzyme.
Third, reaction of inorganic pyrophosphatase with labelled cyanide has been carried out. The result was a labelled protein which was isolated by gel chromatography (Fig. 7).
The labelled protein was then subjected to hydrolysis with 6N HCl (105°C, 24 hours). In the course of the hydrolysis all the radioactivity concentrated in an amino acid. At present we are studying the structure of the amino acid obtained. We hope that the reaction with cyanide will serve as a specific test for the acylphosphate linkage in phosphorylated enzymes.

REFERENCES

(1) KUNITZ, M., J. Gen. Physiol., 35, (1952), 423

(2) SCHACHMANN, H., J. Gen. Physiol. 35, (1952), 451

(3) AVAEVA, C., E. BRAGA, A. EGOROW, Biophysika, 13, (1968), 1126

(4) AVAEVA, C., P. BEISEMBEWA, A. EGOROW, Biochimia, 35, (1970), 257

(5) AVAEVA, C., G. ACHMEDOW, Biochimia, 35, (1970), 31

(6) AVAEVA, C., H. RASKOWA, A. MECH, Westnik, 1, (1970), 100

(7) AVAEVA, C., A. MECH, Biochimia, 35, (1970), 555

(8) ALBRECHT, G., Biochemistry, 9, (1970), 2462

(9) SCHLIESELFELDT, L., T. CONOVER, M. BARANY, Biochemistry, 9, (1970), 1133

(10) COTTON, G. L., P. A. SRERE, Biochem. Biophs. Res. Commun., 35, (1969), 895

(11) SUZUKI, F., K. FUKUNUSHI, G. TAKEDA, J. Biochem., 66, (1969), 767

(12) AVAEVA, C., T. NASAROWA, H. Worobewa, Biochimia, 35, (1970), 1113

(13) EIFLER, R., R. GLAESMER, S. RAPOPORT, Acta Biol. Med. Germ., 17, (1966), 716

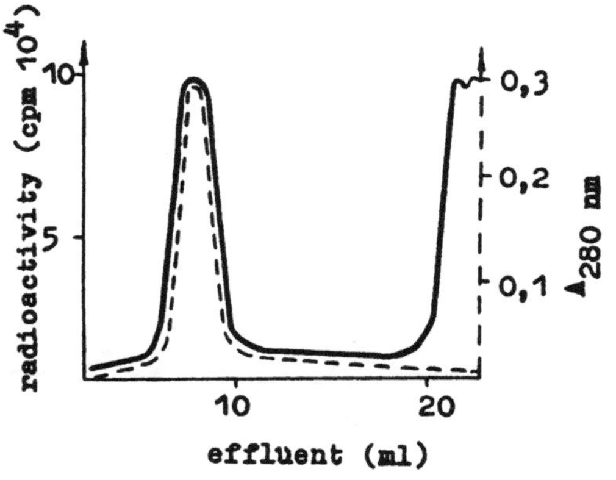

Fig. 1. Gel filtration of phosphorylated protein through a Sephadex G-75 column.

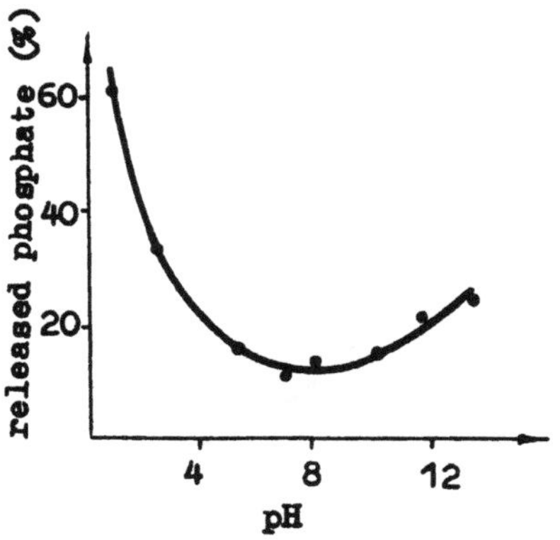

Fig. 2. Effect of pH on ^{32}P-protein hydrolysis (2 hrs, 37°).

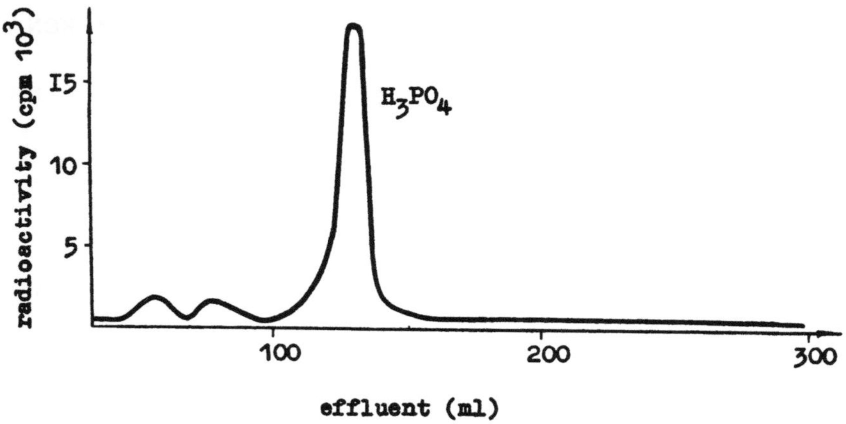

Fig. 3. Dowex 1 x 8 chromatography of phosphorylated protein after hydrolysis by 3 N NaOH (4 hrs, 100°)

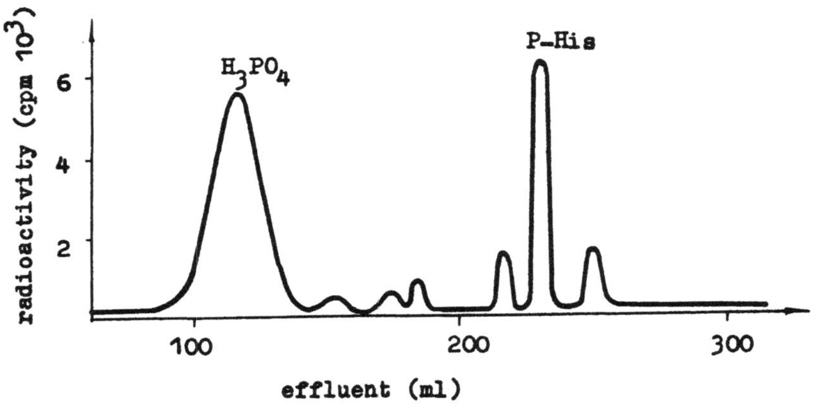

Fig. 4. Dowex 1 x 8 chromatography of phosphorylated protein after incubation at pH 10 (24 hrs, 20°) and hydrolysis by 3N NaOH.

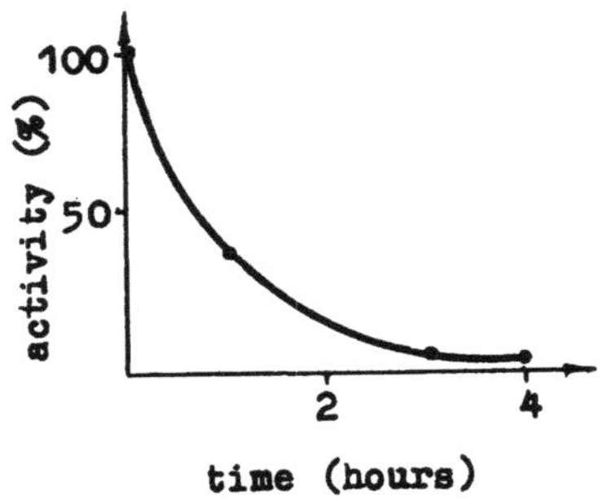

Fig. 5. Inactivation of yeast inorganic pyrophosphatase by 0.4M KCN (pH 7.0, 37°).

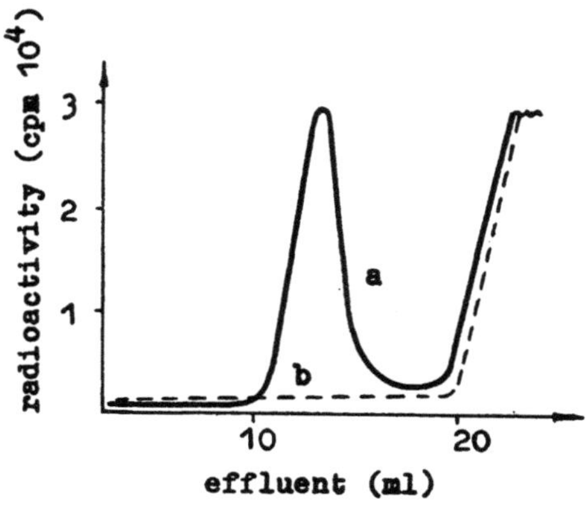

Fig. 6. Dephosphorylation of ^{32}P-protein by KCN. a) before treatment with KCN; b) after treatment with 0.4M KCN

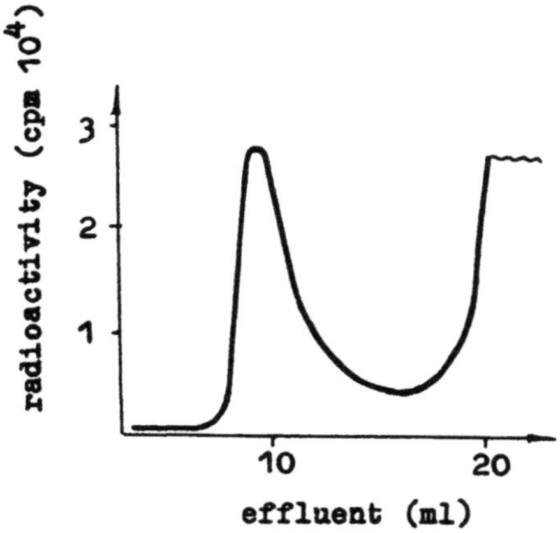

Fig. 7. Gel Filtration of ^{14}C-protein through a Sephadex G-75 column after treatment with 0.4M KCN.

Institut für Physiologische und Biologische Chemie der
Humboldt-Universität zu Berlin

V. HAHN, R. EIFLER, I. HERMANN

ON THE ORIGIN OF INORGANIC PYROPHOSPHATASES OBTAINED FROM BAKER'S YEAST

The inorganic pyrophosphatase prepared from baker's yeast by the original method of KUNITZ (1) is in accordance with the criteria of purity given by KUNITZ. The specific activity and the behaviour of sedimentation are identical, but the endgroup analysis and the disc-electrophoresis show inhomogeneities. Therefore the enzyme was purified further by two chromatographic steps on DEAE-cellulose with a linear NaCl-gradient from 0 to 0, 1 M.
After rechromatography the first peak of the chromatography gave one band only in the disc-electrophoresis as shown in Fig. 2. Peak II and III remainded inhomogenous after rechromatography. All the three peaks showed pyrophosphatase activity.
However the ratios of the three peaks show considerable differences from preparation to preparation. On the other hand the sedimentation behaviour and the specific activity are identical and the N-terminal amino acid is threonine in all cases.
The original preparation procedure of KUNITZ includes an autolysis step of four days and so the question arises: Do pyrophosphatase isozymes exist in a genetically sense or are the different enzymes artefacts formed by autolysis? To give an answer we looked for molecular differences between the proteins in the three peaks. Differences were found in the C-terminal amino acids. The pure enzyme obtained from the rechromatography of the first peak shows only valine after digestion with carboxypeptidase A and traces of serine and glycin. These findings are in opposition to the results obtained by AVAEVA and ACHMEDOV (2). In the products II and III we found either traces of valine or no amino acids at all.
Furthermore the pyrophosphatase containing fractions were tested for activity by disc-electrophoresis after each step of the preparation. In the first case

examinations of the preparation steps by disc-electrophoresis were carried out by staining with amidoblack 10 B with and without addition of the pure enzyme. I. In this way we could see which protein band is identical with the pure pyrophosphatase I. This is shown in Fig. 3. The band which is more intense by the addition of the pure enzyme can be seen clearly.

In the second case the gels were examinated for activity by the method of TONO and KORNBERG (3) with a slight modification. The modification consisted in the addition of tetrasodiumpyrophosphate to the gels before polymerization and also to the buffers in order to improve the activity staining on the gels. The results of these experiments are shown in Fig. 4.

As it can be seen, from the beginning of the preparation the "isozyme" pattern shows two bands. In the first step, the plasmolysis with toluene, the minor band is very small compared with the main band. After the autolysis step this ratio is reversed and an additional third band appears. Unfortunately the third band is not visible in the figure. The first main band is electrophoretically identical with the pure isolated enzyme I.

Therefore we suppose, that the pyrophosphatase from baker's yeast obtained after chromatography are artefacts produced by proteases. The lack of C-terminal amino acids after digestion of pyrophosphatases II and III with carboxypeptidase A can perhaps be explained by the assumption that the artefacts are the result of yeast carboxypeptidase activity.

REFERENCES

(1) KUNITZ, M., Arch. Biochem. Biophys. 92, (1961), 270

(2) AVAEVA, S., and G. ACHMEDOW, Biochimia 35, (1970), 31

(3) TONO, H. and A. KORNBERG, J. Biol. Chem. 242, (1967), 2375

(4) MAURER, H. R., "Analytische Disk-Elektrophorese" de Gruyter Verlag, Berlin 1968

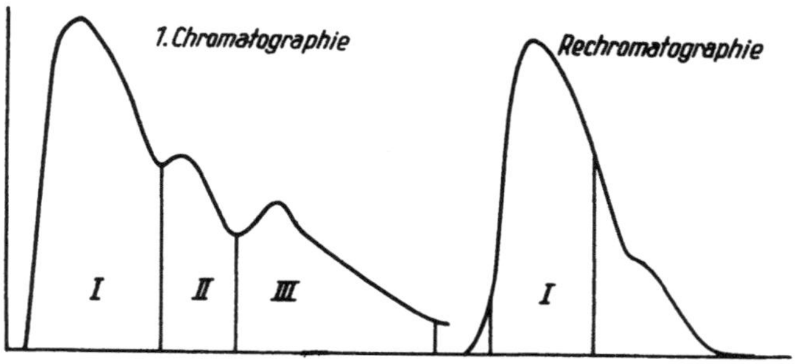

Fig. 1. The left part shows the chromatography of the crystallized pyrophosphatase with a NaCl-gradient (0 - 0, 1 M in 0, 025 M Tris-HCl-buffer pH 7, 6). The fractions were divided into three samples I, II and III. The right part shows the rechromatography of peak I under the same conditions as before. The fractions between the two lines were combined and lyophilized. This part is called pure pyrophosphatase I.

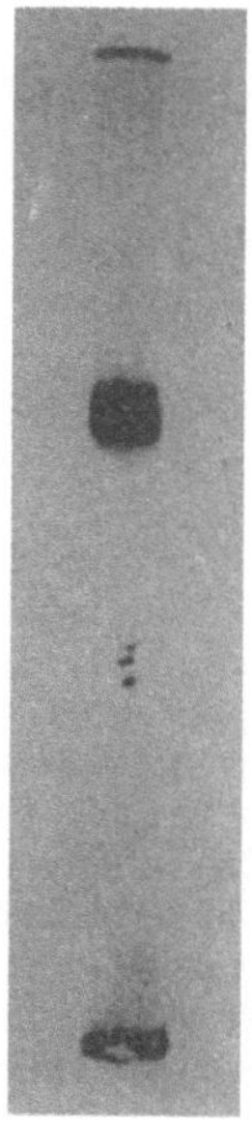

Fig. 2. Disc-electrophoresis of the pure pyrophosphatase I Separation system 1 after Maurer (4) Staining with amidoblack 10 B, 100 μg protein.

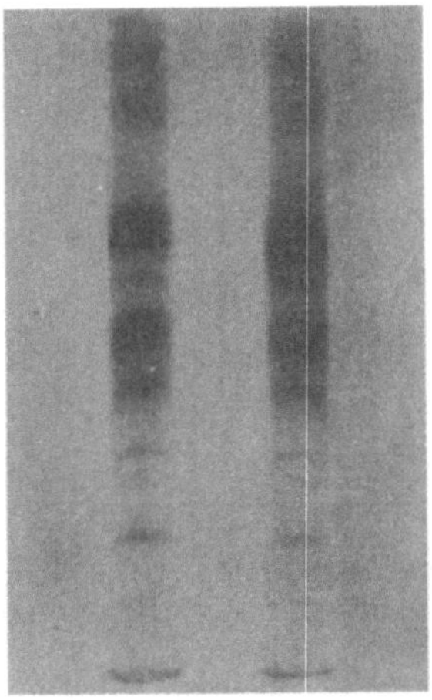

Fig. 3. Disc-electrophoresis of the crude extract after plasmolysis with toluene. Separation system 1 after Maurer (4). On the left gel the crude extract, on the right the crude extract with addition of 30 μg pure pyrophosphatase I.

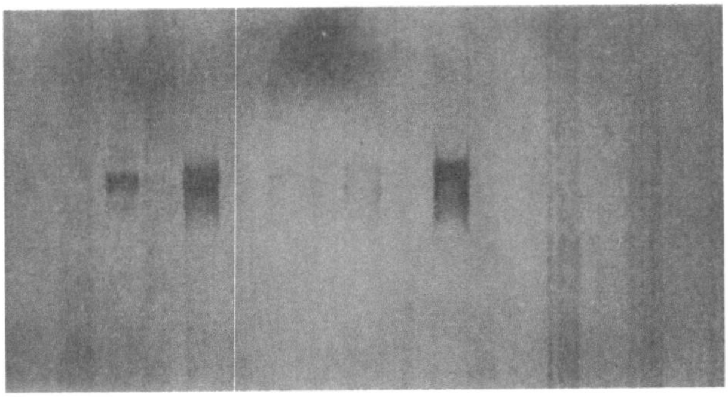

Fig. 4. Disc-electrophoresis of various preparation steps Separation system 1 after Maurer (4) under addition of sodiumpyrophosphate (final concentration 10^{-3} M). The gels were stained by using the enzymatic activity of the pyrophosphatase.
From the left to the right:

1. Crude extract after plasmolysis
2. Crude extract with addition of 1 μg of pure pyrophosphatase I
3. Pure pyrophosphatase I
4. Pyrophosphatase preparation after autolysis
5. Pyrophosphatase preparation after autolysis with addition of 1 μg pyrophosphatase I

Institut für Physiologische und Biologische Chemie
der Humboldt-Universität zu Berlin

G. HANSEN, P. HEITMANN, Ch. SCHOLZ

DISSOZIATION AND ASSOZIATION BEHAVIOR OF INORGANIC PYROPHOSPHATASE FROM BAKER'S YEAST

In 1966 EIFLER and RAPOPORT reported that inorganic pyrophosphatase (pyrophosphate phosphohydrolase, E. C. 3. 6. 1. 1.) from baker' s yeast can be split into subunits by treatment with 8. 0 urea (1). Similar results were obtained by AVAEVA and coworkers with other reagents (2). From experiments on the association-dissociation behavior of the enzyme indications were obtained that the molecular weight of the native enzyme is somewhat higher than the tentative value of 60, 000 originally reported by SCHACHMAN (3). Therefore, attempts were made to determine the correct value by aid of several independent methods. They were: 1. a modified Archibald procedure, 2. density gradient sedimentation, and 3. disc gel electrophoresis.

Fig. 1 shows the apparent molecular weight obtained by the first methods as a function of the protein concentration. The extrapolation to zero concentration yields a molecular weight of 67, 000.

From the concentration dependency of the sedimentation coefficient of the enzyme demonstrated in Fig. 2 $s^{o}_{20,w}$ was determined to be 4. 7. Only a slight alteration of this value is observed down to a protein concentration of 0. 09 mg/ml. Therefore, it can be concluded that no dissociation occurs in this concentration range. The results of the density gradient centrifugation are shown in Fig. 3. The experiments were performed according to the procedure of MARTIN and AMES (4). The position of the protein peaks were determined spectrophotometrically at 280 nm.

In the lower concentration range of about $1\text{-}40\cdot 10^{-4}$ mg/ml an enzymatic assay was used to locate the position of the pyrophosphatase peak. By extrapolation to zero concentration (Fig. 4) a molecular weight of 69, 200 was obtained.

The lowest concentrations employed in the density gradient sedimentation experiments corresponds to those usually used in the pyrophosphatase activity assay.

It is evident that the enzymatically active species of inorganic pyrophosphatase is the undissociated molecule.

The experiments on polyacrylamide gel electrophoresis were performed according to the procedure of ZWAAN (5). A value of 70, 600 was obtained by this method. Based on the three types of independend measurements we propose to use in the future the value of 70, 000 for the average molecular weight of native inorganic pyrophosphatase from baker's yeast.

The enzyme may be split into subunits by treatment with sodium dodecyl sulfate and guanidine hydrochloride. The products have been characterized by sedimentation analysis, polyacrylamide gel electrophoresis, and N-terminal group analysis.

The sedimentation equilibrium analysis in 6. 0 M guanidine hydrochloride was performed according to ARCHIBALD. In Fig. 5 the reciprocal value of M_{app} $(1-\nu^*\rho)$ is plotted against the protein concentration according to the method of ULLMANN(6). Extrapolation to zero concentration yields a molecular weight of 36, 000.

The sedimentation velocity experiments in 6. 0 M guanidine hydrochloride with treatment of the data according to TANFORD (7) yielded a value of 35, 000 for the molecular weight of the split products.

The effect of sodium dodecyl sulfate on the molecular weight of inorganic pyrophosphatase was studied by polyacrylamide gel electrophoresis. The detergent was used in a concentration of 1%. The results were treated according to the procedure described by SHAPIRO (8). The value for the molecular weight was determined to be 37, 000.

It should be mentioned that in no experiment any indication for a heterogeneity of the split products could be detected. Therefore, it is reasonable to assume that inorganic pyrophosphatase is a dimer composed of two identical or very similar subunits. This conclusion is supported by the result of the N-terminal analysis, which yielded threonine as the only amino acid. Its amount of 1. 7 moles per 70, 000 g enzyme approaches the expected value of 2 (9).

Attempts were made to restore the catalytic activity of the enzyme after treatment with guanidine hydrochloride. For this purpose the denaturant was removed by dialysis against 1, 000 volumes of a Tris· HCl buffer pH 7. 2, in the presence of 0. 1 M NaCl, 0. 01 M β -mercaptoethanol and 0. 002 M $MgSO_4$. Under these conditions 70% of the enzymatic activity were recovered. The sedimentation coefficient of the renatured product was determined to be s_{20}^{c} = 4. 8, a value very near to that of 4. 7 for the native enzyme.

Dilution experiments indicate that the renaturation is a fast process. Under optimal conditions more than 50% of the enzymatic activity are restored in less than one minute.

Inorganic pyrophosphatase contains two SH-groups. They possess only a low reactivity against modifing reagents. The treatment with disulfides, i. e. DTNB or GSSG which must be employed in high concentrations, causes a decrease of the SH content of the enzyme due to the formation of mixed disulfides. The catalytic activity is only influenced to a small extent by this mdification. The sedimentation coefficient of the modified enzyme is lower than that of the native protein.

Further investigations are necessary to clarify this effect.

REFERENCES

(1) EIFLER, R., S. RAPOPORT, Third FEBS Meeting, Warsaw 1966, Abstract F168, p. 264

(2) AVAEVA, S.M., E.M. BRAGA, A.M. EGOROV, Biofizika 13, (1968), 1126

(3) SCHACHMAN, H.K., J. Gen. Physiol. 35, (1952), 451

(4) MARTIN, R.G., B. N. AMES, J. Biol. Chem. 236, (1961), 1372

(5) ZWAAN, J., Analyt. Biochem. 21, (1967), 155

(6) ULLMANN, A., M.E. GOLDBERG, D. PERRIN, J. MONOD, Biochemistry 7, (1968), 261

(7) TANFORD, C., K. KAWAHARA, S. LAPANJE, J. Amer. Chem. Soc. 89, (1967), 729

(8) SHAPIRO, A.L., E. VINUELA, J.V. MAIZEL, Biochem. Biophys. Res. Commun. 28, (1967), 815

(9) EIFLER, R., unpublished results

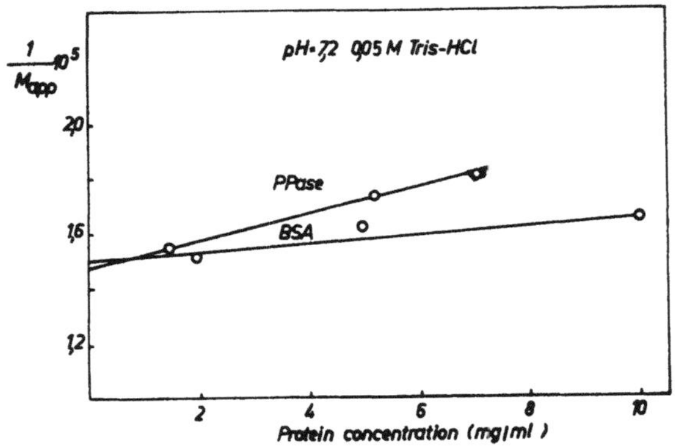

Fig. 1. Dependency of $1/M_{app}$ of inorganic pyrophosphatase and bovine serum albumin on the protein concentration.

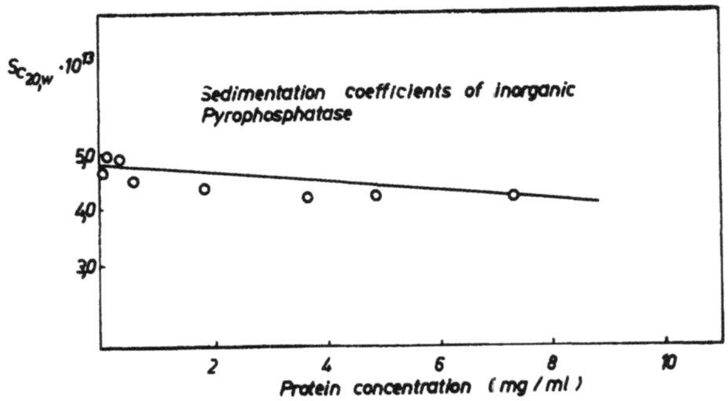

Fig. 2. Sedimentation velocity studies of inorganic pyrophosphatase in NaCl-Tris·HCl buffer pH 7.2.

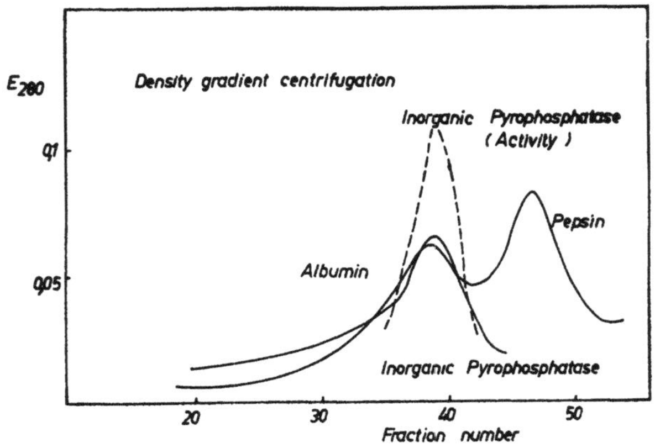

Fig. 3. Sedimentation studies in a 2 to 17 percent sucrose gradient at 40,000 rev/min, 15 h, and 4°.

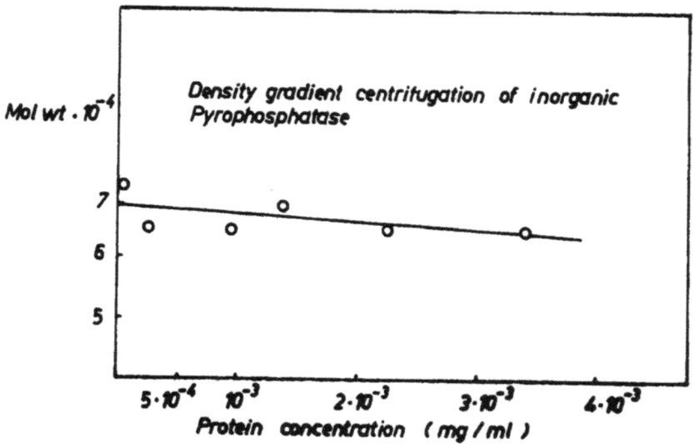

Fig. 4. Determination of molecular weight of inorganic pyrophosphatase by sucrose density gradient centrifugation at different protein concentrations.

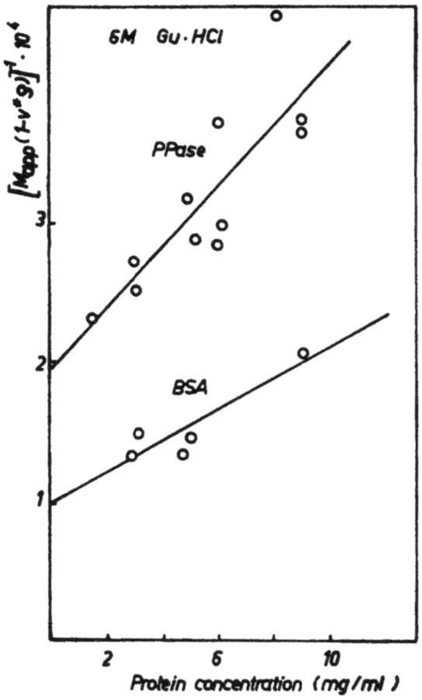

Fig. 5. $1/M_{app}\ (1-v^{*}\varrho)$ values in 6.0 M guanidine hydrochloride as a function of protein concentration.

Institut für Physiologische und Biologische Chemie
der Humboldt-Universität zu Berlin

P. HEITMANN, Ch. MÖLLERKE, H.-J. UHLIG

ON THE STATE OF AROMATIC AMINO ACIDS OF INORGANIC PYROPHOSPHATASE FROM BAKER'S YEAST

For a detailed understanding of the mechanism of an enzymatic it is important to know the spatial arrangement of the amino acids in the enzyme molecule. Great progress was made in this field by means of X-ray crystallography. But other techniques are also useful to approach this problem. Among them the characterization of the state of the amino acids in the protein by spectroscopic methods and modification experiments plays an important part.
The present paper deals with the state of the tyrosyl and tryptophyl residues of inorganic pyrophosphatase (pyrophosphate phosphohydrolase, E.C. 3.6.1.1) from baker's yeast. Both the phenolic group as well as the indole residue are powerful chromophores. It is possible to calculate the tyrosine and tryptophan content of a protein by comparison of its ultraviolet spectrum with those of corresponding low molecular weight model compounds. In order to ensure a great structural correspondence of the chromophores in the protein and in the model system Nα-acetyl-L-tyrosine methylamide and Nα-acetyl-L-tryptophan methylamide were used for this purpose. In the range of 250-300 nm the absorption properties of both compounds were found to be nearly identical with those described in the literature for the corresponding ethyl esters or amides. In order to obtain precise results it is necessary to normalize the absorption of the protein chromophores. According to the procedure suggested by EDELHOCH (1) the spectroscopic measurements were performed in 6.0 M guanidine hydrochloride. This powerful denaturant strongly reduces or eliminates the effects of specific interactions within the protein molecule on the spectroscopic properties of the chromophores. Using this method the total numbers of moles tyrosine and tryptophan per mole inorganic pyrophosphatase were determined to be 24 and 14 on a molecular weight basis of 70,000. The tyrosine value agrees well with that obtained by chromatographical analysis (2).

It is well known that the spectral properties of the tyrosyl and tryptophyl side chains are sensitive to changes of their environment. As a rule, a decrease of the polarity of the surrounding medium leads to a slight red shift of the absorption peaks and a small increase of their intensity. By means of the difference spectroscopic technique this effect can readily be used to study processes which are accompanied by an alteration of the environment of the chromophoric groups. Especially the solvent perturbation method developed by HERSKOVITS and LASKOWSKI (3, 4) has been employed to examine the location of tyrosyl and tryptophyl residues in globular proteins. In this method mild non-aqueous additives, so-called perturbants, for instance polyhydroxy compounds are used to change the polarity of the solvent. In proteins only those chromophores which are accessible to the perturbant possess the ability to "recognize" the change of the solvent composition.

Fig. 1 shows the solvent perturbation spectra of the model compounds here described. They have been recorded with a Cary 14 double-beam spectrophotometer. The observed perturbant effect was produced by 20% ethylene glycol.

Fig. 2. indicates how a chromophoric group can be located in a macromolecule. The chromophores can be classified in four groups: fully exposed (I), partially exposed (II), buried (III), and those in crevices (IVa and IVb).

The solvent perturbation spectrum of inorganic pyrophosphatase due to 20% ethylene glycol is presented in figure 3 (solid line). From a comparison of these data with those obtained with the model compounds it was calculated that 15-16 tyrosyl and 2. 3 tryptophyl residues are exposed to the solvent in this enzyme. One should point out that these values represent an average exposure. For instance, a value of 1 may mean that one residue is fully exposed or 2 residues are 50% exposed. The dashed line in Fig. 3 is a solvent perturbation spectrum of a mixture, which contains the model compounds in Tyr-Try ratio corresponding to the values calculated for pyrophosphatase.

Fig. 4 demonstrates the effect of 8. 0 M urea on the solvent perturbation spectrum of inorganic pyrophosphatase. As expected the accessibility of the chromophoric groups to the solvent is found to be considerably enhanced by urea. 19-21 tyrosyl residues and 11-12 tryptophyl groups are exposed under these conditions.

The conclusions about the location of the tyrosyl residues in inorganic pyrophosphatase drawn from the solvent perturbation experiments are confirmed by the results of a spectrophotometric titration of the enzyme. The titration curve presented in Fig. 5 shows that most of the tyrosyl residues exhibit a normal

titration behaviour with a pK_a value of about 10. Approximately 16 groups are titrated at pH 11. The remaining residues (6-8) require a higher pH value to become ionized. At pH 13 about 21 tyrosyl residues exist in the ionized form. The data of the modification studies are in accordance with the results obtained from solvent perturbation and titration experiments. Tetranitromethane has been used frequently as a very powerful reagent for the nitration of phenolic groups in proteins (5, 6). Fig. 6 indicates that approximately 10 tyrosyl residues of inorganic pyrophosphatase are readily nitrated. Even when a very high excess of reagent is employed no more than 15 residues can be modified. This result corresponds to data obtained in modification studies with a 200-400 fold molar excess of N-acetylimidazole (7), in which about 13 tyrosyl residues could be acetylated. In the presence of 8.0 M urea the number of tyrosyl residues nitrated by an excess of tetranitromethane amounts to 19. From this result and the corresponding data of the solvent perturbation experiments the conclusion can be drawn that the enzyme molecule ist not fully unfolded in 8.0 M urea. An analogous behaviour is known for many other proteins.
The effect of the nitration of the tyrosyl residues on the enzymatic activity is demonstrated in Fig. 7. A total loss of the enzymatic activity is observed even when only 2-3 tyrosyl residues are modified. Therefore it may be possible that tyrosyl groups are involved in the catalytic process.

Table

NUMBER OF ACCESSIBLE TYROSYL AND TRYPTOPHYL RESIDUES IN INORGANIC PYROPHOSPHATASE

	Tyrosine (total number 24)	Tryptophan (total number 14)
Solvent perturbation	15-16 (19-21)	2.3 (11-12)
Titration		
pH 11	16	
pH 13	21	
Reaction with TNM	15 (19)	
Reaction with Ac-Im	13	

The values in parantheses are determined in the presence of 8.0 M urea.

Another explanation of this effect might be that these residues are necessary for maintaining the correct spatial structure of the active center. Further studies are necessary to decide between these two possibilities.
The table summarize the data regarding the state of tyrosyl and tryptophyl residues in inorganic pyrophosphatase from baker' s yeast.

REFERENCES

(1) EDELHOCH, H., Biochemistry 6, (1967), 1948

(2) EIFLER, R., R. GLAESMER, S. RAPOPORT, Acta Biol. Med. German. 17, (1966), 716

(3) HERSKOVITS, T. T., M. SORENSEN, Biochemistry 7, (1968), 2523

(4) HERSKOVITS, T. T., M. LASKOWSKI, Jr., J. Biol. Chem. 235, (1960), 56

(5) RIORDAN, J. F., M. SOKOLOWSKY, B. L. VALLEE, Biochemistry 6, (1967), 3609

(6) BEAVEN, G. H., W. B. GRATZER, Biochim. Biophys. Acta 168, (1968), 456

(7) RIORDAN, J. F., W. E. C. WACKER, B. L. VALLEE, Biochemistry 4, (1965), 1758

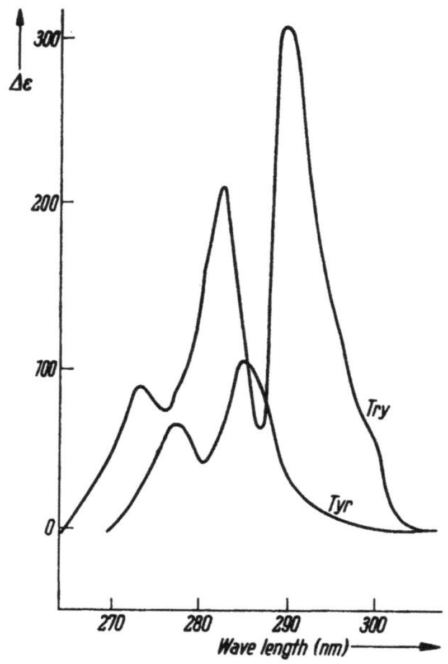

Fig. 1. Solvent perturbation experiments with Nα-acetyl-L-tyrosine methylamide (Tyr) and Nα-acetyl-L-tryptophan methylamide (Try) at pH 6.8.

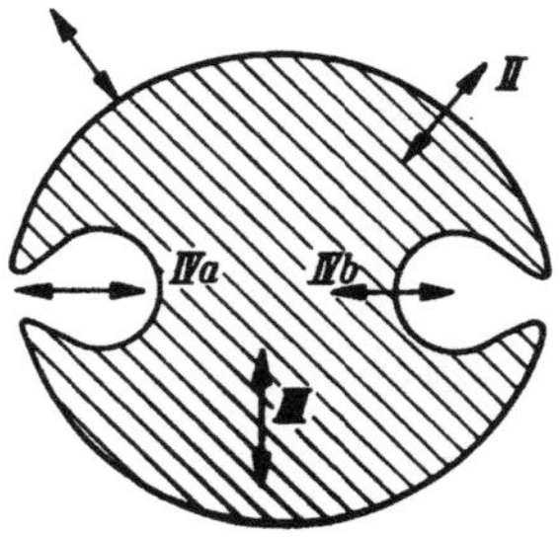

Fig. 2. Classes of chromophoric groups in a macromolecule.

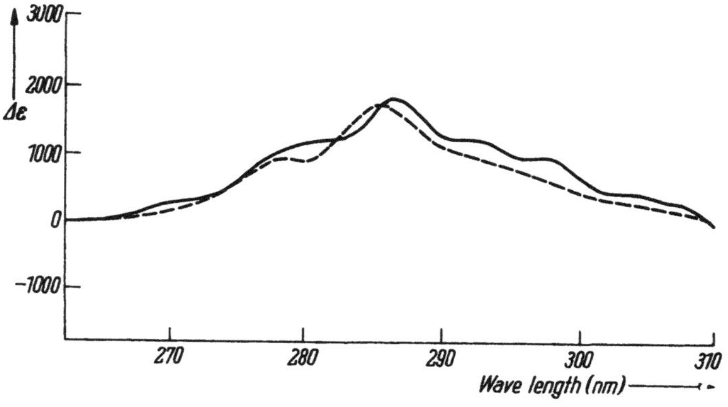

Fig. 3. Solvent perturbation experiments with inorganic pyrophosphatase and a synthetic mixture of model compounds at pH 6.8.

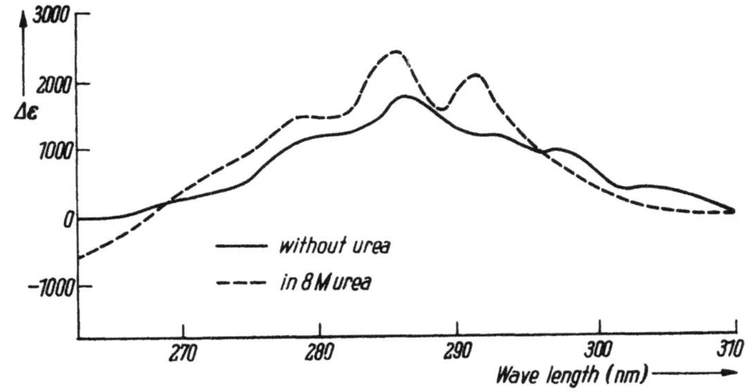

Fig. 4. Influence of 8.0 M urea on the solvent perturbation spectrum of inorganic pyrophosphatase at pH 6.8.

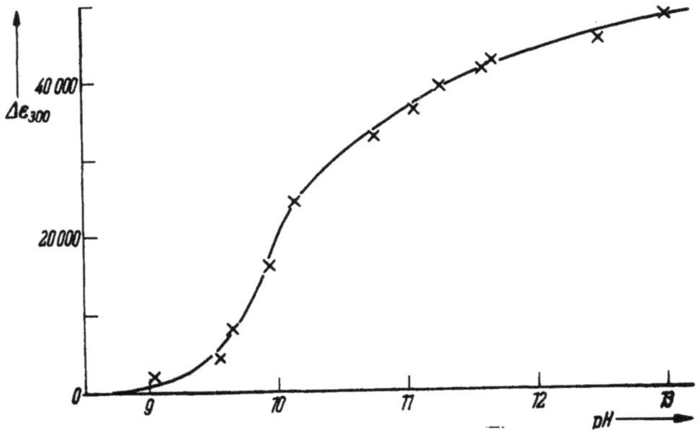

Fig. 5. Titration curve of the tyrosyl residues of inorganic pyrophosphatase.

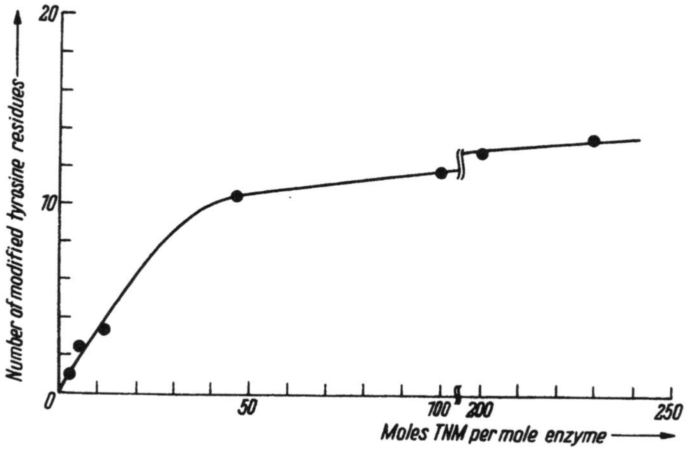

Fig. 6. Modification of the tyrosyl residues of inorganic pyrophosphatase with tetranitromethane (20 °C, pH 8, time 1 h).

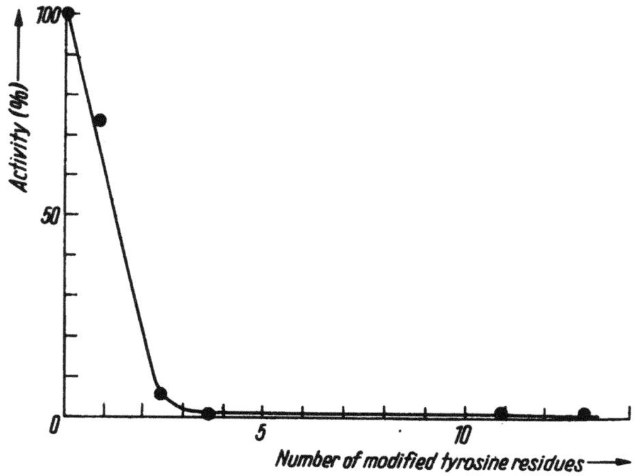

Fig. 7. Dependence of the catalytic activity of inorganic pyrophosphatase on the number of nitrated tyrosyl residues.

Faculty of Biology and Soil Science, Moscow Lomonosov State University,
Institute of Biochemistry and Physiology of Microorganisms,
USSR Academy of Science

I. S. KULAEV

PHOSPHOTRANSFERASE REACTIONS IN THE INORGANIC POLYPHOSPHATE METABOLISM

High polymeric polyphosphates have been found in living organisms as far back as at the close of the last century. The intensive study of their structure and metabolism has however been started only in the middle of this century. The most important results (1) have been obtained in the laboratories of J. P. EBEL, A. N. BELOZERSKY, K. LOHMANN, O. HOFFMANN-OSTENHOF and A. KORNBERG. Studies of J. P. EBEL (3) showed that the high polymeric polyphosphates isolated from the biological materials were the linear polymers, in which the residues of orthophosphates are bound between themselves by the energy-rich bonds as in the case of terminal residues of phosphates in ATP molecule.
The high polymeric polyphosphate structures are shown in Fig. 1.
The molecule of such polyphosphates can contain as much as 300-500 residues of the orthophosphate.
These compounds are rather widely distributed among the living organisms, especially in microorganisms, where they can sometimes account for 10-20% of the dry weight of the cells. But in small quantities they were found in different tissues of both higher plants and animals. Besides information about polyphosphate structure, evidence for enzymes of polyphosphate metabolism was step by step, accumulated.
The first isolated enzyme was ATP: polyphosphate phosphotransferase (4, 5) which catalyzed transfer of energy-rich phosphate residue from ATP to polyphosphates and from polyphosphates to ADP with the formation of ATP according to reaction, presented in Fig. 2.
The most purified form of the enzyme was isolated from E. coli by A. KORNBERG. After the discovery of this enzyme high polymeric polyphosphates have become regarded (6) as some kind of "phosphagens" compounds similar to creatine-

phosphate and argininephosphate. From this point of view polyphosphates may be synthesized and utilized only through the ADP-ATP system. Since polyphosphates are accumulated in substantial quantities in microorganisms they were called microbial phosphagenes. Besides ATP:polyphosphatephosphotransferase another enzyme was isolated, in EBEL's laboratory i. e. polyphosphate: AMP phosphotransferase (7). The enzyme performed the transfer of phosphoryl residues from polyphosphate to AMP with the ADP formation or vice versa.

The discovery of this enzyme seemed to have confirmed once again the thesis that the polyphosphate functions first of all were closely connected with the maintenance of the required level of intra-cellular ATP and ADP. That was one more reason to ascribe them the role of "microbial phosphagenes" (8).

But shortly afterwards, the view-point on physiological role of polyphosphates was changed.

This followed primarily as a result of M. SZYMONA's discovery (9). He found a new enzyme-polyphosphate: glucose-phosphotransferase which (Fig. 3) catalyzes the transfer reaction of phosphoryl group from polyphosphate to glucose followed by glucose-6-phosphate formation in Mycobacteria. The discovery of this enzyme proved on the one hand the possibility of polyphosphate utilisation without ATP-ADP system participation and on the other that in some cases polyphosphates could play the role of ATP itself.

Indeed, the reaction, reported by Dr. SZYMONA, was analogous to the well-known hexokinase reaction, in which glucose phosphorylation proceeds at the expense of ATP. It is interesting to note that it has been shown in that same laboratory recently, that there is a number of adaptive enzymes, which carry out reactions of transfer of phosphate group from polyphosphates also by passing adenylic system not to glucose but to other monosaccharides and their derivatives (10, 11).

Besides M. A. BOBYK, O. SZYMONA and I have recently discovered another enzyme (12, 13) which catalyzed the reaction of polyphosphate synthesis in Neurospora crassa (Fig. 4).

This reaction proceeds at the expense of energy-rich phosphate 1, 3 - diphosphoglyceric acid. The new enzyme - 1, 3-diphosphoglycerate: polyphosphatephosphotransferase, carried out the reaction analogous to the well known reaction of ATP formation in the process of glycolytic phosphorylation.

Thus, in the case of the reaction, discovered by us, high polymeric, polyphosphates seems to play the role usually characteristic of ATP-ADP system. These data present a great interest in the light of the assumption, put forward by A. N. BELOZERSKY in 1957 concerning the evolution of the phosphorous metabolism (14).

In his report made at the I. International Symposium on the problems of origin of life A. N. BELOZERSKY assumed that high polymeric polyphosphates in protobionts could carry out the functions which in contemporary organisms were mainly performed by ATP.

To test this assumption we investigated the distribution of enzymes carrying out the reactions utilizing the polyphosphates and ATP for phosphorylation of the same substrate in different microorganisms which are at different stages of the evolutionary scale.

For this purpose the study of distribution of two enzymes - polyphosphate: glucosephosphotransferase and ATP: glucosephosphotransferase (hexokinase) has been widely carried out in different microorganisms (15-17). Many species of different classes of microorganisms - bacteria, actinomycetes, fungi, green and blue-green algae, have been studied. It was found that hexokinase is present in all the microorganisms studied (Fig. 5), polyphosphate: glucose-phosphotransferase is found only in the organisms, which belong to the class of actinomycetes according to the classification scheme of N. A. KRASILNIKOV (18) and which are included in this class by the common origin according to KLUYVER and van NIEL (19).

Fig. 5 shows that not all the species studied have polyphosphate; glucosephosphotransferase even among the microorganisms which are usually included in the class of actinomycetes by N. A. KRASILNIKOV (18).

It is interesting to point out that almost all the species which do not posses this enzymatic activity are classified as actinomycetes only formally and by far not by all taxonomists.

When the specific activities of the two enzymes of the organisms which have both the enzymes were compared a very interesting fact was established.

Table 1 shows that the ration between the specific activities of these enzymes changes from one species to another.

The specific activity of polyphosphate: glucose-phosphotransferase is considerably higher than that of hexokinase only in the organisms which according to KLUYVER and van NIEL are the most ancient from the phylogenetic point of view in this group. The hexokinase activity is considerably higher in the species which are phylogenetically younger.

Table 1

RATIO OF POLYPHOSPHATE GLUCOKINASE AND ATP-GLUCOKINASE SPECIFIC ACTIVITIES IN VARIOUS MICROORGANISMS
(μM P/hour per mg of protein)

Organism	poly-P-glu-cokinase	ATP-glu-cokinase	poly-P-glucokinase / ATP-glucokinase
Micrococcus lisodeikticus	2	0.4	5
Sarcina lutea	2	0.5	4
Propionbacterium shermanii	6	0.6	10
Mycobacterium tuberculosis	5.2	0.6	9
M. scotochromogenus	14.2	7.1	2
M. phlei	10.1	7.8	1.3
M. smegmatis	16	14	1.2
M. friburgensis	10.6	9.3	1.1
Corynebacterium xerosis	7.5	3.7	2
Nocardia turbatus	3	0.9	3.3
N. gardneri	4	3.4	1.2
N. brasiliensis	5	6.6	0.8
N. madureae	9	1.3	7
N. minima	10.3	9.5	1.1
N. asteroides	4.7	11.6	0.4
Actinomyces olivaceus	3.6	7.6	0.5
A. globisporus	4.8	7.3	0.6
A. fradiae	1.8	16	0.1
A. aureofaciens	2.3	15.1	0.1

Some conclusions can be drawn from these data. First, the presence of some constitutive enzymes, the enzymes of polyphosphate metabolism in particular can be considered as an additional taxonomic feature helping to obtain more specific information about the evolutionary system of microorganisms. Second, if it is true, then the results obtained clearly show that in certain reactions polyphosphates can almost completely replace ATP in the more ancient evolutionary forms, pertaining to the class of actinomycetes. One may suppose that at the very beginning of the origin of life on Earth polyphosphates could really perform functions which further were completely transferred to ATP. It is of great importance that the activity of the enzymatic system which catalyzes the synthesis of polyphosphates at the expense of glycolytic phosphorylation without the participation of ATP is the highest in the phylogenetically ancient microorganisms (20). Thus, on the basis of all the facts mentioned above one can suppose that high-polymeric polyphosphates could have really been the evolutionary precursors of ATP. We are inclined to believe that in protobionts they could be the main compounds (if not the only ones), coupling exergonic and endergonic reactions (see the reactions in Fig. 6).

High polymeric polyphosphates could be synthesized at the expense of 1,3-diphosphoglycericacid formed at the glycolytic splitting of glucose, and on the other hand, they could be used for its primary phosphorylation (21, 22).

Returning again to the enzymes of polyphosphate metabolism we can say that besides the above mentioned enzymes having strongly pronounced phosphotransferase activity in some microorganisms other very interesting enzymes were found, whose action mechanism and role in polyphosphate metabolism are not sufficiently clear. First of all these are different polyphosphatases.

All the polyphosphatases known can be divided into two groups. The polyphosphatases of the first group split highpolymeric polyphosphates to lower molecular fragments. These are the so called polyphosphate-polyphosphohydrolases. They catalyze the reaction, presented in Fig. 7.

The polyphosphatases of the other group carry out the cleavage of orthophosphate molecules one by one at the end of the chain of a polyphosphate.

These enzymes may be called polyphosphate-phosphohydrolases. They are resumed to catalyze the reaction, shown in Fig. 8.

The data accumulated on high molecular polyphosphates permit to assume that physiological role of these two groups of polyphosphatases is basically different. First of all I should like to dwell upon the first group of the above enzymes, namely, polyphosphate-polyphosphohydrolases. It is long known that polyphosphates presented in the cells of microorganisms have different chain lengths. It was found that in microorganisms the higher polymeric fractions of polyphosphates play a role which essentially differs from that of the lower polymeric ones.

Our work performed in the laboratory of A. N. BELOZERSKY (23) and also the work of P. LANGEN in laboratory of K. LOMANN (24) showed that lower molecular polyphosphates are mainly formed from the higher polymeric ones by means of gradual degradation. Further in our laboratory I. A. KRASHENNIKOV and T. P. AFANASYEVA working on the cells of Neurospora crassa and Endomyces magnusii established (25, 26) that high polymeric polyphosphates were localized on the outside of the cells, near the external cytoplasmic membrane and lower polymeric polyphosphates - within the cells, in the nuclei and granules of volutine, in particular. One may suppose that the highest polymeric polyphosphates play some very important role in the functioning of the cytoplasmatic membrane, in the transport of ions and substances through it in particular. The lower polymeric, polyphosphates, which are localized in the volutine granules, may be considered to be a reserve of macroergic phosphate.

Thus, polyphosphatases, depolymerizing high polymeric polyphosphates to lower polymeric fragments, perform apparently a very important function in the cells, namely intracellular translocation of polyphosphates from the outside of the cells inside.

In our laboratory M. S. KRITZKY, E. K. CHERNYSHOVA and G. I. KONOSHENKO have studied polyphosphatases, depolymerizing high polymeric polyphosphates in Neurospora crassa. Two very interesting facts have been established: first, these enzymes split the polyphosphate molecule in the middle of the chain, and second, these enzymes were localized in the cell just with the highest polymeric polyphosphates, i. e. outside of the cell, between the cell wall and cytoplasmic membrane.

These results are in good agreement with the fact that polyphosphate polyphosphohydrolases take part first of all in the transfer of polyphosphates through cell membrane with simultaneous splitting of their molecules.

One may assume that the very long chains of polyphosphates can not penetrate into the cell and a partial fragmentation of them is necessary. From the point of view of Dr. Kritzky it is very likely that the energy of phosphoanhydride bond, released in the course of the polyphosphate splitting, can be used for the translocation of the obtained fragments through the cell membrane.

The other group of polyphosphatases, i. e. polyphosphate-phosphohydrolases, gradually chipping off one molecule of orthophosphate at a time from high polimeric polyphosphates at the end of the chain, play quite a different role in metabolic processes. Together with G. I. KONOSHENKO we studied Neurospora crassa and found that these enzymes as well as depolymerases had been localized mainly or exclusively outside the cell near the external cytoplasmic membrane (27) together with the more physiologically active high polymeric polyphosphates. It seems very likely that polyphosphate phosphohydrolases function in vivo first of all as phosphotransferases releasing at the gradual splitting polyphosphate to orthophosphate the energy for nutrient and ion transport through the cell membrane.

It seems very likely that the action of the enzymes of this kind may be analogous to this of pyrophosphatase and ATPase.

As for pyrophosphatase M. R. STETTEN (28) as well as R. C. NORDLE and W. J. ARION (29) established that it can not only hydrolize pyrophosphate but also at certain conditions transfers one of the chipped phosphates to one or the other substrate. On the other hand H. BALTSCHEFFSKY and M. BALTSCHEFFSKY (30) showed that this enzyme could take part in reverse electron transport along the respiratory chain and also in transport of ions through membrane.

As to ATPase it is well known that ATPase takes part in transport of ions and substances through different cell membranes.

Not very long ago van STEVENINCK et al. (31, 32) have experimentally established the participation of polyphosphates in sugar transport through cytoplasmic membrane of yeasts. At the same time they found that polyphosphates took part in phosphorylation of phosphatidylglycerate to phosphatidylglycerophosphate in membrane, the latter being probably the donor of phosphate for the formation of an active complex of the carrier and the transported sugar.

The reaction postulated by van STEVENINCK et al. can be seen in Fig. 9.

From our point of view it seems very likely that polyphosphate phosphohydrolases take part in such phosphotransferase reactions connected with the transport of substances through the external cytoplasmic membrane.

It is very interesting to note that another specific enzyme of the same kind has been found in a number of microorganisms: tripolyphosphate-phosphohydrolase (or tripolyphosphatase) (33-35) which hydrolizes tripolyphosphate to orthophosphate and pyrophosphate according to the scheme, presented in Fig. 10.

According to our results this enzyme, obtained from Neurospora crasse, is mainly localized in mitochondria (27). It is necessary to underline that all the enzymes of high molecule polyphosphate metabolism (as well as polyphosphates themselves) are completely absent in these organelles. At the present time the role of this enzyme in mitochondria is quite obscure but it is very likely that this enzyme is similar to a pyrophosphatase taking part in the reverse transport of electrons along the respiratory chain as well as in ion transport through the mitochondrial membrane. However this supposition requires further investigations.

One of the works of our laboratory, made jointly with E. I. VOROBYOVA, deals with the detection of the potential ability of different enzymes, participating in polyphosphate metabolism, to biosynthesis and utilization of these phosphorus compounds.

This work was performed on one organism - Propionibacterium shermanii. Moreover, it was shown that the accumulation of polyphosphates at different culture growth stages is in good agreement with the potential possibilities of the action of ATP: polyphosphatephosphotransferase, 1. 3-diphosphoglycerate; polyphosphatephosphotransferases and polyphosphate - phosphohydrolases. The increase of the two former enzymes activities coincided with the polyphosphate accumulation rate observed at the early stage of P. shermanii development. The polyphosphate-phosphohydrolase activity, arising at the stationary phase of growth correlated very well to the termination of polyphosphate accumulation in cells observed at this stage of growth.

Thus, it is the activity of these enzymes only that explains the character of the polyphosphate accumulation curve in the studied organisms. In addition, the very active polyphosphate: glucosephosphotransferase, taking part in phosphorylation of glucose to glucose-6-phosphate, has been revealed in the same cells. Its activity was found to be two orders higher than the activities of the enzymes mentioned above and it failed to agree with the accumulation curve of high polymeric polyphosphates in cells of P. shermanii. It is very difficult to explain the results obtained for the present.

It may be connected either with the latent condition of polyphosphate: glucosephosphotransferase in cells or with our poor knowledge of the main enzymatic mechanism of polyphosphates biosynthesis which could supply this enzyme with the substrate required. The second suggestion seems to be more likely. If it is true, then there must exist in P. shermanii as well as in some other organisms, some yet unknown pathways of polyphosphate biosynthesis, working about a hundred times more actively than that of pathways known at present.

It seems very probable and tempting that this pathways may turn out to be biosynthesis of high polymeric polyphosphates at the expense of pyrophosphate according to the reaction presented in Fig. 11.

Availability of the close correlative relationship between accumulation of certain fractions of polyphosphates, RNA and polysaccharides in fungi cells (36-38) gives us indirect evidence for the existence of this kind of reaction. This hypothetic assumption concerning the possibility of biosynthesis of high polymeric polyphosphates at the expense of nucleic acids and polysaccharides at the expense of pyrophosphate released during the biosynthesis reactions is the only one which can explain the possibility of such correlation. Usually it is considered that pyrophosphates formed in different biosynthesis reactions is split by pyrophosphatase to orthophosphate (39, 40). It seems to us that in some cases there may take place not the hydrolysis of pyrophosphate but the transfer of one of the phosphates from pyrophosphate to polyphosphate with the lengthening of its molecule and reservation of energy of pyrophosphate linkage.
Such a way of pyrophosphate removal from the reaction seems to be far more advantageous from the point of view of cell economy. At present we are working on the proof to this possibility and this idea seems rather realistic.
In conclusion I would like to remind you that at present there is a certain progress in studying phosphotransferases and other enzymes participating in polyphosphate metabolism. However one can hope that the next few years will bring us many new facts which will help us to understand clearly the physiological role of high polymeric polyphosphates and the participation of various phosphotransferases in polyphosphate metabolism.

REFERENCES

(1) KULAEV, I.S., A.W. BELOZERSKY, Izv. Akad. Nauk USSR Ser. biol. 3, (1962), 354 and 4, (1962), 502

(2) HAROLD, F.M., Bacteriol. reviews 30, (1966), 772

(3) EBEL, J.P., Bull Soc. Chem. Biol. 34, (1952), 330

(4) HOFFMANN-OSTENHOF, O., J. KENDY, K. KEEK, O. GABRIEL, K. SCHONFELLINGER, Biochim. Biophys. Acta 14, (1954), 285

(5) KORNBERG, A., S. KORNBERG, E. SIMMS, Biochim. Biophys. Acta, 20, (1956), 215

(6) HOFFMANN-OSTENHOF, O., Colloq. Intern. Centre, Natl. Rech. Sci. (Paris) 106, (1962), 640

(7) DIRHEIMER, G., J.P. EBEL, Compt. Rend. Soc. Biol. 260, (1965), 3787

(8) LANGEN, P., Biol. Rundsch. 2, (1965), 145

(9) SZYMONA, M., M. OSTROWSLI, Biochim. Biophys. Acta, 85, (1964), 283

(10) SZYMONA, O., T. SZUMILO, Acta Biochim. Polon. 17, (1966), 129

(11) SZYMONA, O., H. KOWALSKA, M. SZYMONA, Annales Universitatis M. Curie-Sklodowska (Lublin) 24, (1969), 1

(12) KULAEV, I.S., O. SZYMONA, M.A. BOBYK, Biochimia, 33, (1968), 419

(13) KULAEV, I.S., M.A. BOBYK, Biochimia 36, (1971), 426

(14) BELOZERSKY, A.N., In book "Origin of Life on the Earth", "Nauka", Moscow 1959, p. 370

(15) SZYMONA, O., S.O. URYSON, I.S. KULAEV, Biochimia, 32, (1967), 495

(16) URYSON, S.O., I.S. KULAEV, Dokl Akad. Nauk, 183, (1968), 957

(17) URYSON, S.O., I.S. KULAEV, Biochimia 35, (1970), 601

(18) KRASILNIKOV, N.A., "Mannual of Bacteria a Actinomycetes", "Nauka", 1942

(19) KLUYVER, A.J., C.B. van NIEL, Zentr. Bacteriol. Abt., 94, (1936), 369

(20) KULAEV, I.S., M.A. BOBYK, N.N. NIKOLAEV, N.A. SERGEEV, S.O. URYSON, Biochimia 36, (1971), 5

(21) KULAEV, I.S., In book "Abiogenesis and Initial Stages of the Evolution of Life" "Nauka", 1968

(22) KULAEV, I.S., In book "Molecular Evolution" vol. 1, "Chemical Evolution and Origin of Life". Amsterdam 1971

(23) KULAEV, I.S., A.N. BELOZERSKY, Biochimia 22, (1957), 587

(24) LOHMANN, K., P. LANGEN, Biochem. Z., 328, (1956), 1

(25) KULAEV, I.S. T.P. AFANASIEVA, I.A. KRASHENINNIKOV, S.E. MANSUROVA, Proc. of 2nd Intern. Symp. on Yeast protoplasts, Brno, 1970

(26) KULAEV, I.S., T.P. AFANASIEVA, Autonie vanLeeuwenhock, Suppl. Yeast Symposium, 35, (1969), 13

(27) KULAEV, I.S., G.I. KONOSHENKO, A.M. UMNOV, Biochimia, 36, (1971), 6

(28) STETTEN, M.R., J. Biol. Chem. 239, (1964), 3576

(29) NORDLE, R.C., W.J. ARION, J. Biol. Chem. 239, (1964), 1680

(30) BALTSCHEFFSKY, H., M. BALTSCHEFFSKY, L.V. von STEDINGK "Progress in Photosynthesis Research", 3, (1969), 1313

(31) STEVENINCK, J. van, H.L. BOOIJ, J. Gen. Physiol. 48, (1964)

(32) DEIERKAUF, F.A., H.L. BOOJ, Biochim. Biophys. Acta, 150, (1968), 214

(33) HAROLD, F.M. R.L. HAROLD, J. Bact. 89, (1965), 1862

(34) FELTER, S., G. DIRHEIMER, J.P. EBEL, Bull. Soc., Chim. Biol. 52, (1970), 437

(35) KULAEV, I.S., G.I. KONOSHENKO, Biochimia 36, (1971), 4

(36) KRITSKY, M.S., T.A. BELOZERSKAJA, I.S. KULAEV, Dokl. Akad., Nauk, USSR, 180, (1968), 746

(37) KRITSKY, M. S., E. K. CHERNYSHOVA, I. S. KULAEV, Dokl. Akad. Nauk, USSR, 192, (1969), 1166

(38) KULAEV, I. S., V. H. VAGABOV, A. B. ZYOMENKO, Biochimia in press (1972)

(39) HOFFMANN-OSTENHOF, O., L. SLECHTA, 1958 Proc. Intern. Symp. Enzym. Chem., Tokyo, Kyoto, 2, (1957), 180

(40) KORNBERG, A., In book "Horisons in Biochemistry, Ed. M. KASHA and B. PULLMAN, Acad. Press, N. Y. a London 1962

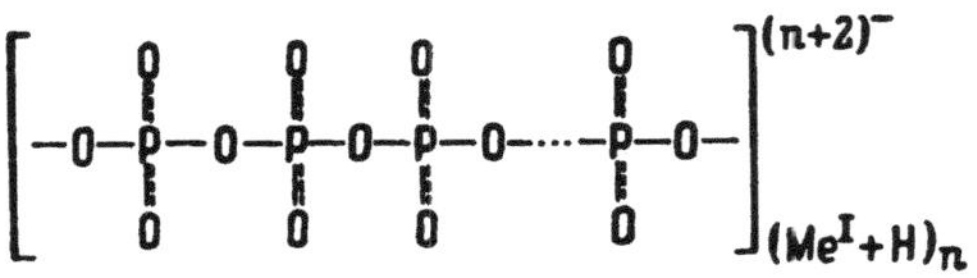

Fig. 1. Inorganic polyphosphate.

$$\text{ATP} + (\text{poly}-\text{P}) \longrightarrow \text{ADP} + (\text{poly}-\text{P})_{n+1}$$

Fig. 2. The scheme of ATP: polyphosphate phosphotransferase reaction.

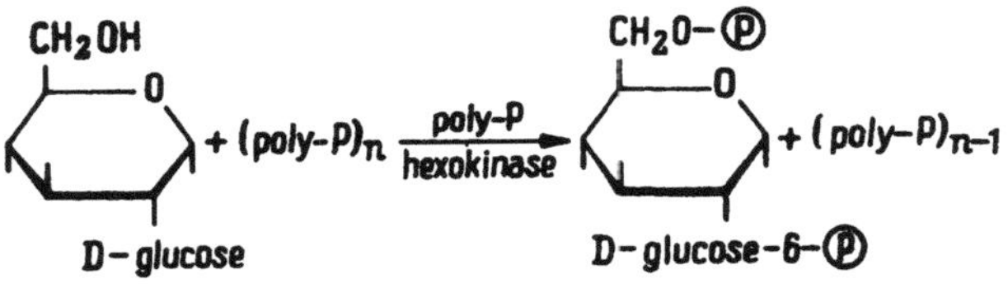

Fig. 3. The scheme of polyphosphate hexokinase reaction.

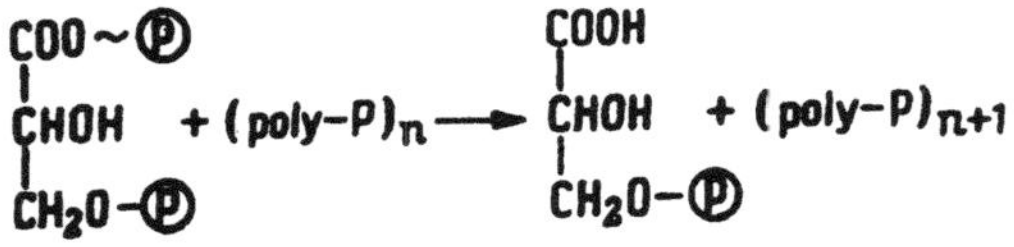

Fig. 4. The scheme of 1, 3-diphosphoglycerate: polyphosphate phosphotransferase reaction.

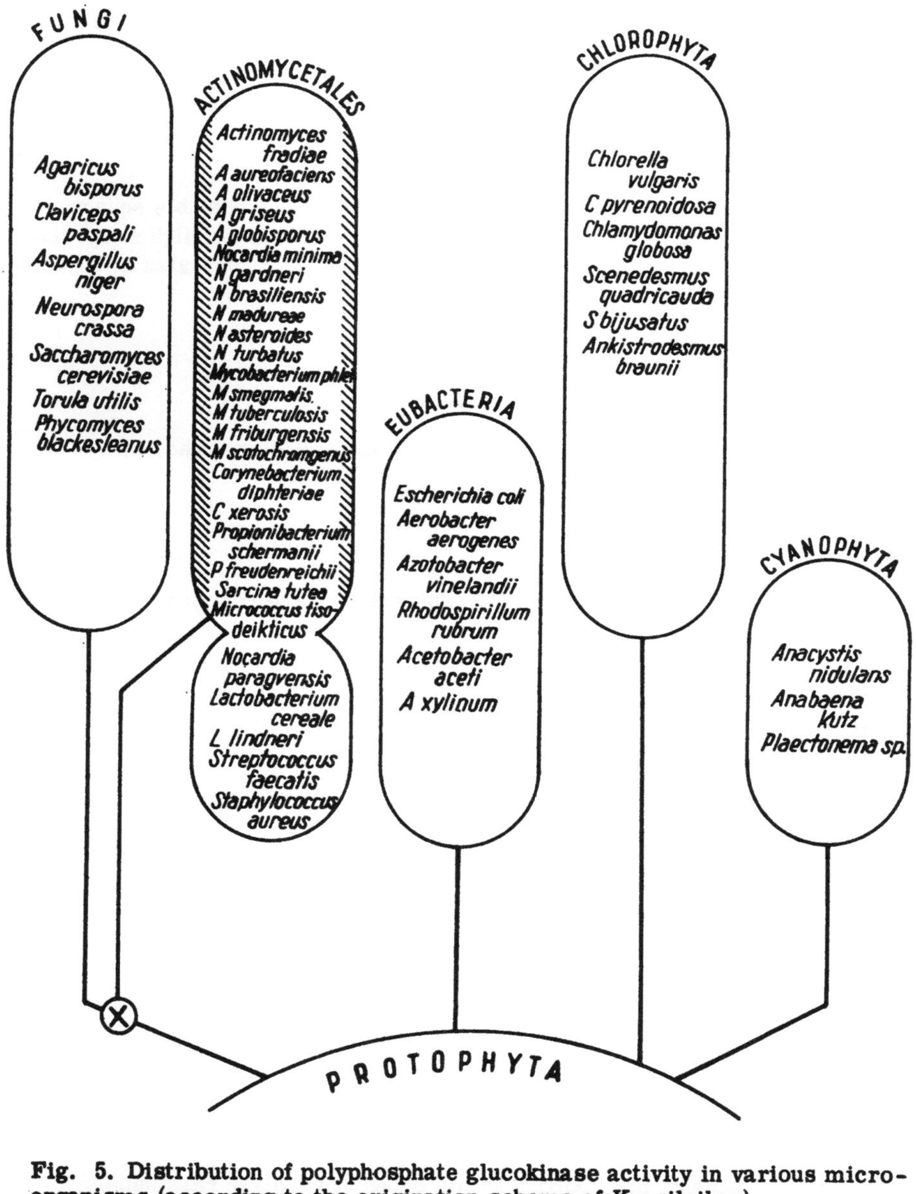

Fig. 5. Distribution of polyphosphate glucokinase activity in various micro-organisms (according to the origination scheme of Krasilnikov).

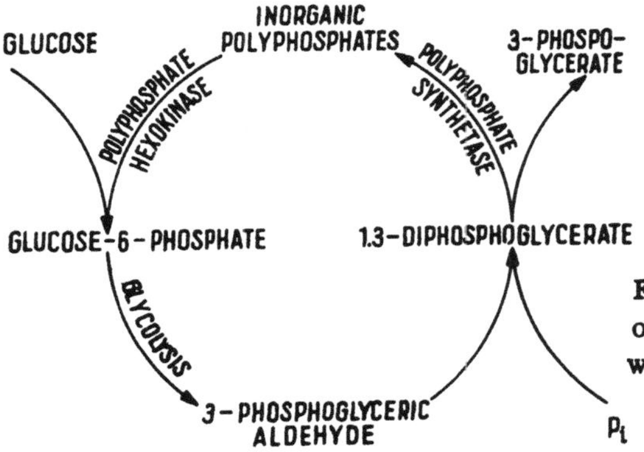

Fig. 6. The possible scheme of metabolism of high molecular weight polyphosphates in Protobionts.

$$(\text{poly-P})_n + H_2O \longrightarrow (\text{poly-P})_{n-x} + (\text{poly-P})_x$$

Fig. 7. The scheme of polyphosphate polyphosphohydrolase reaction.

$$(\text{poly-P})_n + H_2O \longrightarrow (\text{poly-P})_{n-1} + P_i$$

Fig. 8. The scheme of polyphosphate phosphohydrolase reaction.

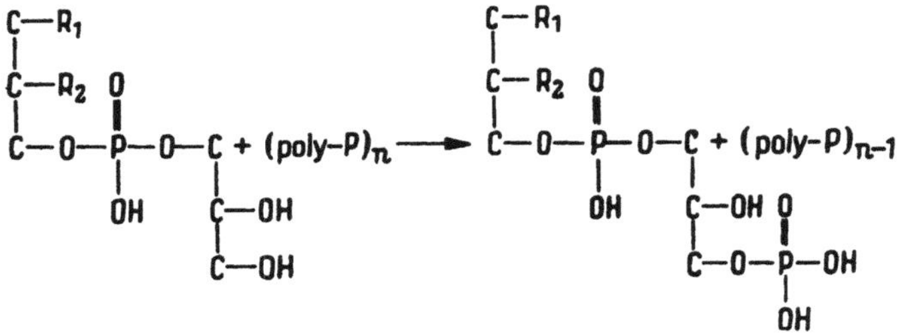

Fig. 9. The possible scheme of phosphatidylglycerol: polyphosphate phosphotransferase reaction.

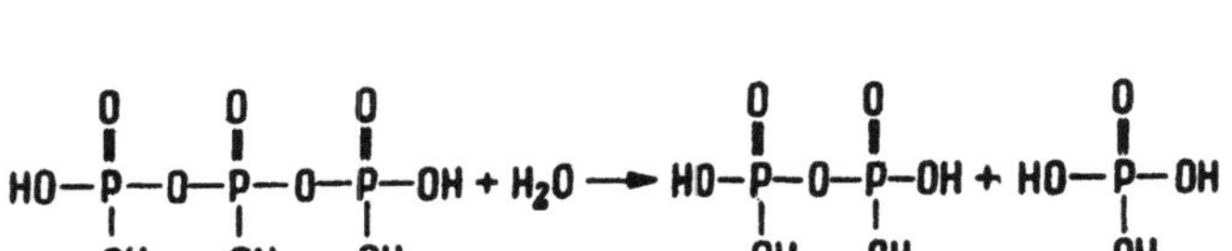

Fig. 10. The scheme of tripolyphosphate phosphohydrolase reaction.

$$PP + (\text{poly-P})_n \longrightarrow (\text{poly-P})_{n+1} + P_i$$

Fig. 11. The possible scheme of pyrophosphate: polyphosphate phosphotransferase reaction.

Bach Institute of Biochemistry of the USSR Academy of Sciences, Moscow

M. S. KRITSKY, E. K. CHERNYSHOVA

THE STUDY OF POLYPHOSPHATE DEPOLYMERASE ACTIVITY OF NEUROSPORA CRASSA

There are two distinct groups of enzymes catalyzing hydrolytic degradation of linear inorganic polyphosphates: polyphosphate orthophosphohydrolases, splitting off terminal phosphates residues, and polyphosphate polyphosphohydrolases, or polyphosphate depolymerases, hydrolyzing P-O-P linkages in the nonterminal sites of the molecule. Up to now our knowledge about the latter group was based on few communications only (INGELMAN, MALMGREN, 1949; KRISHNAN, 1952; NISHI, 1960). In the present work we have demonstrated polyphosphate polyphosphohydrolase activity (E. C. 3. 6. 1. 10) in the cells of Neurospora crassa and studied some its properties.

The enzyme preparation have been isolated from N. crassa 28610 as mycelium in the end of exponential phase of growth. The cells have been disrupted in French press and extracted with 0. 1 M Na-acetate buffer, pH 3. 3. As the substrates, synthetic preparations of Na-polyphosphates with $\bar{n}$ 290; 180; 72; 47 have been used. The standard incubation mixture contained in a final volume 3 ml: Na-acetate, pH 3, 3, 300 micromole; polyphosphate, 25 micromole of acid labile phosphate; cell free extract (17. 000 g supernatant), 0. 5 - 1. 0 mg of protein. When the effect of some additions has been studied, the extract was previously dialyzed for 24-48 hours against water and 0, 1 M Na-acetate, pH 3, 3. The samples have been incubated at 30°C. Degradation of polyphosphates has been assayed viscosimetrically.

In the course of incubation of polyphosphate preparations with a cell free extract of N. crassa some decrease of specific viscosity of the incubation mixture has been observed. This decrease has been sufficiently high, if polyphosphates with initial $\bar{n}$ values 290 and 180 have been used, but when studying the substrates with $\bar{n}$ 72 and 47, we failed to show any fall of specific viscosity, but negligable. This failure has been probably caused by a very low specific viscosity of these polyphosphate preparations, that make impossible a reliable measurement of its change. For this reason, all further experiments have been done, using substrates with $\bar{n}$ 290 and 180. Since we have observed no decrease of the specific viscosity in control experiments, when either polyphosphate or enzyme preparation were taken separately, we have suggested, that a polyphosphate degrading enzyme is present in the extract from N. crassa. Since orthophosphate and low molecular condensed phosphates (n 2-5) were absent from the reaction products, thus we concluded, that P-O-P linkages sufficiently remote from the ends of polyphosphate molecule were hydrolyzed. Basing on viscosimetric data, we have estimated a mean polymerization degree of the products after 30 minute incubation as varying from 60 to 100, if polyphosphates with $\bar{n}$ 290 or 180 were taken as substrates. To express the enzyme activity as the amount of P-O-P linkages splitted, we have suggested the following formula:

$$A = \left(\frac{\bar{n}_o}{\bar{n}_t} - 1\right)\left(\frac{C_{PolyP}}{\bar{n}_o}\right),$$

where A - microequivalents of P-O-P linkages splitted; C_{PolyP}-polyphosphate concentration (micromole of acid labile phosphate); $\bar{n}_o$ and $\bar{n}_t$ - mean polymerization degrees of polyphosphate preparations prior incubation and after incubation for time t. The value of $\bar{n}_t$ could be estimated from a curve of dependence of specific viscosity of polyphosphate under the conditions of incubation mixture from its $\bar{n}$.

The effect of enzyme concentration on the rate of depolymerization of polyphosphate:
The plot in Fig. 1 demonstrates, that the rate of enzymatic depolymerization of polyphosphate is a direct function of enzyme concentration.

The effect of pH:
As it is shown in Fig. 2, a sharp pH-optimum of enzymatic depolymerization of polyphosphates is located in the acidic zone, at pH 3, 2-3, 4. In addition to it, some decrease of specific viscosity of polyphosphate solution in the presence of cell free extract from N. crassa is observed under alkaline conditions (0, 1 M glycylglycine buffer). Thus, polyphosphate depolymerase markedly differs in its pH-dependence from polyphosphate orthophosphohydrolase of N. crassa KONOSHENKO, 1971) which has a pH-optimum in a neutral interval. It is interesting, that the same pattern of pH-dependence was observed for the non-enzymatic hydrolysis of polyphosphates (MC CULLOUGH, VAN WAZER, GRIFFITH, 1956): at neutral pH values terminal

phosphates were splitted off mainly, and under acidic or alkaline conditions, hydrolysis in nonterminal sites of the molecule prevailed.

The effect of divalent metal ions:
Polyphosphate depolymerase activity of N. crassa is stimulated by Zn^{2+}, Mg^{2+}, Ba^{2+} and Mn^{2+} (1.10^{-5} - 1.10^{-3} M). Ca^{2+} has no effect, or in low concentration (1.10^{-5} M) to some extent inhibits its activity (Table 1).

Table 1

THE INFLUENCE OF DIVALENT IONS ON THE POLYPHOSPHATE DEPOLYMERASE ACTIVITY (PER CENT OF ACTIVITY OF DIALYZED PREPARATIONS)

Concentration, M	Zn^{2+}	Mg^{2+}	Ba^{2+}	Mn^{2+}	Ca^{2+}
1.10^{-5}	200	170	140	140	60
1.10^{-4}	330	270	240	230	120
1.10^{-3}	650	390	200	-	-

Substrate: PolyP$_{\bar{n}}$ 290, $0.85 \cdot 10^{-2}$ M

The effect of substrate concentration:
It can be seen in Fig. 3, that the excess of substrate inhibits the rate of polyphosphate depolymerization by a non-dialyzed enzyme preparation. The inhibition may be most likely explained by the presence in the cell free extract of metal cations, activating the reaction. The increase of substrate concentration, probably, leads to the formation of metal-polyphosphate complexes which is resulted in disappearance of free metal ions from the reaction and the decrease of reaction rate.
After dialysis of the enzyme preparation, no inhibition of the reaction rate by the substrate is observed. It can be seen, however, that the dependence of reaction velocity from the Michaellis-Menten hyperbola. This deviation is especially obvious when plotting the same data on a double reciprocal plot. Here, instead of a straight line characteristic for classical Michaellis-Menten kinetics, we find the curve concave upward. The curves of such a shape are observed when two or more active sites with positive cooperativity between them participate in the process (LEVITZKI, KOSHLAND, 1969).

The effect of orthophosphate on the reaction rate:

Since orthophosphate concentrations both inside and outside the cell, are the key factors in the regulation of the activity of enzymes participating in phosphorous metabolism, we have studied the effect of orthophosphate on polyphosphate depolymerase activity. At the concentration of substrate (PolyP$_{\bar{n}290}$) 8 x 10^{-3} M the permanent diminishment of the activity as a function of increased orthophosphate concentration is observed, and at 5 x 10^{2} M orthophosphate completely inhibits the hydrolysis.

However, further investigation has revealed the influence of orthophosphate on the depolymerase activity is more complex, than mere inhibition. When studying the effect of orthophosphate (1, 3 x 10^{-3} M), under various concentrations of substrate, we have found, that inhibitory effect is manifested only with sufficiently high substrate concentration, not less than 0, 85 · 10^{-2} M. Under the smaller concentration of polyphosphate in the incubation mixture, orthophosphate enhances the rate of polyphosphate depolymerization (Fig. 4).

Such double effect of orthophosphate may point to its participation in regulation of polyphosphate depolymerase activity, probably, as an effector of allosteric type. However, the complicated character of this process makes rather difficult the interpretation of its physiological significance in the regulation of polyphosphate metabolism.

Polyphosphate depolymerase activities in the course of mycelium growth:

The data mentioned above, has given no answer, whether various polyphosphate depolymerases exist, specifically adapted for the substrates with various $\bar{n}$, or degradation of polyphosphate fractions with various n is carried out by the same enzyme molecules. Some light onto this problem could be thrown from the comparison of the changes of activities splitting polyphosphates with different $\bar{n}$ during the growth cycle (Fig. 5). The activity, depolymerizing polyphosphate with initial $\bar{n}$ 290, gradually increases from the moment of condia germination, reaches a maximal value to the end of exponential growth, and falls down after that. But in the experiments with polyphosphate $\bar{n}$ 180 taken as a substrate, in addition to the peak at the end of exponential phase, observed with the substrate $\bar{n}$ 290, the other one at the beginning of the exponential phase was revealed. Such different kind of behaviour of the activities, splitting polyphosphate preparations with $\bar{n}$ 290 and 180, may evidence for a possible existence of polyphosphate depolymerase hydrolyzing in aspecific manner the certain polyphosphate fractions.

Intracellular localization of polyphosphate depolymerase activity:
The study of intracellular localization of polyphosphate depolymerase activity carried out in collaboration with G. Konoshenko, has revealed virtually all activity (either splitting polyphosphate with $\bar{n}_0 290$, or $\bar{n}_0$ 180) to be located in a periplasmic space, i. e. between cell wall and outer membrane. The removal of a cell wall by digestion with an enzyme complex of digestive sap of Helix pomatia leads to almost complete loss of activity (Table 2). No activity has been found in mitochondria and in hyaloplasm. On the same time, a minute activity could be observed in other components of the cell, in nuclei and microsomes. It is, however, so small, that entirely might be a result of contamination of these fractions with fragments of the outer cell membrane.

Table 2

INTRACELLULAR LOCALIZATION OF POLYPHOSPHATE DEPOLYMERASE ACTIVITY (m U/mg of protein)

Substrate	Whole cells	Proto-plasts	Nuclei	Mito-chondria	Micro-somes	Hyalo-plasma
$PolyP_{\bar{n}290}$	7.8	0.6	0.15	0.0	2.2	0.0
$PolyP_{\bar{n}180}$	9.4	1.3	0.20	0.0	-	0.0

It should be mentioned here, that polyphosphate depolymerases splitting polyphosphates with $\bar{n}$ 290 and $\bar{n}$ 180, i. e. with a polymerization degrees close to that of the most high molecular polyphosphate fractions of N. crassa (CHERHYSHOVA, KRITSKY, KULAEV, 1971), have been found in the vicinity of their substrates, high molecular polyphosphate fractions which are also localized in a periplasmic space (KRASHENINNIKOV et al., 1971).

By present views, polyphosphates are synthesized in the cell de novo, as the most high molecular fraction and then splitted by depolymerases to form various fractions with smaller $\bar{n}$ values (KULAEV, BELOZERSKY, 1962). If so, it would be reasonable to suggest, that polyphosphate depolymerases present in periplasmic region, participate in the involvment of the polyphosphates, formed out of cell membrane, into the intracellular metabolism. We can believe now, that these enzymes are related with the transfer of newly synthesized polyphosphate molecules through the cell membrane into the protoplasm. In this case the energy produced by enzymatic splitting of P-O-P linkage might be utilized in the transfer of the polyphosphate molecules through the cell membrane into the inner zone of the cell.

REFERENCES

CHERNYSHOVA, E.K., M.S. KRITSKY, I.S. KULAEV, Biokhimia, 36, (1971), 138 (in Russian)

INGELMAN, B., H. MALMGREN, Acta Chem. Scand., 3, (1949), 157

KRASHENINNIKOV, I.A., G.I. KONOSHENKO, S.E. MANSUROVA, A.M. UMNOV, I.S. KULAEV, This volume, 10, (1971)

KONONSHENKO, G.I., This volume, communication, 9, (1971)

KRISHNAN, P.S., Arch. Biochem. Biophys., 37, (1952), 224

KULAEV, I.S., A.N. BELOZERSKY, Izvestia Akademii Nauk SSSR (Seria biologicheskaja), 4, (1962), 402 (in Russian)

NISHI, A., J.Biochem., (Japan), 48, (1960), 758

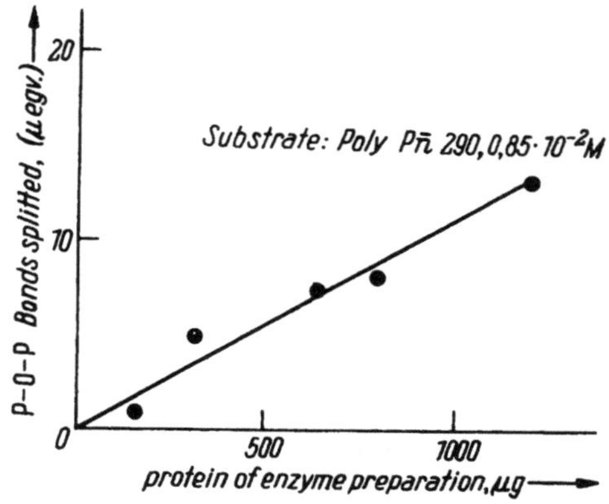

Fig. 1. The effect of enzyme concentration on the reaction rate.

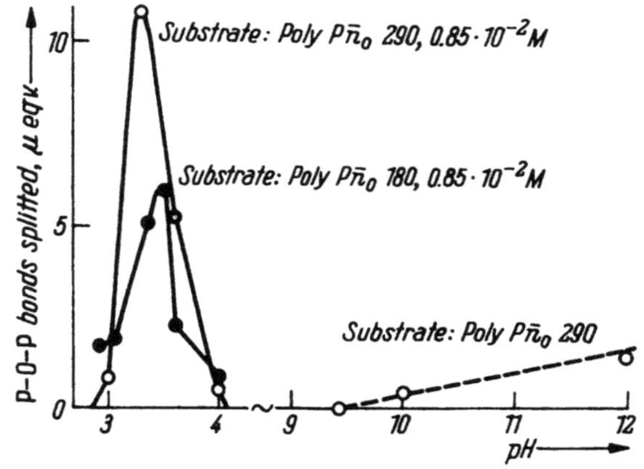

Fig. 2. The effect of pH on polyphosphate depolymerase activity.

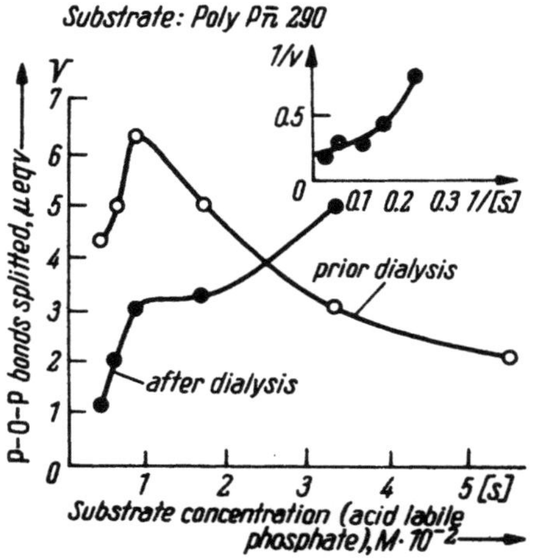

Fig. 3. The effect of substrate concentration on the reaction rate.

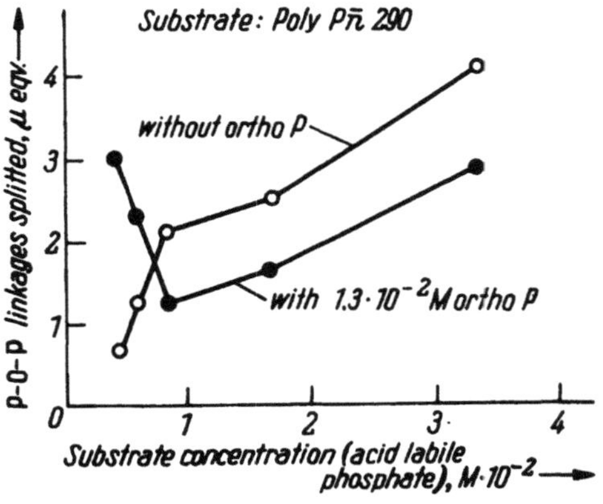

Fig. 4. The effect of orthophosphate upon the rate of polyphosphate depolymerase activity under different substrate concentrations.

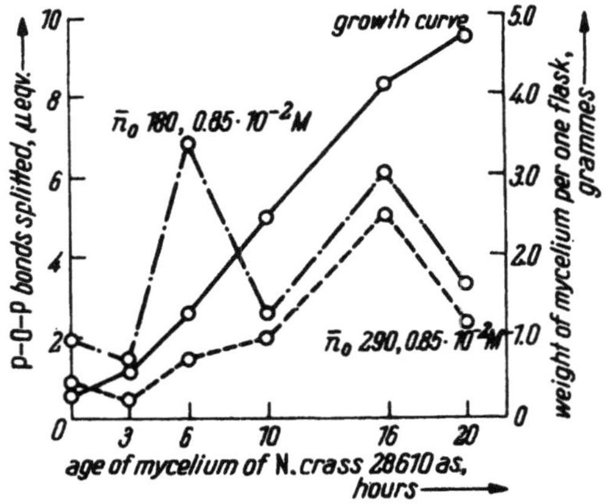

Fig. 5. The changes of polyphosphate depolymerase activities splitting polyphosphates with $\bar{n}_0$290 and 180 in the course of mycelium growth.

Bach Institute of Biochemistry of the USSR Academy of Sciences, Moscow, and Lomonosov State University, Moscow

TATIANA P. AFANASIEVA, V.V. ROTSCHANETZ

SOME PROPERTIES OF POLYPHOSPHATASE FROM Endomyces magnusii

The present communication is concerned, in part, with the study of inorganic polyphosphate metabolism. It was headed by I. S. KULAEV who summarized its results and reported them to this Symposium.

In studying the polyphosphates from Neurospora crassa and Endomyces magnusii it has been found that any destruction of their cell structure results in fast hydrolytic splitting of the high polymer polyphosphates on the outer cytoplasmic membrane to form orthophosphate. This process is likely to proceed under the action of the polyphosphatases which are also bound with the outer cell membrane, and are in the close vicinity of their substrates (KULAEV et al., 1970a).

The polyphosphatases which hydrolyze high molecular polyphosphates to form orthophosphate were found in some microorganisms by a number of workers (MUHAMMED et al., 1959; HAROLD, HAROLD, 1965; FELTER et al., 1970). But the data available show that the properties of polyphosphatases from different biological sources such as optimal pH or the nature of activators and inhibitors for the polyphosphatase reaction, vary considerably.

The object of this work was to study the properties of the polyphosphate phosphohydrolase from E. magnusii. Previously we have investigated the conditions for accumulation and specific intracellular localization of various polyphosphate fraction (KULAEV et al., 1967; AFANASIEVA et al., 1968; KULAEV et al., 1970b).

The yeast cells used in the experiment were harvested at the end of the logarithmic phase. The cells were crushed in a French-press. The crushed material was suspended in 0.05 M tris+maleate buffer pH 7.1 and then centrifuged at 20.000g and 0° for 30 min. The supernatant was dialyzed against buffer to remove salts and low molecular organic compounds. The resulting preparation could be stored for a long time at -15° without any appreciable loss of activity. The enzyme is rendered completely inactive by heating for 20 min. at 45°.

The enzyme may be partly purified by precipitation with ammonium sulphate. The specific activity of the fraction appearing between 60 and 80 % saturation with ammonium sulphate was more than twice as high as the initial one.
A synthetic linear polyphosphate with a chain length of n=290 was used as a substrate for the enzyme reaction. The activity of the polyphosphatase was estimated from the increase of orthophosphate in the reaction mixture. The temperature of incubation mixture was 37°. The linear dependence of the reaction rate on time was recorded during the first 10 minutes.
The effect of pH on the enzyme activity was studied in 0. 1 M tris+maleate buffer at pH 5. 5 - 8. 5 (Fig. 1). The pH optimum was found to be about 7. 1. A high ion strength of the incubation mixture containing KCl (or NaCl) considerably hinders the polyphosphatase reaction. The salt concentration of 0. 4 M almost completely inhibited the reaction (Fig. 2).
Next, the effect of divalent metal ions on the polyphosphatase activity was studied. With such cations added to the preparation after its dialysis, the enzyme activity becomes 10 - 20 times as high. Fig. 3 shows the effect of various metal ion concentrations on the polyphosphatase activity. Mn^{++} at a final concentration of 2×10^{-3} M is the best activating agent. The activity of Co^{++} is almost as high as that of Mn^{++}. The effect of Mg^{++} is practically the same as that of Zn^{++} and is about 60 % of that of Mn^{++}. Ni^{++} promotes the reaction but only slightly. The effect of the above cations decreases with increase of their concentration from 3×10^{-3} to 10^{-2} M. Fe^{++} is an activator of polyphosphatase when its concentration is 10^{-3} M and is an inhibitor at 10^{-2} M. Fe^{3+} does not show any activating properties. The optimal concentration ratio of the activating ion and substrate phosphate is equal to unity. EDTA completely removes the effect of cations.
The composition of the incubation medium optimal for the determination of the polyphosphatase activity is: 2×10^{-2} M tris+maleate buffer pH 7. 1; 2×10^{-3} M $MnCl_2$; 2×10^{-3} M (P) polyphosphate, n=290. The K_m value for the polyphosphatase from E. magnusii as determined according to LINEWEAVER and BURK (1934) is 3×10^{-4} M of phosphate in its polymer form with n=290 (Fig. 4).
Since inorganic polyphosphates are polyanions the cations of the surrounding medium should be responsible for the behaviour of the molecules of these compounds. The effect of some metal ions on the polyphosphatase reaction which has been found may be due to cation binding to a substrate moleculs. It results in specific changes of polyphosphate conformation and affects the following formation

of the enzyme-substrate complex and subsequent hydrolysis. The activating effect of the cations under study allows the following sequence:

$$Mn^{++} > Co^{++} > Mg^{++} = Zn^{++} > Fe^{++} > Ni^{++}$$

It should be pointed out that if another activating cation is added to the incubation mixture containing Mn^{++} the effect of the latter either increases or these cations are incompatible. Fig. 5 shows how simultaneous action of a number of cations of various concentration affects the polyphosphatase reaction rate. It has been found that Co^{++} increases the Mn^{++} effect, Mg^{++} does not almost affect the reaction, Ni^{++} and Zn^{++} decrease the Mn^{++} effect. Fe^{++} stops the reaction completely. Cu^{++}, Ca^{++}, Hg^{++}, Al^{3+}, Ag^{+} (Fig. 6) and also Cd^{++} and UO_2^{++} show inhibition effect independent of the presence of Mn^{++}.

According to WILLIAMS (1959) the cations which form less stable complexes with some compounds and inhibit their non-enzymatic hydrolysis activate the enzymatic decomposition of these compounds and vice verse. According to van STEVENINCK (1966) Mg^{++}, Zn^{++}, Co^{++}, and Ni^{++} inhibit the nonenzymatic hydrolysis of polyphosphates. At the same time we have shown that it is these cations which have an activating effect on the polyphosphatase reaction. Whereas uranyl which superimposes on the H^{+}-catalyzed hydrolysis of polyphosphate hinders enzymatic decomposition of these compounds. Thus, our findings are in accord with the Williams rule and enable definition of the polyphosphatase from E. magnusii as an enzyme which shows its activity only in the presence of Mn^{++}, Co^{++}, Mg^{++} or Zn^{++}. The metal ions seem to play an important role both in specific binding of the substrate and its combining with an enzyme molecule. This is evidenced by the fact that the preparation loses its activity after dialysis or in the presence of EDTA. It is also supported by the interchangeability of the ions and non-stechiometric ratio metal/protein. However, one should not rule out the possibility of metal presence in the active centre of the polyphosphatase as well.

The inhibition effect of p-Cl-Hg-benzoate on polyphosphatase and protecting properties of cystein during storage of the enzyme preparation and its incubation with a substrate provide evidence for a significant effect of sulphhydryl groups of protein on the polyphosphatase activity. This is also supported by the inhibiting effect of Ag^{+}, Hg^{++} and Cd^{++}.

Among the inhibitors of the polyphosphatase reaction are NaF and KCN, their inhibition effect being 35 % and 30 %, respectively.

Nucleoside-phosphates, ATP and UTP, also decrease the reaction rate. But this mechanism is still to be clarified.

Table 1 shows the influence of the enzyme pre-incubation for 10 min. with various detergents on the polyphosphatase activity of dialysed preparation.

Table 1

Detergent	Concentration %	Inhibition %
Na-Dodecyl sulfate	0.05	100
Tween-80	0.10	100
Triton X-100	0.50	80
Deoxycholate	0.50	30

SDS shows the greatest inhibition effect. In 10 min. it completely supressed the polyphosphatase reaction. Tween-80 is also a strong inhibitor. Triton X-100 and deoxycholate, in particular, are less effective. The data obtained suggest that polyphosphatase is involved into the composition of lipoprotein systems of cell membranes and certain native conformation of these complexes is responsible for its activity. This is also evidenced by the localization of the enzyme on the outer cytoplasmic membrane.

REFERENCES

AFANASIEVA, T. P., I. S. KULAEV, S. E. MANSUROVA, V. Ju. POLJAKOV, Biokhimija 33, (1968), 1245

FELTER, S., A. J. C. STAHL, Bull. Soc. Chim. biol. 52, (1970), 75

HAROLD, F. M., R. L. HAROLD, J. Bacteriol. 89, (1965), 1262

KULAEV, I. S., T. P. AFANASIEVA, M. P. BELIKOVA, Biokhimija 32, (1967), 539

KULAEV, I. S., I. A. KRASHENINNIKOV, T. P. AFANASIEVA, Doklady AN USSR 190, (1970a), 1238

KULAEV, I. S., T. P. AFANASIEVA, Doklady AN USSR 192, (1970b), 668

LINEWEAVER, H., D. BURK, J. Am. Chem. Soc. 56, (1934), 658

MUHAMMED, A., A. RODGERS, D.E. HUGHES, J. Gen. Microbiol. 20, (1959), 482

STEVENINCK, J. van, Biochemistry 5, (1966), 1998

WILLIAMS, R.J.P., "The Enzymes", Acad. Press, N.Y., 1, (1959), 391

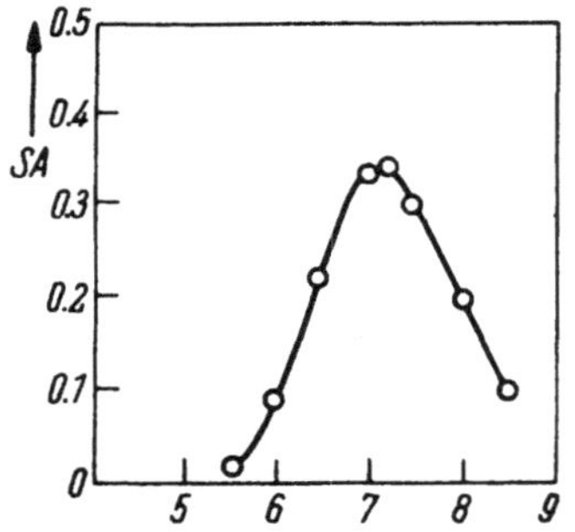

Fig. 1. The influence of pH on the enzyme activity. Specific activity (SA) is expressed in μmoles of orthophosphate formed/min./mg of protein. The reaction mixture contained: tris+maleate buffer, $2x10^{-2}$M; $MnCl_2$, $3x10^{-3}$M; polyphosphate, n=290, $2x10^{-3}$(P); 0. 1mg of protein; water to 0. 5ml. The enzyme activity was measured after 5min. incutation at 37^0.

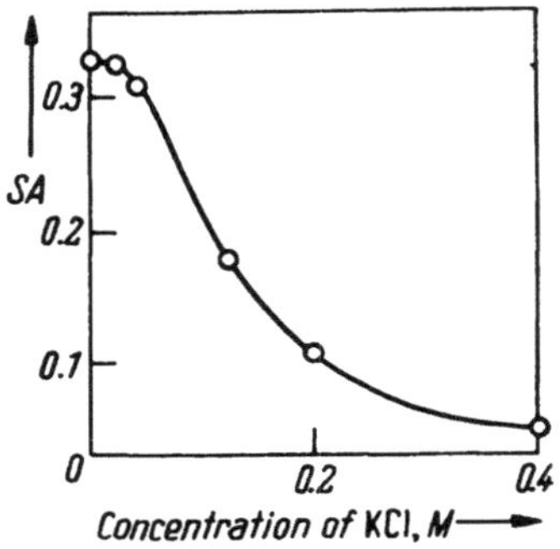

Fig. 2. The effect of KCl on the activity of polyphosphatase. The reaction mixture contained: tris+maleate buffer pH 7. 1, $2x10^{-2}$M; $MnCl_2$, $3x10^{-3}$M; polyphosphate, $2x10^{-3}$M (P); KCl in various concentrations; protein, 0. 1mg; water to 0. 5ml.

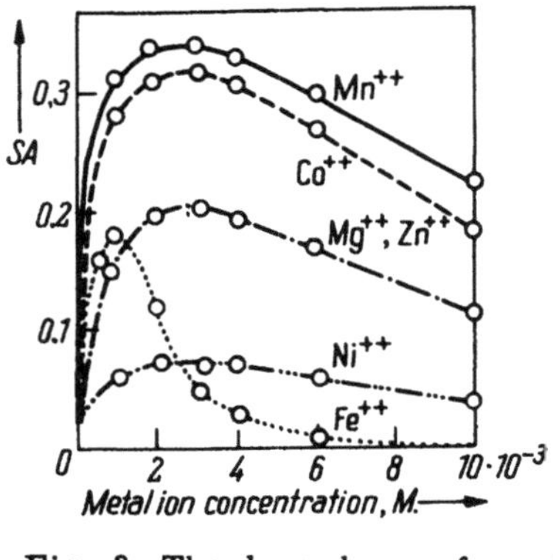

Fig. 3. The dependence of reaction rate on polyphosphate concentration. The reaction mixture contained: tris+maleate buffer pH 7, 1, $2x10^{-2}$M; $MnCl_2$, $3x10^{-3}$M; protein, 0. 1mg. The total volume was 0. 5ml.

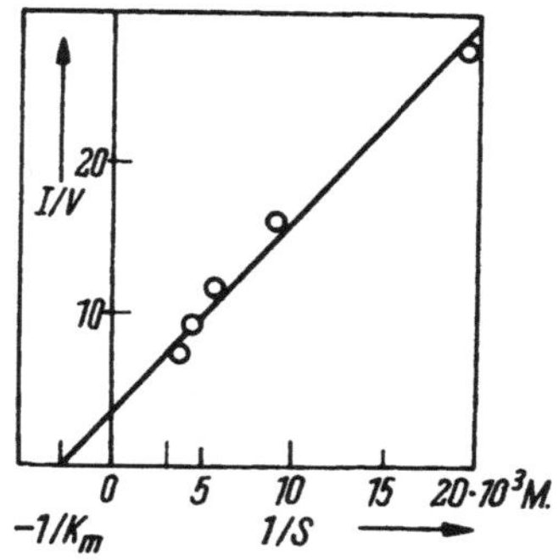

Fig. 4. The effect of various metal ion concentrations on the enzyme activity. The incubation mixture contained: tris+maleate buffer pH 7, 1, $2x10^{-2}M$; polyphosphate, $2x10^{-3}M(P)$; protein, 0. 1mg; water to 0. 5ml.

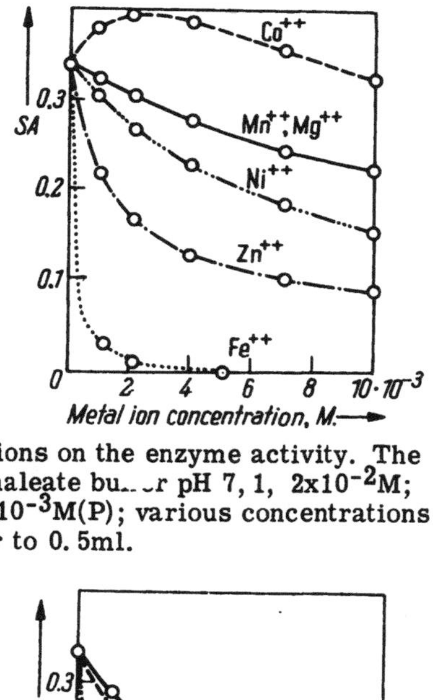

Fig. 5. The effect of various metal ions on the enzyme activity. The incubation mixture contained: tris+maleate buffer pH 7, 1, $2x10^{-2}M$; $MnCl_2$, $2x10^{-3}M$; polyphosphate, $2x10^{-3}M(P)$; various concentrations of metal ions; protein, 0. 1mg; water to 0. 5ml.

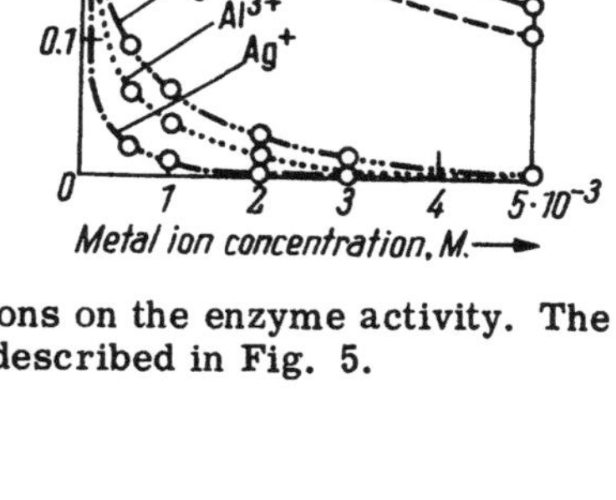

Fig. 6. The effect of various metal ions on the enzyme activity. The incubation mixture was the same as described in Fig. 5.

Lomonosow State University, Moscow

G. I. KONOSHENKO, A. M. UMNOV, I. S. KULAEV

CHARACTERIZATION OF NEUROSPORA CRASSA POLYPHOSPHATE-PHOSPHOHYDROLASES

Polyphosphatases hydrolyzing high-polymeric inorganic polyphosphates to orthophosphate have been reported in some microorganisms (1-6). However, the enzyme properties were studied in details only in Corinebacterium xerosis (3), Aerobacter aerogenes (4) and more recently in yeasts (6). In addition it was shown (4) that cells of A. aerogenes contain also an enzyme splitting tripolyphosphate specifically. No comparable data appear to be available for any polyphosphatase in Neurospora crassa. Indirect data on existence of polyphosphatase activity in this microorganism were obtaind recently in our laboratory (8).
This paper deals with the occurence of the polyphosphatases in the wild type of Neurospora crassa Abbott 4A. It presents also some properties of the enzymes studied. It was shown that cell-free extract of N. crassa has the activities of enzymes which catalyse the hydrolysis of high-polymeric polyphosphates ($\bar{n}$=30-290) and tripolyphosphate to orthophosphate (7). The pH optimum for both reactions was 7,2. The studies of the effects of various agents on the reaction rate revealed the necessity of Mg^{2+}-ions.
The addition of EDTA into incubation medium appreciably decreased the activity of the enzymes. Experiments were performed to determine the effect of various ions on the activities studied.
The optimal concentration of Mg^{2+} for both reactions is approximately 1 μM, as shown in Fig. 1.
Moreover, hydrolysis of high-polymeric polyphosphates is stimulated by K^+-ions (Fig. 2).
The optimal concentration of K^+ is about 100 μM. This fact is in accordance with the properties reported by HAROLD (4) for polyphosphatase from A. aerogenes. The experiments with equal molar concentrations of different polyphosphates demonstrated that the chain length of polyphosphate does not affect the stimulatory action of K^+ and Mg^{2+} on the polyphosphatase activity. The effect of K^+ is

sufficiently specific. For instance ions of Na^+ cannot replace the K^+-ions. Furthermore, ions of K^+ do not stimulate the tripolyphosphatase activity and in some cases they even inhibit the enzyme. The two enzymes studied-tripolyphosphatase and polyphosphatase-also differ in the sensitivity to the inhibitory action of NaF. So, fluoride (100 M) completely inhibited the former and more moderately affected the latter enzyme which retained 52 % of the initial activity.

The results presented here clearly demonstrate that mycelial extracts of N. crassa contain two enzymes catalyzing the splitting of polyphosphates to orthophosphate. The two enzymes differ in substrate specificity. Recent papers of FELTER et al. (6) concerning the occurence of polyphosphatases in yeast Saccharomyces cerevisiae reported that the yeast cells contained only one enzyme catalyzing the hydrolysis of both high-polymeric polyphosphates and tripolyphosphate. It appears, that the authors did not select the proper conditions allowing to divide the two activities.

The existence of two enzymes in N. crassa is confirmed also by the fact that these two activities differ in their behaviour in the course of culture growth. The results are presented in the Fig. 3 and indicate that the maxima of the two activities do not coincide.

As it was shown recently in our laboratory by means of analytical disc electrophoresis on polyacrylamide gel, tripolyphosphatase and polyphosphates represent two different enzyme proteins. The localization of enzyme activities on the gel was assayed by means of procedure described by TONO and KORNBERG (9). The results of such reparation are reproduced in Fig. 4. The polyphosphatase band (A) does not coicide with the bands of the tripolyphosphatase (B) and pyrophosphatase (C). Joint development (D) of the three activities reveals three separate bands.

The studies of localization of the enzymes investigated in the cells of N. crassa (the results will be reported elsewhere) demonstrated that polyphosphatase is localized on the outside of plasma membrane together with the high-polymeric polyphosphates.

The tripolyphosphatase is localized mainly insiae the mitochondria. The question arises about what role is played by tripolyphosphatase in mitochondria. This matter is under investigation. Based on all the data obtained we concludes that the mycelia of N. crassa, contain two enzymes hydrolyzing polyphosphates to orthophosphate. The K^+-activated polyphosphatase localized on the outside of plasma membrane hydrolyses high-polymeric polyphosphates alone. The tripolyphosphatase localized inside the mitochondria hydrolyses tripolyphosphate specifically and is not identical to pyrophosphatase. The existence of the latter enzyme in microorganisms suggests a specific role of tripolyphosphate in the transformation of phosphorus and energy in these organisms.

REFERENCES

(1) KRISHNAN, P. S., Arch. Biochem. Biophys. 37, (1952), 224

(2) KRISHNAN, P. S., V. BAJAJ, Arch. Biochem. Biophys. 42, (1953), 174

(3) MUHAMMED, A., A. RODGERS, D. E. HUGHES, J. Gen. Microbiol. 20, (1959), 482

(4) HAROLD, R. L., F. M. HAROLD, J. Bacteriol. 89, (1965), 1262

(5) SOUZU, H., Arch. Biochem. Biophys. 120, (1967), 338

(6) FELTER, S., G. DIRHEIMER, J.-P. EBEL, Bull. Soc. Chim. Biol. 52, (1970), 437

(7) KULAEV, I. S., G. I. KONOSHENKO, Biochimya (in press) 1971

(8) KULAEV, I. S., I. A. KRASHENINNIKOV, T. P. AFANASJEVA, Dokl. Akad. Nauk USSR 190, (1970), 1238

(9) TONO, H., A. KORNBERG, J. Biol. Chem., 242, (1967), 2375

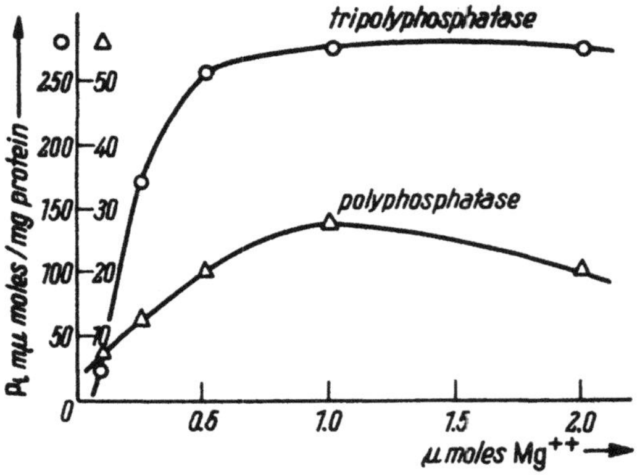

Fig. 1. Effect of Mg^{2+}-ions on the activity of polyphosphatase and tripolyphosphatase.

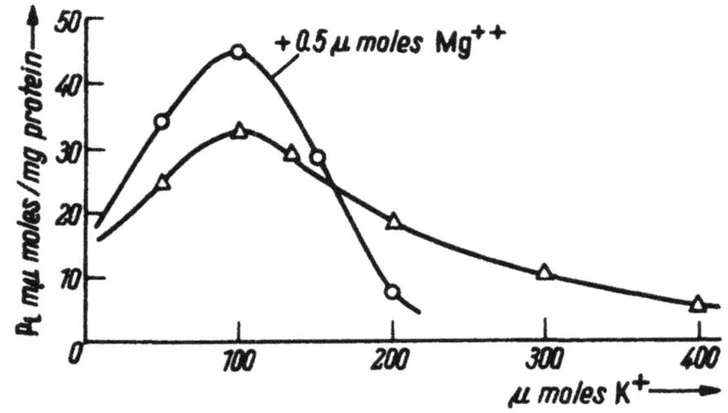

Fig. 2. Effect of K^{+}-ions on the polyphosphatase activity.

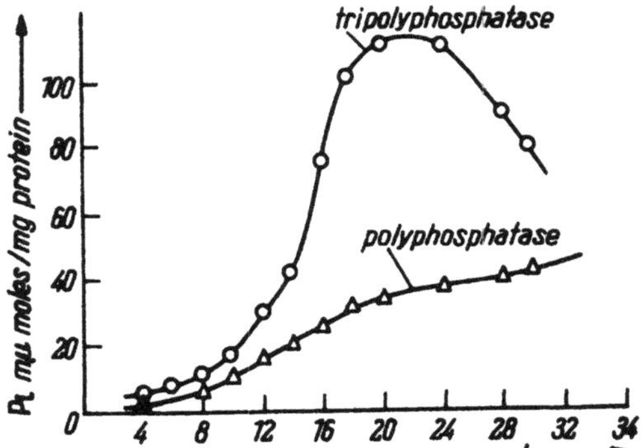

Fig. 3. Influence of culture age on the polyphosphate hydrolyzing activities of N. crassa.

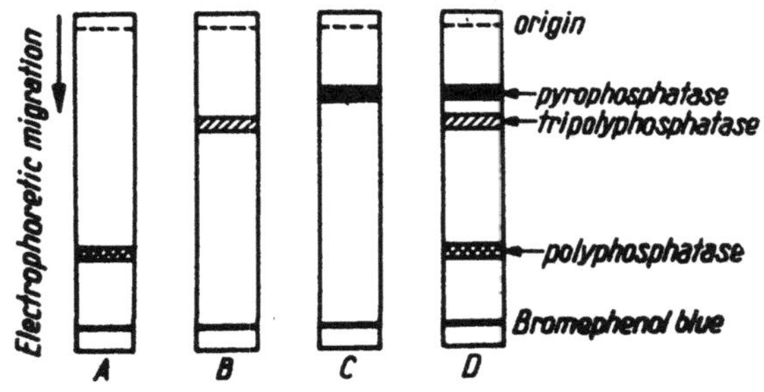

Fig. 4. Disc electropherogram of cell-free extracts from N. crassa.

Faculty of Biology and Soil Science, Lomonosov State University, Moscow

I. A. KRASHENINNIKOV, G. J. KONOSHENKO, S. E. MANSUROVA, A. M. UMNOV, I. S. KULAEV

LOCALIZATION OF POLYPHOSPHATES AND POLYPHOSPHATASES IN NEUROSPORA CRASSA

Phosphorous compounds and enzymes that carry out their transformation occupy central place in metabolic pathways of all living organisms. Study of their localization in different cellular structures may provide a method for elucidation of topography and, probably, mechanism of biochemical processes in the cells of different organisms. At present time a number of facts indicates that essential biochemical and physiological processes in the cell are going on different membrane structures. However, great successes are achieved in this field only in respect of cells of animal tissues; one cannot say the same about plant cells and particularly about cells of fungi. Up to now these organisms were rarely used for the study of localization of substances and processes in the separate cellular structures, and the main reason of it was obviously the absence of methods for gentle destruction of fungal cell walls. Being microorganisms with all specific features of microbial metabolism, yeast and moulds nevertheless have complicated membrane structures that brings them together with higher organisms. That is why fungi are unique objekts for elucidation of the roles of cellular structures in microbial metabolism.

Presence of inorganic polyphosphates of high molecular weight in the cells of microorganisms is the most characteristic feature of their phosphorus metabolism. Polyphosphates make a considerable part of all phosphorus compounds of the cell. Microbial cells contain a large number of polyphosphates with different length of chains - 3 to 500.

We have studied polyphosphates and some enzymes of phosphorous metabolism in different cellular structures of mycelium Neurospora crassa (1).

Data presented in Table 1 show that the removal of the cell walls does not entail alterations in the content of essential phosphorous compounds - nucleic acids and phospholipids that form cellular structures. At the same time the amount of most physiologically active phosphorous metabolites such as nucleoside

Table 1

DIFFERENT PHOSPHORUS COMPOUNDS IN CELLS AND PROTOPLASTS OF NEUROSPORA CRASSA (μg of P per g of dry weight of mycelium)

Compounds	Cells	Protoplasts
Nucleic Acids	6700	6900
Phospholipids	2100	2100
Low molecular weight Polyphosphates	2800	3000
High molecular weight Polyphosphates	2800	0
Orthophosphate	1100	3000

polyphosphates and inorganic polyphosphates considerably decreases. Inorganic polyphosphates of highest molecular weight in this case almost disappear. Amount of polyphosphates of lower molecular weight alters insignificantly; they are mainly localized within the cells. Particularly, nuclei contain a small amount of these compounds. Meantime, using purified preparations of mitochondria from the cells of Neurospora crassa it was found out that mitochondria do not contain polyphosphates. However, these data could not yet solve the question of potential ability of these structures to synthesize polyphosphates and, subsequently, to accumulate them in mitochondria. Then we tried to find enzymes of polyphosphate synthesis in mitochondria of N. crassa.

Data of Fig. 1 show the absence of polyphosphate-synthesizing activity in mitochondria. These data convince us of absence of relations between metabolism of high molecular weight polyphosphates in fungi and processes of oxidative phosphorylation. At the same time, in the volutine-like granules containing polyphosphates they can be synthesized by means of 1, 3 - diphosphoglycerate: polyphosphate transferase (2).

Fig. 2 illustrates the alterations in the content of different fractions of polyphosphates in the process of disruption of N. crassa cells. It was pointed out in the preparations of homogenates from N. crassa that even at 0°C considerable changes occur in the content of different polyphosphate fractions. In first few seconds of disruption amount of polyphosphate of highest molecular weight decreases sharply; simultaneously, amount of polyphosphates of lower molecular weight increases. Data obtained indicate that depolymerizing polyphosphatases

begin to function actively in the first seconds after rupture of cells; these enzymes split polyphosphates of high molecular weight to lower weight polyphosphates. Then the latter ones gradually undergo hydrolysis to orthophosphate.
T. P. AFANASIEVA obtained similar results when she studies the content of separate fractions of polyphosphates in homogenates of Endomyces magnusii cells (3). On the basis of data presented one could assume that polyphosphases and splitting them enzymes are localized in the same cellular structures - that is, outside of cytoplasmic membrane.
We have found that the microorganism studied contains depolymerase splitting polyphosphates to large fragments and at least two enzymes hydrolyzing polyphosphates to orthophosphate. One of them is specific to the polyphosphates of high molecular weight with chain length from 40 to 290 and the other specifically hydrolyzes tripolyphosphate. Data of study of polyphosphatases localization in the cellular structures of N. crassa are presented in Table 2.

Table 2

DISTRIBUTION OF SPECIFIC ACTIVITIES OF POLYPHOSPHATASE AND TRIPOLYPHOSPHATASE IN THE CELLULAR STRUCTURES OF NEUROSPORA CRASSA (mE per mg of protein)

Cellular Structures	Polyphosphatase	Tripolyphosphatase
Whole cells	30-50	90-150
Protoplasts	15-30	100-160
Nuclei	0	0
Mitochondria	0	150-200
Membranes	5	6
Microsomes	2	20
Hyaloplasma	40	160

This table shows that the main part of tripolyphosphatase is localized inside the cell. Study of distribution of its activity in the cellular structures indicates that considerable amount of this enzyme is associated with mitochondria. Polyphosphatase hydrolyzing polyphosphates of high molecular weight to orthophosphate is localized mainly outside of cytoplasmic membrane, since specific activity of this enzyme considerably decreases with removal of cell wall and formation of protoplasts. It was found that bond of polyphosphatase with cytoplasmic membrane is rather labile and enzyme can easily be washed from protoplast. Data of insignificant activity of this enzyme in the preparations of isolated membranes also support the idea of weak bond between polyphosphatase and cytoplasmic membrane (4).

Table 3

EFFECT OF TRITON X-100 ON THE CONTENTS OF POLYPHOSPHATASE IN THE PROTOPLASTS OF NEUROSPORA CRASSA (Data expressed in mE)

	Protoplasts	Medium	Total
Omit Triton	210	20	230
0.02 % Triton	10	160	170

Medium: 0.05 M maleate buffer pH 6.8, containing 0.6 M mannitol and 0.1 % magnesium sulfate

As shown from Table 3, treatment of protoplasts with Tris-buffer at room temperature causes removal of 10 per cent of total activity of polyphosphatase from protoplasts. Addition of 0.02 % Triton X-100 to this medium causes almost perfect removal of enzyme protoplast while its morphological integrity is preserved.

Study of depolymerase activity distribution showed that it is almost exclusively localized at the periphery of the cell. These data can well explain the fact observed by us that removal of cell wall accompanied by destruction of membrane structures of the cell, immediately causes contacting of polyphosphatases with their substrate, that results in complete hydrolysis of polyphosphates. So extreme sensitivity of polyphosphates of high molecular weight to integrity of cell structure can be explained by peripheric localization of these compounds. This supposition agree quite well with the data of van STEVENINCK (5) who showed that inorganic polyphosphates take part in the processes of assimilation of glucose and it transport through cellular membrane.

REFERENCES

(1) KULAEV, I. S., T. P. AFANASJEVA, I. A. KRASHENINNIKOV, S. E. MANSUROVA. In: "Yeast protoplasts". - Proc. of the 2nd Intern. Symp. on Yeast protoplasts. August 1968, Brno, Ed. O. NEČAS and A. SVOBODA. Acta facultatis medicae universitatis brunensis, 1970, 81

(2) KULAEV, I. S., G. I. KONOSHENKO, 1971. Doklady Akad. Nauk. USSR., in press

(3) KULAEV, I. S., I. A. KRASHENINNIKOV, T. P. AFANASJEVA, Doklady Akad. Nauk. USSR, 190, (1970), 1238

(4) KULAEV, I. S., G. I. KONOSHENKO, A. M. UMNOV, Biokhimiya, (1971), in press

(5) STEVENINCK, J. van, H. L. BOOIJ, J. Gen. Physiol., 48, (1964), 43

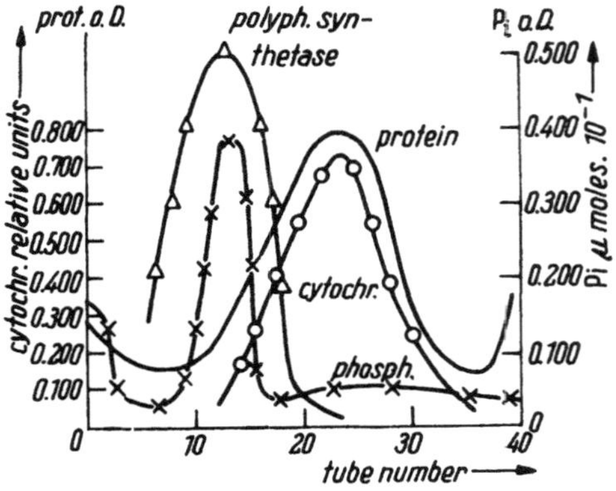

Fig. 1. Distribution of protein, cytochromes, labile phosphorous and polyphosphate-synthesizing activity in the preparation of mitochondria - sucrose gradient centrifugation.

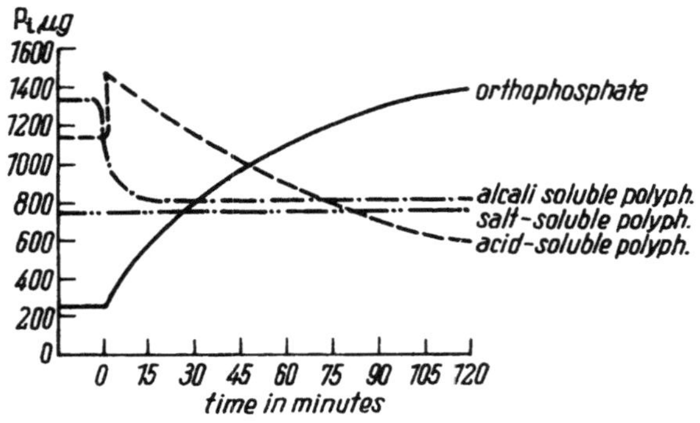

Fig. 2. Changes in content of polyphosphates and orthophosphate with disruption of cells of N. crassa.

Laboratory of Bioorganic Chemistry and Faculty of Biological Sciences,
Lomonosov State University, Moscow

S. E. MANSUROVA, I. S. KULAEV, K. G. SKRYABIN

ON A POSSIBILITY OF NON-ENZYMATIC TRANSPHOSPHORYLATION BY HIGH POLYMERIC POLYPHOSPHATES

This work appeared as a result of studying of pyrophosphate synthesis from polyphosphate and orthophosphate according to equation:

$$(\text{Poly-P})_n + P^* \xrightarrow{\text{Me}} P^*\text{-}P + (\text{Poly-P})_{n-1}$$

The possibility of such enzymatic reaction was described for yeast by IPATA and FELICIOLI in 1964 (1). In our experiments neither with cell free extracts from S. cerevisiae, Endomyces magnusii and Neurospora crassa nor with crude fractions of nuclei and mitochondria from S. cerevisiae this reaction has not been detected. Moreover we have observed a non-enzymatic formation of inorganic $[^{32}P]$ - pyrophosphate in 0. 6 N $HClO_4$ in the presence of labelled orthophosphate and polyphosphate (n=40).

It is well known that enzymes catalyzing the hydrolysis of organic and inorganic polyphosphates in the vast majority of cases require divalent metal ions. This brings up the question of the precise mechanism of cation action. These can explain the great attention to the problem of catalysis of non-enzymatic phosphoorganic compounds hydrolysis and transphosphorylation reactions by divalent cations. As an example we can draw attention to a well known work on the acceleration of acylphosphate hydrolysis by calcium ions by LIPMANN, one of the pioneers in this field (2).

Outstanding role of ATP in bioenergetic reactions explaines the great interest to the catalytic effects of the metal ions in non-enzymatic reactions with adenosine triphosphate and tripolyphosphate (its inorganic analog). LOWENSTEIN et al. studied the reaction of the ATP with nucleophylic agents (3, 4, 5). Very similar data with tripolyphosphate were obtained by BUVET et al. (6). In the course of these non-enzymatic processes catalyzed by divalent cations the terminal phosphate group was transfered either to water (in the case of hydrolysis) or to a labelled orthophosphate followed by the formation of pyrophosphate. Lowenstein studied the formation of hydroxamates by double transfer through the action of ATP and free carboxylic acid in the presence of hydroxylamine and cations.

In our subsequent work we further studied the effect of pH on the rate of pyrophosphate formation in the reaction discovered. The result presented in Fig. 1 show that optimum of transphorylation by polyphosphate (n=40) in the alkaline region was observed in pH 8.9. A reaction is undetectable in the divalent metal ions absence. The specificity of cation action was clear from the influence of these catalytic agents on the non-enzymatic transphosphorylation. As appears from Table 1, 15-30 % of polyphosphates was utilized as phosphoryl group donors in the given condition of the experiment. It is essential to bear in mind that in the case of organic and non-organic triphosphates only 0,2 - 2 % of ones functioned as donor of phosphoryl groups.

It is necessary to emphasize the fact of specific barium catalytic action. It was the sole instance when we observed the formation of not only radioactive pyrophosphate, but also labelled polyphosphates.

These compounds may be synthesized in the following reaction:

$$(Poly\text{-}P)_n + P^* \xrightarrow{Ba^{++}} P^*\ (Poly\text{-}P)_x + (Poly\text{-}P)_{n-x}$$

The obtained data corroborate the assumption that with divalent cation assistance the electronic structure of high polymeric polyphosphate changes in such a way that a partial positive charge is formed on the terminal phosphorous atoms. The nucleophylic attack at these terminal phosphates becomes feasible, which leads to nucleophylic agent phosphorylation. In the case of enzymatic catalysis one of the active center amino acid residue can play the role of nucleophylic agent followed by the formation of phosphorylated enzyme. The forming of these covalently bound phosphorylenzyme as intermediates of phosphohydrolyzing-transphosphorylating reactions is well known at present.

As it was shown in a number of research for the polyphosphatase activity the assistance of divalent metal ions is necessary (7, 8). One may assume that in the course of polyphosphatase reaction cations changed the electronic structure of polyphosphate molecule the same way as in the case of our model experiments. The formation of phosphoryl enzyme-phosphorylpolyphosphatase - may follow as a result of such nucleophylic attack.

It should be also pointed out that the analogy between the non-enzymatic transferase

Table 1

ACTIVATION OF TRANSPHOSPHORYLATION BY BIVALENT METAL IONS

Cations	Pyrophosphate formed (mM)	$\frac{PP}{Poly\text{-}Pn} \cdot 100\ \%$
-	0	0
Mg^{++}	147 ± 11	16
Ca^{++}	205 ± 16	22
Mn^{++}	219 ± 11	23
Cd^{++}	246 ± 24	26
*Ba^{++}	304 ± 32	33

*"Polyphosphate" formed 532 ± 128 mM.
The complete reaction mixture contained:
0. 935 mM poly-P_{40} (pH 8. 17), 50 mM diethanolamine-HCl buffer (pH (pH 9. 0-9. 1), 2, 98 mM [^{32}P] -orthophosphate, 45 mM metal salt.
The final volume was 1. 0 ml, temp. 37°C and the time of incubation 15 h.

reaction catalyzed by Ba^{+2} which we had discovered and enzymatic depolymerization of polyphosphates in N. crassa may happen to be a subject of eventual discussion. It's necessary to outline that the further investigation of catalytic effects of cations on the rate of transphosphorylation reaction will allow to throw more light on their role in the enzymatic process.

REFERENCES

(1) IPATA, P. C., R. A. FELICIOLI, Boll., Soc., ital. biol. sperium, 39 (1964), 85

(2) LIPMANN, F., L. C. TUTLE, J. Biol. Chem., 155 (1944), 571

(3) LOWENSTEIN, J. M., Biochem J., 70 (1958), 222

(4) LOWENSTEIN, J. M., Nature, 187 (1960), 570

(5) TETAS, M., J. M. LOWENSTEIN, Biochemistry, 2 (1963), 350

(6) LEPORT, L., E. ETAIX, F. GODIN, P. LEDUC, R. BUVET, and C. PONNAMPERUMA, 1971, 197. North-Holland Publishing Company

(7) KONOSHENKO, G. I., A. UMNOV, I. S. KULAEV, This Symposium
(8) AFANASIEVA, T. P., V. V. ROTSCHANETZ, This Symposium
(9) KRITSKY, M. A., E. K. CHERNYSHOVA, This Symposium

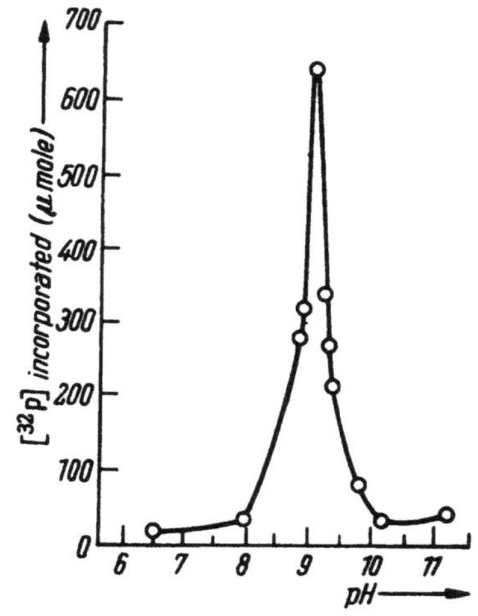

Fig. 1. Effect of pH on transphosphorylation
The complete reaction mixture contained: 0, 935 mM $(Poly\text{-}P)_{40}$ pH 0. 17; 50 mM diethanolamine-HCI buffer; 2, 98 mM [^{32}P]-orthophosphate; 45 mM $MnCl_2$. The final volume was 1. 0 ml, temp. 37 °C and the time of incubation 15 h.

The Institute of Biochemistry and Physiology of Microorganisms of the Academy of Sciences, Puschino, Lomonosov State University, Moscow

M. A. NESMEYANOVA, A. D. DMITRIEV, M. A. BOBYK, I. S. KULAEV

ON THE REGULATION OF SOME PHOSPHOROUS METABOLISM ENZYMES IN ESCHERICHIA COLI

The level of ortho-phosphate and ATP-ADP system plays an important role in the metabolism of bacterial cell. This level can be controlled at least by two reasons. On the one hand the cell is able to produce an alkaline phosphatase (phosphomonoester:phosphohydrolase), which possesses a broad substrate specificity and in conditions of phosphate starvation is able to supply the cell from every possible exogenous sources (1). On the other hand in bacterial cells inorganic, high polymer, polyphosphates are established, which by means of enzymes of the synthesis and degradation of polyphosphates are involved in the control of ATP, ADP and ortho-phosphate level (2). It was proposed, for example, that terminal phosphoric group of ATP can be transferred to polyphosphate by means of ATP-polyphosphate phosphotransferase under excess of ATP, arising in conditions, when growth and nucleic acid synthesis were blocked by any nutrient deficiency (sulfur starvation) (3). The phosphorus of 1, 3-diphosphoglycerol acid can transferred to the polyphosphates by means of 1, 3-diphosphoglycerate-polyphosphate-phosphotransferase under excess of ortho-phosphate and inhibition of ATP synthesis (4). In addition, enzymes, catalyzing the hydrolysis of polyphosphate to orthophosphate (polyphosphate:phosphohydrolase) have been detected in bacterial cells (5, 6). Probably, these enzymes can supply the cell with orthophosphate at the expense of intracellular storage of ortho-phosphate-high polymeryzed polyphosphates, upon phosphate starvation. Unfortunately, at present, the function of these enzymes in the cell are not finally established. The mechanisms, regulating the activity and synthesis of these enzymes are also unknown, but the two latter factors determine the degree of participation of these enzymes in the phosphoric metabolism of the cell.

In the present comunication we are reporting on certain data about the regulation by ortho-phosphate of the biosynthesis of alkaline phosphatase and of some enzymes of polyphosphate metabolism, such as polyphosphatase, three polyphosphatase, ATP-polyphosphate-phosphotransferase and 1, 3-diphosphoglycerate-polyphosphate-phosphotransferase in the cells of E. coli. The alkaline phosphatase from this strain was isolated and purified, its properties and macromolecular structure and its genetic regulation was well established. Ortho-phosphate is the most powerful and specific regulator of this enzyme. It does not only inhibite competitionally the activity of this enzyme, but also represses its synthesis (1). Having studied the biosynthesis of alkaline phosphatase by several wild strains of E. coli, we acknowledged once more the above-said statment (Fig. 1). Indeed, the significant amount of this enzyme is formed only under the conditions of limitting inorganic phosphate. In the presence of ortophosphate the enzyme formation (derepression) began at about the time, when the orthophosphate concentration had fallen to a minimum (curve 2, Fig. 1). By the excess of orthophosphate one could see intensive growth of biomass, but the synthesis of the enzyme was very negligible (curve 3, Fig. 1).

The repression of alkaline phosphatase, it is known, is under the control of two regulator genes (7) and probably, is going at the level of gene transcription. Besides, some of us, previously, obtained, interesting facts, which allow us to suggest that ortho-phosphate can act also at the level of complementation of the enzyme subunits, which occurs at the ribosome-membrane complex of the cell. The phosphate inhibits the formation of dimers from only those subunits, which are in the ribosome-membrane complex (8).

Nothing is known concerning the regulation of enzymes of polyphosphate metabolism in E. coli. But the possible functional identity of these enzymes with the alkaline phosphatase allows us to consider, that the regulation of the synthesis, at least of some of these enzymes, is going on coordinationally. First of all, the data of HAROLD (9) on Aerobacter aerogenes about co-ordinating regulation of three enzymes: alkaline phosphatase, polyphosphatase and ATP-polyphosphate-phosphotransferase convince us of such possibility. We studied dynamics of formation of enzymes during the growth of E. coli in a complete medium with ortho-phosphate (Fig. 2).

Both alkaline phosphatase and enzymes of polyphosphate metabolism, with we studied, have been demonstrated in wild strain of E. coli. Their amount depends upon the stage of cultural growth. The largest amount of all enzymes per unit of dry weight of biomass are accumulated at the early log stage. It is interesting, that this advance coincides with the maximal amount of polyphosphates and intracellular ortho-phosphate. ATP-polyphosphate:phosphotransferase is an exception and reaches its second maximum at the end loge stage. The significant increasing of activity of this enzyme coinsides, on the contrary, with the degradation of polyphosphates.

In order to study the influence of ortho-phosphate on the formation of these enzymes, the cells were harvested at the end log stage, washed and subjected of phosphate starvation. In the control experiment the culture was transferred into the complete medium with ortho-phosphate. Fig. 3 illustrates this experiment. During the phosphate starvation the activities of alkaline phosphatase, polyphosphatase and three-polyphosphatase increased significantly in comparison with the cells which grow in medium rich in phosphate. But the level of induction of these enzymes is very different: if the activity of alkaline phosphatase induces 400 fold, the activity of two other enzymes increases only 8-12 fold. The induction of enzymes under phosphate starvation is accompanied by an intensive degradation of polyphosphates and decrease of intracellular ortho-phosphate. We ded not establish, however, the induction of ATP-polyphosphate:phosphotransferase and 1, 3 diphosphoglycerat:polyphosphatphosphotransferase. Under phosphate starvation the activity of the second enzyme increases on the contrary a little in the control culture with ortho-phosphate. When ortho-phosphate is added to cells, previously subjected to phosphate starvation, alkaline phosphatase, polyphosphatase and three-polyphosphatase are repressed while their activity continuous to increase in the control culture without phosphate (Fig. 4). The influence of orthophosphate on the ATP-polyphosphate-phosphotransferase and 1, 3-diphosphoglycerate-polyphosphate-phosphotransferase less significant. The activity of these enzymes is increases a little under the addition of ortho-phosphate.

First this is accompanied by the increase of the level of ortho-phosphate and then by the increase of polyphosphates. Our data seem to contradict HAROLD' s data (10) concerning ATP-polyphosphate phosphotransferase, but this may be explained by the fact that we have determined the activity of the reverse reaction of this enzyme.

Summing up the above-said, we can say: the enzymes alkaline phosphatase, three polyphosphatase and polyphosphatase which are localized in periplasme (11, 12), are controled coordinationally by exogenous ortho-phosphate, being induced under phosphate starvation and repressed under the excess of the latter. Inhibition of these enzymes formation by ortho-phosphate adds one more examples known for bacteria, where it is the product of an enzyme reaction which controls the synthesis of the enzyme or related enzymes (13). The first experiments with mutant, carrying the defect in regulator gene, responsing for the induction of alkaline phosphatase, allow us to suggest, that these enzymes have common regulator gene. Their answer to the presence of ortho-phosphate in the medium confirms that their main function is the supply of ortho-phosphate from every suitable substrates. As to the enzymes ATP-polyphosphate:phosphotransferase and 1, 3-diphosphoglycerate-polyphosphate:phosphotransferase we see another pattern in this case. The synthesis of these intracellular enzymes (14) are not regulated by the exogenous ortho-phosphate. But, may be, the intracellular ortho-phosphate can regulate their activity. Some of our data confirm such a possibility. Apparently, these enzymes do not also regulate the level of intracellular ortho-phosphate, but only regulate the ATP-ADP ratio and the level components of the glycolytic system.

REFERENCES

(1) TORRIANI, A., Biochim. Biophys. Acta. 38, (1960), 460

(2) KALTWASSER, H., Arch. Microbiolog. 41 (1962), 282

(3) HAROLD, T.M., S. SYLVAN, J. Bacteriol. 86 (1963), 222

(4) KULAEV, I.S., M.A. BOBYK, Biochimia, 36 (1971), 2

(5) MUHAMMED, A., A. BODGERS, D.E. HUGHUES, J. Gen. Microbiol. 20 (1959), 482

(6) HAROLD, F.M., R.L. HAROLD, J. Bacteriol. 89 (1965), 1269

(7) ECHOLS, H., A. GAREN, S. GAREN, A. TORRIANI, J. Mol. Biol. 3 (1961), 425

(8) NESMEYANOVA, M.A., V.P. VITVITSKY, A.A. BOGDANOV, Biochimia. 31 (1966), 902

(9) HAROLD, F.M., J. Gen. Microbiol. 35 (1964), 81

(10) HAROLD, F.M., Bacteriol. revies 30 (1966), 772

(11) MALAMY, M.H., B.L. HORECKER, Biochemistry, 3 (1964), 1889

(12) KRITSKY, M.S., E.K. CHERNYSHOVA, I.S. KULAEV, in press

(13) GERHART, J.C., A.B. PARDEE, J. Biol. Chem., 237 (1961), 891

(14) KULAEV, I.S., G.I. KONOSHENKO, A.M. Umnov, in press

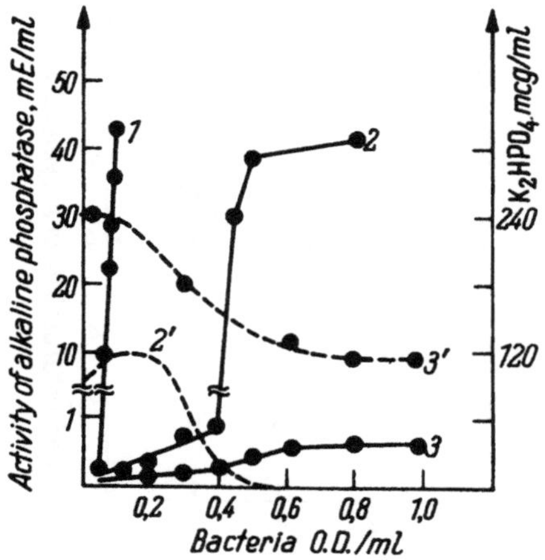

Fig. 1. Dynamics of formation of alkaline phosphatase in E. coli. Curves 1, 2, 3 show the increase of enzyme activity per ml as a function of the bacterial growth in the medium:
1-with 0 mcg/ml K_2HPO_4,
2 - with 120 mcg/ml K_2HPO_4,
3 - with 240 mcg/ml K_2HPO_4,
curves 2'and 3'show the utilization of ortho-phosphate from the medium.

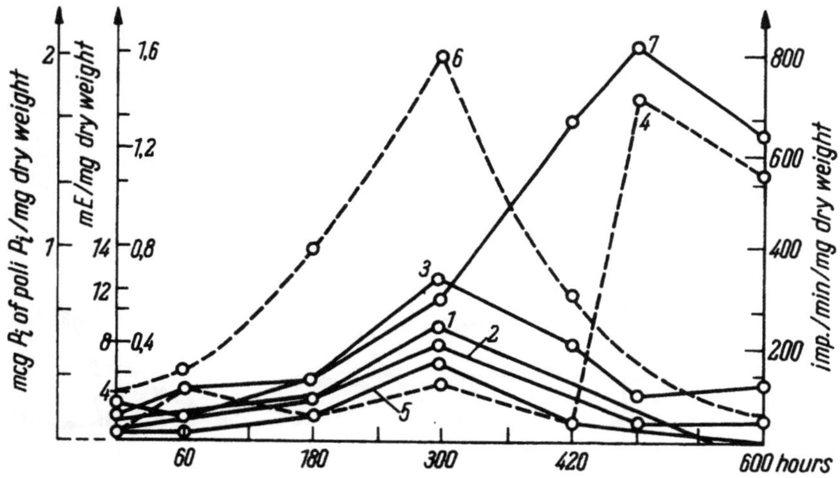

Fig. 2. Dynamics of formation of polyphosphate metabolism enzymes of in the complete medium by E. coli:
1 - alkaline phosphatase,
2 - polyphosphatase,
3 - three-polyphosphatase,
4 - ATP-polyphosphate-phosphotransferase,
5 - 1, 3 diphosphoglycerat-polyphosphate: phospho-transferase,
6 - high polymer polyphosphates,
7 - dry weight of biomass.

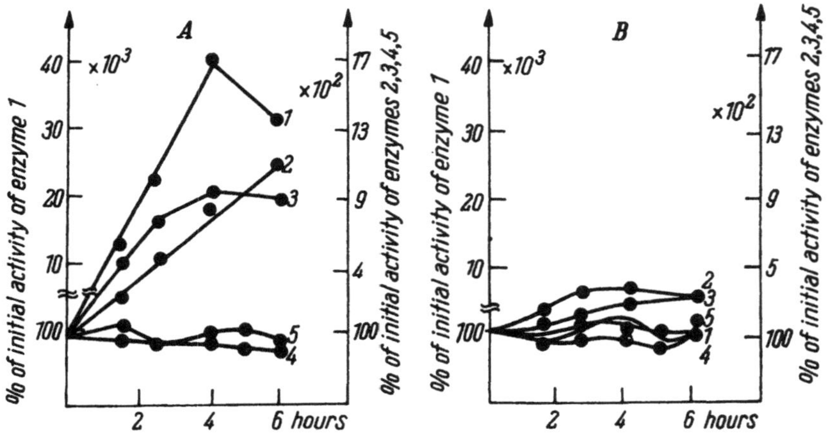

Fig. 3. Influence of ortho-phosphate in the formation of enzymes of phosphorous metabolism in E. coli.
A - The formation of enzymes in conditions of phosphate starvation.
B - The formation of enzymes in control culture with ortho-phosphates.
The indications of enzymes are the same as in Fig. 2.

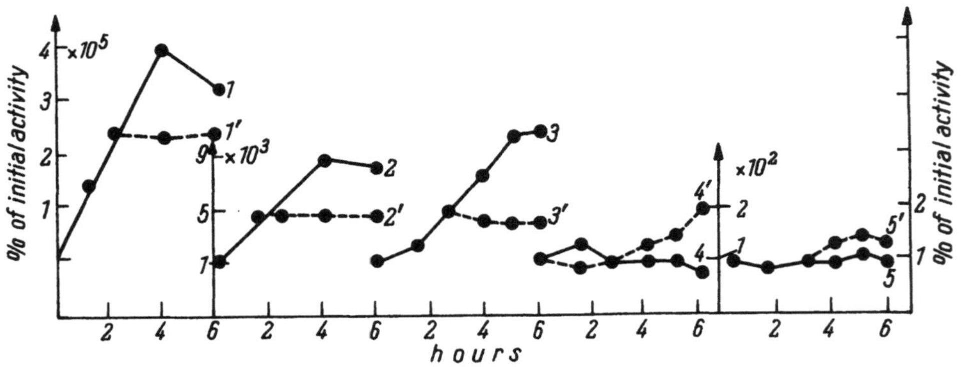

Fig. 4. Influence of ortho-phosphate addition to cells, previously subjected to phosphate starvation in the formation of enzymes in E. coli.
The indications of enzymes are the same as in the Fig. 2.
1, 2, 3, 4, 5-the increase of enzymes activity under phosphate starvation,
1,'2,' 3,' 4,' 5'-under addition of ortho-phosphate. The arrows indicate the addition of 240 mcg/ml K_2HPO_4.

Sektion Biologie - Pflanzenphysiologie der
Friedrich-Schiller-Universität Jena

F. JUNGNICKEL

SIGNIFICANCE OF REPRESSIBLE POLYPHOSPHATE PHOSPHOHYDROLASES IN YEAST AND HIGHER PLANTS

It is well established now, that a rapid synthesis and accumulation of macromolecular inorganic polyphosphates can be observed, especially in unicellular or filamentous organisms. Not so many investigations have been made as yet into the metabolism of these compounds in higher organisms. However, more recent results indicate that macromolecular inorganic polyphosphates do occur in higher plants as well as in animals [MIYACHI (1), GROSSMANN and LANG (2), NIMIERKO (3), PENNIAL and GRIFFIN (4), TEWARI and SINGH (5), VAGABOV and KULAEV (6), GRIFFIN et al. (7, 8), SHIOKAWA and YAMANA (9), GORSKA-BRYLASS (10)]. On the other hand in higher organisms these compounds seem to form only a minor constituent HAROLD (11).
We investigated why an important accumulation of macromolecular inorganic polyphosphates as far as we could see does not take place in higher plants. The cause for this might be high activities of polyphosphate phosphohydrolases. Furthermore we were interested in the consequences of phosphate starvation in the yeast Candida utilis and in sterile cultures of higher plants, especially Lemnaceae, too. Thus it would be possible to compare the effects in unicellular and in multicellular systems.
The yeast Candida utilis has been found to accumulate large amounts of macromolecular inorganic polyphosphates. This may be due to the derepression of the enzyme polyphosphate kinase [HOFFMANN-OSTENHOF et al. (12)].
The metachromatic reaction using toluidine blue indicates that these compounds are located in the so-called volutin granules within the cells. Sixty minutes after start the phosphate uptake is largely reduced by an increased splitting of the macromolecular polyanions (Fig. 1).
According to the polyphosphate cycle of LANGEN and LISS (13) long-chain polyphosphates are transformed into acidsoluble polyphosphates. At that period metachromatic granules also dissapear.

In Candida utilis two phosphohydrolases are derepressed under phosphate starvation conditions. These are: an unspecific orthophosphoric monoester phosphohydrolase E. C. 3. 1. 3. 2. and a polyphosphate phosphohydrolase E. C. 3. 6. 1. 11. which is specific for splitting P-O-P linkages (Fig. 2). Fig. 3 shows that there is only small difference between the splitting rates of living cells and that of partially inactivated cells. These two enzymes are supposed to be located near the cell surface. The are strongly bound to the cell.
Fig. 4 informs about the thermostability of these two enzymes. The polyphosphate phosphohydrolase can be inactivated within 10 minutes at 70°C, the other phosphohydrolase shows more stability. These two enzymes participate in splitting inorganic polyphosphates in different ways: While all inorganic polyphosphates are split by polyphosphate phosphohydrolase, orthophosphoric monoester phosphohydrolase is able to split short-chain polyphosphates only.
Independent of preconditions sterile cultures of higher plants (i. e. Wolffia arrhiza, Lemna trisulca, L. minor, L. gibba) produce an enzyme system which is able to split GRAHAMs salt. The rate of splitting increases after phosphate starvation (Fig. 5) and an interaction between a constitutive and a repressible polyphosphate phosphohydrolase was found. The repressible enzyme is more stable as regard to temperature than the constitutive ones. Both enzymes are not identical with the acid orthophosphoric monoester phosphohydrolase of higher plants which we investigated, too. Nevertheless, short-chain polyphosphates also can be split by this acid phosphatase in the same way as observed in Candida.
Constitutive polyphosphate phosphohydrolases do occur in the potato, in liverworth (Riccia fluitans) and in other axenic plants which we have investigated recently.
We conclude that Candida utilis is able to maintain complete metabolic activity even in media lacking orthophosphate. Derepressed phosphohydrolases degrade macromolecular inorganic polyphosphates as well as orthophosphoric monoesters of the medium, may be these enzymes cause also high turnover rates of endogenic phosphorous compounds. Macromolecular inorganic polyphosphates are mainly deposited in volutin granules within the cytoplasma. They act as trapping centers for phosphate and decrease its outflux from the cells into the medium. This mechanism is a suitable way for cells of unicellular or filamentous organisms to prevent a considerable loss of phosphate. However, in tissues of higher organisms one cell is surrounded by many other cells. Thus a mechanism which

has a favourable effect in single cells can be harmful in tissue.
There may be two ways preventing a considerable accumulation of macromolecular inorganic polyphosphates in tissue:

1. The genetic information for synthesis of macromolecular inorganic polyphosphates has been lost in phylogenesis, or enzymes are repressed in consequence of tissue formation.
2. Enzymes for synthesis of macromolecular inorganic polyphosphates do exist in tissue of higher organisms as well, but no considerable accumulation can be observed in most multicellular organisms, because hydrolases or transferases induce high turnover rates.

Especially the constitutive polyphosphate phosphohydrolase which we find in higher plants seem to confirm the second assumption.
However, the repressible phosphohydrolases which we found, are able to cause increased turnover rates of phosphorous compounds under phosphate starvation conditions in yeast as well as in higher plants.

REFERENCES

(1) MIYACHI, S., J. Biochem. (Tokyo) 50 (1961), 367

(2) GROSSMANN, D., K. LANG, Biochem. Z. 336 (1962), 351

(3) NIMIERKO, W., Colloq. Intern. Centre Natl. Rech. Sci. (Paris) 106 (1962), 615

(4) PENNIAL, R., J.B. GRIFFIN, Biochim. Biophys. Acta 90 (1964), 429

(5) TEWARI, K.K., M. SINGH, Phytochemistry 3 (1964), 341

(6) VAGABOV, V.M., I.S. KULAEV, Dokl. Akad. Nauk USSR 158 (1964), 218

(7) GRIFFIN, J.B., N.M. DAVIDSON, R. PENNIAL, J. Biol. Chem. 240 (1965), 4427

(8) GRIFFIN, J.B., R. PENNIAL, Arch. Biochem. Biophys. 114 (1966), 67

(9) SHIOKAWA, H., K. YAMANA, Exptl. Cell Res. 38 (1965), 180

(10) GORSKA-BRYLASS, A., Acta Soc. Bot. Pol. 35 (1966), 3

(11) HAROLD, F.M., Bacteriol. Rev. 30 (1966), 772

(12) HOFFMANN-OSTENHOF, O., J. KENEDY, K. KECK, O. GABRIEL, H.W. SCHÖNFELLINGER, Biochim. Biophys. Acta 14 (1954), 285

(13) LANGEN, P., E. LISS, Biochem. Z. 330 (1958), 455

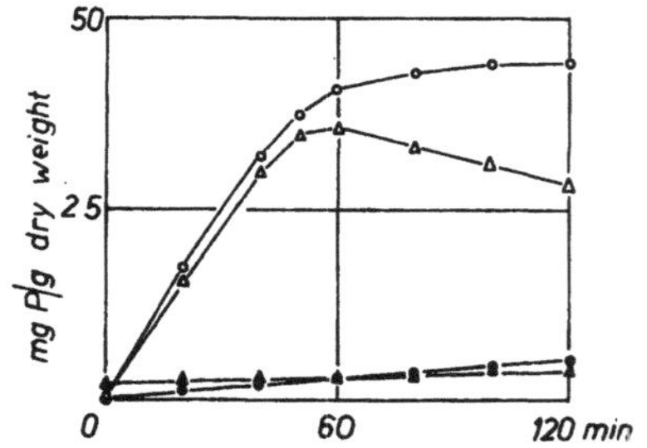

Fig. 1. Phosphate uptake (circles) and acid-insoluble polyphosphates (triangles) in Candida utilis after normal (black) and phosphate starvation (open) conditions. Aerobic incubation, 30 °C. Start: 1 % glucose and 2×10^{-3} M KH_2PO_4 added.

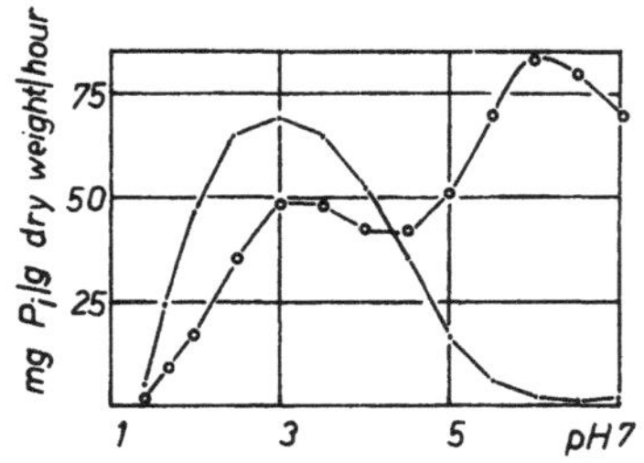

Fig. 2. Splitting of GRAHAMs salt (open circles) and of phenylphosphate (points) by partially inactivated cells of C. utilis. Inactivation: 10 min, 62 °C. Incubation: 30 °C, 0, 04 M buffer (zitrate).

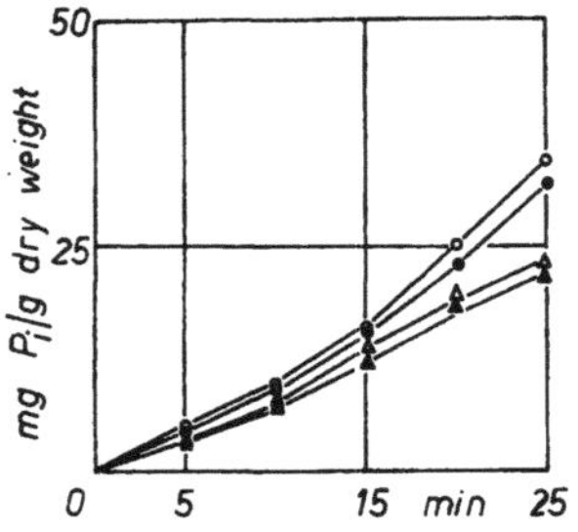

Fig. 3. Splitting of GRAHAMs salt, pH6 (circles), and of phenylphosphate, pH3 (triangles), by living cells (open) and partially inactivated cells (black) of C. utilis. Inactivation: 10 min, 62 °C. Incubation: 30 °C, 0, 04 M buffer (zitrate).

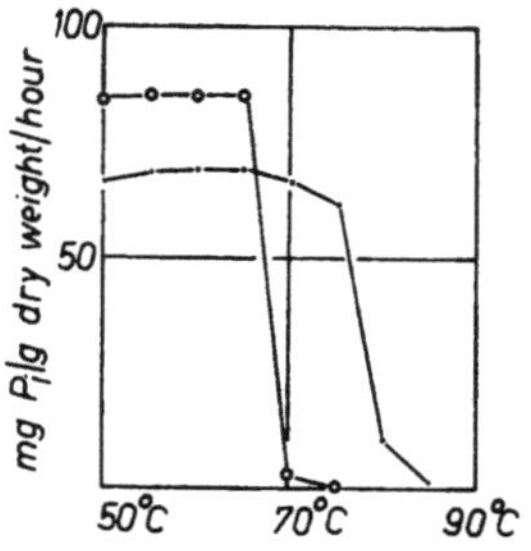

Fig. 4. Splitting of GRAHAMs salt, pH6 (circles), and of phenylphosphate, pH3 (points), by cells of C. utilis after inactivation (10 min) with different temperatures. Incubation: 30 °C, 0, 04 M buffer (zitrate).

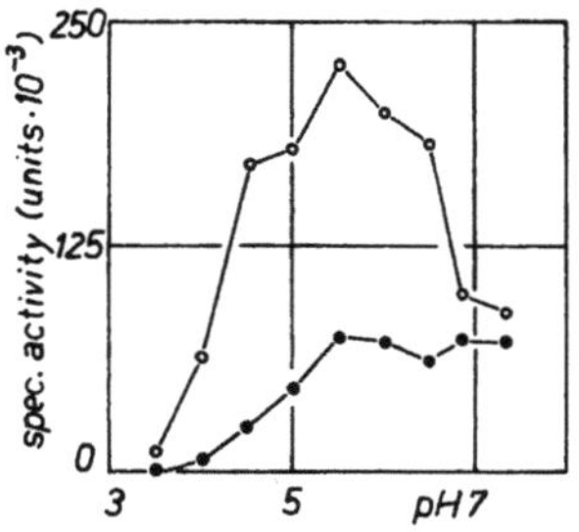

Fig. 5. Splitting of GRAHAMs salt by crude enzyme extracts of Wolffia arrhiza, cultivated under normal (black circles) and phosphate starvation (open circles) conditions. Incubation: 30 °C, 0, 04 M buffer (zitrate or triethanolamine).

2. ADENOSINTRIPHOSPHATASE

Department of Animal Biochemistry, Lomonosov State University, Moscow

A. D. VINOGRADOV, V. K. AKIMENKO, I. B. MINKOV

THE PROPERTIES OF SOLUBLE MITOCHONDRIAL ATPase

In spite of its fundamental importance the mechanism of oxidative phosphorylation could not be described yet in detail in chemical terms. In a simple form oxidative phosphorylation runs as follows:

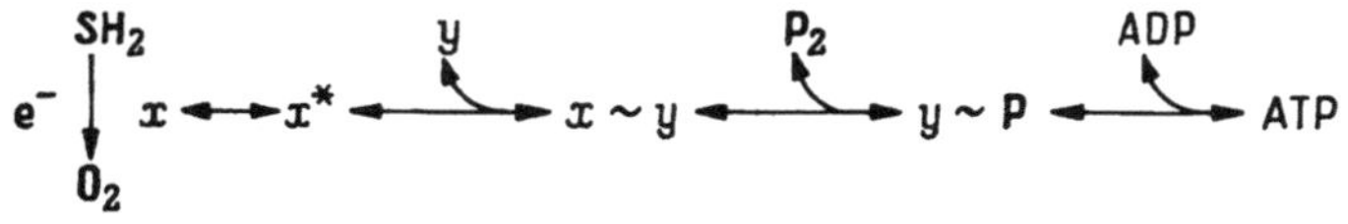

Where SH_2 - oxidable substrate, X and Y - unknown components of energy accumulation mechanism, P_i - inorganic phosphate. Oxidative phosphorylation involves at least three steps:

1. Oxido-reduction in the respiratory chain.
2. Energy conservation in the form of unknown nonphosphorylated intermediate.
3. Energy transfer to ADP involving formation of high-energy phosphate bond.

According to the "chemical" concepts of oxidative phosphorylation (for the review see (1)) oxidation(or reduction) results in activation of X that permits its interaction with Y and increases greatly the potential of component Y transfer to P_i whereas phosphoryltransferase between Y-P and ADP results in a formation of ATP.

According to the "mechanochemical" concepts of oxidative phosphorylation the x~y formation is provided by a special energized conformation of a protein, which arises as a result of electron transfer (2).

The essence of MITCHELL's theory (3) is that the function of the respiratory chain is to separate H^+ and OH^- ions across the inner mitochondrial membrane. The active centre of anisotropic ATPase is located in a membrane in such manner, that this enzyme catalyzes the synthesis of ATP at the expense of electrochemical potential of hydrogen ions.

In any hypothesis the ATP formation is the result of transferase reactions sequence, where the first substrate is the non-phosphorylated high-energy intermediate.
The reversibility of separate steps of energy transfer chain is of a great importance for experimental approach to the mechanism of oxidative phosphorylation. The existance of reversibility may give new tool for the investigation of energy transformation mechanism in mitochondria.
The first indications on the reversibility of energy transfer in mitochondria have been obtained by Boyer and coworkers. They demonstrated mitochondria catalysed P^{32}-ATP exchange (4). The next important step in the study of oxidative phosphorylation was demonstration of several ATPase reactions in intact mitochondria and submitochondrial fragments. These reaction are given below:

Table 1

ATPASE REACTIONS OF MITOCHONDRIA

		Oligomycin	DNP
Energy-linked NAD^+ reduction by succinate	succinate+NAD^+ + ATP ⟶ fumarate+NAD·H+ADP+P_i	inhibits	inhibits
Energy-linked accumulation of ions	$Me_{external}$ + ATP ⟶ $Me_{internal}$+ADP+P_i	inhibits	inhibits
Energy-linked transhydrogenase reaction	NAD·H+$NADP^+$ ATP ⟶ NAD^+ + NADP·H+ADP+P_i	inhibits	inhibits
DNP-stimulated, Mg^{++}-dependent ATPase	ATP + H_2O ⟶ ADP + P_i	inhibits	inhibits

CHANCE et al. (5) and KLINGENBERG et al. (6) demonstrated the energy-linked NAD^+ reduction by flavoproteins of the respiratory chain. It should be noted that thise reaction may occur not only at the expense of energy of respiration but at the expense of ATP also. The existance of stoichiometry between ATP hydrolyzed and NAD^+ reduced allow to consider this data as an evidence of complete reversion of oxidative phosphorylation including electron transport. The further evidence on oxidative phosphorylation reversion were obtained in study of energy dependent ATP-driven reaction in mitochondria - such as active ion accumulation (7) and transhydrogenase (8).

These reactions may be considered as "functional" ATPase, and as an evidence of reversibility of energy transfer chain. The latter can be applied not only to "functional" ones but to DNP - stimulated, Mg-dependent ATPase of intact mitochondria too.

Several authors explain the action of DNP in chemical terms by suggestion that DNP causes the splitting of high-energy intermediate, so that x~y splits without incorporation of P_i and ADP (1). Since the reaction of oxidative phosphorylation is reversible, DNP may induce an appearance of ATPase activity. The effect of DNP may be explained by MITCHELL's theory, if it is assumed that the pair - phenol and phenolate ion can induce proton conductivity through the inner mitochondrial membrane (10).

The comparision of ATPases and oxidative phosphorylation sensitivity to oligomycin and DNP gives an additional evidence of the common reaction-steps in both systems.

All "functional" ATPases are inhibited by oligomycin, which also blocks oxidative phosphorylation, preventing P_i incorporation in the energy-transfer chain. Oligomycin also inhibits DNP - stimulated activity of mitochondrial membrane - bounded ATPase (11, 12). DNP abolishes the reactions depending on ATP - hydrolysis (ion accumulation, transhydrogenase reaction, succinate driven NAD^+ reduction) but does not inhibit the ATP hydrolysis itself.

Even brief analysis of ATPases suggests that ATPase activity represents reversed sequence of oxidative phosphorylation and that the study of ATPase can serve as an useful tool for the studying of oxidative phosphorylation.

Recently much attempts were made to research oxidative phosphorylation as a subject of enzymology. In particular a great effort was made for investigation of mitochondrial ATPase as an individual protein.

However, there are many difficulties in the enzymology study of this system. At first, the proteins participating in energy accumulation are bounded hardly with the inner mitochondrial membrane. The strong physical and chemical influences must be applicated usually for solubilization of such proteins which may result in a modification of native structure and the appearance of artificial activities. The second, essential step of energy transformation is lost when separating of the coupling factors from the mitochondrial membrane take place. On the other hand, the reconstructed systems are still rather complicated for the inhibitory and kinetic analysis.

In spite of these difficulties the study on individual proteins participating in energy transformation seems to be fruitful since the fractionation procedures have been so successful in the study of other biochemical systems.

Two activities may be considered as a criterion of protein' s participating in ATP synthesis in the respiratory chain: ATPase activity and coupling activity - increase of P/O ratio or stimulation of any energy-linked reaction.

The ATPase activity of mitochondrial acetone powder extracts was first demonstrated by LARDY et al. (13). Several years later the soluble ATPase from the acetone powder of mitochondria was obtained in other laboratories (14, 15, 16). In recent works it was shown that the soluble ATPase from acetone powder posses not only ATPase but coupling activities as well (16).

The practically homogenous ATPase of beef heart mitochondria was obtained in RACKER' s laboratory (17, 18). The enzyme was solubilized when beef heart mitochondria had been disrupted by shaking with glass beads in a high speed shaker. The molecular weight of the enzyme as determined by equilibrium ultracentrifugation is 284 000. ATPase activity and coupling factor activity of the purified enzyme are cold-labile. When incubated in a cold, the ATPase dissociated forming an equilibrium mixture of low molecular weight components. The similar protein was isolated in KING' s laboratory from the KEILIN-HARTREE preparation (19). Several proteins featuring ATPase and coupling factor activities were solubilized at mitochondria treatment with detergents (20). The soluble ATPase and the proteins affecting on the energy-linked reactions were also obtained by SANADI et al. (21). It should be noted that in spite of very intensive investigation of the soluble mitochondrial ATPase, the mechanism of it' s action, nature of binding with the membrane and it' s role in energy conservation in mitochondria are still unclear. Moreover, it is not clear at what extent the different soluble mitochondrial ATPases obtained by different methods are similar.

In this paper we shall discuss some properties of soluble ATPase obtained in our laboratory.

Preliminary results and an experimental methods have been described elsewhere (41).

Purification procedure:

The soluble ATPase from beef heart mitochondria was obtained after treatment mitochondrial suspension with acetone at low temperature. The Fig. 1 demonstrates the diagram of purification procedure. After acetone removing, the powder was extracted by distilled water and crude extract containing ATPase activity was purified by ammonium sulphate fractionation. The protein obtained after two precipitation by ammonium sulphate was dissolved in buffer solution containing ATP and was treated by heat exactly 2 minutes at 65°. The clear supernatant obtaining after centrifugation was applied to a column of Sephadex G-200, which has been equilibrated with KCl-Tris buffer solution. Elution was carried out with the same buffer.

The final preparate was 160 times as high as the DNP stimulated ATPase activity of intact mitochondria. The enzyme is stable at the room temperature and inactivates rapidly at low temperature.

It should be noted that solubilization of ATPase becames possible after the treatment mitochondria with acetone. The mixture of acetone with water is known to be a good solvent for the mitochondrial phospholipids (22). One had assumed that solubility of ATPase has taken place as a result of removing of phospolipids from the membrane. We have compared the distribution of the activities in extracts and residues after the treatment of mitochondrial suspension with the different solvents, removing different quantity of phospholipids. These results is demonstrated in the Table 2.

It was found that there is no correlation between the amount of phospholipids removed and the degree of ATPase solubilization. For example, the treatment of mitochondria with chloroform-methanol mixture results in a near completely removing of phospholipids, but the ATPase activity of residue is not solubilized by water. The content of phospholipids in different samples of powder ranges 10-20% from total weight. More than 40% of total ATPase activity may be extracted from such preparations. It seems that the treatment with acetone in our condition rather results in a strong changes in mitochondrial structure and does not result in removing of phospholipids. Moreover it was found that solubilization of ATPase depends on procedure of preparing of mitochondria before the treatment with acetone. The washing of the initial mitochondrial suspension by KCl solution increases extraction of ATPase from acetone powder.

We have compared the structures of mitochondria before and after treatement with acetone in different conditions. One can see an Fig. 2 that the washing

Table 2

THE SOLUBILITY OF ATPASE AND AND CONTENT OF PHOSPHOLIPIDS IN DIFFERENT MITOCHONDRIAL PREPARATE

Sample	Phospholipids content, % of total weight	Distribution of ATPase activity after extraction by distilled water, %	
		Supernatant	Residue
Mitochondria	~20	-	-
The powder after treatement with acetone	10 - 20	86	50
The powder after treatement with chloroform-methanol mixture	~ 3	24	76

mitochondria by KCl solution causes swelling intramitochondrial space. This changes are reversible: the second washing mitochondria by isotonic sucrose solution results in restoration of initial structure. The structures of acetone powder, obtained from these mitochondrial preparation is demonstrated on Fig. 3. It is seen that the powder, prepared from "KCl - mitochondria" practically does not consists a membranes. While the structure of intact mitochondria after treatement with acetone is considerable changed, one can see that such preparations still contain many membrane structures. The solubility of ATPase in water for the "salt mitochondria" is near twice as much as such one of "sucrose preparation". FLEISHER et al. have found a high solubility of mitochondrial phospholipids in mixture of 90% acetone with 10% water. As they pointed out the structure of mitochondria after treatements with such mixture does not change. On the other hand we found that inner mitochondrial membrane in our condition is considerable destroyed. The simpliest explonation of thise facts seems to be the different temperature of treatment with acetone in FLEISHER's and our's experiments. It seems possible that acetone effect at low temperature can be attributed resolves itself to rapid freering of intramitochondrial water which brings about to desintegration of the inner membrane. It seems to be case, because in our experiments the incrase of acetone temperature up to 0° results in an increase of removing phospholipids and in inhibition of ATPase. Moreover a good preparation for extraction of the enzyme may be obtained if acetone treatement is substituted by for such one with butanol at the low temperature. It may be pointed out that the soluble mitochondrial ATPase was isolated in Slater' s laboratory from the powder prepared by treatement with 95% acetone (23) although under such condition the removing of phospholipids is poor (22).

Reconstitution of the ATPase activity of submitochondrial particles:

As it was pointed out above the solubilization of ATPase is concerned with application of influences which may change a properties of the enzyme. Therefore the question arises at what extent is the purified protein intact. For instance oligomycin does not change the activity of our enzyme (as well as RACKER' s F_1) but inhibits the membrane - bound ATPase.

It was shown that F_1 binding with the membrane of submitochondrial fragments accompanies with the restoration of ATPase sensivity to oligomycin (24). We

Table 3

THE BINDING OF SOLUBLE ATPASE WITH APO-PARTICLES

N	Sample	Specific activity, µmol/min/mg	Inhibition by oligomycin %
1	Soluble ATPase	1,41	0
2	ATPase preincubated with $AgNO_3$	1,41	-
3	ASU-particles	0, 03	-
4	ASU-particles+ 1	0, 17	100
5	ASU-particles + 2	0, 05	

The soluble ATPase (4 mg of protein) was incubated with ASU-particles (14 mg of protein) obtained according Racker (25) in the reaction mixture containing 0. 25 M sucrose, 5 mM tris-HCl (pH 7, 5), 5 mM $MgCl_2$ in a final volume 2 ml. After incubation (15 min) the mixture was diluted to 6 ml and centrifugated for 30 min at 145 000 g. The pellets was washed twice with 0, 25 M sucrose. In the experiments 2 and 5 an ATPase was preincubate with AgNO (50 m mol/mg) for 5 min. The activities of samples 3-5 is calculated per mg of particles.

have examinated an ability of our preparation to bind with the mitochondrial membranes. ASU-particles (25), devoided of ATPase was choosen as apo-particles.

The results of such experiments is demonstrated in Table 3.

One can see that the incubation of soluble ATPase with the particles depriving ATPase activity results in a 5-fold increase of specific activity of particles. This increasing is not due to the adsorption, because the new activity is completely sensitive to oligomycin. It is interesting that the blocking of all SH - groups of soluble ATPase prevents the reconstitution but does not change the activity itself.

It should be noted that mitochondrial ATPase is stimulated by low concentration of SH-inhibitorsand is inhibited by high concentration of it (35).

The participation of SH-groups in the mechanism of energy transfer in mitochondria had been discussed by several authors. Flucharty and Sanadi postulated the participation of SH-groups in coupling mechanism (26, 27). It is known that SH-reagents inhibits of oxidative phosphorylation (29, 30) and some of energy linked reactions (31). The activity of coupling factor B is very sensitive to reagents binding thiols (28). The quantity of free SH-groups in mitochondria was found to be correlated with the steady-state concentration of NAD under the additon of substrate and ADP (33).

We can discuss at least two possibilities of SH-group participation in energy transfer in mitochondria.

At first, the absence of inhibition of soluble ATPase may be explained when assuming that the pathways of ATP hydrolysis by the soluble enzyme and by intact mitochondria are quite different. In other words the hydrolysis of ATP by the soluble enzyme does not includes high-energy bond of sulphur. Alternative hypothesis is that SH-groups takes part in the binding of the enzyme with mitochondrial membrane. According the latter, inhibition of oxidative phosphorylation by SH-reagents is the result of coupling mechanism structural changes and of appearance of anomal availability of high-energy compound to water.

The result indicating participation of SH-groups in binding of ATPase with the membrane is in a good agreement with data on stimulating effect of SH-reagents on the ATPase activity of intact mitochondria (35).

Catalytic properties of the enzyme:

The enzyme hydrolysis of ATP, ITP, UTP. The hydrolysis of all substrates requires of the metal-ions. ATPase has a weak specifity for the metal-activator:

ATP is hydrolyzedby the enzyme in the presence of Mg^{++}, Ca^{++}, Co^{++}, Fe^{++}, Ni^{+}, although the rate of hydrolysis is different for these metal. As far as Mg^{++} is the most possible ion, participating in ATP formation in intact mitochondria, we studied the dependence of initial rate of the hydrolysis on the concentration of this ion. The result is presented on Fig. 4.

It seems that ATP- Mg^{++} complex is the real substrate of the enzyme. The free magnesium and free ATP inhibits the reaction. Putting in view this fact, the kinetic constants of the enzyme was measured under constant ATP/Mg ratio equal two. The double reciprocal plot gives a straight line and shows (Fig. 5), that K_m value is $5,2 \cdot 10^{-4}$M. ADP is a competitive inhibitor of the enzyme (Fig. 5, curves 2 and 3).

AMP, adenosine, adenine, pyrophosphate and phosphate does not change activity of the enzyme.

Kinetics of the enzyme was studied for the protein after second ammonium sulphate fractionation.

It was noted that the rate of hydrolysis of ATP by such preparation decreases in course of the reaction. Similar data was obtained by SELWIN (15), however, the nature of this phenomenon was not understanded. It is shown on Fig. 6 the continous trace of acidification of the reaction mixture in course of ATP hydrolysis. The rate of hydrolysis decreases twice already one minute after the initiation of the process. If we put in view the H^{+}/P_i ratio value near one and the buffer capacity of the reaction mixture it is easy to calculate the concentration of ATP, ADP and P_i at any point of this curve. To examinate whether the decrease of rate is due to changes of the reagent concentration the following experiment was carried out. In the reaction vessel ATP, ADP, and P_i was added in concentration exactly equal to one' s at the moment, pointed as A. (Fig. 6)

It was found that initial rate under such condition was much greater than the rate in point A. Therefore the decrease of enzyme activity does not result in a change ob substrate and product concentration. Moreover in a separate experiment it was found there is not inactivation of the enzyme at the course the reaction. The activity falls also with GTP, UTP and ITP although products these nucleoside triphosphates hydrolysis does not inhibits the enzyme.

The enzyme purified to the second $(NH_4)_2SO_4$ fractionation was found to be no homogenous in ultracentrifuge. On the other hand, in Racker' s laboratory the low-molecular weight protein was obtained, which inhibits ATPase (36, 37). We proposed that the purification of ATPase changes the kinetic of the reaction.

Really it turned out that purified enzyme hydrolysed ATP with a constant rate. To have more indications for inhibition of our preparate by low-molecular protein we prepare this inhibitor according HORSTMAN et al. (37). It was found that its addition to the reaction mixture changes the character of hydrolytic activity of the enzyme. The rate of hydrolysis now is not linear with the time. Therefore inhibition of the enzyme activity is due to contaminant protein which inhibits ATPase. It is interesting that the inhibitory effect of this protein manifests only in the presence of the substrates of reaction - under the condition when enzyme is in "working" state. It may be proposed that the high hydrolytic activity of the enzyme does not represent real mechanism of hydrolysis of ATP by intact mitochondria in the presence of DNP and is result of anomal availability to water of some reactive group (for example acylphosphate).

The reaction between ATPase and natural inhibitor results in binding of the latter with activated group of the enzyme and in the appearance of stable product. In this case the rate of inhibition must be dependent on the rate of hydrolysis of different nucleoside triphosphates.

Alternative explanation consists in existion of phosphokinase activity of inhibitor, so that phosphorylated form of ATPase is not active. At last it is possible that ATP-Mg complex is required for the binding of the inhibitor with the enzyme. Now we have no data in favour to any proposed mechanism and this phenomenon requirs of additional experiments.

The effect of uncouplers on soluble ATPase:

The mechanism of action of uncouplers on mitochondrial soluble ATPase is of fundamental interest.

As it was mentioned above according to the "chemical" hypothesis of the oxidative phosphorylation, mechanism of the action of uncouplers is a catalysis of hydrolytic decomposition of non-phosphorylated intermediate of oxidative phosphorylation (1).

According to Mitchell' s hypothesis the free energy of reactions of the respiratory chain is accumulated as a gradient of proton concentration across the inner membrane of mitochondria which is impermeable for hydrogen ions (10). In terms of this hypothesis, the effect of uncouplers on the reactions in intact mitochondria consist in an increase of the mitochondrial membrane proton conductance.

Several authors habe reported experimental evidence for increase in proton conductance of artificial phospholipid membranes caused by uncouplers (38-40). The increase on electric conductance of artificial phospholipid membranes under the action of some uncouplers has been shown to exhibit quantitative correlation with the ability of these substances to induce stimulation of respiration in intact mitochondria (39)*. The latter phenomenon is a strong argument in favour of the mechanism of uncouplers action as proton conductors.

On the other hand, Racker's group have reported a small but distinct stimulation of soluble ATPase by DNP (18). Similar results were obtained earlier by PENNIAL who used ATPase of acetone-dried mitochondria (14); they were confirmed by SELWIN (15).

The mechanism of stimulation by DNP of ATPase released from membranes is not clear.

To obtain more detail information about this phenomenon, in a preliminary experiment we measured the rate of hydrolysis of ATP in the presence of DNP and without it. The curves, demonstrating the results of such experiments are presented on Fig. 7.

One can see that after 45 sec from the beginning of the reaction the quantity of protons released in the presence of 1 mM DNP (upper curve) exceeds that in the control by approximately 50% (lower curve). Such an increase in enzyme activity is in a good agreement with the quantitative data of other authors (15, 16, 18).

The reaction rate with and without DNP, estimated by linearization of the curves, 45 sec after the beginning of the reaction was 7 and 4 μg H^+ ion per min per mg of protein, respectively, which indicated activation of the enzyme. However, it is clear from Fig. 7 that the difference between the initial rates of reactions with and without DNP is very small.

These results should have been expected if one assumes that the decrease in reaction rate is due, at least partially, to product inhibition and the effect of DNP is due to removal of this inhibition. To verify this assumption, some experiments were performed the results of which are shown in Fig. 8.

ADP was added to the reaction mixture, and the rate of reaction was measured 15 sec after addition of the enzyme. The results obtained show that DNP prevents the inhibitory effect of ADP.

The simulating effect of DNP has been proved to be specific to the ATPase reaction. Other nucleoside triphosphatase activities of soluble ATPase do not change under

---

* RECENTLY TING et al. Arch. Biochem. Biophys., 141 (1970), 141 found that there is no correlation between the effectiveness of uncouplers on mitochondria and their effects on conductions of lipid membrane.

the action of DNP. This specificity may be understood because ADP is the only nucleoside diphosphate that inhibits the enzyme.
To verify the specificity of the DNP effect, the action of other uncoupler (FCCP) on soluble ATPase has been studied. It was shown that this and other uncouplers had no effect on the activity of the enzyme and on its sensitivity to ADP. These results reported show that there is no correlation between the effects of uncouplers on intact mitochondria and on soluble ATPase and that the stimulation of the enzyme by DNP is due to prevention of the inhibition of ATPase by ADP.
It should be pointed out that PULLMAN et al. found the stimulating effect of DNP in the presence of regenerating system (phosphoenol pyruvate and pyruvate kinase) (18).
However, there is no data on the steady state concentration of ADP under such condition. Such apparent disagreement of our data and PULLMAN' s results requires more detailed analysis.
For the explanation the effect of DNP it may be proposed that DNP affect the hydrophobic area of ATPase, so that ADP does not able to react with the active centre of enzyme.
On the fact that hydrophobic area in ATPase exists and take place in catalytic activity indicates its cold lability. It is quite clear, that so mechanism is very hypothetical and requires further studies.

REFERENCES

(1) SLATER, E. C., in M. FLORKIN ed. "Comprehensive biochemistry", 14 (1966), 327, Elsevier Publishing Company, Amsterdam, London, New-York, 327

(2) BOIER, P. D., Molekularbiologie, Nauka, 227 (1964)

(3) MITCHELL, P., J. MOYLE, Nature, 213 (1967) 137

(4) BOYER, P. D., W. W. LUCHSINGER, A. B. FALCONE, J. Biol. Chem., 223 (1956) 405

(5) CHANCE, B., G. HOLLUNGER, Federat. Proc. 16 (1957) 163

(6) KLINGENBERG, M., W. SLENCZKA, Biochem. Z., 331 (1959), 334

(7) VASINGTON, F. D., J. V. MURPHY, J. Biol. Chem. 237 (1962), 2670

(8) DANIELSON, L., L. ERNSTER, Biochem. Z., 338 (1963), 188

(9) SLATER, E. C., Proc. 5-th Internat. Congr. Biochem., Moscow 1961, 5, Pergamon Press, London, 1963

(10) MITCHELL, P., J. MOYL, In E.C. SLATER, Z. KANIUGA, L. WOJTCHAK (Eds), "Biochemistry of Mitochondria, Acad. Press, London, New-York, (1967), 53

(11) LARDY, H.A., D. JOHNSON, W.C. McMURRAY, Ach. Biochem. Biophys., 78 (1958), 587

(12) LARDY, H.A., W.C. McMURRAY, Federat. Proc., 18 (1958), 269

(13) LARDY, H.A., H. WELLMAN, J. Biol. Chem., 201 (1953), 357

(14) PENNIAL, R., Biochem. Biophys. Acta, 44 (1960), 395

(15) SELWIN, M.J., Biochem. Jl., 105 (1967), 279

(16) VALLEJOS, R.H., E.C. SLATER, Biochem. Biophys. Acta, 143 (1967), 441

(17) PULLMAN, M.E., H. PENEFSKY, E. RACKER, Arch. Biochem. Biophys., 76 (1958), 227

(18) PULLMAN, M.E., H. PENEFSKY, A. DATTA, E. RACKER, J. Biol. Chem., 235 (1960), 3322

(19) KETTMAN, J., T.E. KING, Biochem. Biophys. Res. Communs., 11 (1963), 255

(20) TZAGOLOFF, A., J. Biol. Chem., 243 (1968), 2405

(21) ANDREOLI, T.E., K.W. LAM, D.R. SANADI, J. Biol. Chem., 240 (1965), 2644

(22) FLEISHER, S., B. FLEISHER, W. STOECKENIUS, J. Cell. Biol., 32 (1967), 193

(23) VALLEJOS, R.H., S.G. VANDEN BERGH, E.C. SLATER, Biochem. Biophys. Acta, 153 (1968), 509

(24) KAGAWA, Y., E. RACKER, J. Biol. Chem. 241, (1966), 2467

(25) RACKER, E., L.L. HORSTMAN, J. Biol. Chem., 242 (1967), 2547

(26) FLUHARTY, A.L., D.R. SANADI, J. Biol. Chem., 236 (1961), 2772

(27) FLUHARTY, A.L., D.R. SANADI, Proc. Nat. Acad. Sci. USA, 44 (1960), 609

(28) LAM, K.W., Arch. Biochem. Biophys., 123 (1968), 642

(29) JACOBS, E.E., M. JACOBS, D.R. SANADI, L.B. BRADLEY, J. Biol. Chem. 223 (1960), 147

(30) FLETCHER, M.S., A.L. FLUHARTY, D.R. SANADI, Biochem. Biophys. Acta, 60 (1960), 425

(31) KURUP, C.K.R., D.R. SANADI, Biochemistry, 7 (1968), 4483

(32) TYLER, D.D., Biochem. J., 107 (1968), 121

(33) WINOGRADOW et al., Biochimia, 31, (1966), 501

(34) COOPER, C., J. Biol. Chem., 235 (1960), 1815

(35) KILLEY, W.W., Proc. 5-th Internat. Congr. Biochem., Moscow 1961, 5, Pergamon Press., London, 1963

(36) PULLMAN, M.E., G.C. MONROY, J. Biol. Chem., 238 (1963), 3762

(37) HORSTMAN, L.L., E. RACKER, J. Biol. Chem., 245 (1970), 1336

(38) BIELAWSKY, L., T.E. THOMPSON, A.L. LEHNINGER, Biochem. Biophys., Res. Communs., 24 (1966), 943

(39) SKULACHEV, V.P., A.A. SHARA, L.S. YAGUJINSKY, A.A. JASAITIS, E.A. LIBERMAN, V.P. TOPALY, Currents in Modern Biol., 2 (1968), 98

(40) LIBERMAN, E.A., V.P. TOPALY, Biochem. Biophys. Acta, 163 (1968), 123

(41) AKIMENKO, V.K., J.B. MINKOV, A.D. VINOGRADOV, Biochemistry (Moscow), 1971, in press

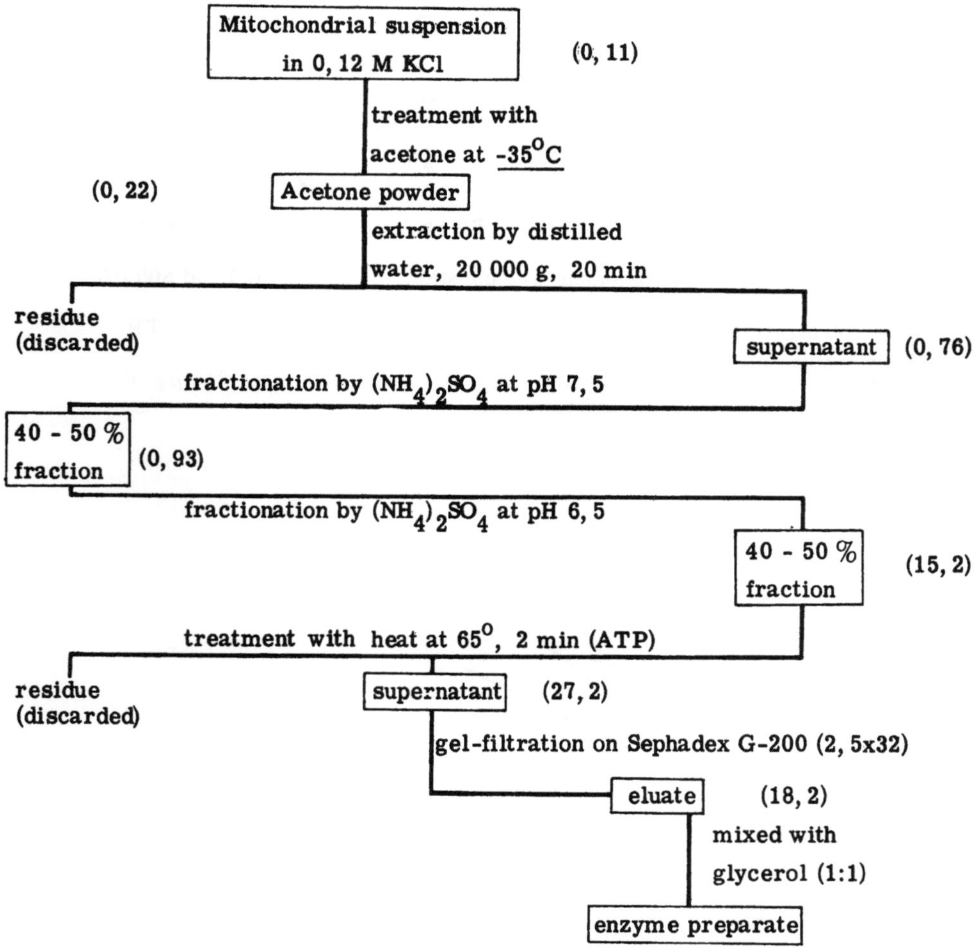

Fig. 1. Diagram of purification procedure.

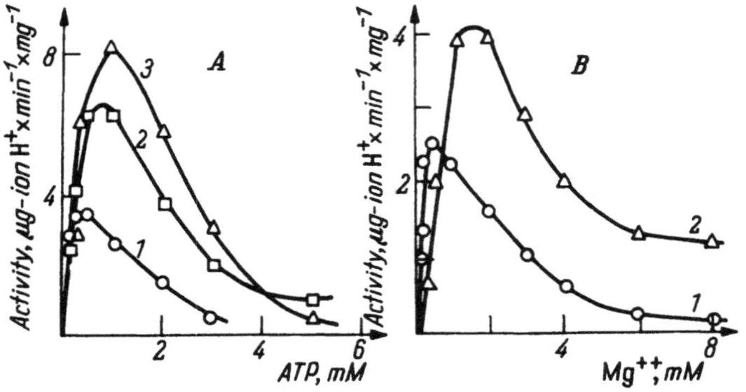

Fig. 4. Dependence of initial rate of ATP hydrolysis on concentration of ATP and magnesium
A - concentrations of magnesium are constant, 1-3 - 50, 300 and 500 μM of Mg^{++} respectively
B - concentrations of ATP are constant, 1 and 2 - 1 and 2 mM of ATP respectively
The reaction mixture: KCl 50 mM, tris-HCl (pH 7, 5) 5 mM, 210 μg of protein, ATP and Mg^{++} as indicated.

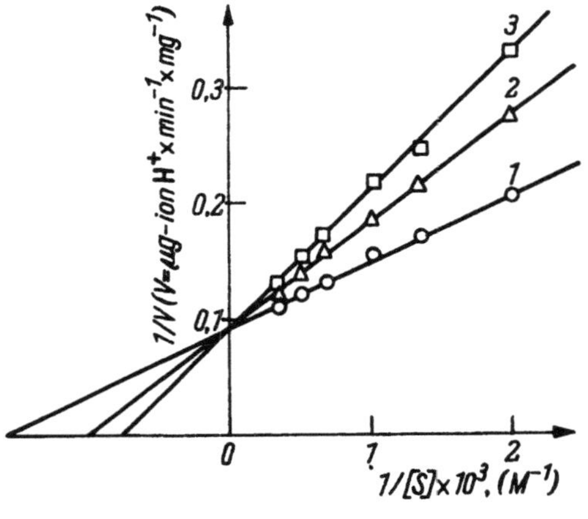

Fig. 5. Kinetic of ATPase reaction
Reciprocal values of initial rates are plotted against reciprocal values of Mg^{++}-ATP complex
The reaction mixture: KCl 50 mM, tris-HCl (pH 7, 5) 5 mM, 200 μg of proteins.

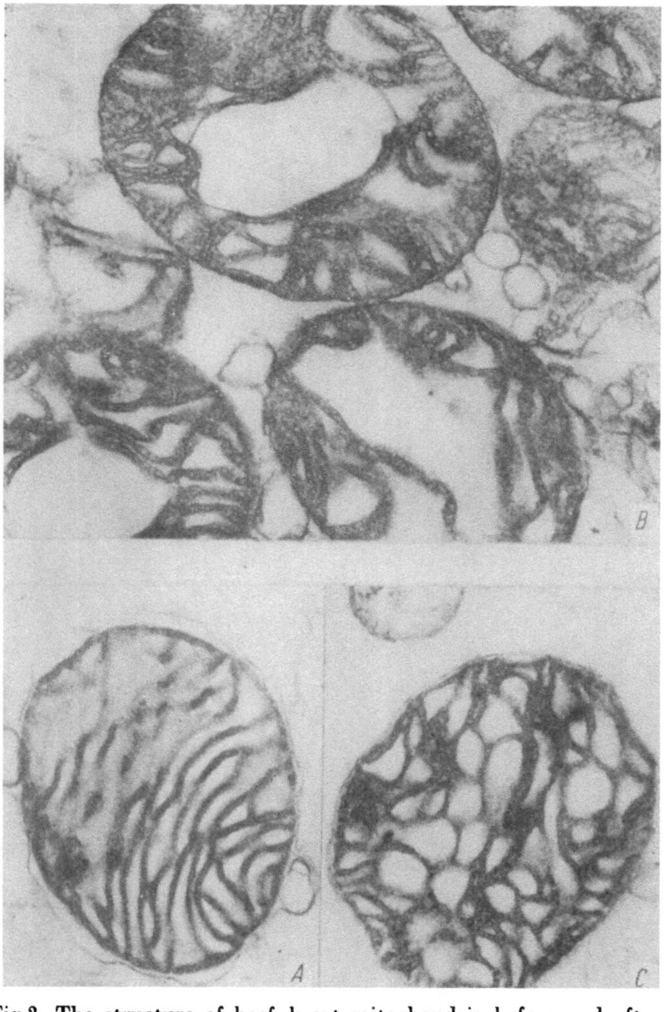

Fig 2. The structure of beef heart mitochondria before and after treatment with acetone
A -- mitochondria suspended in 0,25 M sucrose
B — mitochondria were washed with 0,12 M KCl
C — mitochondria were washed with 0,12 M KCl and after washing with 0,25 M sucrose were suspended in 0,25 M sucrose.

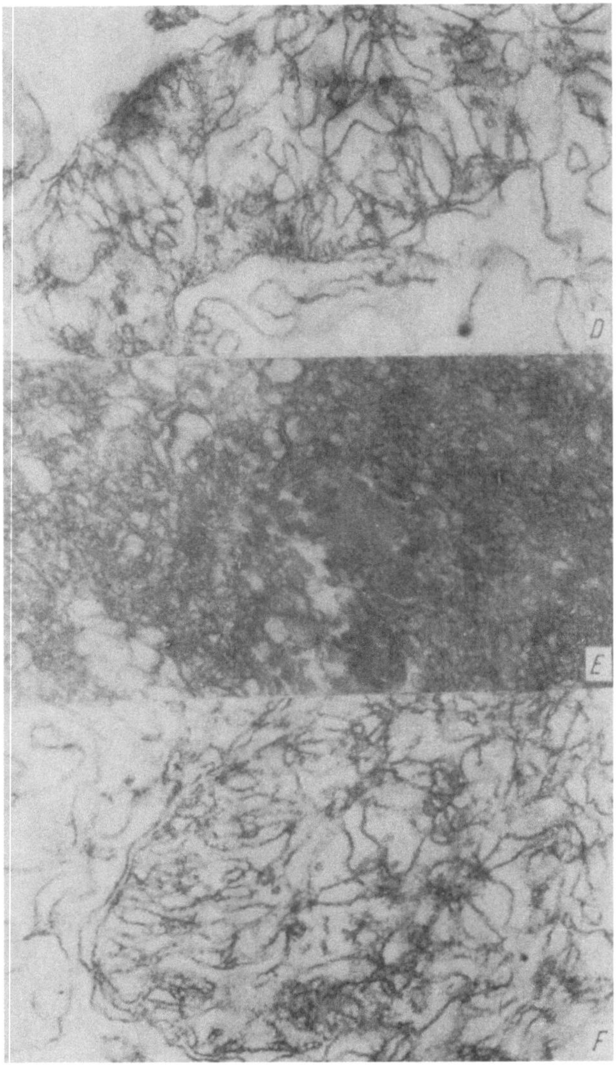

Fig. 3 The structure of acetone powders D—E-preparates A, B and C (see fig. 2), respectively, were treated with acetone at −35 °C.

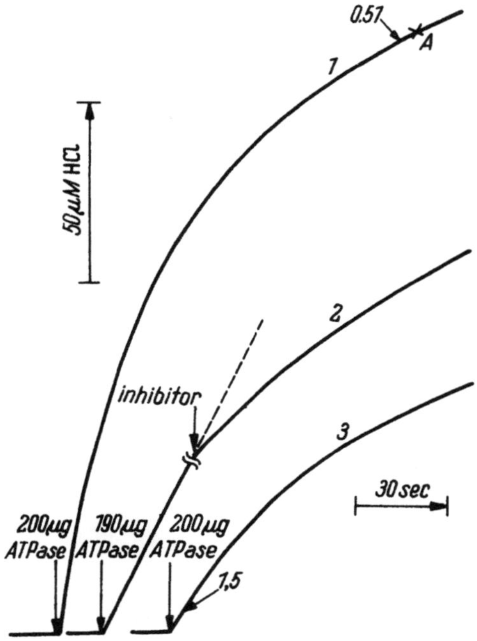

Fig. 6. Acidification of the reaction mixture at the ATP hydrolysis by different enzyme preparations inhib. -addition of 14 μg of natural protein-inhibitor.

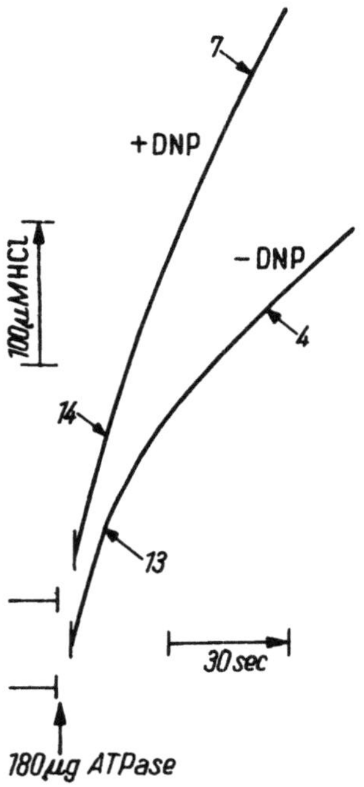

Fig. 7. Acidification of the reaction mixture at the ATP hydrolysis by soluble ATPase
Upper curve - in the presence of 1 mM DNP
Lower curve - control
Figures on arrows indicate the values of reaction rate calculated for the moment indicated
The reaction medium consist: 10 mM tris-HCl buffer (pH 7, 5), 50 mM KCl, 2 mM ATP and 1 mM $MgCl_2$.

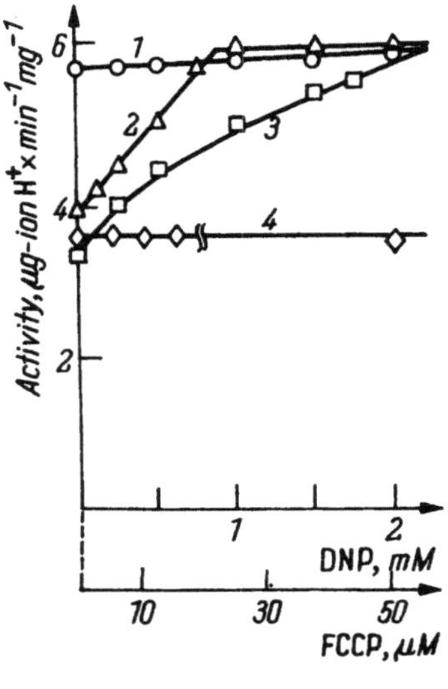

Fig. 8. Effects of DNP and FCCP on ATPase inhibition by ADP
The reaction was started by the enzyme addition 1 - without ADP, 2 and 3 - with 1 and 2 mM of ADP respectively, 4 - with 1 mM ADP, 1-3 - experiments with DNP, 4 - with FCCP.

Animal Biochemistry Department, Lomonosov State University, Moscow

Y. N. LEYKIN and A. D. VINOGRADOV

IDENTIFICATION OF IONIC FORM OF PHOSPHATE, ACCOMPANYING UPTAKE OF CA^{2+} BY RAT LIVER MITOCHONDRIA

Measurements have been made of the pH dependence of respiration rate of isolated rat liver mitochondria in the presence of calcium ions and limited amounts of phosphate. A kinetic analysis of this dependence and study of relationship between oxygen consumption rate and concentrations of different ionic forms of phosphate reveal that phosphoric acid monoanion is transported across the mitochondrial membrane. K_m value for phosphate transport is $7 \cdot 10^{-5} M H_2PO_4^{-1}$.

Investigations on the transport of oxidizable substrates, inorganic phosphate and other anions and cations into the internal mitochondrial compartment demonstrated that the mitochondrial membrane has a very specific permeability to various agents. This has been interpreted to mean that the mitochondrial membrane contains special translocases. Considering that some enzymes are localized in an intramitochondrial compartment, these translocases may play a vital role in mitochondrial metabolism. At present very little is known about the structure and active centers of carriers, as well as of the molecular mechanism of the reactions catalyzed by them. Since many substrates of enzymes localized inside mitochondria at physiological pH values are ions (nucleotides, substrates of cycle of tri-carboxylic acids, phosphates, cations ets.), identification of ionic forms of transported agents deserves special attention. Such data are needed for deeper insight into the mechanism of substrate transport across the membrane.

It has been shown that inorganic phosphate (P_i) in the course of oxidative phosphorylation is transported across the inner membrane as a monoanion of $H_2PO_4^{-1}$ (1). Active uptake of calcium is another process paralleled by phosphate transport across the mitochondrial membrane (2). It is known that the energy-linked massive accumulation of Ca^{2+} occurs only in the presence of anions penetrating through the membrane (3). Under physiological conditions phosphate seems to play the part of such an anion. Present paper deals with the identification of an ionic form of phosphate participating in transport across the membrane in the course of active Ca^{2+} transport by rat liver mitochondria.

Methods

Rat liver mitochondria were isolated by differential centrifuging in an isotonic solution of sucrose, containing $1\cdot 10^{-3}$ M EDTA, pH 7,4. Upon washing isolated mitochondria were suspended in a 0,25 M solution of sucrose and stored at $0^{o}C$. Incubation medium contained 0,15 M sucrose, 0,075 M KCl, 10 mM tris-HCl (pH values are given in legends to figures), $5\cdot 10^{-6}$ M rotenone, 10 mM succinate and 3-5 mg of mitochondrial protein; volume of samples =2ml, temperature 19-21°C. Oxygen consumption rate was determined polarographically with rotating platinum electrode. Protein was measured by biurete method.

Results

Relationship between the ratio of various forms of phosphate and pH of incubation medium has been used as a basic experimental approach to solve the set task. At any pH value and stability of other factors (ionic strength, temperature), concentrations of all ionic forms of phosphate are derived from its overall concentration and are easily calculated (4).

Fig. 1 presents a polarographic record of mitochondrial respiration in medium (see Methods) containing $5\cdot 10^{-4}$ M phosphate with two fixed pH values. It is seen from the figure that under these conditions oxygen consumption rate during Ca^{2+} transport is strongly dependent upon pH of incubation medium, so that change in pH from 7,2 to 8,1 decreases respiration rate five times (0,20 and 0,04 μg = atoms 0_2/min per mg of protein respectively).

Since active transport of Ca^{2+} is coupled with respiratory chain energy consumption, the measured respiratory rate in active state is correlated with the rate of Ca^{2+} ions transport across the mitochondrial membrane.

When analyzing factors aocounting for the marked dependence of the rate of Ca^{2+} transport across the mitochondrial membrane on pH, it should be remembered, that the registered oxygen consumption is a complicated process involving both proper transport system and activity of respiratory chain. Hence, there are several possible explanations for the effect of changes in pH upon the respiration rate in active state. First, it is possible that the energy accumulation system itself (substrate transfer, respiratory chain and coupling mechanism) are dependent on pH in such a manner that acidification of the incubation medium will induce activation of the system. Another possibility is that the experimentally observed effect of the concentration of hydrogen ions is connected with pH-dependence of the activity of carriers of phosphate and Ca^{2+} (dissociation of a group of active centers of carriers). This suggestion is very likely to be true of the phosphate carrier, for it was demonstrated that Ca^{2+} affinity to the mitochondrial membrane had but a slight dependence on pH in the range of 7,4 to 9,0 (5). The third hypothesis is that changes in pH induce alterations in the concentration of the real substrate of phosphate carrier as a result of variations in the ratio fo different ionic forms (concentration of free ions of Ca^{2+} is practically independent of pH in the observed interval).

Formally, the first two reasons are responsible for a decrease in concentration or activity of a enzyme, participating in the accumulation of Ca^{2+}, whereas the third one is associated with changes in concentration of the reacting substrate. Proper choice among these alternatives can be made with kinetic analysis.

Decrease in amount of an active form of enzyme should lead to a change of the V_{max} value and elimination of K_m dependence upon the concentration of added phosphate. In this case the value of half-maximal rate of oxygen consumption (K_m for P_i) remains constant at various pH values.

On the other hand, a change in the concentration of the reacting substrate must lead to the pH dependence of the apparent value of K_m for phosphate and to the stability of the maximal rate of respiration at different concentrations of hydrogen ions in the incubation medium.

Results of experiments on the dependence of oxygen consumption rate upon phosphate concentration during active transport of Ca^{2+} at different pH values are plotted in Fig. 2.

It is clear from the figure, that values of maximal rate are independent of pH of the incubation medium and that the earlier observed dependence (Fig. 1) is responsible for the change of the apparent K_m value for phosphate.

Changes in pH between 7,15 and 7,80 induce a four-fold change of the K_m value from 150 μM at pH 7,15 up to 600 μM at pH 7,80. The obtained data reveal that the activity of Ca^{2+} and phosphate carriers, as well as the activity of the energy accumulation system, do not change in the studied interval of pH. In other words, acidification of the incubation medium causes an increase in the concentration of the really transported form of phosphate.

For identification of ionic form of phosphate, directly participating in transport across the mitochondrial membrane we investigated the quantitative aspect of

the pH dependence of respiration rate during Ca^{2+} transport in the presence of phosphate amounts, limiting the process as a whole ($5 \cdot 10^{-4}$ M). In this study consideration was given to the fact that acidification of the incubation medium leads to an increase in concentrations of non-dissociated phosphoric acid and its monoanion. The observed dependence is plotted in Fig. 3 (curve 1).
One may see that change between 8,4 and 7,2 in pH of the incubation medium containing $5 \cdot 10^{-4}$ M phosphate leads to a ten-fold increase in oxygen consumption rate during transport of 500 μM Ca^{2+}.
Concentrations of phosphate monoanion and uncharged molecules (H_3PO_4), calculated with account of ionic strength of medium in each experimental point of curve 1, are given in the same figure as dotted lines (2-nd and 3-nd lines respectively). It is evident that a tenfold increase in oxygen consumption rate corresponds to a ten-fold increase in concentration of phosphate monoanions and to almost a 200-fold increase in concentration of an uncharged form of phosphate. Complete correlation between concentration of phosphate monoanion and oxygen consumption rate during active accumulation of calcium phosphate proves that exactly this ionic form of phosphate directly participates in transport across the mitochondrialmembrane. As it is seen from the figure, further increase in $H_2PO_4^{-1}$ concentration at pH below 7,2 does not alter respiration rate, which proves that the carrier is substrate-saturated.
To evaluate real kinetic parameters of the transport system, the obtained dependence (Fig. 3) was substituted by the dependence of enzyme reaction rate upon substrate concentration (Fig. 4). For this purpose, pH values plotted against the abscissa in Fig. 3, were substituted by calculated values of $H_2PO_4^{-1}$ concentrations. The resulting slight deviation from hyperbolic dependence seems to proceed from a somewhat erroneous choice of pK values of phosphoric acid in our experiments.
Such an explanation is in good agreement with the earlier obtained dependence of respiration rate upon phosphate concentration (see Fig. 2). On the basis of the obtained data proving the participation of phosphate monoanion in transport across the membrane, it is possible to explain the quantitative aspect of differences between K_m values associated with a change in pH of the incubation medium (Fig. 2). Indeed, at pH 7,15 and 7,80, concentrations of phosphate monoanions are related as 4:1, and the ratio between apparent K_m values at the same pH values equals 4 (600 and 150 μM). Thus real K_m values for $H_2PO_4^{-1}$, upon recount of experimental

values, coincide and are in good accord with calculated data (K_m for $H_2PO_4^{-1}$ is about 70 μM, Fig. 4).

The present communication provides evidence that monoanion is an actively transported form of phosphate, accompanying Ca^{2+} accumulation. In other words, at pH values close to neutral transport of any molecule of phosphate is accompanied by transport of H^+ion across the inner mitochondrial membrane. The existence of a mechanism providing for such a transport correlates with the postulated exchange of phosphate - hydroxyl and corresponds to the experimentally observed ratio of H^+/Ca^{2+} equal to 1 (6, 7).

REFERENCES

(1) EVTODIENKO, Yu. V., L. Yu. KUDZINA, L. V. PESHKOVA, V. N. SCHIPAKIN, Biochemistry (Moscow), 1971,in press

(2) TYLER, D. D., Biochem. J., 107 (1968), 121

(3) LEHNINGER, A. L., C. S. ROSSI, J. W. GREENAWALT, Biochem. Biophys. Res. Communs., 10 (1963), 444

(4) LAITINEN, H. A., Chemical Analysis, McGraw-Hill Book Company, Ink., New-York, Toronto, London, 1960

(5) GEAR, A. R. L., C. S. ROSSI, B. REYNAFARJE, A. L. LEHNINGER, J. Biol. Chem., 242 (1967), 3403

(6) CHAPPELL, J. B., A. R. CROFTS, Biochem. J., 95 (1965), 393

(7) AZZONE, G. F., A. AZZI, C. S. ROSSI, in Mitochondrial Structure and Compartmentation. Ed. QUAGLIARIELLO, E., S. PAPA, E. C. SLATER, J. M. TAGER, Adriatica Editrice, Bari, 1967

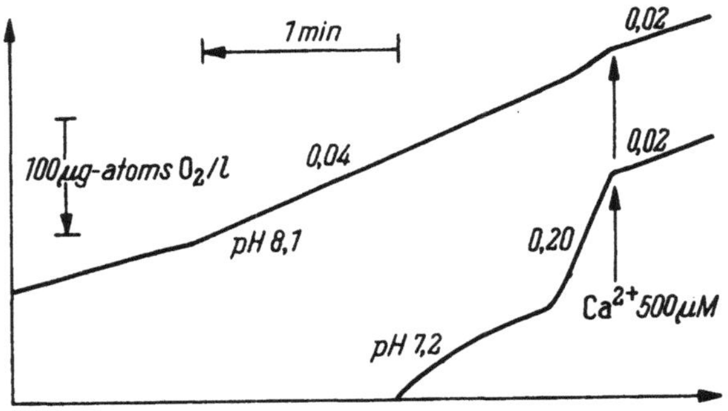

Fig. 1. Oxygen consumption by rat liver mitochondria under $CaCl_2$ addition. Mitochondria (4 mg of protein) were incubated in the medium as described under "Methods" except the 5 10^{-4} M of P_i was present. Initial pH values of the reaction mixture are indicated on the curves. Figures on the curves indicates the rate of respiration, µgatoms 0_2/min/mg.

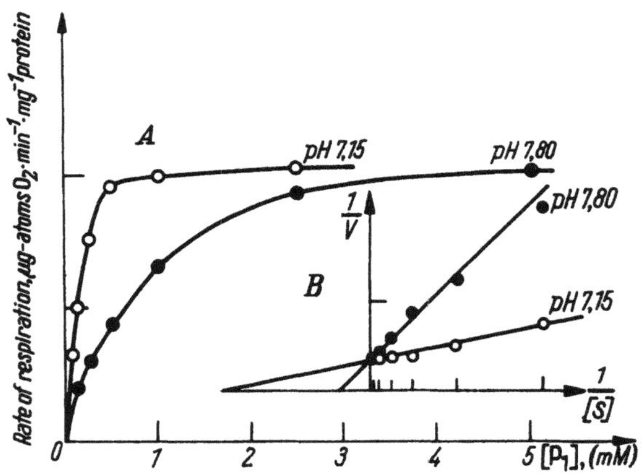

Fig. 2. Effect P_i concentration on the rate of oxygen consumption during Ca^{+2}-ions accumulation. The inset contains a plot of the reciprocal reaction velocity against the reciprocal of P_i concentration. pH values of the reaction mixture are indicated on the curves. Mitochondria (4, 8 mg of protein) were incubated in the medium as described under "Methods" except varying P_i concentration were present. Reaction was initiated by 500 µM $CaCl_2$ addition.

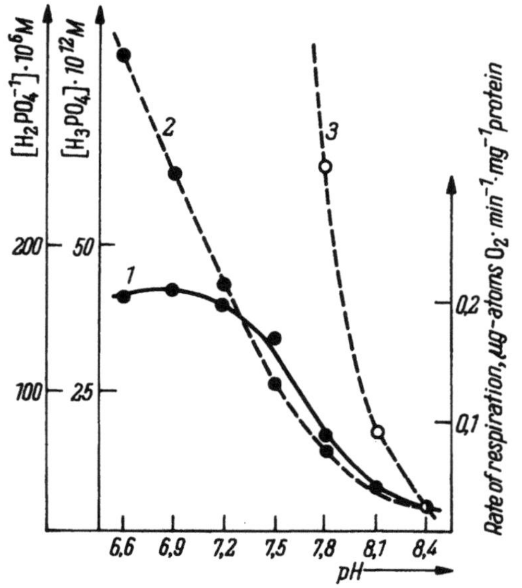

Fig. 3. pH-dependence curve of the rate oxygen consumption during Ca^{+2}-ions uptake. Mitochondria (4 mg of protein) were incubated in the medium as described under "Methods" except the $5\cdot 10^{-4}$ M of P_i was present. Reaction was initiated by 500 μM $CaCl_2$ addition. Curve 1 - the rate of respiration, curves 2 and 3 - the calculated $H_2PO_4^{-1}$ and H_3PO_4 concentrations respectively.

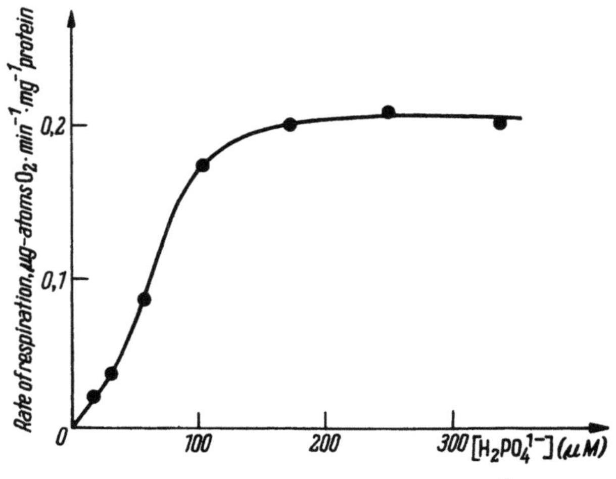

Fig. 4. The respiration rate plotted against the calculated $H_2PO_4^{-1}$ concentration. Conditions were the same as those described in Fig. 3.

Institute of Biological Physics, Academy of Sciences of the USSR,
Pushchino, Moscow Region

V. J. DESHCHEREVSKY

MECHANISM OF ACTIVATION OF MYOSIN ATPase BY ACTIN

At present all enzymes can be considered to some degree as mechanoenzymes. Thus, the studies of functional conformational rearrangements are necessary to understand the enzymatic mechanisms. This conclusion concerns to a large extent such a specific mechano-chemical system as actomyosin. A native quarternary structure of actin and myosin in striated myofibrills has been studied rather well. But its transformation during the contraction can not be detected as the structural analytical methods are not rapid enough for this purpose. Nevertheless, the simulation of muscle contraction gives the chance to understand some features of elementary conformational rearrangements of actomyosin during its enzymatic action. It should be noted that all conclusions about the nature of actomyosin ATPase obtained in this work are valid as much as the kinetic model of striated muscle contraction (1, 2) from which these conclusions result.

The model is based on the totality of facts which is traditionally named as "sliding-filament hypothesis". It results from the sliding-filament concept that myosin cross-bridge must have at any rate two essentially distinct states:free and closed. Each cross-bridge must act cyclically, that is, make a link with the actin receptor, develop moving force and then break the link. The elementary working cycle of the cross-bridge which necessarily results from the sliding-filament concept is shown in Fig. 1a.

But this working cycle is not sufficient enough to explain contractile properties of striated muscles. For this purpose it is necessary to assume that a closed cross-bridge transits at least through two essentially distinct states. In the first state which can be called as a "pulling" one the cross-bridge develops a positive force which initiates the sliding of filaments in the direction corresponding to the muscle shorting. In the second state which can be named as a "hindering" one the cross-bridge develops a negative force which hinders the contraction.

The chance of the splitting of the hindering cross-bridge is many times higher than that of the pulling one. The rate of the transition from the pulling state to the hindering one is proportional to the speed of filament sliding, i. e. the speed of the contraction. Such an elementary working cycle of the cross-bridge which is the basis of the kinetic theory of striated muscle contraction (1, 2) is shown in Fig. 1b. This theory allows one to calculate practically each regime of the contraction. From this analysis it follows: Hill's equations result from the theory automatically; calculated rates of force development and redevelopment after quick release coincide with experimental data; the theory accounts for the behaviour of insect flight muscles (2) and sceletal muscles of vertebrates (3) during driven oscillations of the lenghth; it permits one to predict some new facts, for instance, auto-oscillations of the force developed by a frog sartorius muscle under isotonic conditions; these auto-oscillations have been registered under the predicted experimental conditions (4).
The existence of the hindering state of closed cross-bridge is, as a matter of fact, the single assumption introduced in the theory in addition to the principles of sliding-filament concept. The coincidence of the theory with the experimental data gives the grounds to maintain that this assumption reflects real situation in the contractile system of striated muscles. If so, some conclusion may be done about the nature of actomyosin ATPase, because the speed of circulation of cross-bridge through the elementary cycle gives the rate of ATP splitting reaction. Really, the introduced postulate can be formulated in other terms. We could assume, that cross-bridge splitting is a slow stage in the elementary cycle and that this stage is accelerated when cross-bridge is subjected to the deformation. There is only one source of external deformational force for linked cross-bridge and this is the relative sliding of actin and myosin filaments. At the first stage (pulling) just after the cross-link formation cross-bridge transforms the energy of ATP splitting into the work of conformational rearrangement, which is the source of filament sliding, and at the second stage (hindering) the sliding of filaments (at the expense of other pulling cross-bridge) is the source of cross-bridge deformation, which causes its splitting. Thus, we should conclude that the mechanical connection between active sites, i. e. the "crystalloid" organization of actomyosin contractile system makes possible to use energy of ATP hydrolysis for the acceleration of this reaction, more accurately, for the lowering of activation barrier of the slowest stage of the reaction.
Actomyosin solutions in vitro have the structures, which are similar to thin and

thick filaments of striated myofibrils (5). It is evident that the mechanism under discussion can be performed in this case also, but can't take place in the case of pure myosin system. This can explain the lower ATPase activity of this protein in comparison with the actomyosin.

The utilization of a part of the reaction energy for acceleration of the same reaction probably takes place in the RNA synthesis on DNA by RNA-synthetase (6). Similar idea may be applicable to the working mechanism of other enzymes which have quarternary structure. Protein oligomers can use the energy which is liberated at one of the active sites during the exothermic stage of the reaction for the acceleration of the endothermic stage of the coupled reaction at the adjacent active site.

REFERENCES

(1) DESHCHEREVSKY, V.I., Biofizika, 13 (1968), 928

(2) DESHTSHEREVESKII, V.I., Bioreology, 7 (1971), 147

(3) DESHCHEREVSKY, V.I., Transactions of the symposium on oscillations in chemistry and biology. Pushchino, 1970

(4) BURAVTSEV, V.N., V.I. DESHCHEREVSKY, Biofizika, 15 (1970), 541

(5) MORALES, M.F. In: Molecular biology, Moscow, Nauka, 202 (1964)

(6) IVANOV, V.I., V.L. FLORENTJEV, Molecular biology, 4 (1970), 866

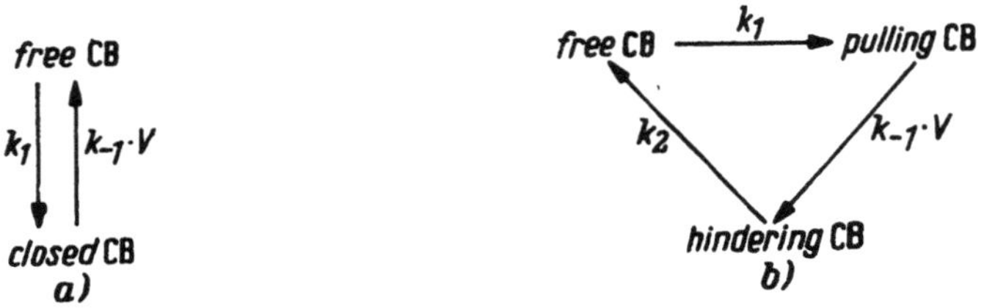

Fig. 1. The diagram of the elementary working cycle of the cross - bridge:
1a - the cycle resulting from the "sliding - filament" concept.
1b - the postulated cycle which is sufficient for the complete description of the contractile properties of the striated muscles.

Animal Biochemistry Department, Lomonosov State University, Moscow

A. A. BOLDYREV, V. B. RITTOW

TRANSPORT ATPases OF MUSCLE MEMBRANES

Two transport ATPases of muscle membranes - Na^{+}, K^{+} - activated ATPase (Na^{+}, K^{+} - ATPase) of sarcolemma and Ca^{2+} - activated ATPase (Ca^{2+}-ATPase) of sarcoplasmic reticulum - provide energetically the active transport of corresponding cations. The accumulating capacity of transport system determines the rate of relaxation and the duration of refractory state, what is essential for intensity of repeated contraction cycle, especially in slow muscles. Ion transporting capacity seems to be controlled by two factors: rate of ATP hydrolysis and its coupling with cation active transport. In this report some examples of the two types of the regulation are discussed.

Ion regulation of ATPase activity: Transport ATPase activity depends on transporting ions. Thus Na^{+}, K^{+}-ATPase (Fig. 1) is controlled both by Na^{+}/K^{+} ratio and by calcium concentration. It is also known the requirement for Mg^{2+} ions of transport ATPases (1-3).

Influence of biologically active compounds: Acetylcholine and caffeine prolongate the Ca^{2+}-uptake by fragmented sarcoplasmic reticulum without changing the ATPase activity (Fig. 2A). The natural result is the decrease of Ca/ATP ratio and Ca^{2+}-uptake efficiency what may be due to the increased outflow of Ca^{2+} from vesicles to medium. The opposite effect was observed in the presence of imidazole compounds which are of interest because of their chemical and biological properties (4). In their presence Ca/ATP ratio and thus efficiency of Ca-uptake rise significantly (Fig. 2B), but ATPase activity ist not affected in this case. However ATPase uncoupled with production of ion gradient is activated by these compounds (Fig. 3).

pH influence: pH of muscle cells is known to change markedly at muscle function. The contribution of transport ATPase into supporting of ion gradients depends on pH medium. Optimum for Na^{+}, K^{+}-ATPase is in slightly alkaline region. At the state of Ca^{2+}-uptake pH optimum for Ca^{2+}-uptake lies at pH 6. 5 (Fig. 4A, I) and for Ca^{2+}-ATPase at neutral pH (Fig. 4A, II). At this pH region Ca^{2+}-response of myofibrillar suspension is rather week (Fig. 4B, I). In the state of so called "exchange" between Ca^{2+} in and Ca^{2+} out (5), when only outflow of Ca^{2+} from reticulum is stabilized by Ca^{2+}-ATPase, its activity rises with the increase of pH (Fig. 4A, III). That results from acceleration of Ca^{2+} outflow (6) and corresponds at the same time to enhancement of Ca-response of myofibrillar suspension (Fig. 4B, II).

It is interestingt to note that imidazole compounds do not influence these Ca^{2+}-curves (Fig. 4B). However in a simple contracting model (myofibrillar suspension plus fragmented sarcoplasmic reticulum) imidazole and anserine accelerate sufficiently the "relaxation" (Fig. 5). That is accounted for their influence on Ca^{2+}-uptake by sarcoplasmic reticulum.

On the mechanism of cation transport: The investigation of transport mechanism may be conveniently carried out on fragmented sarcoplasmic reticulum. Both Ca^{2+} inflow (7), and Ca^{2+} outflow at this system follows to Michaelis kinetics and fits to the equations:

$$E + Ca^{2+}_{out} \underset{K-1}{\overset{K+1}{\rightleftharpoons}} E - Ca_{out} \underset{K-2}{\overset{K+2}{\rightleftharpoons}} E - Ca_{in} \underset{K-3}{\overset{K+3}{\rightleftharpoons}} E + Ca^{2+}_{in}$$

$$E - Ca_{in} + Ca^{2+}_{in} \underset{K-4}{\overset{K+4}{\rightleftharpoons}} E - Ca_{2in}$$

where E is a Ca - transporting unit, $\frac{K-1}{K+1} = 10^{-7}$, $\frac{K+3}{K-3} = 10^{-5}$ (estimated experimertally), Ca^{2+} is a substrat for transfer and V_{max} of inflow equal to V_{max} of outflow of Ca^{2+}. Equation I describes the Ca^{2+}-transport through membrane and equation II - inhibition of this process by Ca^{2+}_{in} (5). In such a way calcium is at the same time the substrate and the modifier of the system. The excess of Ca^{2+}_{in} inhibits significantly the Ca^{2+}-uptake. The analysis of this inhibition gives a straight line in coordinates ATP/Ca versus $(Ca^{2+})_{in}$ (Fig. 6). The point of its intersection with ordinate corresponds to reciprocal value of maximal coupling of ATP hydrolysis with Ca^{2+} transport. H^{+} and Mg^{2+} eliminate competitively Ca^{2+}_{in}-inhibition of transport process. This fact accounts well an increase of Ca/ATP ratio in the presence of H^{+} or Mg^{2+}, observed by many authors (5, 6).

If maximal coupling is less than 2, we may estimate the per cent of disrupted vesicles in this reticulum preparation. If the preparation is characterized by maximum coupling equal to 2, reduction of this value under experimental conditions indicates the Ca^{2+} outflow. In this way:

$$V_{Ca\ inflow} = V_{ATP\ hydr.} \times \frac{Ca}{ATP_{max}}$$

$$V_{Ca\ outflow} = V_{ATP\ hydr.} \times \left(\frac{Ca}{ATP_{max}} - \frac{Ca}{ATP_{exptl.}} \right)$$

$V_{Ca\ inflow} / V_{Ca\ outflow}$ versus reciprocal $[Ca^{2+}]_{in}$ is a straight line, which intersects with ordinate at 1 (Fig. 7). That means that V_{max} for Ca^{2+} inflow and for Ca^{2+} outflow are equal to each other. Thus our data confirm experimentally HASSELBACH' s idea about the unified mechanism for Ca^{2+} transfer through membrane in both directions (8, 9).

REFERENCES

(1) HEXUM, T., F. SAMPSON, R. HIMES, Biochim. Biophys. Acta 212 (1970), 322

(2) YAMAMOTO, T., Y. TONOMURA, J. Biochem. (Tokyo), 62 (1967), 558

(3) INESI, G., E. MARING, A. MURPHY, B.H. McFARLAND, Arch. Biochem. and Biophys. 138 (1970), 285

(4) SEVERIN, S.E., In "Proceedings of the plenary sessions of sixth international congress of Biochemistry", New-York, 1964

(5) WEBER, A. Current topics in Bioenerg. 1 (1966), 203

(6) DUGGAN, P.F., A. MARTONOSI, J. gen. Physiol. 56 (1966), 203

(7) WORSFOLD, M., J.B. PETER, J.Biol. Chem. 245 (1970), 5545

(8) BALZER, H., M. MAKINOSE, W. HASSELBACH, Arch. Pharm. und exptl. Pathol. 260 (1968), 444

(9) MAKINOSE, M., W. HASSELBACH, FEBS Lett. 12 (1971), 271

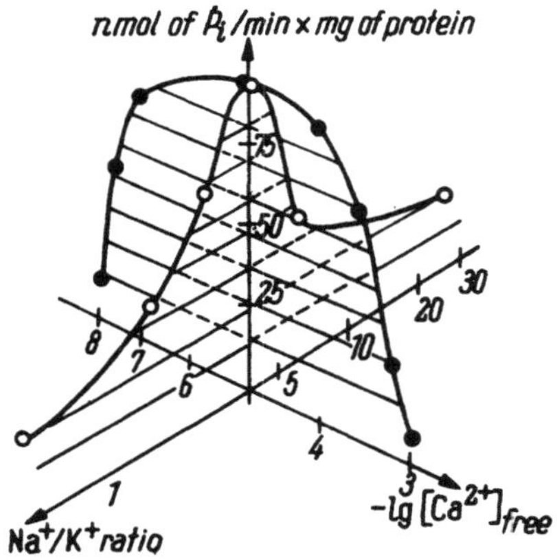

Fig. 1. The dependence of Na^+, K^+-ATPase of rabbit skeletal muscle sarcolemma from Na^+/K^+ ratio (o—o—o $Na^+ + K^+ = 140$ mM) or free Ca^{2+} concetration (●—●—●).

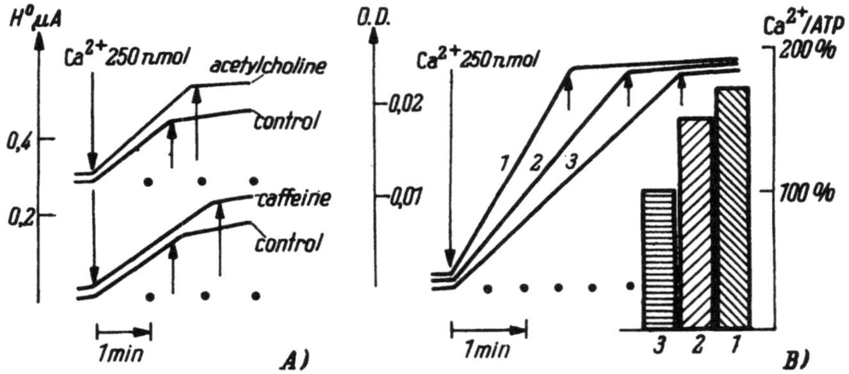

Fig. 2. The duration of Ca^{2+}-uptake in the presence of acetylcholine (10 mM), caffeine (2 mM), imidazole (10 mM), anserine (10 mM) or tris-buffer (20 mM). The start of uptake is denoted ↓ and the finish- ↑ . Incubation medium: A. pH 7.5, 25 °C, NaCl 100 mM, Na-oxalate 5 mM, Mg x ATP 2 mM; bars denote the Ca/ATP ratio with imidazole (1) or anserine (2) in comparison with tris-buffer (per cent to 3).

Fig. 3. The influence of imidazole (▲, △), carnosine (■, □), histidine (▼) and tris-buffer (○, ●) on Na^+, K^+-ATPase (black symbols) and Mg^{2+}-ATPase (empty symbols) of rabbit skeletal muscle sarcolemma.

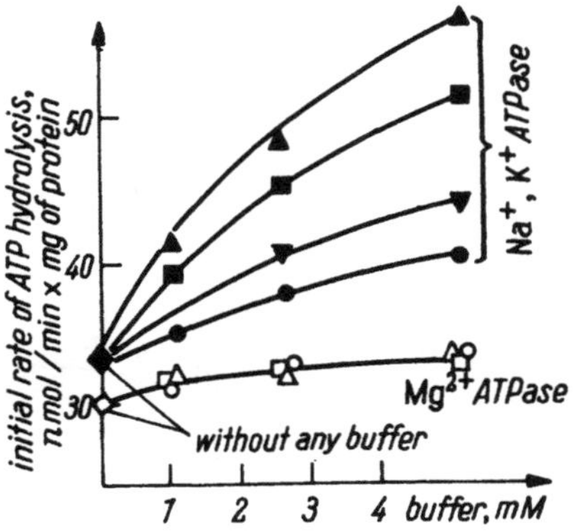

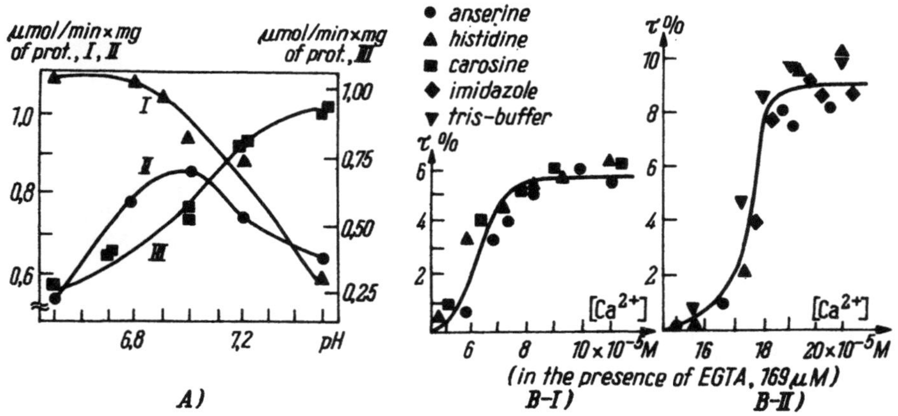

Fig. 4. A. pH dependence of Ca^{2+}-uptake (I, ▲) and Ca^{2+}-ATPase in the state of Ca^{2+}-uptake (II, ●) or in the state of Ca^{2+}-exchange (III, ■) of fragmented sarcoplasmic reticulum. B. Ca^{2+}-dependence of transmittens (τ %) of myofibrillar suspension in different buffers (5-25 mM) at two pH (I - 6, 8. II - 7, 8). 20-22 °C, μ 0, 11 - 0, 12.

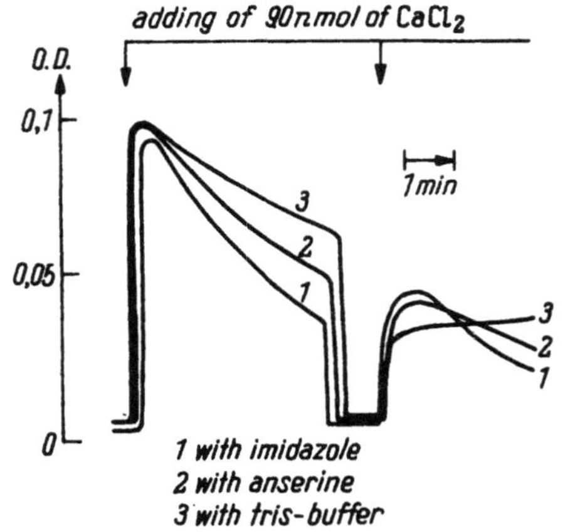

Fig. 5. Optical density variation in mixture of myofibrillar suspension and fragmented sarcoplasmic reticulum at the Ca^{2+} - adding. 28 °C, pH 7, 10 ± 0, 07 K-oxalate 5 mM, KCl 90 mM, Mg x ATP 2 mM, buffer concentration 10-20 mM, myofibrillar protein/microsomal protein ratio.

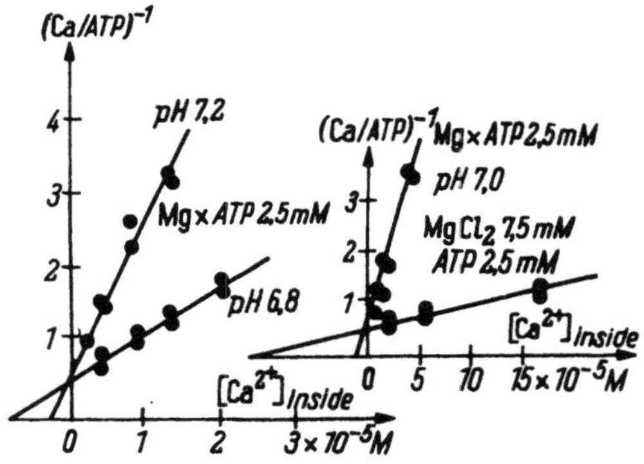

Fig. 6. $(Ca/ATP)^{-1}$ dependence from free $[Ca^{2+}]$ inside (at different oxalate concentration: 0, 5 - 10 mM) at different pH (A) and Mg^{2+} concentrations (B).

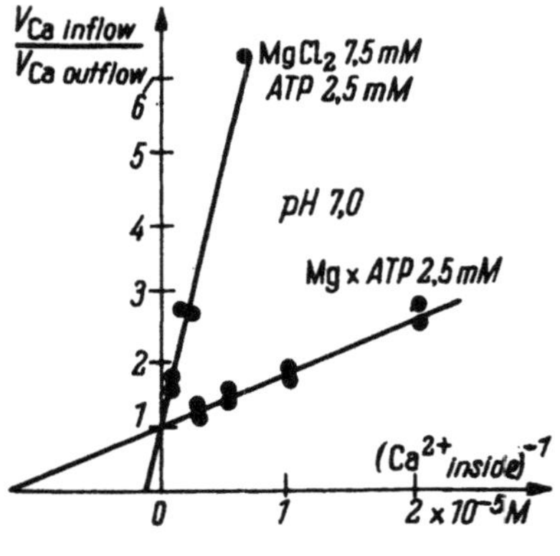

Fig. 7. V_{Ca} inflow/V_{Ca} outflow dependence from (free Ca^{2+} inside)$^{-1}$ at two different Mg^{2+} concentrations.

Department of Biochemistry, State University Leningrad

N. S. PANTELEEVA

O^{18} EXCHANGE REACTION AS A METHOD IN THE STUDY OF THE TRANSPORT OF PHOSPHORYL GROUPS FROM ATP TO MYOSIN

Introduction

During ATP hydrolysis by purified myosin, H-meromysin and actomyosin in the medium, enriched by H_2O^{18} exchange reaction takes place between the P_i released and the H_2O^{18} of the medium (1-4). There is good evidence that this reaction occurs at an intermediate stage in hydrolysis, supposedly in the stage of phorphorylated protein. This reaction has been termed the "intermediate O^{18} exchange". Simultaneously, "direct exchange" or "medium exchange" occurs when P_i is added to the medium or released as a result of ATP hydrolysis. The proportion of direct exchange under similar conditions is negligible (5, 6).
In this communication, the method of O^{18} exchange reactions has been used for studying the role of myosin functional groups in the formation of the ATPase reaction intermediate.

Methods

Myosin was prepared from rabbit skeletal muscle (7); H-meromyosin, by tryptic digestion of myosin (8). Myosin modification was carried out by treating the protein with specific reagents. For the binding of sulfhydryl groups, pCMB and NEM[1], (9) were used, for aminogroups - TBS (10), for carboxyl groups - pNTP (11) and for imidazol groups - DEPC (12). As a criterion for the modification, an alteration of Mg^{++} - ATPase activity served. Reagent concentrations leading to activation of ATPase ("activating concentrations") and to its inhibition ("inhibiting concentrations") were used. Since myosin precipitates when treated with DEPC, H-meromyosin was used for binding imidazol groups.
The ATP hydrolysis was carried out in an enriched H_2O^{18} medium under conditions favourable to the intermediate O^{18} exchange (6, 13). P_i was isolated in the form

---

1) Abbreviations: pCMB, p-chloromercuribenzoate; NEM, N-ethylmaleimide; TBS, 2. 4. 6-trinitrobenzenesulphonate; pNTP, p-nitrothiophenol; DEPC, diethylpyrocarbonate.

of Ba phosphate. The 0^{18} content was determined mass spectrometrically; the oxygen of the phosphates and medium was transformed into CO_2 form by the use of the guanidine hydrochloride reaction (14).

Results and discussion

As seen in the table, the 0^{18} exchange reaction decreases against control with both the activating and inhibiting concentrations of the reagents binding the different functional groups of myosin molecule. As a rule, 0^{18} exchange was decreased even when the highest activation of Mg^{++}-ATPase was obtained.
Thus, first of all, we should note the presence of a nonspecific uncoupling of the hydrolytic and the 0^{18} exchange reactions after treatment of myosin with modifiers of different nature. However, the mechanisms of such uncoupling may be essential different for each case.
If we work from the hypothesis that the phosphorylmyosin intermediate is capable of exchanging oxygen with the medium, then the sequence of ATP hydrolysis and 0^{18} exchange may be presented as the following simplified scheme:

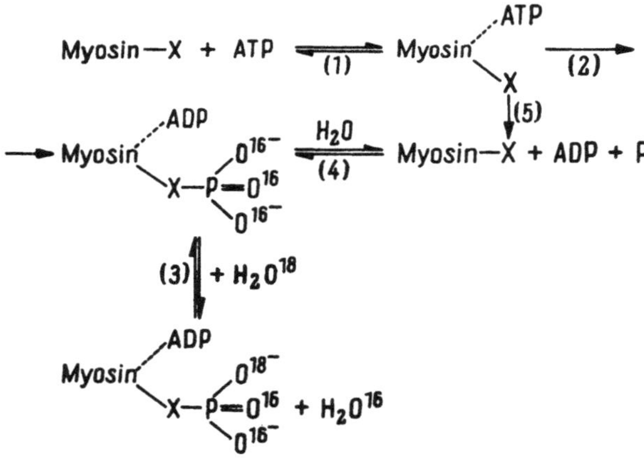

At point X we may put any group (- SH, - NH_2, - COOH, imidazol) to which binding of phosphoryl residues may be expected (reaction 2). According to this scheme, the P_i labeling should be determined by the time of the reversal of reaction 3. The change in the value of 0^{18} exchange should reflect the change in the properties of the phosphorylated protein. Let us consider the effects of pNTP and TBS on myosin. It is known that the nucleophilic reagent pNTP attacks

Table

THE INFLUENCE OF THE BINDING OF ACTIVE GROUPS OF MYOSIN ON MG^{++}-ATPASE ACTIVITY AND ON 0^{18} EXCHANGE REACTION IN "ACTIVATING" AND "INHIBITING" CONCENTRATIONS OF REAGENTS

Active Group	Reagent	Amount of Bound Reagent Mole · 10^5 g protein	ATPase Activity	0^{18} Exchange Reaction
			% Against Control	
Sulfhydryl	pCMB	2.2	200˙	129
		4.3	400	52
		6.0	40	25
Sulfhydryl	NEM	1.3	100	81
		1.9	200	56
		3.8	50	50
Amino	TBS	1.2	486	69
		2.0	534	59
		2.5	786	48
		12.4	876	24
Carboxyl	pNTP	0.5	300	102
		1.0	126	72
		2.2	78	59
Imidazol	DEPC	50[1)]	328	87
		100	278	56
		200	183	35

1) DEPC equivalents per mole H-meromyosin.

the acylphosphate and acylthioether bonds, forming a covalent complex with carboxyl groups of myosin (15, 16). The amount of NTP-binding is highest in the presence of ATP and Mg^{++} (specific binding), where, as is expected, the reactive myosin-phosphate complex is formed. The inhibition of 0^{18} exchange (by 30-40%) under such specific binding testifies to the participation in the 0^{18} exchange reaction of myosin carboxyl groups phosphorylated in ATP hydrolysis. However, with low concentration of pNTP, an increase of ATP decay, parallel with the 0^{18} exchange inhibition was observed. One may suppose that in the presence of NTP-myosin, direct hydrolysis of ATP by way of binding of water elements to an enzyme-substrate complex predominates (see scheme, reaction 5). The blocking of carboxyl groups may be considered as a depreciation of hydrolysis, uncoupling it from the phosphorylation.

Another mechanism may be expected in the case of the action of TBS on myosin. The myosin molecule in the active center region of ATPase has two lysyl residues, whose aminogroups can rapidly and specifically react with TBS (17, 18). According to our data, the joining of TBS to even one of these groups leads to a fivefold increase in Mg^{++}-ATPase activity and a simultaneous decrease in the 0^{18} exchange reaction by 40 %. With further growth of the bonded NH_2 groups, ATPase activity rapidly increases, and the oxygen exchange under the same conditions decreases. If X in such a scheme is replaced with an NH_2-group, then blocking of the latter will lead to depression of hydrolysis in reaction 4. Since, on the contrary, hydrolysis is accelerated many times, it may be supposed that the aminogroup itself does not take part directly in binding the phosphate residue.

Most probably the blocking of aminogroups in the neighborhood of the active center, or possibly in the allosteric center, forms favourable steric conditions for rapid formation and decay of the phosphorylated intermediate and thus diminishes the time of its existence (18). The latter stimulates hydrolysis in reaction 4 and leads to a decrease in 0^{18} exchange reaction in the 3rd stage (see scheme).

The action of TBS is similar to the action of the physiological modifier of actin. The latter, as is known, increases Mg^{++}-ATPase activity of myosin by 5 - 20 times, simultaneously decreasing the value of intermediate 0^{18} exchange by 30 - 35 % (6. 13). The similarity of the action of these two substances, differing substantially in nature, is caused by the fact that a similar catalytically active conformation of the active center is formed under the influence of both actin and TBS.

The affect of pCMB, NEM and DEPC on myosin ATPase and oxygen exchange in its nonspecific aspect is similar to the effect of pNTP and TBS. So far nothing definite can be said concerning the concrete mechanism of their influence.

Study of intermediate 0^{18} exchange in combination with the method of chemical modification enables us to obtain information concerning the properties of the intermediate formed in the stationary phase of ATP hydrolysis by myosin and concerning the possible paths of energy transport from ATP to contractile proteins.

Summary

The role of functional groups of myosin ATPase (EC 3.6.1.3.) in the formation of the intermediate of ATPase reaction was studied, employing the methods of 0^{18} exchange reaction and chemical modification. Effects were discovered, both specific and nonspecific, of modifiers of various nature on ATPase activity and on "intermediate 0^{18} exchange" catalyzed by myosin in the process of ATP hydrolysis.

REFERENCES

(1) KOSHLAND, D. E., H. M. LEVY, In: The Biochemistry of Muscle Contraction. Ed. J. GERGELY, New York, (1967), 87

(2) BOYER, P. D., In: The Biochemistry of Muscle Contraction. Ed. J. GERGELY, New York, (1964), 94

(3) PANTELEEVA, N. S., Izv. AN USSR, ser. biol. 4 (1968), 565

(4) PANTELEEVA, N. S., Second All-Union Biochem. Confer. Tashkent, 1969. Precis of Reports of Symposium, p. 253. Pub. FAN, Uzbekistan SSR (1969)

(5) SWANSON, J. R., R. G. YOUNT, Biochem Z. 345 (1966), 395

(6) ILIN, L. A., N. S. PANTELEEVA, Tzitologiya 9 (1967), 553

(7) IVANOV, I. I., V. A. YURYEV, Biokhimiya i Patobiokhimiya Myshts, Medgiz, Moscow - Leningrad (1961)

(8) SZENT-GYORGYI, A., Arch. Biochem. Biophys. 42 (1953), 305

(9) KULEVA, N. V., N. S. PANTELEEVA, Vestn. Leningrad State Univ. ser. biol. 3 (1968), 123

(10) KULEVA, N. V., N. S. PANTELEEVA, E. A. KARANDASHOV, Biokhimiya, 35 (1970), 42

(11) KULEVA, N. V., A. MÜHLRAD, N. S. PANTELEEVA, Biokhimiya, in press

(12) PANTELEEVA, N. S., N. V. KULEVA, E. A. KARANDASHOV, in press

(13) LEVY, H. H., N. SHARON, E. LINDEMANN, D. E. KOSHLAND, J. Biol. Chem. 235 (1960), 2628

(14) ILIN, L. A., Vestn. Leningrad State Univ. Ser biol. 3 (1966), 85

(15) KITAGAWA, S., K. CHIANG, Y. TONOMURA, Biochim. Biophys. Acta, 82 (1964), 83

(16) KUBO, S., N. KINOSHITAL, Y. TONOMURA, J. Biochem. (Tokyo) 60 (1966), 476

(17) TOKUYAMA, H., Y. TONOMURA, J. Biochem. (Tokyo) 62 (1967), 456

(18) TOKUYAMA, H., Ann. Rep. Biol. Works, 14 (1966), 1

Animal Biochemistry Department, Lomonosov State University, Moscow

N. P. MESHKOVA

THE PHOSPHORYLATED DERIVATIVES OF HISTIDINE-CONTAINING DIPEPTIDES: SYNTHESIS AND PHYSIOLOGICAL ROLE

Histidine-containing dipeptides, carnosine and anserine are specific components of skeletal muscles. Analysis of their effect on the processes of oxidative metabolism of muscle tissue has revealed these two compounds to facilitate the coupling of oxidation and phosphorylation, the maintenance of the structure of mitochondria and the increase of the energy level of high-energy phosphorous compounds (phosphocreatine). These facts indicate that in the presence of dipeptides the content of ATP in the sample should be higher. One of the reasons for higher ATP content could be a slower rate of ATP hydrolysis in the presence of carnosine and anserine. A possibility of phosphorylation of dipeptides in the course of biological oxidation could not be ruled out either.

In the present report we describe the results of a study of the effect of dipeptides from skeletal muscles, carnosine and anserine, and also imidazole and their N-phosphorylated derivatives on ATPase activity of myosin.

Experiments were carried out with pigeon breast muscle myosin at an optimal ATP concentration of $2 \cdot 10^{-3}$ M and that of Ca^{2+} - $4. 10^{-3}$ M in Tris-buffer pH 7, 5. Quantity of protein depended on the activity of the enzyme and was from 0. 15 to 0. 5 mg/ml; time of incubation - 2-3 minutes. Phosphorylated derivatives of imidazole compounds were obtained by treating the latter with $POCl_3$ in alkali. Depending on alkalinity of the medium mono-and diphosphorylated derivatives of imidazole can be prepared. If phosphorylation is performed at pH 13 mono-phosphorylimidazole is synthetized and under milder conditions (pH 11) - symmetrical diphosphorylimidazole.

As has been reported earlier, free carnosine and anserine as well as imidazole do not affect myosin ATPase activity. Our data do not contradict those in the literatur. PETUSHKOVA and BOCHARNIKOVA reported carnosine to produce an inhibiting effect on myosin ATPase only at substrate inhibition. AVENA and BOWEN observed an activating effect of carnosine during incubation of myosin in the medium of low ionic strength and in the absence of Ca^{2+} ions. The authors used the conditions similar to ours and addition of carnosine did not cause activation of ATPase.

The experiments made by us with the phosphorylated derivatives of imidazole, carnosine and anserine have shown them to have an inhibiting effect.

Trying to clear out the effect of different concentrations of phosphorylimidazole on the activity of the enzyme we have found a direct relationship between the degree of inhibition and the ratio of the quantities of added monopnosphorylimidazole and protein. The higher the phosphorylimidazole/protein ratio the greater is the degree of inhibition (Fig. 1).

A great difference was revealed between the effect of mono- and diphosphorylimidazole on the enzymatic activity of myosin. The inhibiting action of monophosphorylimidazole is always considerably higher than that of diphosphorylimidazole and has a competitive character (Fig. 2). The effect of diphosphorylimidazole is non-competitive (Fig. 3).

It was also noted that the inhibitory effect of phosphorylated derivatives of imidazole depends on the specific activity of the enzyme. At higher specific activity the inhibiting effect was more pronounced. (Fig. 4)

We suggested that for the competitive inhibition of ATPase activity of myosin by phosphorylated derivatives of imidazole the presence of a phosphoryl residue and free tertiary nitrogen in imidazole ring is required. To verify this suggestion experiments were carried out with the phosphorylated derivatives of carnosine-monophorylcarnosine, in which only one nitrogen atom of the imidazole ring was phosphosphorylated, and with triphosphorylcarnosine in which both nitrogen atoms in the imidazole ring and also the amino group of the β -alanine residue were phosphorylated (Fig. 5).

It turned out that it is only monophosphorylcarnosine that produces an inhibiting effect on ATPase activity. ATPase cannot be inhibited by triphosphorylcarnosine. Monophosphorylcarnosine as well as monophosphorylimidazole causes a competitive inhibition of myosin ATPase (Fig. 6).

The detailed mechanism of enzymatic splitting of ATP and the structure of myosin active centre is still the subject of intensive investigation. TONOMURA et. al. offered a model of the myosin-ATP enzyme-substrate complex, having studied the kinetics of enzymatic hydrolysis of ATP by myosin and the chemical structure of the globular portion of myosin. These authors believe that the ATP molecule binds myosin at several sites. The amino group of adenine binds with a unidentified group of myosin, ribose - with the NH-peptide bond of prolyl-lysine, β- and γ-phosphate residue bind via Ca with the asparagine residue. Besides the terminal phosphate group binds with the SH-group, and, possibly, with the arginine residue (Fig. 7).

Taking into consideration that only the monophosphoryl derivatives of imidazole compounds cause a competitive inhibition of the enzyme, that free imidazole compounds have no effect on the activity and that diphosphorylimidazole causes irregular and non-competitive inhibition of the process, the mechanism of inhibition of ATPase by myosin can be presented in the following way.

Monophosphorylimidazole and monophosphorylcarnosine compete with ATP for theactivesite of the enzyme, the free nitrogen atom of imidazole ring binding with active site on the place of amino group of adenine, the phosphoryl residue replaces the terminal phosphate group of ATP. Naturally, the rate of cleavage of ATP thereby decreases.

Inhibition of ATPase of myosin caused by diphosphorylimidazole is much less and of non-competitive nature. This way be associated with its effect on allosteric site of the enzyme. One of such sites may be a fragment of polypeptide chain containing histidine. Diphosphorylimidazole is a potent phosphorylating agent, and it may phosphorylate amino acid residues and first and foremost the imidazole group of histidine. This may lead to the conformational changes in the active site of myosin and thereby brings about changes in its enzymatic activity. This hypothesis agrees with the data on dependence of the degree of inhibitory effect on the specific activity of the enzyme.

Lower inhibitory effect of monophosphorylcarnosine as compared to that of monophosphorylimidazole can be explained by the larger size of its molecule; it can hinder the formation of myosin-monophosphorylcarnosine complex or facilitate its degradation.

In this work several problems are raised both in respect to the proposed mechanism of inhibitory effect of phosphorylated derivatives of imidazole and on the possibility of physiological significance of the phenomenon in question. Further investigation is required for solving these questions.

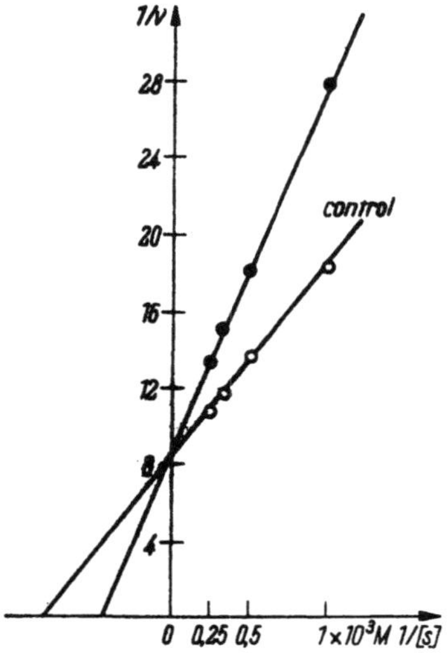

Fig. 1. Competitive inhibition of ATPase activity by MPI.

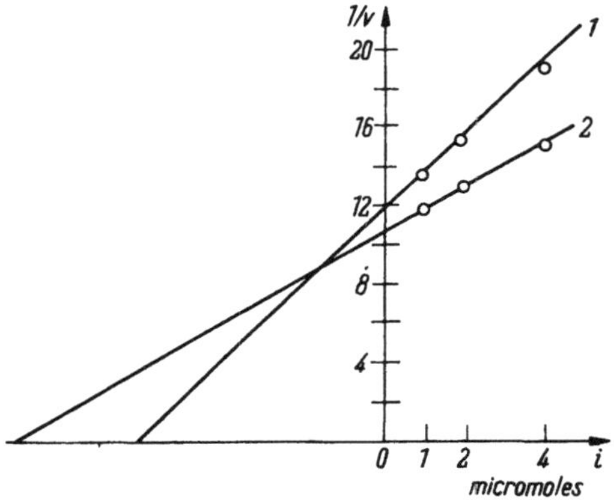

Fig. 2. Kinetics of inhibition of myosin ATPase by MPI at 3 (1) or 6 (2) micromols ATP per tube.

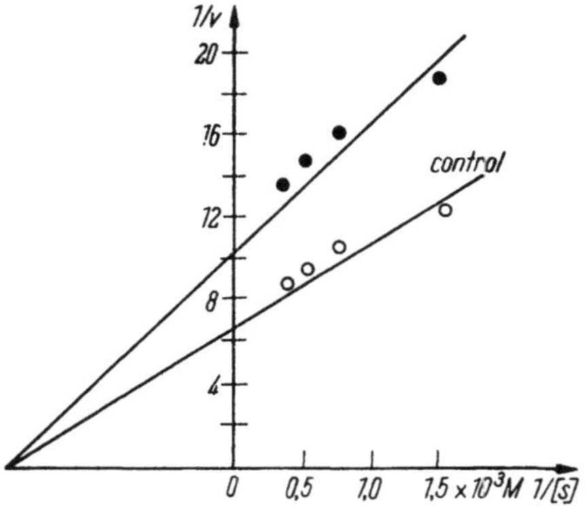

Fig. 3. Non-competitive inhibition of ATPase activity by DPI.

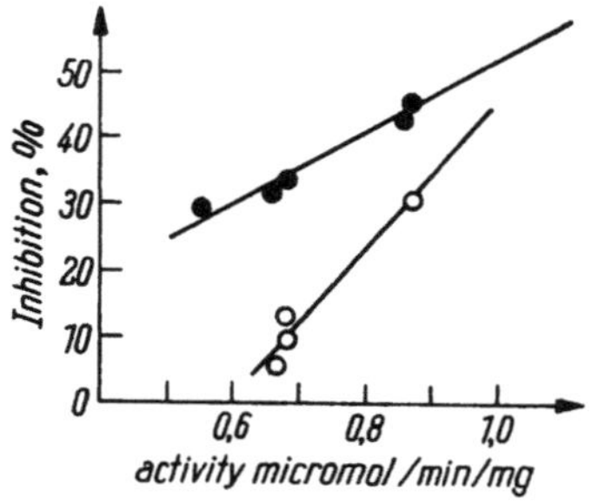

Fig. 4. The dependence of the degree of inhibition by ● MPI and ○ DPI on the level ATPase activity of myosin preparations.

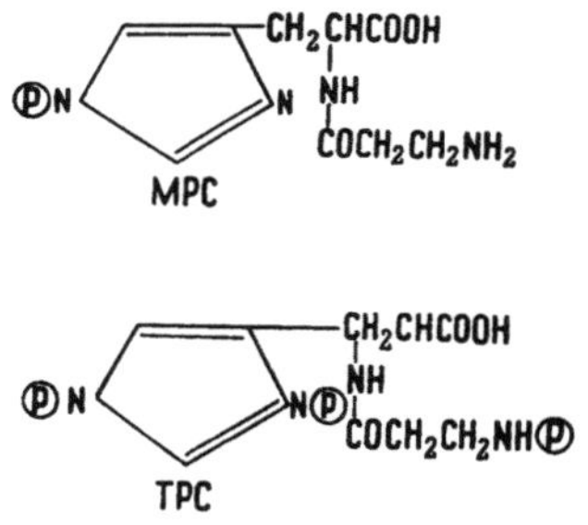

Fig. 5. Monophosphorylcarnosine (MPC) and Triphosphorylcarnosine (TPC).

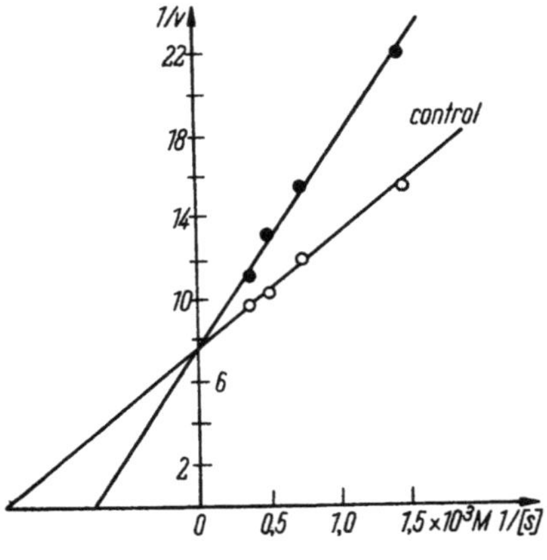

Fig. 6. Competitive inhibition of ATPase activity by MPC.

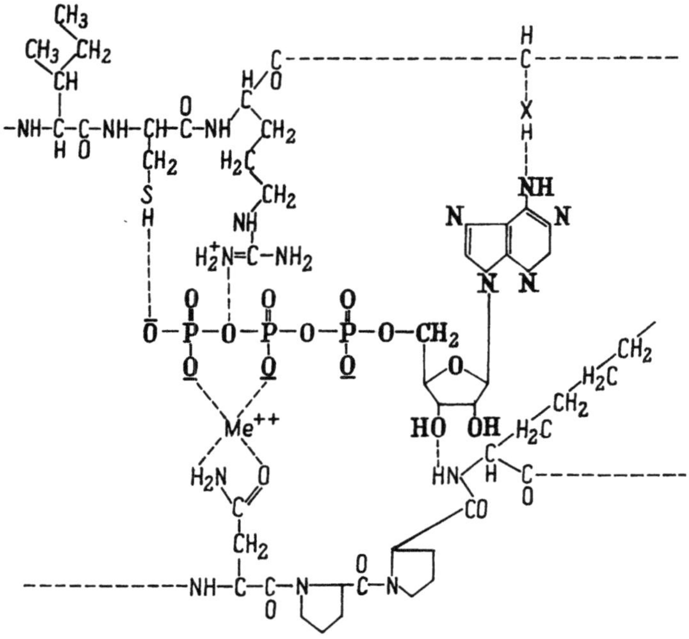

Fig. 7. Scheme of myosin-ATP complex.

Department of Biochemistry, University of Nijmegen, Nijmegen,
The Netherlands

S. L. BONTING

NA-K ACTIVATED ATPASE AND ACTIVE CATION TRANSPORT

Introduction

The search for the cation pump mechanism, which maintains the cation gradients across the outer membranes of most animal cells, first met with success through the work of SKOU. He reasoned that active cation transport requires energy, which will be supplied by ATP. Hence he looked for an ATPase activity, which might be stimulated by Na^+ and K^+, the cations transported. In crab nerve he detected a Mg-dependent ATPase, which markedly increased in the simultaneous presence of Na^+ and K^+ SKOU (1). Subsequently, he demonstrated that the Na-K induced increase in ATPase activity was inhibited by ouabain SKOU (2), which was known to inhibit active cation transport in erythrocytes SCHATZMANN.

In subsequent years a large number of papers has appeared concerning this enzyme system. The present paper attempts primarily to survey the physiological significance of the Na-K activated ATPase (Na-K ATPase system). A more comprehensive review has recently been published by the author BONTING (3).

Assay and properties

Since the Na-K ATPase activity is always accompanied by a non-Na-K activated, ouabain-insensitive Mg ATPase activity, a differential assay method is necessary. The sum of both ATPase activities is determined in a complete medium, containing Na^+, K^+, Mg^{2+}, ATP, buffer (pH 7,4). Mg ATPase activity is determined by media inhibitory to Na-K ATPase, by the omission of either Na^+ or K^+, by the addition of ouabain (10^{-4}M) or a combination of these. When these inhibitory media give the same activity, the difference in activity between the complete medium and the inhibitory media represents the Na-K ATPase activity.

Table 1 shows an example of the results obtained in the case for lens epithelium BONTING, CARAVAGGIO, HAWKINS (4).

Table 1

RELATIVE EPITHELIAL ATPase ACTIVITY IN VARIOUS SUBSTRATE MEDIA[1)]

Medium	Cat	Calf	Rabbit
	%	%	%
A (complete)	100	100	100
B (no K)	28, 9 ± 1, 5	43, 1 ± 2, 0	39, 9 ± 7, 5
C (no Na)	23, 2 ± 1, 4	32, 5 ± 0, 6	19, 2 ± 0, 3
D (10^{-4}M ouabain)	22, 1 ± 2, 9	32, 1 ± 0, 1	28, 0 ± 1, 3
E (no K, 10^{-4}M ouabain	19, 7 ± 0, 8	32, 4 ± 3, 8	20, 0 ± 3, 6

1) ATPase activity in medium A (total ATPase activity) set at 100; figures for media B, C, D and E (means with standard errors from six determinations) indicate the activity remaining upon inhibition of the Na-K ATPase activity.

The Na-K ATPase system is a membrane-bound insoluble complex, hence its purification is a difficult matter.

Partial removal of the Mg ATPase, without seriously affecting the Na-K ATPase activity, is possible by pretreatment with urea (Fig. 1) or sodium iodide. Treatment with detergents in low concentrations (e.g. 0.06% deoxycholate) also removes non-Na-K ATPase material without considerably changing the Na-K ATPase activity. Detergents in higher concentrations, however, cause inactivation of the enzyme, apparently by removal of essential phospholipids. Isolation of a membrane fraction by differential or gradient centrifugation increases the relative Na-K ATPase activity due to removal of mitochondrial Mg ATPase. The optimal method for determination of the absolute Na-K ATPase activity is the use of lyophilized, aqueous homogenates without further treatment. For the study of its properties the use of treated membrane fractions is usually advantageous.

The properties of the Na-K ATPase in a variety of tissues studied so far are remarkably similar. The pH-optimum is approximately 7.3, as shown for lens epithelium in Fig. 2. The Mg-activation curve, e.g. for rat liver BAKKEREN, BONTING (5) presented in Fig. 3, usually reaches maximal activity at a molar Mg/ATP ratio of 1, suggesting that Mg ATP^{2-} is the optimal substrate. An example of the Na-activation curve is shown in Fig. 4 from herring gull salt gland BONTING, CARAVAGGIO, CANADY, HAWKINS (6). Half-maximal activation by Na^+ is usually found at a Na^+-concentration of approx. 10 mM. Fig. 5 shows a K-activation curve, likewise from herring gull salt gland. The slope for K-activation is much steeper than that for Na^+, half-maximal activation usually being found at a K^+-concentration of approx. 1 mM. An example of the ouabain inhibition curve is shown in Fig. 6 for shark rectal gland BONTING (7). The halb-inhibition concentrations for most tissues lies between 10^{-7} and 10^{-6}M. This curve also shows a slight stimulation by 10^{-9}-10^{-8}M ouabain. This so-called biphasic effect has also been found in the functional effect of ouabain (e. g. in pancreatic secretion, Fig. 14). Another typical aspect of the ouabain inhibition is its reversal by increased K^+-concentrations, which is shown in Fig. 7 for rabbit brain BONTING, HAWKINS, CANADY (8).

Occurrence and correlation with the cation pump

The Na-K ATPase has been demonstrated in 33 out of 39 tissues of the cat, the 6 inactive tissues being either cell-free or poor in cells BONTING, SIMON, HAWKINS (9). Currently it can be stated that the enzyme system is common to all animal cells. A weak activity has also been shown in the bacterium Escherichia coli (HAFKENSCHEID, BONTING 10, 11). For plant tissues no clear evidence for the presence of an ouabain-sensitive Na-K ATPase system has ben reported so far.

A correlation between the occurrence of the Na-K ATPase system and an ouabain-sensitive cation transport system was shown by the presence of the enzyme system in all 21 tissues from 10 species for which such a transport system had been demonstrated at the time (BONTING, CARAVAGGIO, HAWKINS (12)).

A quantitative correlation between active cation flux and Na-K ATPase activity over a 25, 000-fold range (Table 2) has been demonstrated for six tissues (BONTING, CARAVAGGIO (13)). The average molar ratio of cation transported to ATP hydrolyzed was 2, 6 $\pm$ 0, 19. This finding clearly demonstrates the role of the enzyme system in the active cation transport involved in maintenance of ion gradients in single cells, transport of salt and water across epithelial membranes and repolarisation processes in muscle, nerve and electric organ. A cation/ATP ratio of nearly 3 has now been found in a great number of tissues, both when calculated from the Na-K ATPase assay and from extra energy requirement for transport (Table 3).

The occurrence and function of the Na-K ATPase cation pump is present in three types of systems: single cell systems, excitatory systems and secretory systems. Examples of each will be discussed in the next three sections.

Table 2

COMPARISON OF CATION FLUXES AND Na ·K ATPase ACTIVITY

Tissue	Temp.	Ion	Cation flux	Na-KATPase activity	Ratio
	°C		10^{-14}M/cm^2/sec	10^{-14}M/cm^2/sec	
Human erythrocytes	37	Na, K	3, 87	1, 38	2, 8
Frog toe muscle	17	Na, K	985	530	1, 86
Squid giant axone	19	K	1, 200	400	3, 00
Frog skin	20	Na	19, 700	6, 640	2, 97
Toad bladder	27	Na	43, 700	17, 600	2, 48
Electric eel, non-innerv. membr., Sachs organ	23	Na	86, 100	38, 800	2, 22
					2, 56
			r=+0, 866, P=0, 017		± 0, 19

Single cell systems

Erythrocytes represent the first thoroughly studied system (POST, MERRITT, KINSOLVING, ALBRIGHT, DUNHAM, GLYNN (15)). It has been shown that the stimulation of the Na-K ATPase system occurs by Na^+ intracellularly and by K^+ extracellularly. ATP has to be supplied internally, while the ouabain inhibition occurs from the outside. The pump system, through its considerable ATP requirement, has a pacemaker effect on the cellular energy metabolism. There are only a few hundred pump sites per erythrocyte.

The eye lens behaves as a single cell system with regard to cation transport (Fig. 8). While passive cation movements can occur on both sides, active transport in the mammalian lens takes place on the anterior side, where the epithelial cell layer is located. The Na-K ATPase activity in the mammalian lens is located entirely in the epithelium (BONTING, CARAVAGGIO, HAWKINS, (4)). Excellent agreement between transport and enzyme characteristics is found (Table 4), indicating that the epithelial Na-K ATPase system represents the lens cation

Table 3

RATIOS OF CATION TRANSPORT TO HYDROLYSIS OF ATP IN VARIOUS TISSUES

Tissue	Cation/ATP (eq/mol)	
	Based on Na-K ATPase assay	
Human erythrocytes	2, 8 (Na)	BONTING and CARAVAGGIO, 1963
Frog toe muscle	1, 9 (Na)	BONTING and CARAVAGGIO, 1963
Squid giant axon	3, 0 (K)	BONTING and CARAVAGGIO, 1963
Frog skin	3, 0 (Na)	BONTING and CARAVAGGIO, 1963
Toad bladder	2, 4 (Na)	BONTING and CANADY, 1964
Electric eel, non-inn. membrane electroplax	2, 2 (Na)	BONTING and CARAVAGGIO, 1963
Calf lens	3, 1 (Na)	BONTING, CARAVAGGIO and HAWKINS, 1963
Rabbit lens	2, 5 (Na)	BONTING, CARAVAGGIO and HAWKINS, 1963
Herring gull nasal gland	2, 0 (Na)	BONTING and COWORKERS, 1964
Human leukocytes	1, 8 (Na)	BLOCK and BONTING, 1964
Rabbit ciliary epithelium	3, 0 (Na)	BONTING and BECKER, 1964
Spiny dogfish rectal gland	2, 2 (Na)	BONTING, 1966
Frog sartorius muscle	2, 2 (K)	CORRIE and BONTING, 1966
Escherichia coli	1, 6-3. 2 (K)	HAFKENSCHEID and BONTING, 1968
Dog pancreas	1, 9 (Na)	RIDDERSTAP and BONTING, 1969a
	Not based on Na-K ATPase assay	
Human erythrocytes	2, 7 (Na)	SEN and POST, 1964
	3, 0 (Na)	GLYNN, 1962a
Frog muscle	2, 5-3. 0 (Na)	HARRIS, 1967
Frog skinn	2, 7-3. 3 (Na)	ZERAHN, 1956
	3, 5 (Na)	LEAF and RENSHAW, 1957
Toad bladder	2, 8 (Na)	LEAF, PAGE and ANDERSON, 1959
Calf lens	3, 0-4. 5 (Na)	KERN, ROOSA and MURRAY, 1962

Table 4

PROPERTIES OF CATION TRANSPORT AND EPITHELIAL Na-K ATPase SYSTEMS IN THE LENS

	Cation transport	Na-K ATPase
Location	Epithelium	Epithelium
Substrate	ATP (?)	ATP
Cation requirements	Mg^{2+}(?), Na^+, K^+	Mg^{2+}, Na^+, K^+
	Rb^+ competes with K^+	Rb^+ can replace K^+
pH optimum (rabbit)	7,5	7,3
Temperature coefficient Q_{10}(rabbit)	2	2,4
Ouabain	Inhibitor	Inhibitor
pI_{50}, calf (5 mM K^+)	7,0-7.4	7,1
pI_{50}, rabbit (5 mM K^+)	6,0	6,0
Rates (µmole/lens/hr)		
Calf	2,6 (K^+), 3,1 (Na^+)	0,78 (ATP)
Rabbit	6,5 (K^+), 3,9 (Na^+)	1,56 (ATP)
	3,9 (Rb^+)	1,34 (ATP)

After BONTING, CARAVAGGIO and HAWKINS (4)

pump system BONTING (16). The primary function of this system appears to be the maintenance of lens transparency through the prevention of osmotic swelling (Duncan, 1968).

The occurrence of the Na-K ATPase system outside the animal kingdom has been established in Escherichia coli (HAFKENSCHEID, BONTING (10, 11). The low activity relative to the Mg ATPase activity required extensive testing in various media, but a significant activity could be demonstrated (Table 5). Besides the dominant K^+-H^+ transport system in these bacteria, a weak ouabain-sensitive K^+-Na^+ transport system could be shown to be present HAFKENSCHEID, BONTING (17). Incubation of the bacteria in a medium containing 10^{-4}M ouabain, 117 mM Na^+ and 5 mM K^+ increases their Na^+ content by 14,7 $\pm$ 4,1 % (17 meq/kg dry wt.) and decreases their K^+ content by 7,4 $\pm$ 2,5% (12 meq/kg dry wt.), both changes being significant and equivalent within the experimental error. The uptake of ^{86}Rb at pH 5, 3 is reduced significantly by 10% in the presence of 0, 8 x 10^{-4}M ouabain (Fig. 9).

Table 5

RELATIVE ATPase ACTIVITIES OF E. COLI IN VARIOUS SUBSTRATE MEDIA

Code	Incubation media				ATPase activities[1)]	
	Additions					
	Mg^{2+}	Na^{+}	K^{+}	Ouab. (10^{-4}M)		
A	+	+	+	-	100	(21)
B	+	+	-	-	91, 5 ± 0, 9	(21)
C	+	-	+	-	89, 7 ± 0, 8	(21)
D	+	+	+	+	93, 1 ± 1, 0	(21)
E	+	+	-	+	91, 9 ± 0, 5	(21)
F	+	-	-	-	83, 8 ± 0, 7	(21)
G	+	-	-	+	83, 8 ± 0, 8	(21)
H	+	-	+	+	87, 0 ± 0, 9	(21)

1) in percent of activity in medium A, with standard error of mean and number of determinations in parentheses.

Excitatory systems

In this category fall tissues which upon stimulation (electrical, chemical, light or auditory stimulus) show passive cation fluxes: nerve, brain, muscle, retina, cochlea, electric organ. In these systems a pump mechanism serving to maintain the cation gradients required for the cation fluxes is needed. As examples muscle and cochlea will be discusses.

Table 6 demonstrates the presence of Na-K ATPase activity in two types of frog muscle BONTING, CARAVAGGIO (13), CORRIE, BONTING (18). Cation/ATP ratios of 1, 9 and 2, 2. were calculated, in fair agreement with the ration of 2, 5-3 obtained by HARRIS (19) from direct determinations of ATP consumption during ion movements in frog sartorius muscle, as well as with the ratios for a variety of tissues in Table 3. The decrease in membrane potential of frog

Table 6

RELATIVE AND ABSOLUTE ATPase ACTIVITY OF FROG MUSCLE IN VARIOUS SUBSTRATE MEDIA

Medium	Toe (%)	Sartorius (%)
A, complete	100	100
B, $-K^+$	88,7	74
C, $-Na^+$	85,3	58
D, +ouabain (10^{-4}M)	85,1	72
E, $-K^+$, +ouabain	80,3	81
Na-K ATPase activity in % of total ATPase	14,9	24,3[1)]
in mmol/kg wet wt/hr	67	29
cation/ATP ratio	1,9	2,2

1) Omitting the result for the Na^+-free medium, which deviates because of Na-sensitivity of Mg ATPase activity.

sartorius muscle after a 4 hr treatment with 10^{-4}M ouabain or erythrophleine is equal to that calculated from the measured decrease in K^+ concentration (Table 7). The gain in Na^+ content is slightly larger than the loss of K^+, and is accompanied by isotonic uptake of Cl^- and H_2O. The activity of the cation pump is increased when the internal Na^+ concentration is raised from 5 to 20 mM SJODIN, BEAUGE (20). Absence of external K^+ reversibly inhibits the pump NODA (21). These findings demonstrate that in muscle the Na-K ATPase system plays a major role in maintaining the cation gradients necessary for muscle function.

In the cochlea we have the interesting phenomenon of an extracellular fluid with high K^+ (150 mM) and low Na^+ (1 mM) concentration: the endolymph present in the scala media (Fig. 10). This fluid borders on the perilymph in scala vestibuli and scala tympani with reverse cation ratio, as in other extracellular fluids. We asked ourselves: is a Na-K ATPase system responsible for the maintenance of these cation gradients, and if so, where is it located? Microdissection of guinea pig cochlear structures and ultramicro enzyme assays gave us the distribution of Na-K ATPase activity indicated in Fig. 10 KUIJPERS, BONTING (22).

Table 7

EFFECT OF OUABAIN AND ERYTROPHLEINE ON FROG SARTORIUS MUSCLE

	ΔH_2O	ΔK^+	ΔNa^+	$\frac{\Delta Na^+ - K^+}{\Delta H_2O}$	ΔE_{calc}	ΔE_{obs}
	mg	µmole	mole	M	mV	mV
Ouabain 4 h., 0,1 mM	17,1	-2,49	4,51	0,12	-22,1	-19,3
S. E.				0,02	1,8	1,6
					P = 0,28	
Erythrophleine, 4 h., 0,1 mM	13,3	-2,80	4,77	0,15	-16,6	-21,0
S. E.				0,06	3,7	3,5
					P = 0,42	

$\frac{\Delta Cl^-}{\Delta H_2O} = 0,15$ M $\qquad \Delta E_{calc} = 59,4 \log \frac{K^+ \text{ exptl}}{K^+_{\text{contr.}}}$ (mV)

The stria vascularis has a high activity, which is about 20 times as high as in the other structures surrounding the endolymph. On a surface area basis the stria vascularis is even 168 times as active as REISSNER's membrane. The cation gradients between endo- and perilymph are accompanied by a steady potential, the endolymphatic resting potential (ERP). Stimulation of the organ of Corti by sound waves generated an alternating potential, the cochlear microphonic potential (CMP), which is dependent on the presence of ERP. Perfusion of the scala vestibuli with ouabain inhibits both potentials simultaneously and to the same extent (Fig. 11). The pI_{50} values for the stria vascularis enzyme and the two potentials are equal KUYPERS, BONTING (23). This shows that the Na-K ATPase system of the stria vascularis plays a primary role in the maintenance of the cochlear cation gradients and in the generation of the cochlear potentials.

Secretory systems

Occurrence and transport function of the Na-K ATPase system have now been demonstrated in kidney, frog skin, toad bladder, marine bird salt gland, elasmobranch rectal gland, liver, intestine, pancreas, gall bladder, formation of aqueous humor and cerebrospinal fluid. The two examples to be illustrated are toad bladder and pancreas.

The toad bladder has been widely studied as a simple model of renal function. By means of an epithelial layer with a sodium pump and a varying water permeability regulated by vasopressin the toad is able to maintain its osmotic equilibrium on land and in water. When a piece of toad bladder is clamped between two short-circuited chambers, containing aerated physiological salt solution, a current flows which is equivalent to the net Na^+ flux from mucosal to serosal side LEAF, ANDERSON, PAGE (24). The Na-K ATPase activity compared to this Na^+ flux gives a Na/ATP ratio of 2, 5 BONTING, CARAVAGGIO (13), in good agreement with the value of 2, 8 derived from the increased O_2 consumption upon stimulation of Na^+ transport LEAF, PAGE, ANDERSON (25). Ouabain inhibits the Na^+ transport when present on the serosal side, but not when added to the mucosal side, indicating that the pump is located on the serosal side. The ouabain inhibition curves for Na^+ transport and enzyme agree very well (Fig. 12), and give pI_{50} values of 4, 6 and 4, 7, respectively BONTING and CANADY (26). Vasopressin and aldosterone have no effect on the Na-K ATPase activity, indicating that they affect the Na^+ transport in the toad bladder indirectly. The low sensitivity of the toad bladder transport system to ouabain (pI_{50} = 4, 6) is not due to rapid detoxication or slow transformation into an active metabolite HERRMANN, PORTIUS, REPKE (27). but to the low sensitivity of the enzyme system (pI_{50} = 4, 7). Most likely, this is related to the fact that the toad produces several bufagins in its venom which are chemically and pharmacologically analogous to ouabain. This conclusion is supported by the much higher sensitivity of the enzyme system to the structurally dissimilar erythrophleine (pI = 6, 0). The Na^+ transport rate is related to the intracellular Na^+ concentration FRAZIER, DEMPSEY, LEAF (28) and is reversibly inhibited by removal of serosal K^+ BOWER (29), while the ouabain inhibition is reversed by high serosal K^+ concentrations BOWER (30). These findings clearly demonstrate the crucial role of the Na-K ATPase system.

The exocrine function of the pancreas consists of isotonic fluid and electrolyte secretion by the acino-ductular cells and of enzyme secretion by the acinar cells. The presence and properties of the Na-K ATPase system in the pancreas of dog and rabbit have been established RIDDERSTAP, BONTING (31, 22). A Na/ATP ratio for the dog pancreas of 1, 8 has been found. Inhibition of fluid secretion by the dog pancreas in vivo after ouabain injection into its artieral blood supply occurs RIDDERSTAP, BONTING (31). More detailed studies are possbile with the isolated rabbit pancreas because of its sheetlike configuration. Addition of ouabain to the bath inhibits flow greatly without affecting the Na^+ concentration as shown in Fig. 13 RIDDERSTAP, BONTING (32). The inhibition curves for enzyme activity and flow rate are greatly similar with pI_{50} values of 5,4 each and a biphasic effect in each (Fig. 14). There is also excellent parallelism between the two inhibitory effects for a variety of transport inhibitors (erythropleine $>$ ouabain $>$ sciallaren A $>$ digoxin $>$ hexahydroscillaren). In all these experiments the Na^+ concentration of the secreted fluid remained unchanged, suggesting that the primary process is a Na-K ATPase mediated Na^+ transport.

Coupled transport

In the case of pancreatic fluid secretion we encounter an example of coupled isotonic water transport. This is further corroborated by the following two experiments. With $^{22}Na^+$ added to the bathing medium the $^{22}Na^+$ secretion and flow rate are equally inhibited by $5 \cdot 10^{-6}$M ouabain (57 and 58%, resp.) and equally stimulated by secretin (39 and 35%, resp.). An 85% decrease in the Na^+ concentration of the bathing fluid lowers Na^+ secretion and flow rate reversibly by 84 and 80%, respectively.

This process of isotonic water transport coupled to active Na^+ transport is due to local osmosis DIAMOND (33, 34). In Fig. 15 the process is illustrated for the gall bladder KAYE, WHEELER, WHITLOCK, LANE (35). In the resting state the intercellular spaces are very narrow. The pump is located mainly on the lateral membranes. On the lumen side (top) the cells are locked together by desmosomes. When the pump begins to function, Na^+ ions are secreted into the intercellular space. The resulting local hypertonicity draws water from the lumen through the cell into the space with dilution to isotonicity and distension of the space. The resulting hydrostatic pressure expels the isotonic fluid into the lamina propria, where it is carried away by the capillaries.

Morphological evidence for the occurrence of this type of process in the pancreas has been found by electronmicroscopic observation of dog pancreas biopsis before and after injection of secretin RIDDERSTAP, STADHOUDERS, BONTING (unpublished observations). Before secretin injection there were no intercellular spaces, after secretin very clear spaces had developed, showing desmosomes at the lumenal side and narrow openings at the basal side. Since in this case Na^+ and water are secreted into the lumen and the desmosomes are at the lumenal side, the process of local osmosis must function here in the opposite way as in the gall bladder. The Na^+ pump must be located mainly on the lumenal side, drawing Na^+ passively across the lateral cell membrane from the intercellular space. The resulting hypotonicity in this space makes water flow with the Na^+ ions. Theoretical calculations by DIAMOND, BOSSERT (36) show this to be feasible. Evidence for coupled isotonic water transport exists also for kidney intestine and frog skin.

In addition to being responsible for coupled isotonic water transport, the Na-K ATPase system also appears to function in the active transport of a number of other substances: sugar, amino acids etc. in kidney and other tissues. Criteria for this type of coupled transport are: 1. the transport is dependent on the presence of Na^+ ions, 2. it is inhibited by ouabain, and 3. the ouabain inhibition is reversible by raising the K^+ level. The next two papers by DETTMER, MÜLLER and GLANDER and by MÜLLER, DETTMER and REMKE deal with the coupled transport of monosaccharides across the intestinal epithelium, and indicate that there is a direct relation between the Na-K ATPase system and the sugar transport probably at the level of the K^+ activated dephosphorylation of the phosphorylated enzyme complex.

Mechanism

Our current knowledge of the mechanism of the Na-K ATPase system will be briefly summarized here. A more detailed discussion of various aspects will be found in the papers of REPKE and of NORBY and JENSEN.

The Na-K ATPase system of crab nerve, electric organ and cerebral cortex catalyses a $Na^+ + Mg^{2+}$ activated ATP-ADP exchange reaction SKOU (2), FAHN, ALBERS, KOVAL (37), SWANSON (38). This suggests that the first step in the reaction mechanism is the formation of an intermediary phosphorylated enzyme complex:

$$E + ATP \xrightleftharpoons{Mg^{2+} + Na^+} E \sim P + ADP \qquad (1)$$

$$E \sim P + H_2O \xrightleftharpoons{K^+} E + P_i \qquad (2)$$

This reaction is inhibited by ouabain in high concentrations only.

The initial step is followed by a K^+ activated, ouabain-sensitive dephosphorylation step POST (39).

A K^+ activated, ouabain-sensitive phosphatase activity, which hydrolyses acetylphosphate and p-nitrophenylphosphate, has been demonstrated in Na-K ATPase preparations from a variety of tissues HASHIUCHI (39a). Since no labelling of ATP occurs on incubation of the system with ^{32}P-acetylphosphate, there is a possibility that this step actually consists of two steps:

$$E \sim P \rightleftharpoons E - P \quad (3)$$

$$E - P + H_2O \xrightarrow{K^+} E + P_i \quad (4)$$

The phosphorylated complex appears to be an acylphosphate, since it is hydrolyzed by NH_2OH, molybdate and acylphosphatase HOKIN et al. (40), BADER, SEN, POST (41). Reaction of peptic fragments of the ^{32}P-phosphorylated complex with tritium-labelled N-(n-propyl) hydroxylamine yields a radioactive hydroxamate, which has been positively identified by comparison with synthetic compounds KAHLENBERG, GALSWORTHY, HOKIN (42). The acylphosphate is L-glutamyl--phosphate.

A model, incorporating current insights into the Na-K ATPase mechanism, is shown in Fig. 16 ALBERS, KOVAL, SIEGEL (43). The circle is the cell membrane carrying six pump sites in different stages of activity. At 1. Na^+ activated phosphorylation takes place with the enzyme in the "cis"-form, i. e. with the cation binding sites pointing to the cytoplasm and binding Na^+ ions. The phosphorylated enzyme is rapidly converted (2) into the "trans"-form, with the cation binding sites pointing extracellularly. In the presence of K^+ exchange of Na^+ for K^+ ions on the binding sites takes place, while reaction 3 activates the hydrolytic step (4). This tep makes the "trans" enzyme less stable, which then reverts to the "cis"-form (reaction 5). The cycle is completed when Na^+ displaces K^+ from the "cis"-enzyme (reaction 6).

This concludes our survey of the properties, occurrence, function and mechanism of the Na-K ATPase cation transport system.

It has a crucial physiological function in three types of system:

1. single cell systems: maintenance of cation gradients and coupled transport of metabolites.

2. excitatory systems: maintenance and restoration of cation gradients, required for passive cation movements during excitation.

3. secretory systems: sodium pump required for isotonic water-electrolyte movement, often with coupled transport of other substances.

In its study the disciplines of biochemistry, physiology, biophysics, morphology and pharmacology all interact to contribute what may be called the molecular biology of active cation transport. Much has been learned since the discovery of the Na-K ATPase system, yet, much still remains to be elucidated!

REFERENCES

(1) SKOU, J.C., Biochim. Biophys. Acta, 23 (1957), 394

(2) SKOU, J.C., Biochim. Biophys. Acta, 42, (1960), 6

(3) BONTING, S.L., In: E.E. BITTAR (Ed.), Membranes and Ion Transport, Wiley, New York, 1, (1970), 286

(4) BONTING, S.L., L.L. CARAVAGGIO, N.M. HAWKINS, Arch. Biochem. Biophys., 101, (1963), 47

(5) BAKKEREN, J.A.J.M., S.L. BONTING, Biochim. Biophys. Acta, 150 (1968), 460

(6) BONTING, S.L., L.L. CARAVAGGIO, M.R. CANADY, N.M. HAWKINS, Arch. Biochem. Biophys., 106, (1964), 49

(7) BONTING, S.L., Comp. Biochem. Physiol., 17, (1966), 953

(8) BONTING, S.L., N.M. HAWKINS, M.R. CANADY, Biochem. Pharmacol., 13, (1964), 13

(9) BONTING, S.L., K.A. SIMON, N.M. HAWKINS, Arch. Biochem. Biophys., 95, (1961), 416

(10) HAFKENSCHEID, J.C.M., S.L. BONTING, Biochim. Biophys. Acta, 151, (1968), 204

(11) HAFKENSCHEID, J.C.M., S.L. BONTING, Biochim. Biophys. Acta, 178, (1969), 128

(12) BONTING, S.L., L.L. CARAVAGGIO, N.M. HAWKINS, Arch. Biochim. Biophys., 98, (1962), 413

(13) BONTING, S.L., L.L. CARAVAGGIO, Arch. Biochim. Biophys., 101, (1963), 37

(14) POST, R.L., C.R. MERRITT, C.R. KINSOLVING, C.D. ALBRIGHT, J. Biol. Chem., 235, (1960), 1796

(15) DUNHAM, E.T., J.M. GLYNN, J. Physiol., 156, (1961), 274

(16) BONTING, S.L., Invest. Ophthalmol., 4, (1965), 723

(17) HAFKENSCHEID, J.C.M., S.L. BONTING, Comp. Ciochem. Physiol. (1971, in press)

(18) CORRIE, W.S., S.L. BONTING, Biochim. Biophys. Acta, 120, (1966), 91

(19) HARRIS, E.J., J. Physiol., 193, (1967), 455

(20) SJODIN, R.A., L.A. BEAUGE, J. Gen. Physiol., 52, (1968), 389

(21) NODA, K., Kurume Med. J., 15, (1968), 51

(22) KUIJPERS, W., S.L. BONTING, Biochim. Biophys. Acta, 173, (1969), 477

(23) KUIJPERS, W., S.L. BONTING, Pflügers Arch., 320, (1970), 348

(24) LEAF, A., J. ANDERSON, L.B. Page, J. Gen. Physiol., 41, (1958), 657

(25) LEAF, A., L.B. PAGE, J. ANDERSON, J. Biol. Chem., 234, (1959), 1625

(26) BONTING, S.L., M.R. CANADY, Amer. J. Physiol., 207, (1964), 1005

(27) HERRMANN, I., H.J. PORTIUS, K. REPKE, Arch. Exptl. Pathol. Pharmakol., 247, (1964), 1

(28) FRAZIER, H., E.F. DEMPSEY, A. LEAF, J. Gen. Physiol., 45, (1962), 529

(29) BOWER, B.F., Fed. Proc., 22, (1963), 445

(30) BOWER, B.F., Nature, 204, (1964), 786

(31) RIDDERSTAP, A.S., S.L. BONTING, Amer. J. Physiol., 216, (1969a), 547

(32) RIDDERSTAP, A.S., S.L. BONTING, Amer. J. Physiol., 217, (1969b), 1721

(33) DIAMOND, J.M., J.Gen. Physiol., 48, (1964a), 1

(34) DIAMOND, J.M., J. Gen. Physiol., 48, (1964b), 15

(35) KAYE, G.I., H.O. WHEELER, R.T. WHITLOCK, N. LANE, J. Cell Biol., 30, (1966), 237

(36) DIAMOND, J.M., W.H. BOSSERT, J. Cell Biol., 37, (1968), 694

(37) FAHN, S., R.W. ALBERS, G.J. KOVAL, Science, 145, (1964), 283

(38) SWANSON, P.D., J. Neurochem., 15, (1968), 1159

(39) POST, R.L., In: J. JÄRNEFELT (Ed.) Regulatory Functions of Biological Membranes, Elsevier, Amsterdam, 1(1968), 173

(39a) HASHIUCHI, Y., Nara Igaku Zasshi, 18, (1967), 469

(40) HOKIN, L.E., P.S. SASTRY, P.R. GALSWORTHY, A. YODA, Proc. Natl. Acad. Sci. U.S., 54, (1965), 177

(41) BADER, H.A., K. SEN, R.L. POST, Biochim. Biophys. Acta, 118, (1966), 106

(42) ALBERS, R.W., G.J. KOVAL, G.J. SIEGEL, Mol. Pharmacol., 4, (1968), 324

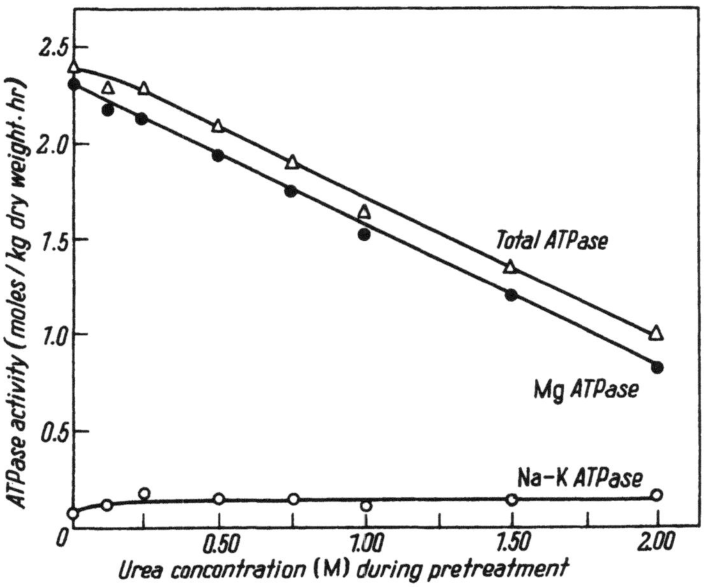

Fig. 1. Effect of pretreatment of urea on ATPase activities in Escherichia coli.

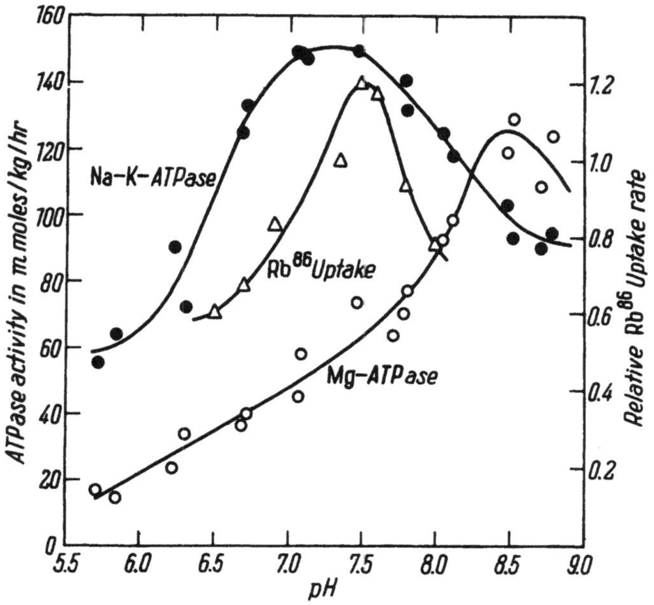

Fig. 2. pH-Activity curves
for Na-K ATPase and Mg ATPase
ATPase for lens epithelium.

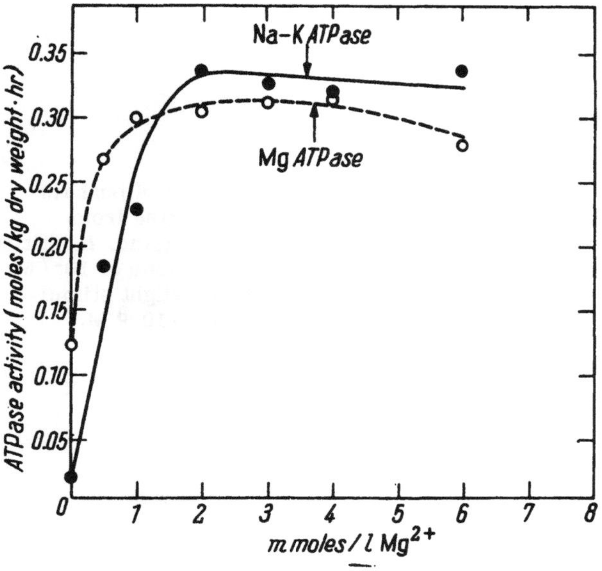

Fig. 3. Mg-activation curves for Na-K ATPase and Mg ATPase for rat liver.

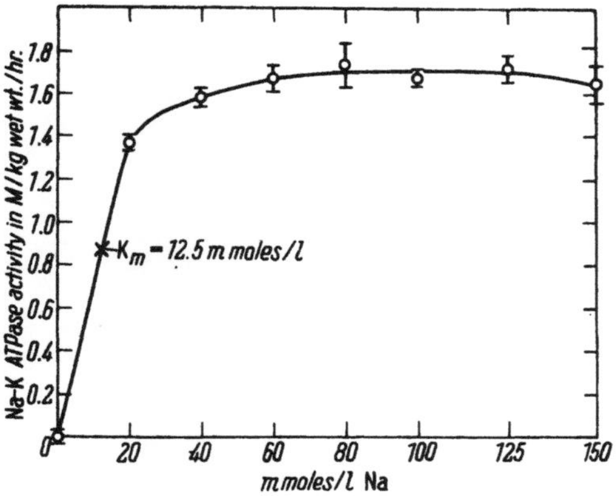

Fig. 4. Na-activation curve for Na-K ATPase from herring gull salt gland.

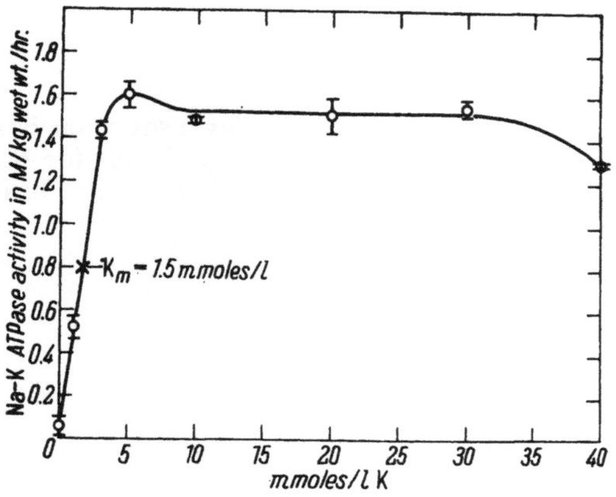

Fig. 5. K-activation curve for Na-K ATPase from herring full salt gland.

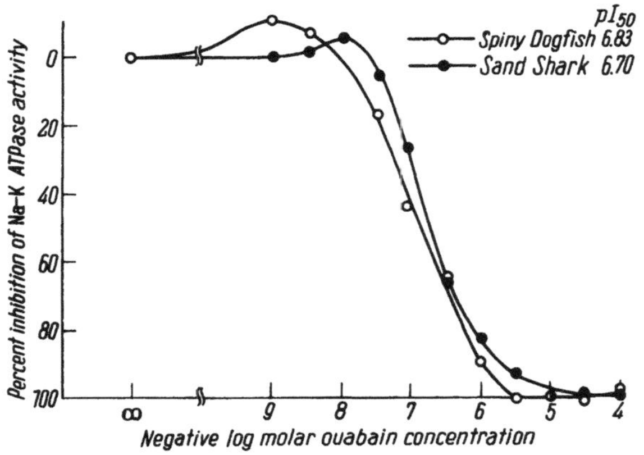

Fig. 6. Effect of ouabain on Na-K ATPase from shark rectal gland. Inhibition at concentrations of 10^{-7} M and higher, slight stimulation at 10^{-9}-10^{-8} M.

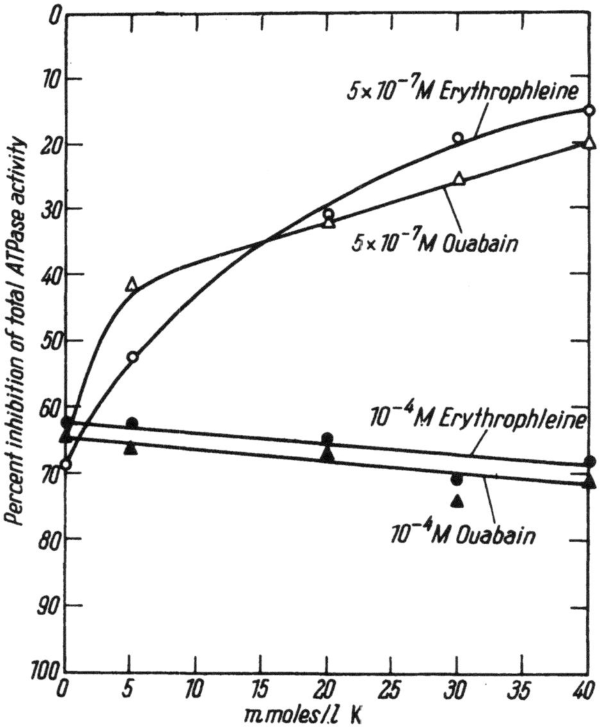

Fig. 7. Reversal of ouabain-inhibition by K^+ions for Na-K ATPase from rabbit brain.

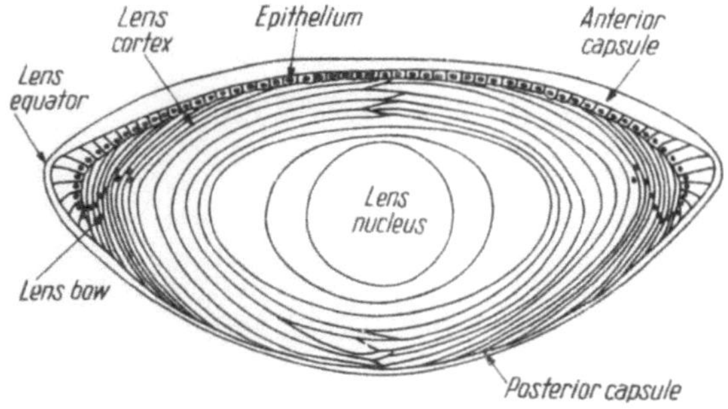

Fig. 8. Anatomy of the lens.

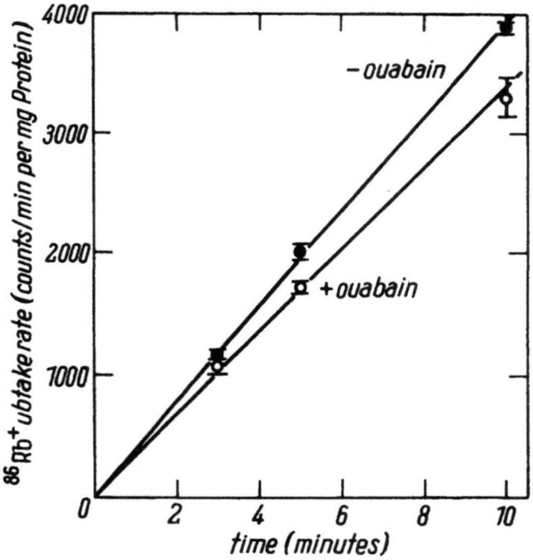

Fig. 9. Inhibitory effect of ouabain (0.8×10^{-4}M) on ^{86}Rb uptake by Escherichia coli.

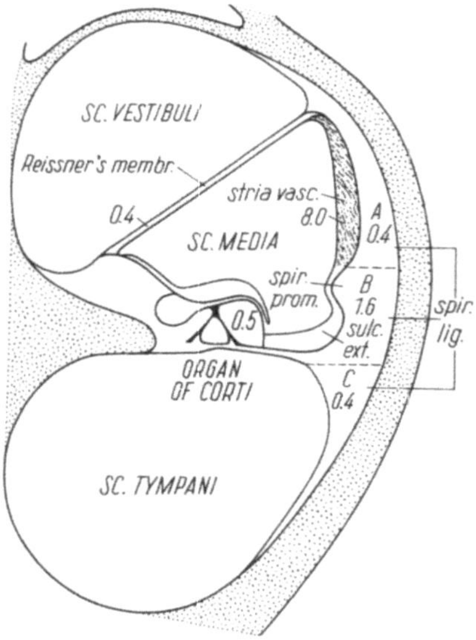

Fig. 10. Schematic cross-section of the cochlea with the Na-K ATPase activity (in moles/kg dry wt/hr) of the various structure indicated by numbers.

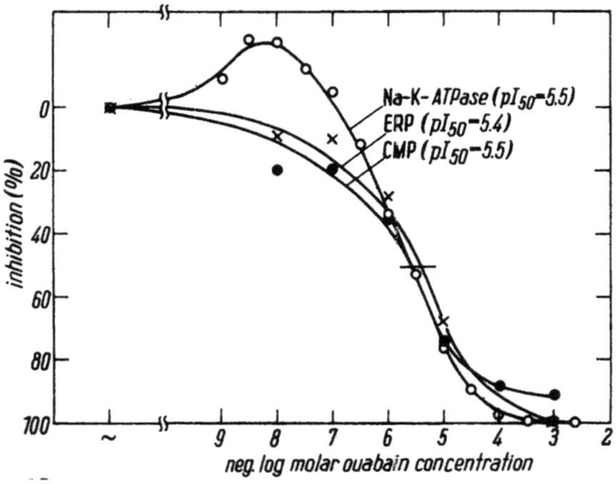

Fig. 11. Inhibitory effects of ouabain on Na-K ATPase activity of stria vascularis and on the cochlear potentials (ERP and CMP).

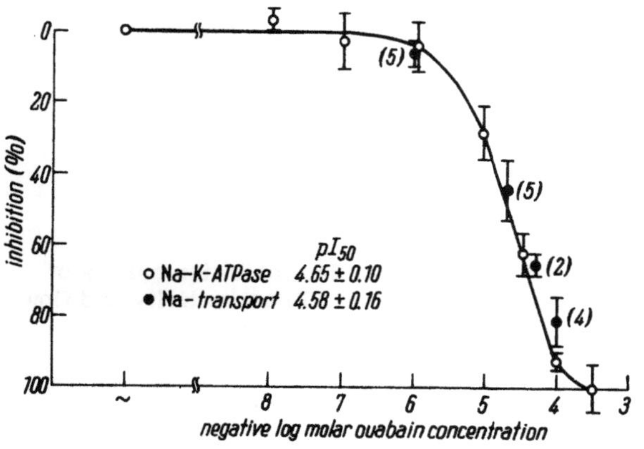

Fig. 12. Effect of ouabain on Na-K ATPase activity and Na^+ transport in toad bladder.

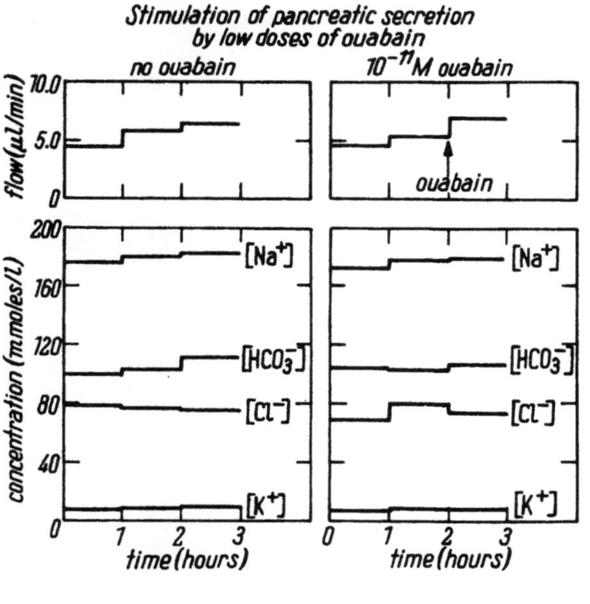

Fig. 13. Effect of ouabain on flow rate and ion levels of pancreatic fluid secreted by the rabbit pancreas in vitro.

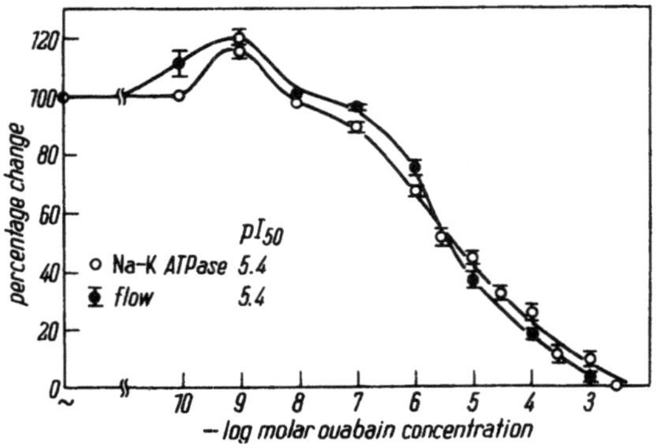

Fig. 14. Inhibition curves of Na-K ATPase activity and flow in rabbit pancreas.

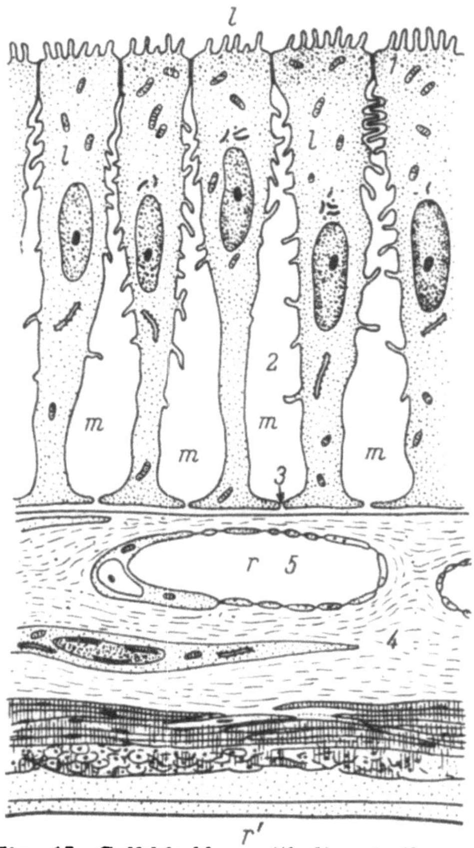

Fig. 15. Gall bladder epithelium in the secreting state.
Explanation: 1. desmosome, 2. intercellular space, 3. narrow opening, 4. lamina propria, 5. capillary.

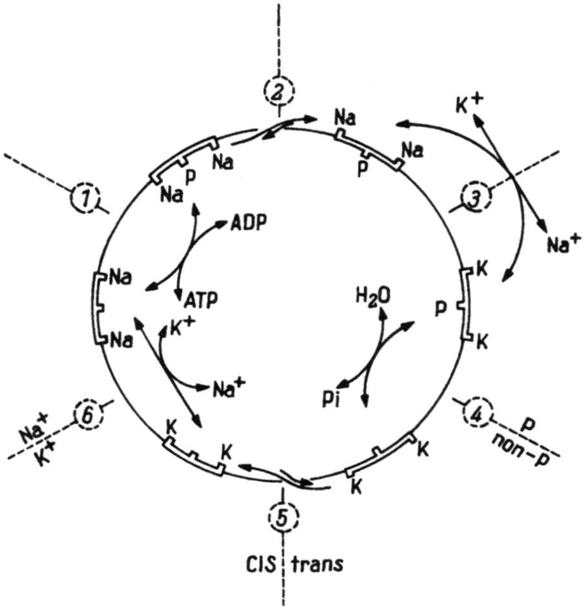

Fig. 16. Model of Na-K ATPase mechanism according to ALBERS, KOVAL and SIEGEL (43).

Department of Biochemistry, University of Nijmegen,
Nijmegen, The Netherlands

S. L. BONTING, J. J. H. H. M. DE PONT, T. D. HANSEN

ANION EFFECTS ON Mg-ACTIVATED ATPase OF LIZARD GASTRIC MUCOSA

A bicarbonate-activated, thiocyanate-sensitive ATPase activity in frog gastric mucosa was reported by KASBEKAR and DURBIN (1). Since thiocyanate also inhibits the gastric acid secretion, these authors suggested that this enzyme participates directly in the sequence of reactions leading to acid secretion. In particular, they DURBIN and KASBEKAR (2) postulated that this enzyme might play a role in a chloride-bicarbonate exchange mechanism analogues to the sodium potassium exchange mechanism of the Na-K ATPase system. The inhibitory effect of SCN^- ions would then be analogous to the action of ouabain in the latter enzyme. In the course of a study of the possible role of Na-K ATPase in gastric secretion HANSEN et al. (3) we made a detailed study of the effects of bicarbonate, thiocyanatean other anions on the Mg-activated ATPase of the lizard gastric mucosa.

Gastric mucosa was isolated from the lizard stomach by dissection, was homogenized and lyophilized. A 0. 02% (w/v, final concn.) homogenate was incubated for 1 hr. at 37°C in a medium of the following composition:
ATP, 2 mM; Na^+, 60 mM; Mg^{2+}, 2 mM; Tris, 100 mM; ouabain, 0, 1 mM. The pH was either 7, 5 or 8, 4. Various anions were present such that the total concentration was always 147 meq/l at pH 7, 5 and 97 meq/l at pH 8, 4. The medium was made inhibitory to Na-K ATPase by the absence of K^+ and the presence of 10^{-4}M ouabain. After termination of the enzyme reaction by addition of trichloroacetic acid, the orthophosphate formed was determined by means of a molybdate-$FeSO_4$ color reagent BONTING (4).

Inhibition curves were determined for thiocyanate in the presence of various anions at a constant total anion concentration. From the results for four anions shown in Fig. 1, it is apparent that in the absence of thiocyanate there are considerable differences in the Mg-ATPase activity for these various anions, the activity decreasing in the order HCO_3^- > acetate > Cl^- > NO_3^-.

Moreover, the half-inhibition concentration for thiocyanate varies considerably with the anion.
Assuming that these effects are to be ascribed to differences in the affinity of the various anions for the anion-binding sites on the enzyme, we calculated the ratio of the thiocyanate concentration to the other anion at which 50% inhibition of the enzyme activity was obtained (Table 1, last column). This ratio, at which presumably half of the anion binding sites are occupied by thicyanate ions, should give a measure of the affinity of the other anion relative to that of thicyanate for the binding site: the larger the ratio, the larger its affinity.
Another approach was applied by determining the enzyme activity in mixtures of chloride and each of the other anions, total anion concentration again being

Table 1

HALF-MAXIMAL INHIBITION CONDITIONS FOR Mg ATPase FROM LIZARD GASTRIC MUCOSA BY THIOCYANATE

Anion	pH	Maximal activity (mol/kg/hr)	Concentrations at 50% inhibition		
			SCN^- (mM)	Anion (meq M)	ratio $\frac{SCN^-}{anion}$ x 10^3
HCO_3^-	8,4	6, 00	8, 0	89	90
Cl^-	8,4	3, 58	6, 0	91	66
glucuronate	7, 5	3, 50	0, 35	146, 6	2, 4
acetate	7, 5	3, 44	0, 5	146, 5	3, 4
SO_4^{2-}	7, 5	3, 55	0, 6	73, 2	8, 2
Cl^-	7, 5	2, 32	6, 6	140, 4	47
NO_3^-	7, 5	1, 57	20	127	157

total anion concn. 147 meq/1 at pH 7, 5; 97 meq/1 at pH 8, 4. Figures in last column indicate affinity of anion to enzyme, value for SCN^- set at 1000.

kept constant. An example of the curves is shown in Fig. 2 for the mixture chloride-acetate. In this case an activity midway between the activities for either anion alone was obtained at 8 meq/1 chloride and 139 meq/1 acetate, indicating a larger affinity of chloride ions for the anion bindings sites.

Table 2

EFFECT OF ANION MIXTURES ON Mg ATPase FROM LIZARD GASTRIC MUCOSA

Anion	pH	Concentrations at 50% activity change		
		Cl mM	Anion mM	ratio Cl / anion
glucuronate	7, 5	5	142	0, 035
acetate	7, 5	8	139	0, 058
SO_4^{2-}	7, 5	9	69	0, 13
NO_3^-	7, 5	117	30	3, 9
HCO_3^-	8, 4	70	27	2, 6

total anion concn. 147 meq/1 at pH 7, 5, 97 meq/1 at pH 8, 4

The results of these experiments are summarized in Table 2. The last column gives the ratio of the concentrations of chloride and the other anion at which the activity was midway between the activities for either anion alone. This ratio should give a measure of the affinity of the other anion relative to that of chloride for the binding site: the larger the ratio, the larger its affinity. A comparison of the results from Tables 1 and 2 is presented in Table 3.

Table 3

AFFINITY OF ANIONS TO ENZYME RELATIVE TO AFFINITY OF CHLORIDE

Anion	from direct substition	from thiocyanate inhibition
glucuronate	0, 035	0, 051
acetate	0, 058	0, 072
SO_4^{2-}	0, 13	0, 17
Cl^-	1, 0	1, 0
HCO_3^-	2, 6	1, 4
NO_3^-	3, 9	3, 4
SCN^-	21, 3	21, 3

There is quantitative agreement between the affinity ratios from the thiocyanate-inhibition and the anion-interaction experiments. When comparing the affinity sequence to the activity sequence appearing from Fig. 1, there also turns out to be qualitative agreement between the affinity sequence and the inverse activity sequence except for the place of bicarbonate: affinity: glucuronate < acetate < SO_4^{2-} < CL^- < HCO_3^- < thiocyanateactivity: HCO_3^- > SO_4^{2-} ≃ glucuronate ≃ acetate > Cl^- > NO_3^- > thiocyanate. With regard to the place of bicarbonate, it should be kept in mind that all assays involving this anion had to be done at pH 8,4, while the other assays were done at pH 7,5.

The following conclusions seem to be warranted. The Mg ATPase of lizard gastric mucosa has anion binding sites. The enzyme activity depends on the anions occuying these sites. The stronger an anion binds to these sites, the lower the enzyme activity becomes, with exception of bicarbonate. The anions may cause a conformational change in the enzyme protein, affecting the active site. These findings suggest that the bicarbonate, thicyanate and chloride effects on this enzyme are of a different nature than those of the cations Na^+ and K^+ and of ouabain on the Na-K ATPase system.

REFERENCES

(1) KASBEKAR, D. K., R. P. DURBIN, Biochim. Biophys. Acta, 105, (1965), 472

(2) DURBIN, R. P., D. K. KASBEKAR, Fed. Proc., 24, (1965), 1377

(3) HANSEN, T. D., J. F. G. SLEGERS, J. J. H. H. M. de PONT, S. L. BONTING, (1971, in preparation)

(4) BONTING, S. L., In: E. E. BITTAR (ed.), Membranes and Ion Transport, Wiley, New York, 1, (1970), 257

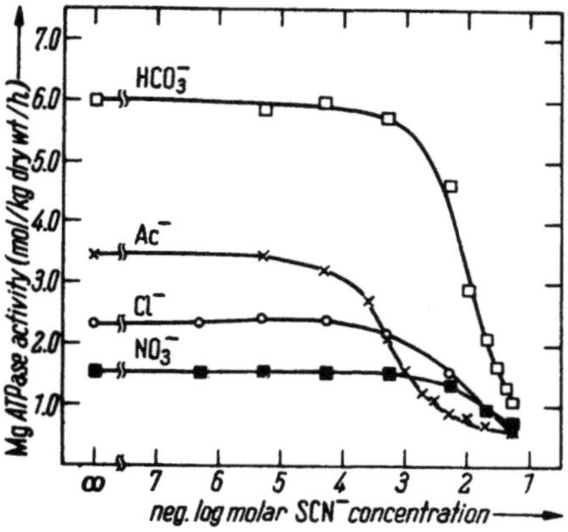

Fig. 1. Inhibition of Mg ATPase from lizard gastric mucosa by thiocyanate in the presence of various anions.

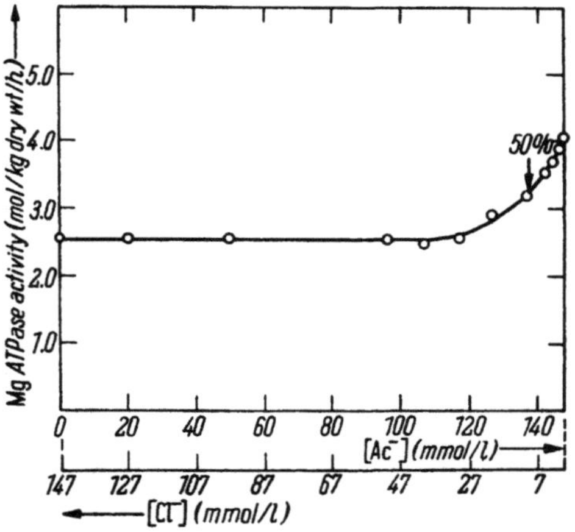

Department of Biochemistry, University of Nijmegen,
Nijmegen, The Netherlands

S. L. BONTING, J. H. M. WÖLTGENS

INORGANIC PYROPHOSPHATASE IN DEVELOPING TEETH: EFFECTS OF MAGNESIUM AND DIPHOSPHONATES

The occurrence of an inorganic pyrophosphatase in mineralizing hamster teeth was demonstrated WÖLTGENS, BONTING, BIJVOET (1). There is considerable evidence that this enzyme plays a key role in the mineralization process by removing from the site of hydroxyapatite deposition inorganic pyrophosphate, which inhibits the latter process. In a study of the properties of the enzyme we determined the effects of magnesium ions and of three diphosphates.

For the pyrophosphatase assay a 0, 02% (w/v, concn.) homogenate of 3-day old hamster molars was incubated for 1 hr. at 37°C in a medium of the following composition: 0, 1 M TRIS pH 8, 7, 3, 3 mM inorganic pyrophosphate (PP_i) and 3, 3 mM $MgCl_2$.

Fig. 1 shows the effect of Mg^{2+} ions, added in a concentration range of 0-10 mM at three different substrate concentrations (1, 6, 2, 2 and 3, 3 mM). In the absence of added Mg^{2+} the enzyme showed marked activity. Addition of Mg^{2+} caused an increased activity, which was always maximal at a Mg^{2+}: $P_2O_7^{4-}$ molar ratio of 1. At higher Mg^{2+} concentrations the activity was less than in the absence of Mg^{2+}. This indicates that the best substrate for the enzyme is Mg $P_2O_7^{=}$, less suitable is $P_2O_7^{4-}$ and least suitable is $Mg_2P_2O_7$.

The chelating agent EDTA strongly inhibits the enzyme activity. When Mg^{2+} was omitted from the medium nearly full inhibition was obtained with only 0, 1 mM EDTA (Fig. 2). This suggests that a bivalent cation acts as a cofactor or activator of the enzyme. The inhibition by EDTA can be largely reversed by adding 5-10 mM Mg^{2+} (Fig. 3). The need for a 50-100 fold excess of Mg^{2+} suggests that a bivalent cation other than Mg^{2+} (possibly Zn^{2+}) acts as cofactor or activator, and that the excess of Mg^{2+} displaces this cation from the EDTA-complex and/or favors its binding to the enzyme.

Diphosphonates are structural analogues of pyrophosphate, in which the P-O-P link is replaced by a P-C-P link. The following three diphosphonates were used in our study:

EHDP - Na ethane-1-hydroxy-1, 1 diphosphonate	Et P - C - P OH
MDP - Na methane-diphosphonate	H P - C - P H
Cl_2MDPMDP - Na dichloro methane diphosponate	Cl P - C - P Cl

All three diphosponates showed an inhibitory effect on the enzyme, which did not hydrolyse them. Fig. 4 shows inhibition curves for the three compounds. The half-inhibition concentrations were about equal to the $MgP_2O_7^{2-}$ substrate concentration used in the assay and were of the order of 1-3 mM. In the case of Cl_2MDP and MDP increase of the Mg^{2+} concn. largely or completely reversed the inhibitory effect of these substances, as shown for MDP in Fig. 5. This suggests that these two diphosphonates inhibit the pyrophosphate activity primarily through complexation of bivalent cations, like in the case of the chelating agent EDTA.

The diphosphonate EHDP displays a more complicated behavior. Fig. 6 shows the occurrence of clear minima in the activity curves for Mg^{2+} concentrations representing Mg: (pyrophosphate + EHDP) ratios of 1. By varying the pyrophosphate concentration in the presence and absence of 1, 65 mM EHDP, while adding Mg^{2+} so that all times a Mg: (pyrophosphate + EHDP) ratio of 1 was maintained, the Lineweaver-Burke plot presented in Fig. 7 was obtained. This plot indicates that under these conditions the EHDP-Mg complex competitively inhibits the enzyme by competing with the $(MgP_2O_7)^{2-}$ substrate complex for the active site. The two complexes have apparently a nearly equal affinity for the active site, since the K_m value (1 mM) and the K_i value (0, 7 mM) derived from Fig. 7 are approximately equal. A likely explanation for this competitive action of EHDP, which is absent in the case of the other two diphosphonates, is the presence of the OH-group on the carbon linking the two phosphoryl groups in EHDP. Apparently the P-C-P configuration (with O on C) is sufficiently similar to the P-O-P link in pyrophosphate to permit binding at the active site. The similarity in the inhibition curves of Fig. 4 suggests that EHDP also can inhibit through complexation of bivalent cations.

The inhibitory effects of the diphosphonates is of practical significance for the therapeutic treatment of conditions of abnormal calcification such as aortic calcification in rats given large dosis of vitamin D_3 FRANCIS et al. (2) and dental deseases such as supragingival calculus formation MÜHLEMANN (3) et al.

REFERENCES

(1) WÖLTGENS, J.H.M., S.L. BONTING, O.L.M. BIJVOET, Calc. Tiss. Res., 5, (1970), 333

(2) FRANCIS, M.D., R.G.G. RUSSELL, H. FLEISCH, Science, 165, (1969), 1264

(3) MÜHLEMANN, H.R., D. BOWLES, A. SCHAIT, J.P. BERNIMOULIN, Helv. Odont. Acta, 14, (1970), 31

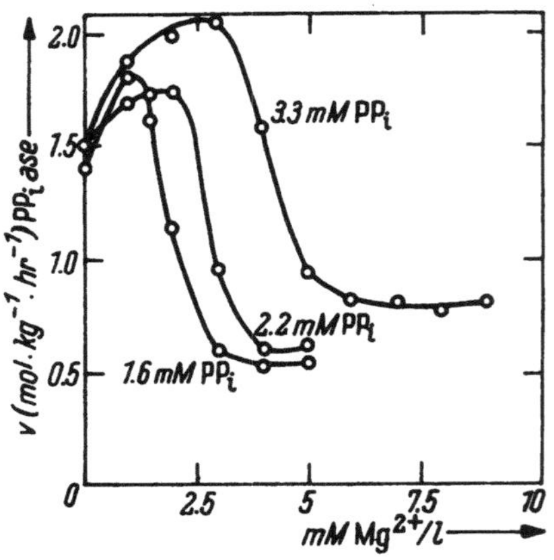

Fig. 1. Effect of magnesium on pyrophosphatase activity. Note optimal activities at Mg : P_2O_7 ratio of 1.

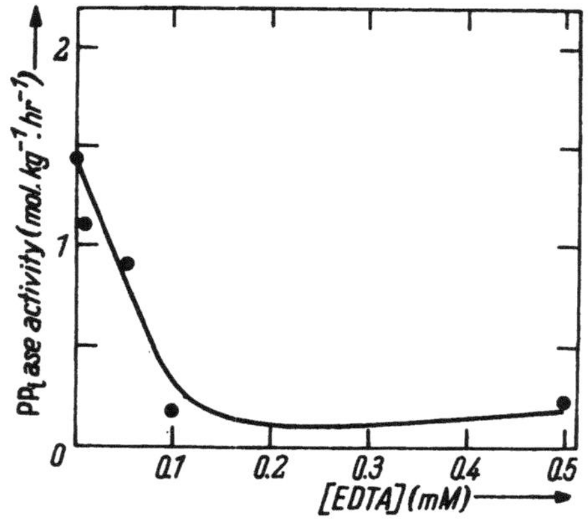

Fig. 2. Effect of ethylene-diamine-tetra acetic acid (EDTA) on pyrophosphatase activity. Concentration of pyrophosphate was 3. 3. mM.

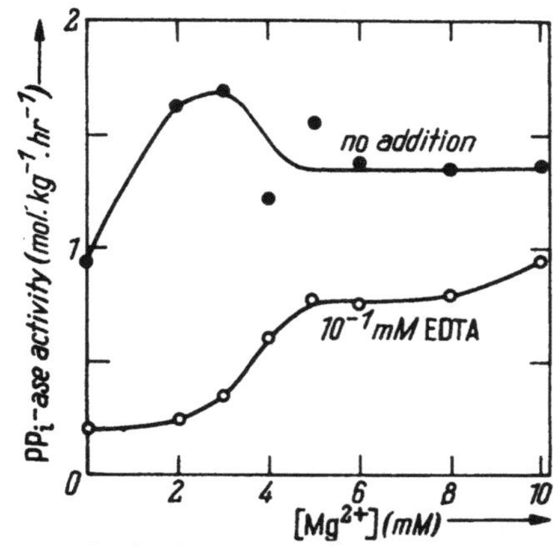

Fig. 3. Reversal of inhibitory effect of EDTA by magnesium. Concentration of pyrophosphate was 3.3 mM, of EDTA 0.1 mM.

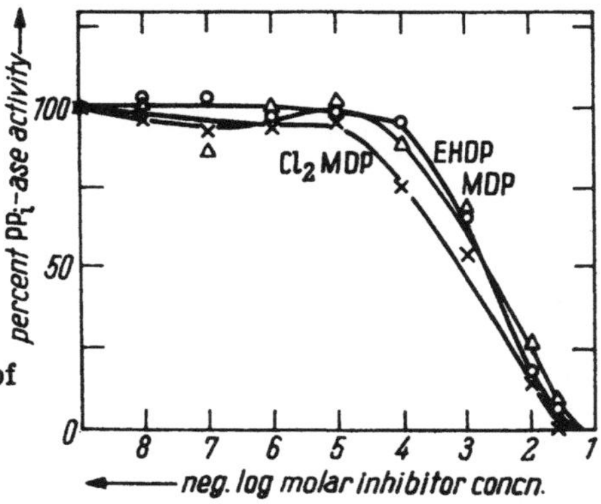

Fig. 4. Effect of three diphosphonates on pyrophosphatase activity. Concentrations of Mg^{2+} and pyrophosphate were 3.3 mM.

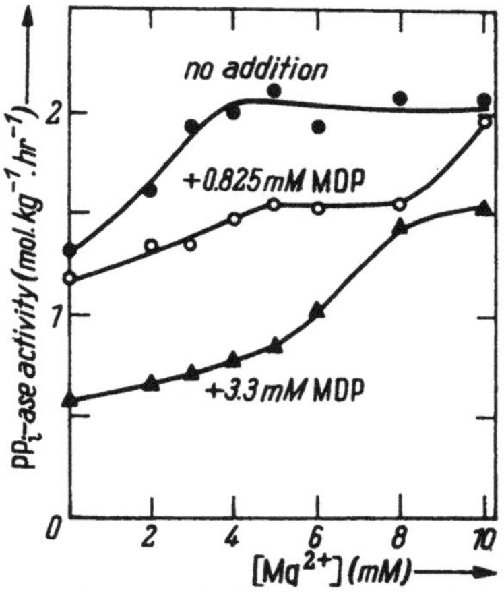

Fig. 5. Reversal of MDP inhibition of pyrophosphatase activity by magnesium. Concentration of pyrophosphate was 3.3 mM.

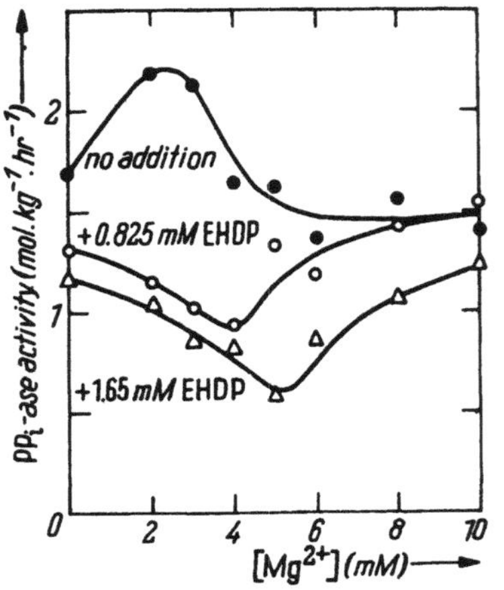

Fig. 6. Effect of magnesium on the inhibition of pyrophosphatase activity by EHDP.

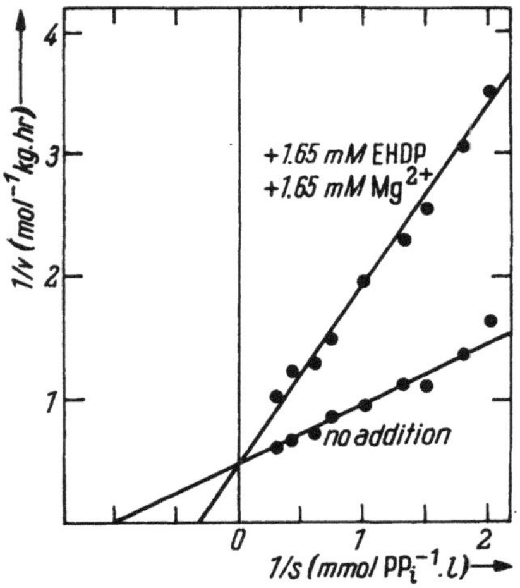

Fig. 7. LINEWEAVER-BURKE plot for the inhibition of pyrophosphatase activity by EHDP.
Molar concentration of magnesium was kept equal to the sum of the pyrophosphate and EHDP concentration.

Forschungszentrum für Molekularbiologie und Medizin der Deutschen Akademie der Wissenschaften zu Berlin, Berlin-Buch

K. R. H. REPKE, R. SCHÖN, W. SCHÖNFELD, H. J. PORTIUS[1)]

DOES Na^+ + K^+-DEPENDENT ATPase CONSIST OF MORE THAN ONE ENZYME ENTITY

Regarding the present state of knowledge on transport ATPase action (for reviews see SKOU (3), ALBERS (4), HOKIN (5), POST and COWORKERS (6)), progress might partially depend on additional ideas about its possible reaction mechanism for the uphill transport of sodium and potassium ions across cell membrane. So, we asked ourselves whether or not the Na^++K^+-dependent ATPase could eventually consist of more than one enzyme entity. This question occurred to us when it became evident by the work of NEUFELD and LEVY (7) and of WOODIN and WIENEKE (8) that a Na^+-dependent ATPase -apparently involved in the uphill transport of sodium ions- and a K^+-dependent phosphatase -apparently involved in the uphill transport of potassium ions- exist as separately and independently working entities in the plasma membrane of certain cells. We thought it possible that in Na^++K^+-dependent ATPase (NaK)-ATPase both enzymes might interact in a concerted manner so that the initial Na^+-dependent transphorsphorylation from ATP to enzyme I:

$$\text{enzyme I+MgATP} \underset{}{\overset{Na^+}{\rightleftharpoons}} \text{phosphoenzyme I+ADP+}Mg^{++} \qquad (1)$$

is more or less tightly coupled by a Mg^{++}-catalyzed transphosphorlation from phosphoenzyme I to enzyme II:

$$\text{phosphoenzyme I+enzyme II} \overset{Mg^{++}}{\rightleftharpoons} \text{phosphoenzyme II+enzyme I} \qquad (2)$$

with the K^+-dependent hydrolytic reaction:

$$\text{phosphoenzyme II} \overset{K^+}{\rightleftharpoons} \text{enzyme II+}P_i \qquad (3)$$

Reaction (1) could be catalyzed by the Na^+-dependent ATPase and reactions (2) and (3) by the K^+-dependent phosphatase. So, the normal phosphoryl donator for the phosphatase would be phosphoenzyme I. As well-known, however, (NaK)-

1) Parts of our work reviewed here were presented in preliminary short communications (1, 2)

ATPase easily splits acetyl phosphate and p-nitrophenyl phosphate, too. As to the supposed transphosphorylation step (2), ROBINSON's suggestion derived from kinetic and chemical evidence may be cited here that after the initial Na^+-activated formation of an acylphosphate (localized in enzyme E_1) the phosphoryl group becomes transferred to a hydrolytic site called E_3:

$$E_1 \xrightarrow[\text{ATP} \curvearrowright \text{ADP}]{Na^+,\ Mg^{++}} [E_1 \sim P] \longrightarrow E_2\text{-}P \xrightarrow{K^+} [E_3\text{-}P] \longrightarrow E_1 + P_i$$

This hydrolytic site could be identic with the supposed enzyme II, although ROBINSON left it open whether or not the site should be localized at the same enzyme protein as the initial acyl phosphate.

In an attempt to check the question about the involvement of two enzyme proteins in transport ATPase action we used as a possibly useful tool the dialdehyde derivative of ATP which, it was hoped, could interact with an ATPase moiety, but not with a phosphatase moiety. In "ATP-dialdehyde", which was synthesized by Dr. SCHÜTT in Professor LANGENs group the vicinal cis-located hydroxy groups of ribose are oxidized to aldehyde groups (Fig. 1).

ATP-dialdehyde does not undergo hydrolysis by (NaK)-ATPase and does not serve as phosphoryl group donator to the enzyme protein. This can be deduced from the observation that, in contrast to ATP, the dialdehyde derivative does not reduce the steady state level of phosphoprotein obtained by incubation of the enzyme preparation with ATP-γ^{32}P, Mg^{++} and Na^+ (Table 1).

Table 1

Nucleotide	^{32}P incorporated (cpm/U)
ATP-γ^{32}P	25, 8
ATP-γ^{32}P + ATP	13, 9
ATP-γ^{32}P + ATP-dialdehyde	24, 3

Lack of effect of ATP-dialdehyde on steady state level of Na^+-induced phosphorylation of (NaK)-ATPase by ATP-γ^{32}P. ATP-γ^{32}P, ATP and ATP-dialdehyde each 0, 005, Mg^{++} 1, Na^+ 80, pH 7, 4, 60 sec, 0° (concentrations in mM).

ATP-dialdehyde inhibits (NaK)-ATPase by a relatively slow reaction. The resulting inhibition is essentially irreversible. The same percentage of inhibition is reached whether ATP-dialdehyde was present during activity determination or present during pre-incubation only (Table 2).

The chemical basis of the observed inhibitory effect could be the formation of a SCHIFF base between ATP-dialdehyde and an amino group shown to be present in the active centre of (NaK)-ATPase by POST and coworkers (6) (Fig. 2). That this amino gruop apparently plays an essential role in enzymic activity is suggested by the observation, originally made by PULL (9) and confirmed by us, that citraconic anhydride, an almost specific reagent for primary amino groups, likewise inhibits (NaK)-ATPase.

Table 2

LACK OF REVERSIBILITY OF (NaK)-ATPase INHIBITION BY ATP-DIALDEHYDE

Special conditions	Inhibition %
1 hr incubation in presence of ATP alone or ATP plus ATP-dialdehyde	90
1 hr pre-incubation in presence of ATP alone or ATP plus ATP-dialdehyde. After removal of ATP-dialdehyde excess 1 hr incubation in presence of ATP	87

Common conditions: ATP and ATP-dialdehyde each 2, Mg^{++} 2 or 4, Na^{+} 135, K^{+} 5, pH 7,4, 37° (concentrations in mM).

Although ATP-dialdehyde is not used as substrate by (NaK)-ATPase, it nevertheless apparently induces a similar change of enzyme conformation as does ATP. This becomes evident when studying the binding of the negative effector ouabain to the enzyme (Fig. 3). While in the absence of nucleotides there is only a very low ouabain binding, ATP-dialdehyde proves almost as effective as ATP in producing ouabain affinity.

On the basis of its apparently specific interaction with the ATPase, ATP-dialdehyde seemed to be suited for answering the question on the existence of a separate,

though normally interdependtly acting phosphatase in (NaK)-ATPase.
However, after ATP-dialdehyde treatment of the enzyme preparation, K^+-dependent phosphatase activity proved to be inhibited to a comparable percentage as the overall (NaK)-ATPase activity (Table 3).
The result does not favour the idea that (NaK)-ATPase consists of two enzyme entities. Indeed, ATP-dialdehyde does not act as specifically as desired, since besides the K^+-dependent phosphatase it also inhibits an unspecified phosphatase activity presumably independent of (NaK)-ATPase activity (Table 4).
In addition to the analysis of ATP-dialdehyde enzyme interactions, we followed (NaK)-ATPase and (K)-phosphatase activities in various isolation and purification procedures and found an essentially constant ratio. Accordant with this observation, in a very recent paper HOKIN et al. 10 described, that the specific activity of

Table 3

COMPARISON OF THE INHIBITORY EFFECT OF ATP-DIALDEHYDE ON (NaK)-ATPase ACTIVITY AND (K)-PHOSPHATASE ACTIVITY

Enzymic activity	Inhibition (%)
Na^+ + K^+-dependent ATPase	68
K^+-dependent Phosphatase (NPPase)	46

After pre-incubation of an enzyme sample with ATP-dialdehyde 2 in presence of Mg^{++} 3 and Ca^{++} 1 at pH 7,4 and 37° for 15 min, the medium was removed and portions of the sample were incubated either with ATP 2, Mg^{++} 2, Na^+ 135, K^+ 5, EGTA 0,1 or p-nitrophenyl phosphate 5, Mg^{++} 5, K^+ 5, EGTA 0,1 at pH 7,4 and 37° for 30 min (concentrations in mM).

K^+-activated p-nitrophenyl phosphatase increased in parallel with that of the (NaK)-ATPase on 30 to 50 times purification. These and other data prompted us to think about possibilities for the understanding of the peculiar ATPase and phosphatase activities of (NaK)-ATPase alternative to the conception as discussed up to now.
As the result of an attempt to integrate the mass of data and ideas from literature and our own experience, we emerged with a model for the Na^++K^+-dependent hydrolysis of ATP and the K^+-dependent hydrolysis of p-nitrophenyl phosphate by

one single enzyme protein (Fig. 4). Its main features are as follows. The basic assumption is that the enzyme protein may exist in form of the two conformational isomers E_i and E_o. The E_i isomer is combined with sodium ions and may therefore

Table 4

INHIBITION OF MEMBRANE-LOCALIZED PHOSPHATASES BY ATP-DIALDEHYDE

Enzymic activity	Inhibition (%)
(K)-phosphatase	86
Unspecified phosphatase[1)]	100

1) K^+ omitted from incubation mixture

Pre-incubation (15 min): ATP-dialdehyde 2, Mg^{++} 5, K^+ 5, pH 7,4, 37°.
Incubation (1 hr): additional presence of p-nitrophenylphosphate(7) (concentrations in mM).

also be called the sodium form; its cation binding site is located at the inside of the membrane. This sodium form preferentially uses ATP, but partially also acetyl phosphate as substrate. The E_o isomer is combined with potassium ions and may therefore also be called the potassium form; its cation binding site is located at the outside of the membrane. This potassium form preferentially uses p-nitrophenyl phosphate as substrate.

These assumptions refer among other things to the following data. Sodium ions, in the presence of saturating K^+-concentrations, accelerate the hydrolysis of acetyl phosphate and grossly depress the hydrolysis of p-nitrophenyl phosphate as shown by ROBINSON (11). Potassium ions decrease the enzyme-ATP dissociation constant NORBY and JENSEN (12). p-Nitrophenyl phosphate competitively inhibits Na^++K^+-dependent ATPase activity and ATP competitively inhibits K^+-dependent phosphatase activity, as described by BADER and SEN (13), ISRAEL and TITUS (14) or FORMBY and CLAUSEN (15).

With both enzymic acitivities, the hydrolysis of the substrates proceeds through the stage of an intermediary phosphoenzyme. However, the phosphorylated group is on the Na^++K^+-dependent ATPase activity a hydroxylamine-sensitive carboxyl phosphate, according to HOKIN (5) a glutamyl-γ-phosphate residue, and on the

K^+-dependent p-nitrophenyl phosphatase activity it is, according to ROBINSON (16), an unknown phosphate-carrying residue largely insensitive to hydroxylamine. Moreover, in (NaK)-ATPase activity there are two phosphorylated stages, according to our denotation the sodium form and the potassium form of the phosphoenzyme. Both presumably carry an acyl phosphate, but are apparently largely different in their conformational energy. Thus, only the sodium form of the phosphoenzyme exhibits an ADP-ATP exchange activity (6).

On the basis of this model the uphill transport of Na^+ and K^+ by (NaK)-ATPase across cell membrane can be thought to proceed as follows. Under the catalytic action of Na^+, the enzyme protein is readily phosphorylated, which imposes a heavy conformational constraint upon the phosphoprotein E_i. This enables the Na^+-loaden protein to undergo a vectorial conformational change carrying the cation binding site and with it sodium from in to out. The phosphoprotein E_o thus formed releases $Na^+$and attracts K^+. K^+ catalyzes the hydrolysis of the phosphoprotein which presumably proceeds through the stages of bond cleavage and separation of phosphate from the enzyme phosphate addition complex. This latter stage involves a backward conformational change, carrying the cation binding site and with it potassium from out to in.

The oscillating, energy-consuming conformational changes of enzyme protein, postulated to effect the vectorial cation transport, should tentatively be described in terms of free energy changes. Thus, it may be permitted to speculate a little further. Fig. 5 shows a comparison of free energy diagrams for the hydrolysis of the phosphate group of those phosphoproteins which are thought to be intermediates of alkaline phosphatase, (K)-phosphatase or (NaK)-ATPase. The diagram for alkaline phosphatase is based on data of WILSON and DAYAN (17), while the diagrams for the two other enzymes are but speculatively constructed in a preliminary attempt to relate conformational and chemical events to free energy changes and transport work.

According to PECK et al. (18), the large positive value of the standard free energy change accompanying hydrolysis of the phosphoserine residue of alkaline phosphatase, +5, 5 kcal, in comparison with the value of -3, 3 kcal for phosphoserine suggests the presence of large positive noncovalent interactions between the phosphate group and the protein in the phosphoform of phosphatase. The relatively small free energy change for bond breaking, -2, 5 kcal, suggests that these interactions are not extensively altered during bond breaking. For the final step, the separation of the enzyme phosphate complex, the free energy change is a high as +8 kcal. As to the phosphoprotein intermediates of (K)-phosphatase and (NaK)-ATPase, the corresponding values for bond cleavage and dissociation of the phosphate group are not available.

As will be shown later, the phosphorylation of the glutamyl acid residue of (NaK)-ATPase is accompanied by a conformational change which might confer on the protein a conformational energy able to do vectorial transport work. As pointed out KLOTZ et al. (19), but minors changes in structure or orientation of small groups in macromolecules are accompanied by substantial changes in free energy. The free energy changes of (NaK)-ATPase by phosphorylation with ATP (formation of ${}^{Na}E_i \sim P$) or p-nitrophenylphosphate (formation of ${}^{K}E_o$-P) are unknown as yet, but the differences in enthalpy between the transition states and the initial reactants were calculated by FORMBY and CLAUSEN (20) to be 12,4 and 7,2 kcal/mole, respectively. Thus, the transport of Na^+ from in to out does not necessarily require bond cleavage, i. e. the potassium form of the phosphoprotein might carry the unchanged acyl phosphate residue. The free energy, released by K^+-catalyzed bond breaking of the acyl phosphate residue with consecutive dissociation of enzyme phosphate complex, could be consumed for a backwards directed conformational change of the protein transporting K^+ from out to in.

Summing up this part of the talk, it appears to us that the conception on the existence of both a Na^+- and a K^+-form of (NaK)-ATPase protein, using preferentially either ATP and acetyl phosphate or phosphate esters as substrates, makes it unnecessary to think that (NaK)-ATPase activity might involve the concerted action of more than one enzyme entity. The separately occuring and independently working Na^+-dependent ATPase and K^+-dependent phosphatase, as described by NEUFELD and LEVY (7) or WOODIN and WIENEKE (8), could rather be related to (NaK)-ATPase as far as having lost either the site for K^+-activation or the site for Na^+-activation.

In our model for the transformation of chemical energy into osmotic work by (NaK)-ATPase, it is postulated that the initial reaction stages impose on the enzyme protein a conformational constraint able to do vectorial transport work. As will be shown now, the Na^+-catalyzed phosphorylation of the enzyme and already the formation of the enzyme substrate complex appear to be involved in the induction of this supposed conformational constraint. As a sensitive and specific tool for evidencing the production of a conformation change accompanying those first steps of enzyme action, we used the appearance of a strong affinity for the negative enzyme effector ouabain LINDENMAYER and SCHWARTZ (21).

As well-known, ouabain can combine with one of the two intermediary phosphoforms of the enzyme and stabilize it against the dephosphorylating action of K^+, thus interrupting the phosphorylation-dephosphorylation cycle. However, the widely accepted view (6) that the potassium form (frequently called E_2-P) attracts ouabain, is presumably not correct. This is one of the conclusions to be drawn from the following experimental observations.

In the absence of K^+ from the medium, Na^+ catalyzes the phosphorylation of the enzyme protein by ATP (6), the steady state level being reached within about 60 sec (Fig. 6). Ouabain at very high concentrations suppresses the phosphorylation so that the steady state level of phosphoenzyme is reduced to that found in the absence of added Na^+. However, ouabain, at the very low concentrations used in the following experiments, solely stabilizes the phosphoenzyme against the hydrolytic action of contaminating K^+. Under such conditions favouring maximal phosphorylation, ATP in concentrations as low as 0, 02 mM induces a change in enzyme conformation as evidenced by production of affinity for the effector ouabain (Fig. 7).[1)]

Compared to the rate of phosphorylation, ouabain binding to the phosphoenzyme proceeds very slowly. Apparently, it is a statistically seldom event that, at the very low effector concentrations used, an ouabain molecule in the right position for combination meets an enzyme molecule in the appropriate conformation.

When Na^+ is replaced by Ca^{++} in the incubation mixture, there is, if any, but a slow phosphorylation of the enzyme by ATP (Fig. 8). Under these slowly or non-phosphorylating conditions, ATP in concentrations as low as 0, 01 mM likewise induces a change in enzyme conformation as evidenced by production of ouabain affinity (Fig. 9).

However, a comparison of the rates of ouabain binding induced by ATP, Mg^{++} and Na^+ or by ATP, Mg^{++} and Ca^{++} (Fig. 7 and 9) reveals that in the Ca^{++} system the rate is markedly lower.

The tempting conclusion that, before phosphorylation, already the formation of the enzyme substrate complex may induce a change of enzyme conformation, is substantiated by the results of our experiments with ATP-dialdehyde mentioned earlier.

---

1) The abrupt levelling of the ouabain binding curves at lower ATP concentrations may be assumed to be due to complete ATP hydrolysis at the time of levelling.

As may be recalled here, although ATP-dialdehyde is unable to phosphorylate the enzyme, it nevertheless induces ouabain binding in presence of Mg^{++} and Na^{+} as effectively as ATP (Fig. 3 and 10). Apparently, in (NaK)-ATPase as in many other proteins quaternary structure is modified when substrate is bound (19). Taken together, our observations suggest that the formation of the enzyme substrate complex induces a similar conformational change of the enzyme as the consecutive phosphorylation does or maintains. Thus, the enzyme substrate formation may contribute to the production of the conformational energy postulated to be important for the osmotic work of the enzyme.

Our studies on the dissociation of the enzyme effector complex (Fig. 10) revealed the important fact, that its stability essentially depends on the presence of Na^{+}, but is independent whether the ouabain binding conformation was induced by formation of enzyme substrate complex (with ATP-dialdehyde) or by phosphorylation (with ATP).

With both nucleotides, Ca^{++} is less effective than Na^{+} in stabilizing enzyme ouabain complex. Furthermore, the ouabain binding induced by ATP, Mg^{++}, Na^{+} is reduced to the rate obtained with Mg^{++} alone, if K^{+} is substituted for Na^{+} (Fig. 11). These observations let us assume that it is the sodium conformation of the phosphoenzyme which preferentially or solely binds ouabain.

One of the puzzling features of (NaK)-ATPase discovered by SCHWARTZ et al. (21), is that ouabain binding to the enzyme can be obtained with P_i and Mg^{++} as effectively as with ATP, Mg^{++} and Na^{+} (Fig. 11). Reduction of incubation temperature reduces the rate of binding in both systems to the same extent (Fig. 12). The activation energy of binding is in both cases 25 kcal/mole. Moreover, after removal of excess ouabain, the enzyme effector complexes independently on way or phase of formation dissociate with the same rate (Fig. 13). The half life time of both complexes is about 120 min.

These results suggest the conclusion that in both media ouabain becomes bound to the same enzyme conformation apparently induced or maintained by phosphorylation. As shown by POST and coworkers (6), ouabain in presence of Mg^{++} effects phosphorylation of the enzyme with inorganic phosphate to about the same level as that obtained with ATP, Mg^{++}, Na^{+}. ALBERS and coworkers (22) described that the electrophoretic mobilities of the phosphopeptides obtained by peptic and pronase digestion of both preparations are indistinguishable and that the chemical stability of the incorporated phosphate is also similar with both phosphoproteins. Thus, we are left with the conclusion that ouabain in presence of Mg^{++} and P_i can effect the esterification of a carboxyl group with phosphate yielding an acyl phosphate of high free energy of hydrolysis. If this conclusion should prove to be correct, the energy for this ester formation might be derived from the removal of a conformational constraint which could result from a moulding effect of ouabain on enzyme protein during combination in presence of Mg^{++} and P_i. This standard free energy change for the interaction of the enzyme-ouabain complex was calculated by TOBIN and SEN (23) to be about 9,5 kcal/mole suggesting a large conformational change of the enzyme. According to ALBERS et al. (24), the resulting conformational potential could eliminate the free energy difference between free enzyme and phosphoenzyme and adjust the free energy potential of the system to the thermodynamic requirements of phosphorylation.

Summing up this final part of the talk, we would say the following (Fig. 14). The formation of the enzyme substrate complex and the consecutive Na^+-catalyzed phosphorylation result in similar changes of enzyme conformation as evidenced by production of similar affinities to ouabain. In this way, the enzyme could gain conformational energy to be used for the vectorial work of Na^+ transport from in to out. In presence of Mg^{++}, ouabain, due to a moulding effect on enzyme protein, is supposed to confer on the enzyme a conformational constraint, of which the energy could be used for the incorporation of inorganic phosphate into the enzyme or possibly for the phosphorylation of the carboxyl residue in the active centre. As may be added here, this synthesis does not work if Na^+ is present (23). Presumably, Na^+ by its relaxation effect on conformational constraint removes that ouabain induced conformational potential of the enzyme protein thought to supply the free energy required.

Note added after the Symposium (May 10, 1971)

In a very recent paper, W. BOOS and A. S. GORDON (J. Biol. Chem. 246, (1971), 621), have shown that the galactose-binding protein of E. coli exists in two forms, representing different conformational states. According to these authors, the active transport of galactose through the cell membrane would involve preferential association of galactose to the highly binding active conformer on the outside of the cell membrane and subsequent dissociation of galactose from the less binding conformer on the inside of the membrane. The energy-dependent accumulation of galactose inside the cell might be explained by an energy-dependent continuous transformation from one state to the other followed by an energy-independent reversal of this reaction. Thus, the conclusion reached in the present lecture that the vectorial coupled transport of Na^+ and K^+ across cell membrane by (NaK)-ATPase is effected by one single enzyme entity oscillating between four anisotropic conformational states of different ionic selectivities and affinities, appears to be of more general significance.

REFERENCES

(1) SCHÖN, R., W. SCHÖNFELD, K.R.H. REPKE, Acta biol. med. germ. 24, (1970), 61

(2) SCHÖN, R., W. SCHÖNFELD, K.R.H. REPKE, Acta biol. med. germ., 25, (1970), 1

(3) SKOU, J.C., Physiol. Rev., 45, (1965), 596

(4) ALBERS, R.W., Ann. Rev. Biochem., 36, (1967), 727

(5) HOKIN, L.E., J. Gen. Physiol., 54, (1969), 327

(6) POST, R.L., S. KUME, T. TOBIN, B. ORCUTT, A.K. SEN, J. Gen. Physiol. 54, (1969), 306

(7) NEUFELD, A.H., H.M. LEVA, J. Biol. Chem., 245, (1970), 4962

(8) WOODIN, A.M., A.A. WIENEKE, Biochem. Biophys. Res. Commun., 33, (1968), 558

(9) PULL, I., Biochem. J., 119, (1970), 377

(10) UESUGI, S., N.C. DULAK, J.F. DIXON, T.D. HEXUM, J.L. DAHL, J.F. PERDUE, L.E. HOKIN, J. Biol. Chem. 246, (1971), 531

(11) ROBINSON, J.D., Arch. Biochem. Biophys., 139, (1970), 164

(12) NORBY, J.G., J. JENSEN, Biochim. Biophys. Acta, 233, (1971), 104

(13) BADER, H., A.K. SEN, Biochim. Biophys. Acta, 118, (1966), 116

(14) ISRAEL, Y., E. TITUS, Biochim. Biophys. Acta, 139, (1967), 450

(15) FORMBY, B., J. CLAUSEN, Z. physiol. Chem., 349, (1968), 909

(16) ROBINSON, J.D., Biochem. Biophys. Res. Commun., 42, (1971), 880

(17) WILSON, I.B., J. DAYAN, Biochemistry, 4, (1965), 645

(18) PECK, E.J., D.S. KIRKPATRICK, J.W. RAY, Biochemistry, 7, (1968), 152

(19) KLOTZ, I.M., N.R. LANGERMAN, D.W. DARNALL, Ann. Rev. Biochem., 39, (1970), 25

(20) FORMBY, B., J. CLAUSEN, Z. physiol. Chem., 350, (1969), 973

(21) LINDENMAYER, A. SCHWARTZ, Arch. Biochem. Biophys., 140, (1970), 371

(22) SIEGEL, G.J., G.J. KOVAL, R.W. ALBERS, J. Biol. Chem., 244, (1969), 3264

(23) TOBIN, T., A. K. SEN, Biochim. Biophys. Acta, 198, (1970), 120

(24) ALBERS, R.W., G.J. KOVAL, G.J. SIEGEL, Mol. Pharmacol., 4, (1968), 324

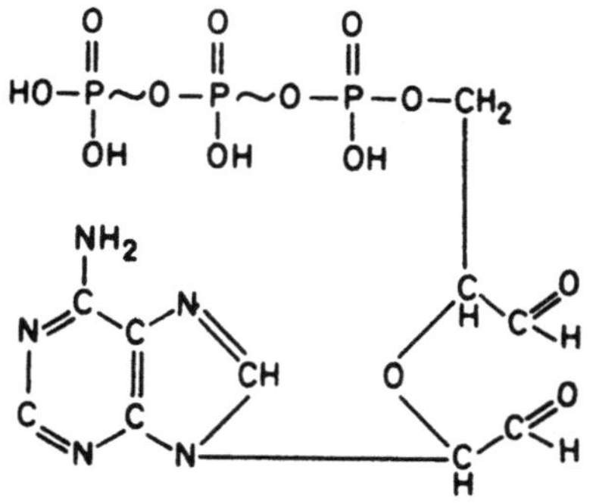

Fig. 1. "ATP-dialdehyde".

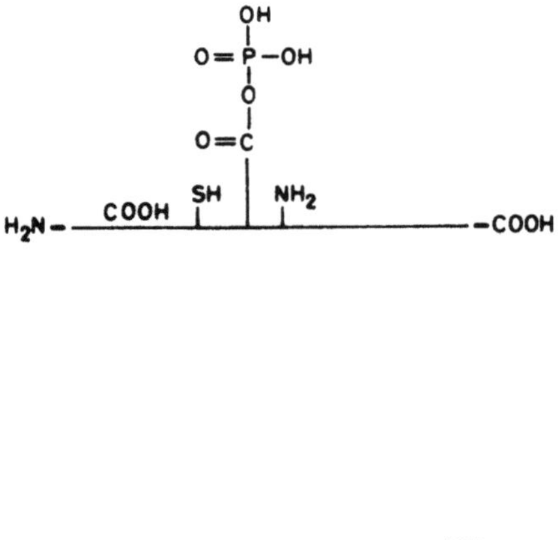

Fig. 2. Sequence of reactive groups and bonds in the primary structure of the phosphorylated third peptic fragment of the phosphorylated enzyme intermediate (POST et al.: J. Gen. Physiol. 54, 306s (1969)).

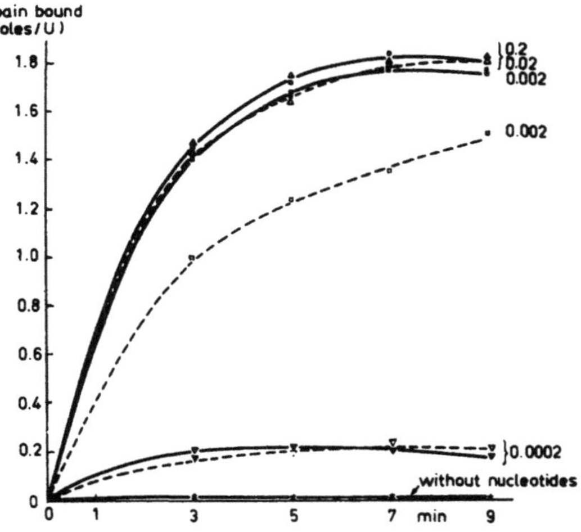

Fig. 3. Comparison between ATP (filled signs) and ATP-dialdehyde (open signs) as to the effectiveness for ouabain binding to (NaK)-ATPase. Concentrations (in mM) of nucleotides as indicated, $Mg^{++}4$, $Na^{+}80$, ^{3}H-Ouabain 0.0001, pH 7.4, 37°.

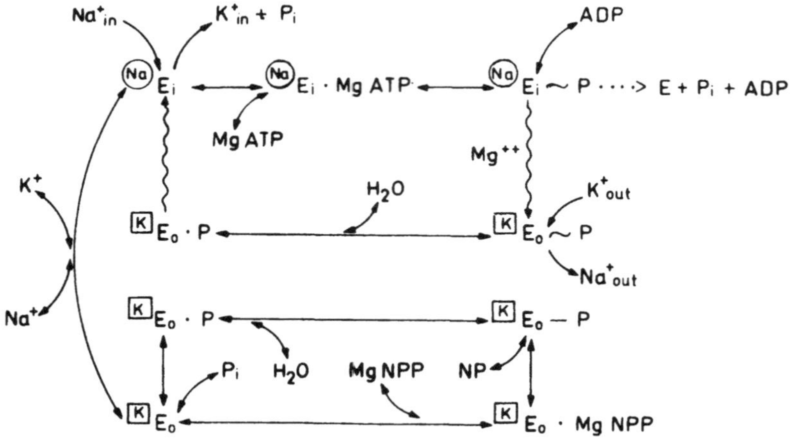

4. Scheme for hydrolysis of ATP or p-nitrophenyl phosphate (NPP) by one single enzyme protein existing as conformational isomers E_i (Na^+-form) or E_o (K^+-form) which use either ATP or NPP as substrate: reaction sequences for pH 7.4 and 37°.

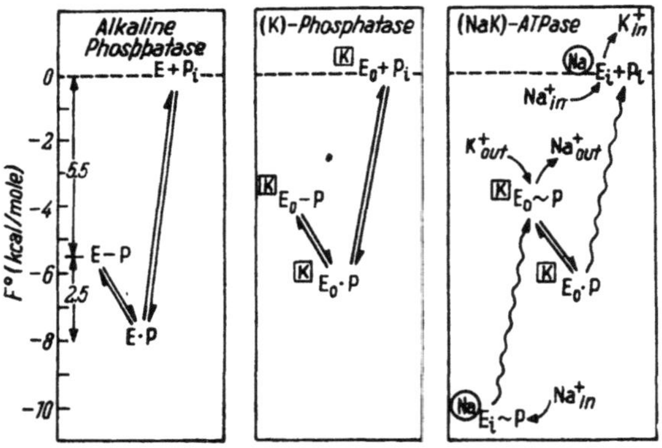

5. Comparison of free-energy diagrams for the hydrolysis of the phosphate group of three phosphoproteins.
Alkaline phosphatase: Figure taken from PECK et al. (Biochemistry 7, 152 (1968)). (K)-Phosphatase and (NaK)-ATPase: analogous, but speculative diagrams (ordinates in arbitrary units).

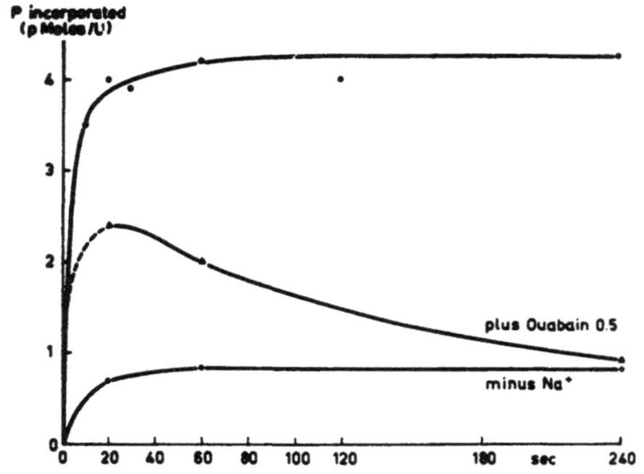

6. Effect of ouabain on Na^+-catalyzed phosphorylation of enzyme protein by ATP. Basic conditions: ATP-γ^{32}P 0.005 Mg^{++} 1, Na^+80, pH 7.4, 0° (concentrations in mM).

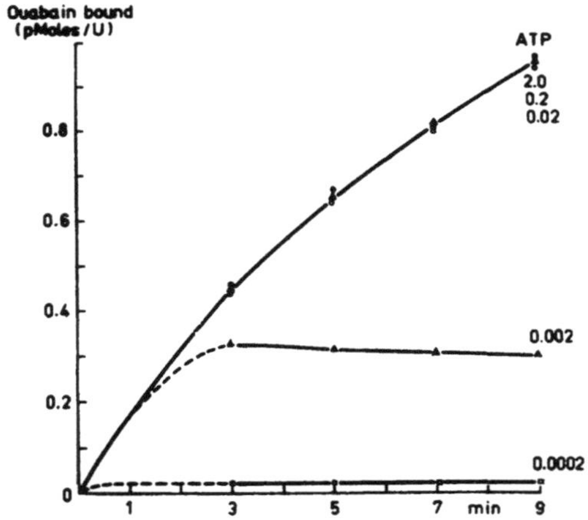

Fig. 7. Induction of a change in enzyme conformation by ATP under phosphorylating conditions, evidenced by production of affinity for the effector ouabain. ATP as indicated, Mg^{++}4, Na^+80, ^{3}H-ouabain 0.00002, pH 7.4, 37° (concentrations in mM).

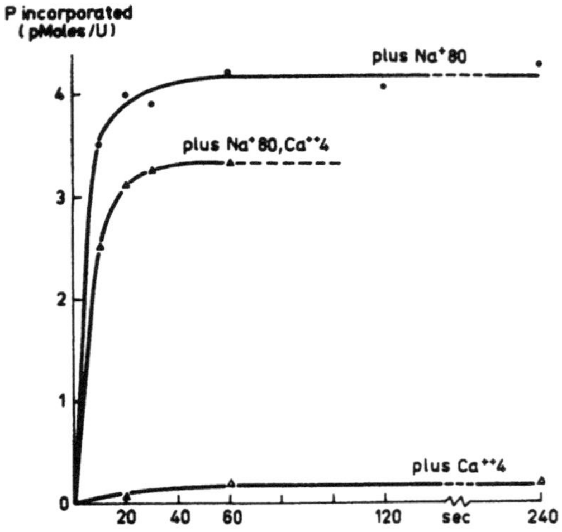

Fig. 8. Inability of Ca^{++} when replacing Na^{+} to catalyze phosphorylation of enzyme protein by ATP. Basic conditions: ATP-γ^{32}P 0.01, Mg^{++}1, pH 7.4, 0° (concentrations in mM).

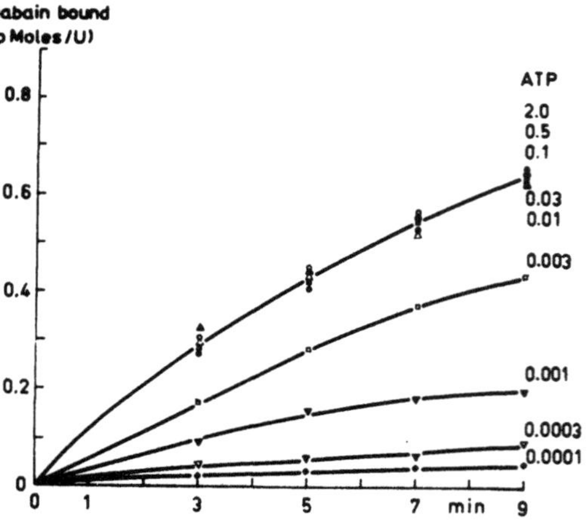

Fig. 9. Induction of a change in enzyme conformation by ATP under non-phosphorylating conditions, evidenced by production of affinity for the effector ouabain. ATP as indicated, Mg^{++}4, Ca^{++}1, ^{3}H-ouabain 0.00002, pH 7.4, 37° (concentrations in mM).

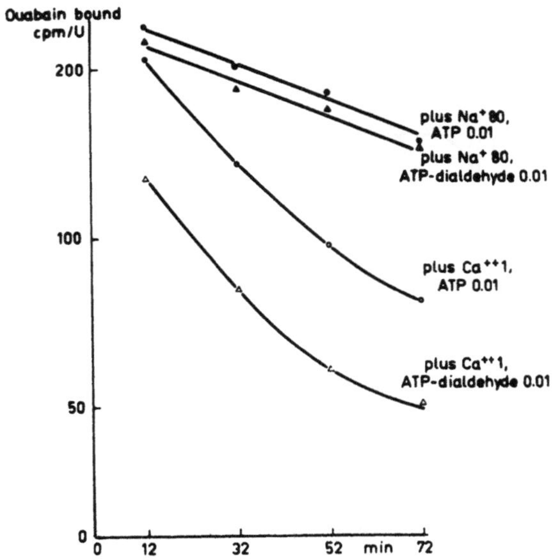

Fig. 10. Differences in exchange abilities of bound ^{3}H-ouabain against ouabain in the enzyme effector complexes involving either Na^+ or Ca^{++}. Binding of ^{3}H-ouabain 0.0001 to the enzyme protein was effected in the presence of Mg^{++}4 and the indicated other components at pH 7.4 and 37° during a 12 min period. After the twelfth minute ouabain 0.056 was added and the release of ^{3}H-ouabain followed (concentrations in mM).

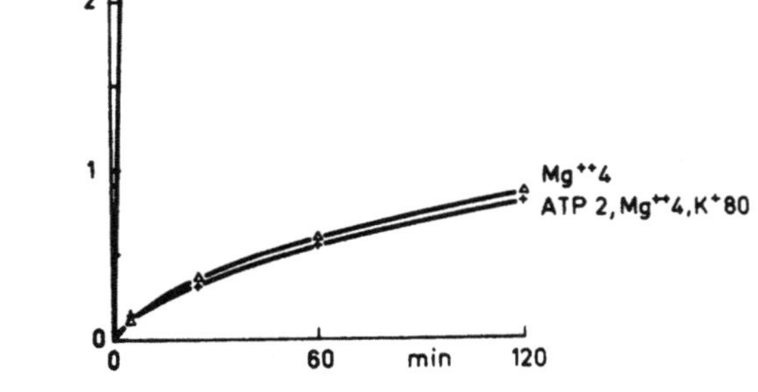

Fig. 11. Induction of a change in enzyme conformation under different conditions evidenced by production of affinity for the effector ouabain. Basic conditions: ^{3}H-ouabain 0.0001, pH 7.4, 37° (concentrations in mM).

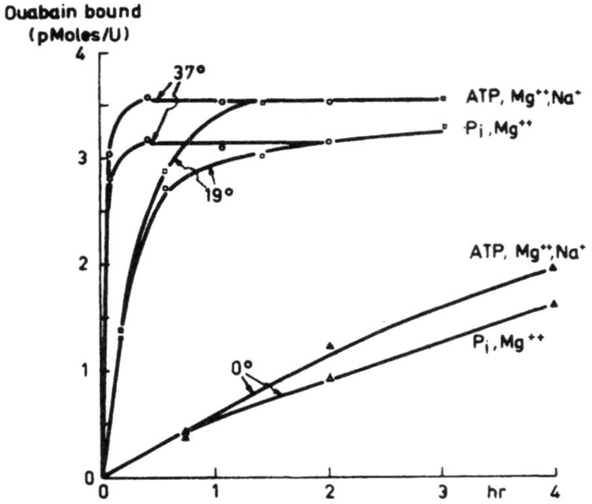

Fig. 12. Temperature dependence of ouabain binding to enzyme protein under different conditions. ATP 2, $Mg^{++}4$, $Na^{+}80$ or P_i 2, $Mg^{++}4$; ^3H-ouabain 0.0001, pH 7.4 (concentrations in mM).

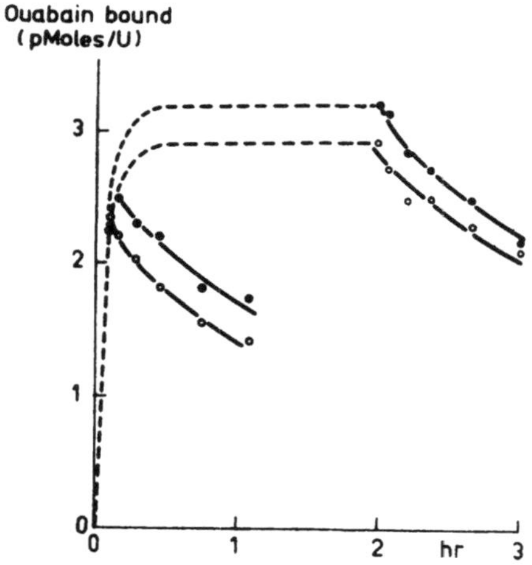

Fig. 13. Reversibility of ouabain binding to enzyme protein obtained under different conditions. Binding of ^3H-ouabain 0.0001 in presence of ATP 2, $Mg^{++}4$, $Na^{+}80$ (filled signs) or P_i 2, $Mg^{++}4$, (open signs) at pH 7.4 and 37^{0}. After removal of ^3H-ouabain excess, the reversibility of binding was followed when incubating the samples with ATP 2, $Mg^{++}4$, $Na^{+}80$, $K^{+}5$ at pH 7.4 and 37^{0} for periods indicated (concentrations in mM).

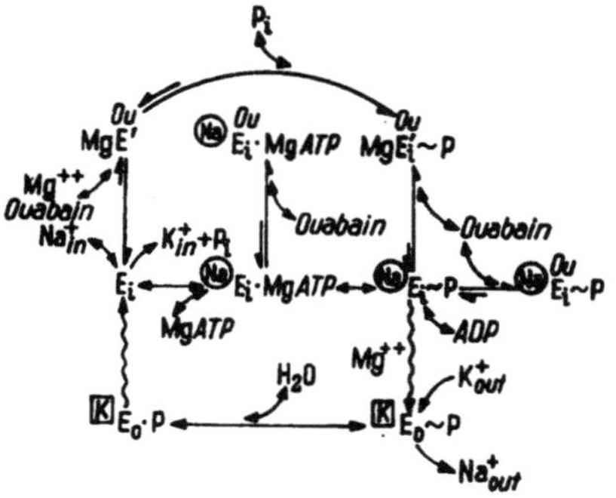

Fig. 14. Scheme on the combination of ouabain (Ou) with (NaK)-ATPase protein in the initial stages of activity in consequence of a conformational change of protein either by formation of enzyme substrate complex or by phosphorylation on the one side and in consequence of a moulding effect of ouabain on enzyme protein during combination in presence of Mg^{++} and P_i on the other side.

Institute of Physiology, University of Aarhus, Denmark

G. NORBY, J. JENSEN

BINDING OF ATP TO $(Na^+ + K^+)$-ACTIVATED ATPase, NUMBER OF BINDING SITES AND ENZYME-ATP DISSOCIATION CONSTANT

Since much evidence exists for relating $(Na^+ + K^+)$-activated ATP hydrolysis to active cation transport it is of interest to study the intermediate steps involved in this ATP hydrolysis in order to understand the mechanism of active cation transport SKOU (1), ALBERS (2), HEINZ (3) and WHITTAM, WHEELER (4).
In the following we report the results of some studies concerned with the binding of the ATP molecule to ox brain microsomal ATPase prepared as described by SKOU, HILBERG (5). The binding was investigated by a rapid dialysis rate technique (COLOWICK, WOMACK (6)) at 2° C, pH 7.4, ionic strength 0.073-0.084 M in the presence of 10 mM EDTA (free $Mg^{++} <$ 5 nM, total ATP from 0.1 to 3 μM). Details of the general experimental procedure have been published elsewhere (7).
The results of the ATP-binding studies were interpreted according to the following simple model:

$$E + ATP \rightleftharpoons EATP$$

At equilibrium,

$$\frac{[E]' \cdot [ATP]}{[EATP]} = K'_{diss} \qquad (1)$$

or

$$\frac{([E_t] - [EATP]') \cdot [ATP]}{[EATP]'} = K'_{diss} \qquad (2)$$

in which $[E_t]$, $[EATP]'$ and $[ATP]$ denote the concentration of total binding sites, bound ATP (B) and free ATP (F), respectively. K'_{diss} is the apparent dissociation constant.
Equation (2) is transformed to:

$$B = [E_t] - K'_{diss} \cdot \frac{B}{F} \qquad (3)$$

and the results are plotted in a Scatchard-type plot (Fig. 1). K'_{diss} is obtained from the slope and E_t from the intercept with the ordinate axis.

Relationship between ATP-Binding Capacity (E_t) and ATPase Activity (V):
The microsomal ATPase preparations used in present and previous studies (7) exhibit hydrolytic activity in the presence of Mg alone (Mg-ATPase) as well as ($Na^+ + K^+$)-stimulated activity (NaK-ATPase). In our previous investigations (7) we demonstrated that E_t is proportional to the NaK-ATPase activity when the latter was varied by partial heat denaturation. In these preparations the ratio V_{NaK}/V_{Mg} varied from 3.2 to 6.7.
The above mentioned studies have now been extended to include enzyme preparations in which the NaK- and/or Mg-ATPase is partially inhibited with ouabain, p-chloromercuribenzoate (PCMB), N-Ethylmaleimide (NEM) or ethacrynate (ETHA).
The Mg-ATPase activity was either unaffected or decreased by the inhibitors and the ratio V_{NaK}/V_{Mg} varied from about 2 (ouabain-inhibited) to about 50 (NEM- or ETHA-inhibited preparations). It is apparent from Fig. 2 that the results obtained support our previous suggestion that the ATP-binding measured is largely or exclusively confined to NaK-ATPase. The observation that the ouabain-binding capacity of native and heat denatured enzyme preparations was almost equal to the ATP-binding capacity (8) further demonstrates the selectivity of the ATP binding measured.
The molecular activity of NaK-ATPase calculated from the slope of the straight line in Fig. 2 is 6700 molecules of ATP transformed per binding site per minute.

Specificity of the ATP-Binding Site of NaK-ATPase:
Information about the specificity of the ATP-binding site was obtained by investigating the influence of ATP congeners on ATP^{32} binding. Pyrophosphate, adenine, adenosine, and AMP in concentrations up to about 3 orders of magnitude higher than that of free ATP were not able to displace ATP from the ATP-binding site. For those able to displace ATP from the enzyme apparent dissociation constants were calculated (Table 1) according to a model in which it is assumed that the substance in question competes with ATP for the same binding site (9).
The nucleotide specificity of the ATP-binding site appearing from Table 1 is similar to the substrate specificity of ($Na^+ + K^+$)-activated ATPase JENSEN, NORBY (9) thereby lending further support to the suggestion that this binding site is identical with the substrate site of ($Na^+ + K^+$)-activated ATPase.
Moreover, we feel justified to conclude that the 6-amino group on the purine or pyrimidine ring as well as the β-phosphate group are essential for binding to

Table 1

APPARENT DISSOCIATION CONSTANTS OF ADP-, GTP-, ITP- AND CTP-ENZYME COMPLEXES EXPRESSED RELATIVE TO THAT OF THE ATP-ENZYME COMPLEX

K'_{ATP}	K'_{ADP}	K'_{GTP}	K'_{ITP}	K'_{CTP}
1	5	575	865	250

the substrate site of $(Na^+ + K^+)$-activated ATPase, and that binding of ATP also involves a link between the γ -phosphate group and the enzyme.

The Effect of K^+ and Na^+ on the Apparent Dissociation Constant, K'_{diss}:
Determination of K'_{diss} on the native enzyme preparations revealed a variation of this value between 0.17 and 0.35 μM. Experiments in which K^+ was added to washed enzyme preparations (contaminated with about 40 μM K^+ and 100 μM Na^+) revealed that K'_{diss} was dependent upon the K^+-concentration (Fig. 3). These observations led us to propose the formation of a K^+-enzyme complex which bind ATP with a lower affinity than the potassium-free enzyme (7). According to this model the following dissociation constants were calculated:
$K^{E\text{-}ATP} = 0.12\ \mu M$, $K^{KE\text{-}ATP} = 0.7\ \mu M$, $K^{E\text{-}K} = 87\ \mu M$ and
$K^{EATP\text{-}K} = 500\ \mu M$ (ref. 7).
The effect of Na^+_{diss} was studied in a similar way and the results obtained are shown in Fig. 4. In contrast to what was observed for K^+, an increase in Na^+ concentration tends to promote a decrease in K^+_{diss}, the effect being highly dependent on the K^+ concentration. This, as well as the tendency for K'_{diss} to reach a level independent of the K^+ and Na^+ concentrations, can be explained by assuming 1) that there is a binding of Na^+ to a site on the enzyme, 2) that Na^+ on this site is without influence on the affinity for ATP or, in other words, that the enzyme and enzyme-ATP complex have the same affinity for Na^+, and 3) that binding of Na^+ excludes binding of K^+ to the site at which it exerts its effect on the binding of ATP. The Na^+ - and K^+ - site might be the same site.

Conclusion

The results reported here are confined to what may be called the first step in the hydrolysis of ATP by $(Na^+ + K^+)$-activated ATPase. The observations that ATP will bind to the enzyme with a high affinity, in the absence of Mg, and that K^+

affects the binding process, require a reconsideration of earlier conclusions drawn from kinetic studies concerning the enzyme-substrate-cation interaction.

REFERENCES

(1) SKOU, J.C., Physiol. Rev., 45, (1965), 596
(2) ALBERS, R.W., Ann. Rev. Biochem., 36, (1967), 727
(3) HEINZ, E., Ann. Rev. Physiol., 29, (1967), 21
(4) WHITTAM, R., K. P. WHEELER, Ann. Rev. Physiol., 32, (1970), 21
(5) SKOU, J.C., C. HILBERG, Biochim. Biophys. Acta, 185, (1969), 324
(6) COLOWICK, S.P., F.C. WOMACK, J. Biol. Chem., 244, (1969), 774
(7) NORBY, J.G., J. JENSEN, Biochim. Biophys. Acta, 233, (1971), 104
(8) HANSEN, O., J. Jensen, J.G. NORBY, Nature, submitted for publ.
(9) JENSEN, J., J.G. NORBY, Biochim. Biophys. Acta, 233, (1971), 395

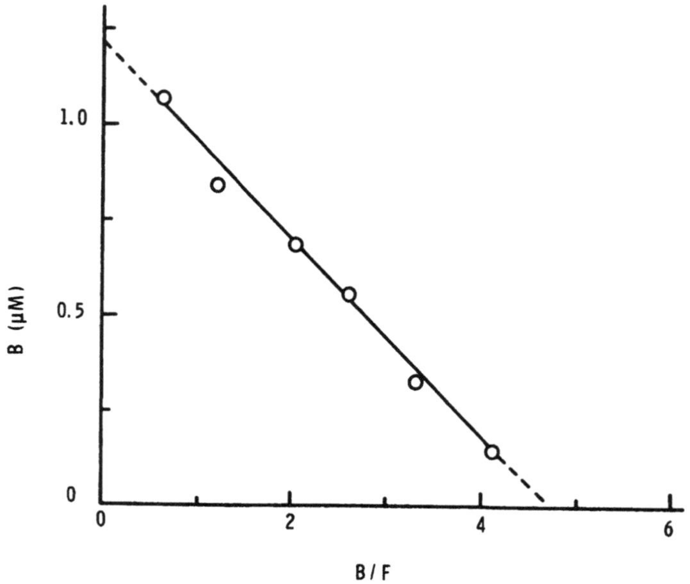

Fig. 1. Scatchard-type plot of a typical ATP-binding experiment (7).

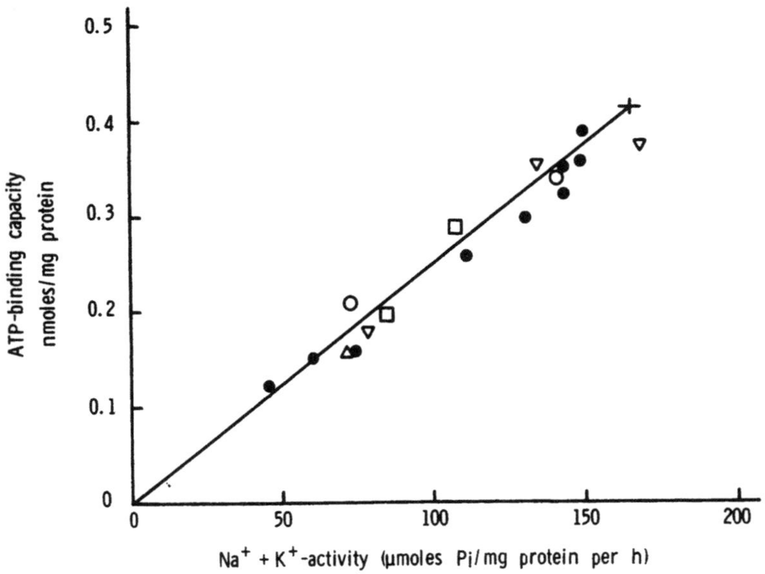

Fig. 2. Relationship between binding capacity and (Na^+ + K^+)-activity for heat-denatured preparations (•), and preparations partially inhibited with ouabain (o), PCMB (△), NEM (□) or ethacrynate (▽). The straight line is drawn through the origin and the point (+) representing the average ± SEM of 14 native preparations (unpublished).

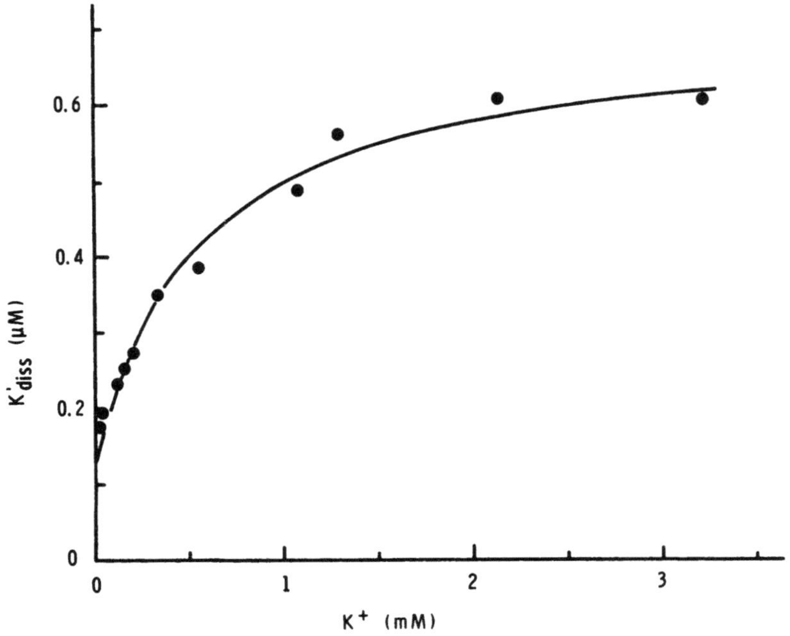

Fig. 3. The effect of K^+ on the apparent dissociation constant of enzyme-ATP complex. The curve is calculated according to the model(7) and dissociation constants mentioned in the text.

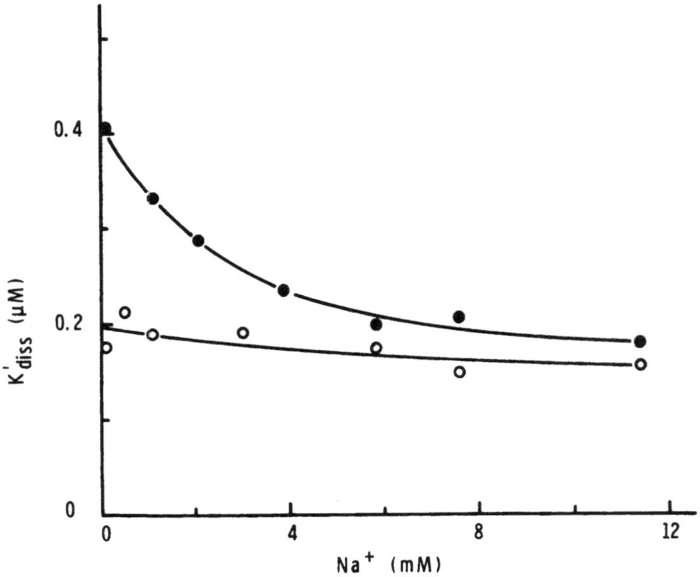

Fig. 4. Relationship between the apparent dissociation constant of enzyme-ATP complex and Na^+ concentration at $K^+ = 72\,\mu M$ (o) and $560\,\mu M$ (•) (unpublished).

Timiriazev Institute of Plant Physiology, USSR Academy of Sciences, Moscow

M. S. KRASAVINA, E. I. VYSKRIBENTSEVA

SOME PROPERTIES OF ATPase OF PLANT TISSUES

Unlike animal tissues the study of ATPase activity of plants has not been practised until recently so that its properties have not been agreed upon heretofore. In this connection there has been started the study of ATPases specific to functionally different plant tissues. The most of our experiments have been conducted on Cucurbita roots. An enzyme preparation was obtained by homogenation of the tissue in a five - fold volume of tris-sucrose buffer, precipitation by ammonium sulphate and dialysis against tris-buffer or distilled water. ATPase reaction of such a preparation has an optimum activity at pH 4.5-5 and shows less pronounced increase in activity at pH 10. To lay open the processes capable of regulating the intracellular enzyme activity, we assumed it to be inexpedient to carry out experiments under conditions of maximum enzyme activities; it is why the majority of experiments were carried out at pH 7.2 which is approximate to the physiological conditions of the root tissues.

A response of the ATPase reaction to the activators and inhibitors has been studied (Table 1).

The common inhibitor of all phosphatases, viz., NaF is found to suppress enzyme activity almost completely. Of interest is also the stimulation of ATPase in the presence of DNP in the reaction mixture, as well as a weak inhibitory effect in the presence of NaN_3. These results suggest that the hydrolysis of ATP is similar to that in mitochondria of animal cells and the utilization of energy of ATP in Cucurbita roots may occur by inverting the reaction of oxidative phosphorylation.

Strophanthin - sensitive ATPase has been found in Cucurbita roots. Introduction of this inhibitor into the incubation mixture results in a 20-25 per cent suppression of ATPase reaction. Such a dependence of ATPase on strophanthin is of special interest for plant physiologists because the presence of a strophanthin-sensitive ATPase in plant tissues is so far called in question (1, 2, 5). Meanwhile a strophanthin - sensitive ATPase in animal tissues is closely coupled with the

Table 1

ATPase ACTIVITY OF CUCURBITA ROOTS

Treatment	μmoles P_i / mg protein· h	Treatment[1)]	μmoles P_i / mg protein· h
-	2.10		1.62
NaF 10 mM	0.20	Mg^{2+} 3 mM	3.89
DNP 0.25 mM	3.31	Ca^{2+} 3 mM	3.17
NaN_3 10 mM	1.82	Mg^{2+} + Na^+ 50 mM + K^+ 10 mM	4.67
strophanthin 0.05 mM	1.71	Mg^{2+} + Na^+ 50 mM + K^+ 10 mM + stroph.	4.04

1) Homogenation and dialysis with 5 mM EDTA.

The tissues were homogenized in 5 volumes of icecold tris-HCl buffer pH 7.2 (20 mM tris, 0.25 M sucrose) precipitated by ammonium sulphate (sat. 0.7) and dialysed against 3 mM tris buffer pH 7.2 Incubation mixture contains 1.2 mM ATP, 20 mM tris -HCl pH 7.2 and other components shown in the table. The incubation time was 1 hr at 30°. During this period the reaction was linearly with time.

transport of K and Na cations through cellular membrane. And if the mechanisms of cation active transport in animal and plant tissues are the same, transport ATPases must have been especially specific to the root as an organ specialized in uptake, transport and excretion of mineral elements. As it is evidenced by the data of Table 1, a highest activity of ATPase develops in the concurrent presence of magnesium, sodium and potassium ions. An ATPase activated by Na^+ and K^+ is responsible for about 17-20 per cent of this activity. It is of interest that strophantin inhibits by 80 per cent just the above part of the enzyme activity. Bivalent cations stimulate ATPase considerably. Thus, a 3 mM solution of Mg^{++} increases about twice the activity while the effect of the same concentration of Ca^{++} is weaker. Higher cation concentrations inhibit the enzyme.
The ability of ATPase to be stimulated or inhibited by cations is environment-dependent. Ph is one of the factors affecting the response of ATPase to cations (Table 2).

Table 2

THE EFFECT OF pH ON THE ATPase SENSITIVITY TO CATIONS

Treatment	μmoles Pi / mg protein· h pH 5.1	pH 9.0
-	2.52	1.08
Mg^{++} 3 mM	3.06	0.96
Mg^{++} + Na^{+} 50 mM + K^{+} 10 mM	3.52	0.62
Ca^{++} 3 mM	2.36	1.35

The reaction mixture: 3 mM ATP, 50 mM tris and different cation.

The optimum activity of Mg^{++} or (Na^{+} - K^{+})-stimulated ATPase was found to be at lower pH values than that of Ca^{++}-stimulated ATPase. Thus, at pH 5.1 both Mg^{++} and Na^{+} plus K^{+} activate the ATPase, while at pH 9 these cations in the same concentration even inhibit the enzyme activity. Conversely, Ca-cation stimulates the reaction occurring at alkaline pH.

The ion status of a tissue also affects the dependence of ATPase reaction on cations. Thus, when the 5 mm-segments of Cucurbita stem have been preliminarily held in a solution of KCl, a surface ATPase is stimulated by potassium and sodium cations whereas the enzyme activity is badly inhibited by strophanthin (Table 3). The above cations exhibit but quite different effect on the ATPase activity of segments pretreated with a solution of NaCl. In this case the "sensitivity" of the enzyme to monovalent cations must have been raised so that the same concentrations of Na^{+} and K^{+} are inhibitory. Strophanthin probably hinders the interaction between the potassium excess and the enzyme, prevents its inhibitory effect and reactivates ATPase.

Thus, the response of ATPase reaction to various substances probably depends on both the intracellular pH and ion composition and on those of the environmental medium. The fact that ATPase in Cucurbita roots may be stimulated by a joint action of K^{+} and N^{+} and inhibited by strophanthin suggests that the ATPase of plant tissues resembles to the animal transport ATPases described by SKOU (7). These evidences are of great interest since they help to understand the mechanism of uptake and transport of monovalent cations in a plant cell.

Table 3

THE ATPase ACTIVITY OF STEM SEGMENTS TREATED WITH KCl OR NaCl

Treatment	ATPase activity $\left[\frac{\mu moles\ P_i}{g\ tissues \cdot h}\right]$ after preincubation in	
	KCl	NaCl
-	1.62	2.42
Mg^{++} 3 mM	1.60	2.40
Mg^{++} + Na^{+} 50 mM + K^{+} 10 mM	2.78	1.80
Mg^{++} + Na^{+} + K^{+} + strophan. 0.05 mM	1.26	2.28

500 mg of 5 mm stem segments were preincubated in a 10 mM solution of KCl or NaCl at 0° C during 18 hours. A surface ATPase activity was then determinated by placing segments in an incubation mixture containing 1.2 mM ATP, 20 mM tris pH 6.0 and different cations.

ATPases participating in a transmembrane transport of cations are as a rule characteristic of the cell membrane. We isolated the fractions of cell walls, plastids, mitochondria and microsomes by differential centrifugation (Table 4).

Table 4

THE ATPase ACTIVITY IN DIFFERENT SUBCELLULAR FRACTIONS

Fractions	μmoles P_i/mg protein · h	
	Mg^{++} 3 mM	Mg^{++} 3 mM Na^{+} 50 mM K^{+} 10 mM
1000 g (nucleiand cell walls)	4.3	5.2
4000 g (plastids)	1.8	2.1
20000 g (mitochondria)	2.2	2.5
105 000 g (microsomes)	1.3	2.7

Incubation with 1.2 mM ATP, 20 mM tris-HCl pH 7.2.

The protein of a fraction precipitating at 1000 g is found to possess the highest specific ATPase activity. In the main, this activity probably belongs to fragments of the cellular membranes that permeate the cell walls. This enzyme activity is stimulated by sodium and potassium ions. ATPase considerably stimulated by monovalent cations is found also in a microsomal fraction. An addition of Na^+ and K^+ to this fractions results in a two- or threefold increase in enzyme activity. Since the (Na^+-K^+)-ATPase is bound up with the membrane elements, in particular with the external cytoplasmic membranes, it has been found expedient to carry out experiments on an intact root system. These experiments have been carried out to study the participation of the ATPase in cation transport. This problem is at the very point of discussion by plant physiologists, inasmuch as some investigators failed to find any strophanthin influence upon the fluxes of K^+ into plant cells (4, 6). We also failed to disclose any strophanthin influence upon the influx of K^+ into the roots of intact plants grown on the Knop solution containing 0. 5 mM K^+ and 0. 5 mM Na^+ (Fig. 1).

However, if the roots are grown under conditions of a sharply decreased potassium concentration and an increased sodium content in the medium, the influx of potassium into the cells is strophanthin-dependent (Table 6).

Table 6

THE ESTIMATED AND PREDICTED K^+-CONCENTRATION IN THE ROOTS OF CUCURBITA PLANTS GROWN WITH LOW POTASSIUM ENVIRONMENTAL CONTENT

experimental condition	E[mV]	K^+-content $\left[\frac{\mu \text{eqiv}}{\text{ml tissue water}}\right]$	
		K^+pred	K^+ estim
initial	-140	4.6	32.2
after 2 hrs			
in 2 mM K^+			50.7
in 2 mM K^+ + stroph.			39.0

5-day seed germination in dark was followed by seedlings grown for 5 days on nutrient solution containing 0. 02 mM K^+ and 1. 0 mM Na^+. Other conditions were the same as in Table 5.

A similar phenomenon is observed for Na^+ when it is transported from the cells containing a low sodium concentration (Table 5). Therefore an assumption was made that the strophanthin influence is develop in transport of the cations under conditions of their low concentration only, when the transport probably occurs against the gradient of electrochemical potential.
To estimate the gradient of electrochemical potential in the root cells the membrane potential was measured and intracellular concentration of potassium and sodium ions was calculated by the Nernst equation. These concentrations should be attained if these cations are passively distributed between the cell and the medium. It was found that potassium distribution is close to electrochemical equilibrium. At the same time a high Na gradient was found to be directed from the medium into the cell; therefore a "pump" must have to be in the cell so as to transfer Na uphill. It is in this case that the effect of strophanthin is exhibited. A contrary phenomenon takes place for potassium - depleted plants. In this case a gradient of electrochemical potential for K^+ is directed from the cell into the medium; therefore, a K^+ - influx mechanism is required to maintain an intracellular potassium level. And this is inhibited by strophanthin.
Thus, the strophanthin inhibition effect develops only when the ionic fluxes are directed uphill. This fact is in strict accordance with the USSING criterion of active transport (9). Proceeding from this the (Na-K)-activated ATPase in the root tissues may be considered as a mechanism of cation active transport.

REFERENCES

(1) ATKINSON, M.R., G.M. POLYA, Austr. J. Biol. Sci., 20, (1967), 1069

(2) BROWN, H.D., A. ALTSCHUL, Biochem. Biophys. Res. Comm., 15, (1964), 479

(3) BONTING, S.L., In E.E. BITTAR, ed., Membranes and Ion Transport, London, (1970), 286

(4) CRAM, W.J., J. Exp. Bot., 19, (1968), 611

(5) FISHER, J., T.K. HODGES, Pl. Physiol., 44, (1969), 385

(6) HODGES, T.K., Nature, 209, (1966), 425

(7) SKOU, J.C., Biochim. Biophys. Acta, 23, (1957), 394

(8) SKOU, J.C., Prog. Biophys. Mol. Biol., 14, (1964), 134

(9) USSING, H.H., Physiol. Rev., 49, (1949), 127

experimental condition	cation content µeq/g tissies	
	K^+	Na^+
initial	68.0	4.0
after 2 hrs		
in 2mM K^+	77.5	2.7
in 2mM K^+ + strophanthin	77.0	6.8

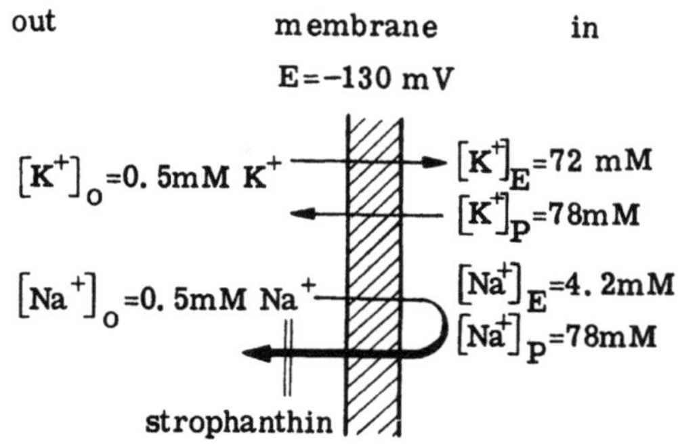

Fig. 1. Effect of strophanthin on the cation content in plant root.
Seeds germinated vor 5 days at 25 °C in dark on the filter paper moisted with tap water. The following 5 days the young seedlings were grown on Knop nutrient solution containing 0.5 mM K^+ and 0.5 mM Na^+. Cation contents - $[C]_E$ - were determined by flame photometry in cold-water extract of roots immediately or after K^+-uptake from the nutrient solution containing 2 mM K^+ and 0.5 mM Na^+ with or without strophanthin. The electropotential measurements were made with glass microelectrodes by usual technique. The predicted cation concentration - $[C]_P$ - was calculated by the Nernst equation.

Physiologisch-chemisches Institut der Karl-Marx-Universität Leipzig

D. DETTMER, F. MÜLLER, H.-J. GLANDER

THE INTERACTION OF MONOSACCHARIDES WITH INTESTINAL (Na^+-K^+)-ATPase

In 1958 RIKLIS and QUASTEL (1) detected effects of monovalent cations on glucose absorption. They concluded that sodium is essential for glucose transport. Since that time this effect was intensively investigated. With the accumulation technique CRANE (2) could demonstrate the to-day well known effect of Na^+ on the affinity of the transport system for its substrate (sugar). For 6-deoxy-glucose as substrate the apparent K_M increased about 200-fold by reducing the Na^+-concentration from 145 Equ/l to zero, while V_{max} was always constant. CRANE postulated a ternary complex consisting of Na^+, glucose and the carrier molecule which passes through the membrane. The affinity of the carrier to the monosaccharide is regulated by the Na^+-concentration and vice versa. The Na^+-sugar-carrier-complex is built up on the mucosal surface of the brush border, passes through the membrane and dissociates at the intracellular face. The obligatory asymmetry of the Na^+-distribution is maintained by the active extrusion of Na^+ by the Na^+-pump. The inhibition of active sugar transport by cardiac glycosides like ouabain was interpreted as an effect on this Na^+-pump. Later SCHULTZ and ZALUSKY (3) localized this pumping on the serosal membrane of the epithelial cell.

This model suggested a twofold involvement of Na^+ in the active sugar transport:

1. Direct coupling between sodium and sugar influx,
2. active Na^+-extrusion for maintenance the low intracellular Na^+-concentration.

Thus the enzymatic basis of active Na^+-transport - the (Na^+-K^+)-ATPase - is involved only in an indirect manner.

In an earlier publication we have described experimental results which can hardly be explained by this model DETTMER et al. (4). The activity of the (Na^+-K^+)-ATPase was decreased under conditions of an increased sugar transport. In accordance with other authors we also found an inhibition of the (Na^+-K^+)-ATPase by glucose. Besides there exists no simple correlation between sugar transport and (Na^+-K^+)-ATPase activity. On the basis of these results we discussed the possible role of the phosphorylated intermediate of the (Na^+-K^+)-ATPase for active sugar transport.

Also some results in the literature can be interpreted in the same manner - factors which dephosphorylize the intermediate, show inhibitory effects on the sugar transport. BOSACKOVA and CRANE (5) found an inhibition of sugar transport by Li^+, Rb^+, Cs^+, NH_4^+, i. e. ions which all dephosphorylate the intermediate. On the other hand, extracellular Na^+ is essential for active sugar transport but inhibits the ATPase SCHATZMANN (6).

KIMMICH (7) investigated the sugar accumulation in isolated intestinal cells. He concluded, that the Na^+-gradient is not a condition for uphill monosaccharide transport. On the basis of inhibition experiments with ouabain, oligomycin and dinitrophenol he discussed a direct involvement of the (Na^+-K^+)-ATPase in active sugar transport.

The key to this problem is to investigate interactions of glucose with the (Na^+-K^+)-ATPase.

The first step was the isolation and purification of a (Na^+-K^+)-ATPase from intestinal mucosa of rats. We have got a preparation of good stability (one month and more at -20°), good activity (about 2 I. U. per mg protein, measured at 37°) and a ratio of Mg^{++} to $Mg^{++}-Na^+-K^+$ activity of 1:1. The basic enzymatic properties of our preparation were similar to those published by other authors. Our (Na^+-K^+)-ATPase was inhibited by ouabain, phloridzin and SH-reagents. Of special interest was the glucose effect. Glucose inhibits intestinal (Na^+-K^+)-ATPase (Fig. 1). The inhibitory effect is dependent on the sodium and potassium concentration and the sodium-potassium concentration ratio. The greatest inhibitory effect is detectable at low Na^+- and K^+-concentrations. Further we observed an other glucose effect: the Na^+-activation curve was altered by increasing K^+-concentrations to a sigmoid shape. If glucose is present simultaneously the Na^+-activation curve becomes sigmoidal at lower K^+-concentrations. Galactose, mannose, and 3-O-methyl-glucose act principally in the same manner, whereas arabinose is without any influence. The kinetic effect is most clear demonstrable in the Eadie plots (Fig. 2). Without glucose, the HILL-coefficients for the Na^+ activation curve increase linearly between 10 - 100 mM K^+, but with 10 mM glucose the increase is steeper with nearly maximal values between 40 - 60 mM K^+.

PRIESTLAND and WHITTAM (8) studied the alteration of the K^+-activation curve with red cell ghosts. With this preparation a strict separation of the K^+-

and Na^+-activating sites was possible. The sigmoidicity occurs only, when Na^+ acts on the extracellular K^+-activating site. In this sense our results can be interpreted as effects of extracellular sodium and glucose on the ATPase. Extracellular Na^+ and glucose can be regarded as negative allosteric effectors which increase the cooperativity. In order to study the molecular basis of the glucose effect it is necessary to analize the action of glucose on the several steps of the (Na^+-K^+)-ATPase. Fig. 3 shows a model of reaction sequences for the phosphorylation of (Na^+-K^+)-ATPase Post et al. (9). There are various steps: ATP:ADP exchange - Mg^{++}, Na^+-dependent building of a phosphorylated intermediate - its conversion to a second phosphorylated intermediate- and a K^+-dependent dephosphorylation. The Na^+-dependent phosphorylation is reversible but the interconversion of the two intermediates is irreversible. As possible phosphorylated products KAHLENBERG et al. (10) isolated γ-glutaminyl phosphate and AHMED AND JUDAH (11) serine phosphate.

The first step, the Na^+-dependent phosphorylation can be studied with labelled ATP. There are two kinds ob labelling - an unstable and a stable one SKOU (11a). It is most likely that the unstable labelling represents an acyl phosphate like γ-glutaminyl phosphate. Possibly serine phosphate causes the stable labelling. The inhibition of the (Na^+-K^+)-ATPase by diisopropylfluorophosphate SACHS (12) is in accordance with this assumption.

The K^+-dependent dephosphorylation step can be investigated separately as K^+-stimulated p-nitrophenyl-phosphatase. This is a further evidence for the presence of an hydroxyl group in the active center.

Inhibition experiments with SH-reagents allow to postulate a SH-group being involved in the catalytic process. In our experiments the K^+-stimulated phosphatases were inhibited by mersalylic acid, while is has only little effect on the Na^+-dependent phosphorylation. SIEGEL and ALBERS (13) described similar effects by NEM and localized the SH-group between the two phosphorylated intermediates. Cardiac glycosids, such as ouabain, inhibit the dephosphorylation at low concentrations SKOU (11b), HOKIN (14) investigated the inhibition of (Na^+-K^+)-ATPase by glycosides and haloderivatives of glycosides. Strophanthidin-3-haloacetates caused inhibition being irreversible to about 70 %. HOKIN discussed alkylation of the enzyme at a SH-group near the site, where the cardiac glycosides act.

WEISS (15) gives a plausible explanation for the reaction between the functional groups in analogy of the molecular mechanism of 3-glycerol aldehyde phosphate dehydrogenase. He discussed a thioacyl bond between the γ-carboxyl group of glutaminic acid and a SH-group leading to conformational changes of the whole enzyme. In our studies glucose had little influence on the total amount of phosphate incorporated in the first reaction step, the forming of a phosphorylated intermediate, but the break down was clearly retarded. (Fig. 4). Thus we could demonstrate that glucose has a stabilizing effect on the intermediate. On the other hand glucose inhibits the phosphatase activity, and the K-activation curve of the phosphatase is influenced by glucose in the same manner as the Na^+-activation curve of the (Na^+-K^+)-ATPase. (Fig. 5). Both glucose effects - the inhibition of the overall activity and the altering of the kinetic behaviour can be localized at the phosphatase site which represents the dephosphorylation step and the extracellular site of the (Na^+-K^+)-ATPase.

Most likely (Na^+-K^+)-ATPase is present in most membranes. ROSENBERG et al. (16) could demonstrate it also in the intestinal brush border membrane. Na^+ and glucose could act as negative allosteric effectors on the mucosal face of the brush border membrane.

All this facts and comments we have summarized in the following hypothetical model of (Na^+-K^+)-ATPase (Fig. 6).

ATP will be split if Mg^{++} and Na^+ are present. The terminal phosphate is transferred to a γ-carboxyl group glutaminic acid. Thereby this group becomes able to react with a thiol group. The phosphate from acyl phosphate is transferred simultaneously to the OH-group of serine. This ester phosphate is then hydrolized in a K^+-dependent step.

ASKARI et al. (17) investigated the effect of oligomycin on the ATPase. Oligomycin inhibits the (Na^+-K^+)-ATPase, though it is without any influence on the K^+-stimulated phosphatases. But it inhibits the activation of the phosphatases by Na^+ and ATP. ASKARI et al. discuss an effect of oligomycin on the interconversion of the two phosphorylated intermediates. According to the model described above oligomycin acts on the conformational state caused by the thioacylation bond. We also found inhibitory and cooperative effects of Na^+ and glucose on the phosphatase, which may be regarded as allosteric effects at the same site. The conformation state induced by the thioacylation bond and influenced by Na^+, glucose, and oligomycin could be discussed as a special energized state of membrane essential for ion and sugar transport.

REFERENCES

(1) RIKLIS, E., J.H. QUASTEL, Canad. J. Biochem. Physiol., 36, (1958), 347

(2) CRANE, R.K., Fed. Proc., 24, (1965), 1000

(3) SCHULTZ, S.G., R. ZALUSKY, J. Gen. Physiol., 47, (1964), 1043

(4) DETTMER, D., F. MÜLLER, E. KUHFAHL, Acta Biol. Med. German., 18, (1967), 555

(5) BOSACKOVA, I., R.K. CRANE, Biochim. Biophys. Acta, 102, (1965), 423

(6) SCHATZMANN, H.J., Biochim. Biophys. Acta, 94, (1965), 89

(7) KIMMICH, G.A., Biochemistry, 9, (1970), 3669

(8) PRIESTLAND, R.N., R. WHITTAM, Biochem. J., 109, (1968), 309

(9) POST, R.L., S. KUME, T. TOBIN, B. ORCUTT, A.K. SEN, J. Gen. Physiol, 54, (1969), 306

(10) KAHLENBERG, A., P.R. GALSWORTHY, L.E. HOKIN, Science, 157, (1967), 434

(11) AHMED, K., J.D. JUDAH, Biochim. Biophys. Acta, 104, (1965), 112

(11a) SKOU, J.C., Physiol. Rev., 45, (1965), 596

(11b) SKOU, J.C., C. HILBERG, Biochim. Biophys. Acta, 185, (1969), 198

(12) SACHS, G., E.Z. FINLEY, T. TSUIJ, I.B. HIRSCHOWITZ, Arch. Biochem. Biophys., 134, (1969), 497

(13) SIEGEL, G.J., R.W. ALBERS, J. Biol. Chem., 242, (1967), 4972

(14) HOKIN, L.E., M. MOKOTOFF, S.M. KUPCHAN, Proc. nat. Acad. Sci. U.S.A., 55, (1966), 797

(15) WEISS, D.E., Austral. J. Biol. Sci., 22, (1969), 1373

(16) ROSENBERG, I.H., L.E. ROSENBERG, Comp. Biochem. Physiol., 24, (1968), 975

(17) ASKARI, A., D. KOYAL, Biochem. Biophys. Res. Commun., 32, (1968), 227

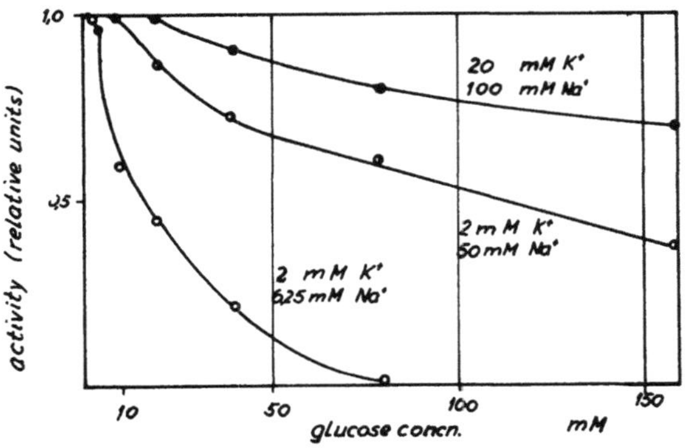

Fig. 1. Inhibition of the (Na^+-K^+)-ATPase by glucose under several conditions.
●——● with 20 mM K^+ and 100 mM Na^+
○——○ with 2 mM K^+ and 50 mM Na^+
◑——◑ with 2 mM K^+ and 6, 25 mM Na^+

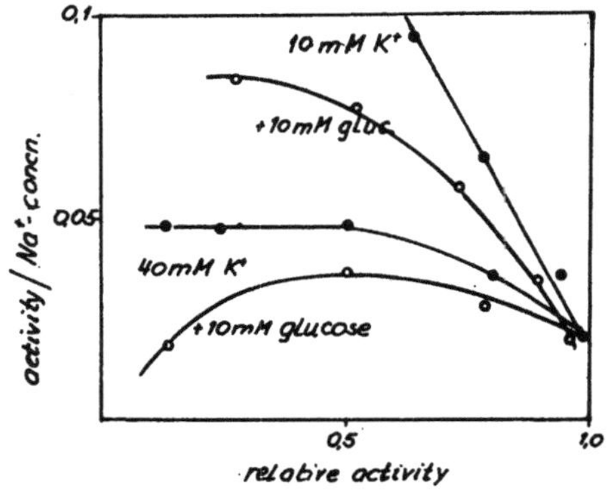

Fig. 2. Eadie-plots of the Na^+-activation curve of the (Na^+-K^+)-ATPase at different K^+-concentrations and with and without glucose.
●——● with 10 mM K^+
◑——◑ with 10 mM K^+ and 10 mM glucose
●——● with 40 mM K^+
◑——◑ with 40 mM K^+ and 10 mM glucose

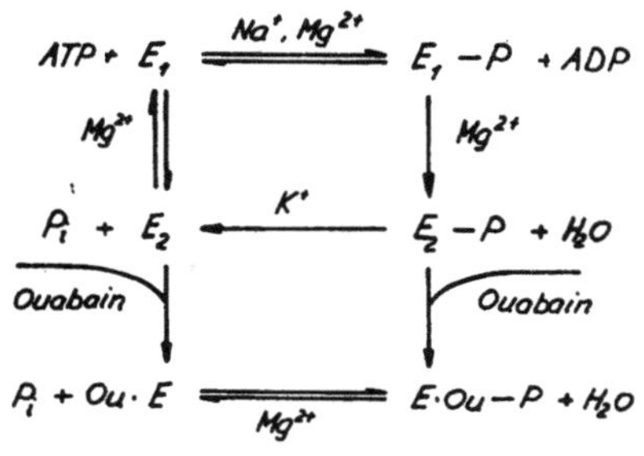

Fig. 3. Reaction scheme of the phosphorylation steps of the $(Na^{+}-K^{+})$-ATPase (POST et al.)

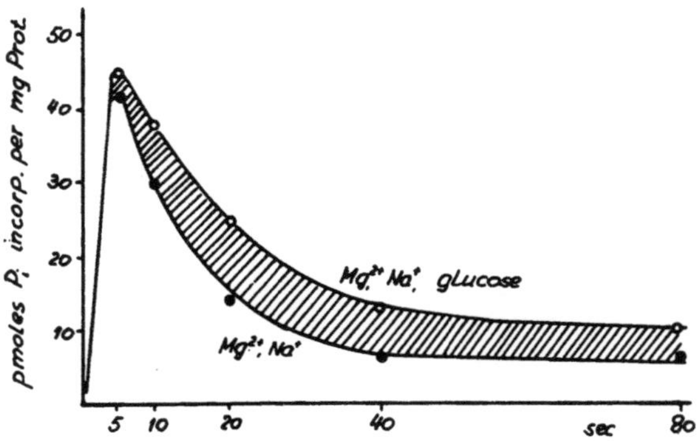

Fig. 4. Influence of glucose on building and break down of the phosphorylated intermediate.

o——o with 3 mM Mg^{++}, 100 mM Na^{+} and 100 mM glucose

●——● with 3 mM Mg^{++}, 100 mM Na^{+}

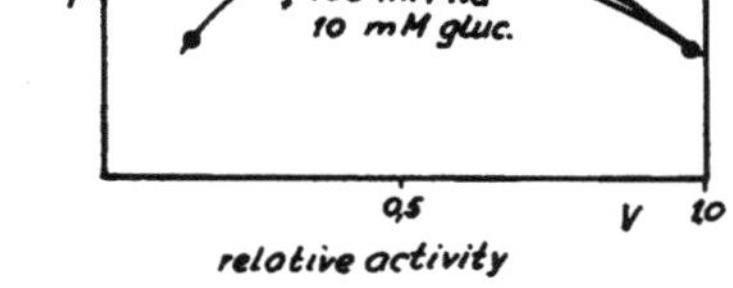

Fig. 5. Eadie-plots of the K^{+}-activation curve of the K^{+}-stimulated p-nitrophenylphosphatase at several conditions.

◐——◐ with 100 mM Na^{+} and 10 mM glucose

o——o with 100 mM Na^{+}

●——● without Na^{+} and glucose.

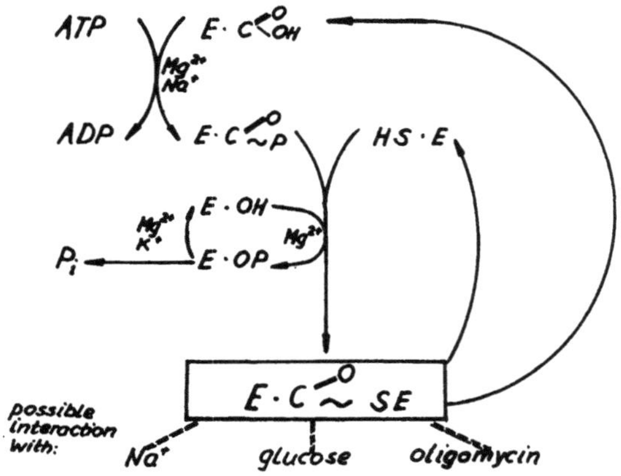

Fig. 6. Hypothetical model of the mechanism of the $(Na^{+}-K^{+})$-ATPase and the interaction with glucose.

Physiologisch-chemisches Institut der Karl-Marx-Universität Leipzig

F. MÜLLER, D. DETTMER, H. REMKE

MONOSACCHARIDE DEPENDENT Na^+-FLUXES ACROSS INTESTINAL EPITHELIUM AND THEIR POSSIBLE RELATIONS TO (Na^+-K^+)-ATPase

The problem we deal with is the mechanism of the coupled Na^+- and monosaccharide transport across the microvilli membrane of the mucosal cell of small intestine. Fig. 1 demonstrates the CRANE hypothesis about the coupling mechanism. This model (1, 2) envisions a mobile membrane carrier, which has binding sites for both, the sodium ion and the sugar molecule. Na^+ increases the affinity of the carrier to the sugar and vice versa. Therefore in the high sodium enviroment a sodium-sugar-carrier-complex is readily formed, while at the relative sodium poor enviroment (intracellular) the complex dissociates. The energy necessary for the accumulation of an actively transported monosaccharide against its concentration gradient is dependent on an inwardly directed sodium gradient. The model includes an obligatory link and high degree of correlation between the unidirectional fluxes of sodium and monosaccharide.

In order to test this model, we have investigated the unidirectional sodium influx into the cells of rat jejunum in the presence of various monosaccharides. Two kinds of methods are available for measurements of unidirectional Na^+-fluxes across the cell membrane.

1. The tracer is added to the mucosal compartment and transported into the epithelial cells, and the tracer appearence - it equals the net flux across the intestinal wall - can be measured continuously in the serosal compartment (Fig. 2). The fluxes across the microvilli membrane may be calculated from the tracer appearance rate and the intracellular sodium concentration (3).
2. The second method is a short-time incubation method (4, 5) for direct determination of the sodium influx. Fig. 3 shows that the uptake of labelled Na^+ is directly dependent on the time of incubation up to 50 sec, and gives a straight line, which indicates that only influx occurs, which may be calculated from the increment of the intracellular tracer content.

Results

Na^+- net fluxes: Fig. 4 demonstrates, that the sodium net fluxes in the presence of nonmetabolizable sugars, for instance actively transported galactose and passively transported arabinose, are nearly identical and do not differ from the control values without sugar. On the other hand in the presence of a metabolizable sugar (glucose) the Na^+-net flux is increased by about 50 % (possibly a secondary effect caused by lactate production). These results are not in agreement with the CRANE hypothesis, because there is no distinction between actively and passively transported sugars.

Na^+ - influx: Fig. 5 shows the unidirectional Na^+-influx, measured with the short-time incubation method, across the microvilli membrane under the influence of various monosaccharides. Again, the value in the presence of 10 mM of arabinose does not differ from the control. The surprising result is the decreased influx of sodium under the influence of glucose.

In the presence of galactose there is only a small increase.

Intracellular Na^+concentrations: Fig. 6 demonstrates experiments on isolated mucosal sheets, incubated in 10 mM galactose and 145 mM Na^+. During the accumulation of sugar (initially up to 6 fold in relation to the medium) the intracellular Na^+-concentration increases. After reversion of the Na^+-gradient (intracellular Na^+-concentration $>$ 145 mM), there is still an accumulation of galactose in the cells. That means, the sodium gradient could not be the only energy source for uphill sugar transport. A reasonable explanation would be that a direct energy input into the sugar transport system exists, for instance an ATP-driven carrier. It was suggested by KIMMICH (6), that monosaccharide transport system is directly linked to the energy supply system by (Na^+-K^+)-dependent ATPase.

Fig. 7 represents his conception for the reactions and equilibria between ATP, the phosphorylated intermediates and the monosaccharide transport.

It provides an explanation especially for the accumulation of monosaccharides against the sodium gradient by direct energy supply by ATP.

But it is known that the ATPase reaction is principally reversible, i. e. even the sodium influx caused by an inwardly directed Na^+-gradient is able to produce ATP from ADP (7, 8, 9, 10). Therefore we have modified this reaction scheme by introducing ion fluxes in the following manner (Fig. 8). We propose that the Na^+, K^+-gradients, resp. the corresponding fluxes, influence the ATPase-equilibria

by a mechanism, where Na^+ or K^+-intra- and extracellular-act like substrates on the ATPase. The free energy of ATP-hydrolysis could be utilized for the vectorial reaction, which lead to monosaccharide- and K^+-influx and to Na^+-efflux against their gradients, but also the free energy of the inwardly directed sodium gradient could be converted for the same purpose.
This would explain an increase in the intracellular Na^+- and a decrease in the K^+-concentration during galactose transport, because the equilibria are shifted to the left.
On the other hand raising of the intracellular ATP level (by metabolism of glucose) would shift the reactions to the right and lead to a decreased Na^+-influx.

REFERENCES

(1) CRANE, R. K., Fed. Proc. 24, (1965), 1000
(2) LYON, I., R. K. CRANE, Biochem. biophys. Acta, 112, (1966), 278
(3) CURRAN, P. F., F. C. HERRERA, C. F. FLANIGAN, J. Gen. Physiol. 46, (1963), 1011
(4) SCHULTZ, S. G., P. F. CURRAN, R. A. CHEZ, R. E. FUISZ, J. Gen. Physiol. 50, (1967), 1241
(5) GOLDNER, A. W., S. G. SCHULTZ, P. F. CURRAN, J. Gen. Physiol., 53, (1969), 362
(6) KIMMICH, G. A., Biochemistry, 9, (1970), 3669
(7) PRIESTLAND, R. N., R. WHITTAM, J. Physiol., 204, (1969), 49
(8) LANT, A. F., R. WHITTAM, J. Physiol., 207, (1969), 291
(9) GLYNN, I. M., V. L. LEW, U. LÜTHI, J. Physiol., 207, (1969), 371
(10) WHITTAM, R., K. P. WHEELER, Ann. Rev. Physiol., 32, (1970), 21

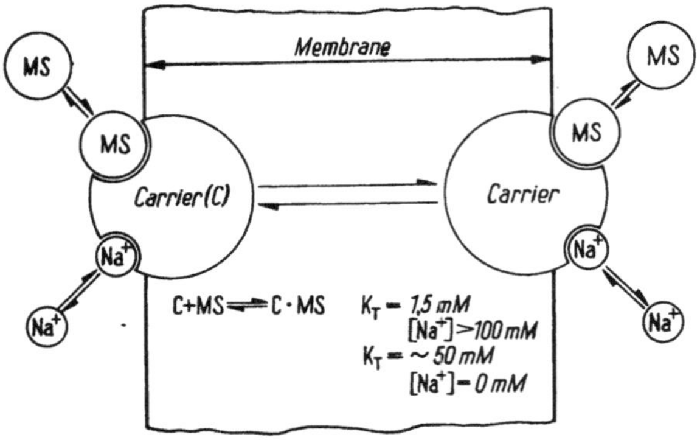

Fig. 1. The Crane model for a mobile membrane carrier with binding sites for sodium and sugar. (K_T = sugar concentration for half maximal transport velocity).

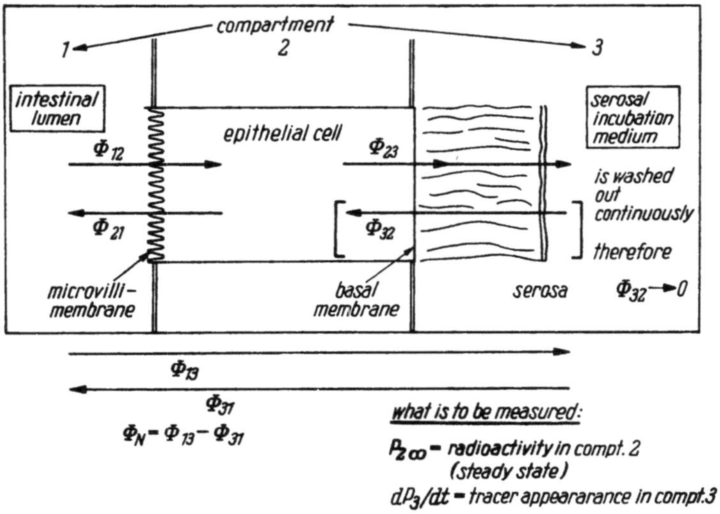

Fig. 2. Compartment system of isolated intestinal wall. Labelled Na^+ is transported from the mucosal compartment (1) across the mucosal cells (2) to serosal compartment (3).

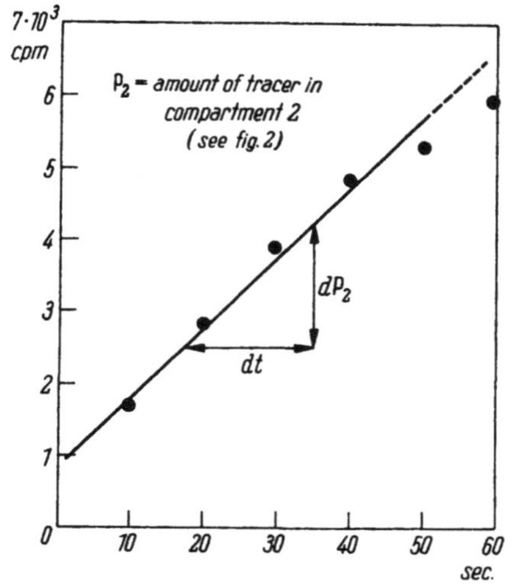

Fig. 3. Linear uptake of labelled sodium from the mucosal compartment into the cells during short-time incubation.

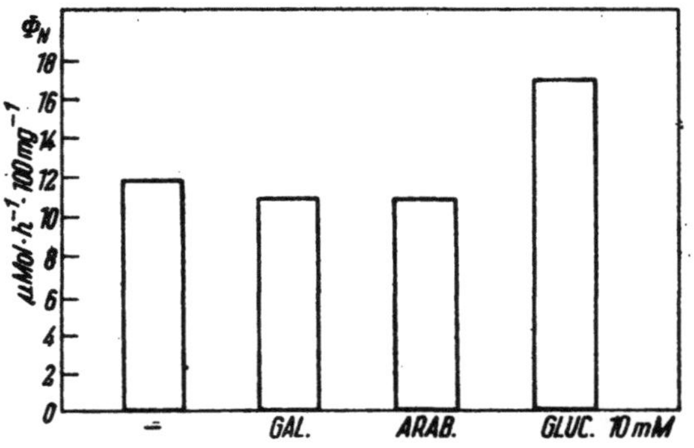

Fig. 4. Steady state Na^+-net fluxes in the presence of galactose, arabinose and glucose. Incubation medium: Krebs-Henseleit-buffer, monosaccharide: 10 mM, Na^+: 145 mM, tracer: $^{22}Na^+$.

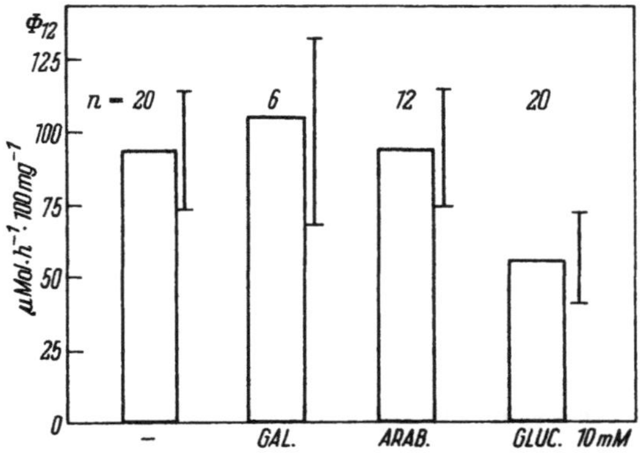

Fig. 5. Steady state sodium influxes (mucosal compartment ⟶ epithelial cell) of glucose, galactose and arabinose. Incubation medium: Krebs-Henseleit-buffer, monosaccharide: 10 mM, Na^+: 145 mM, tracer: $^{22}Na^+$.

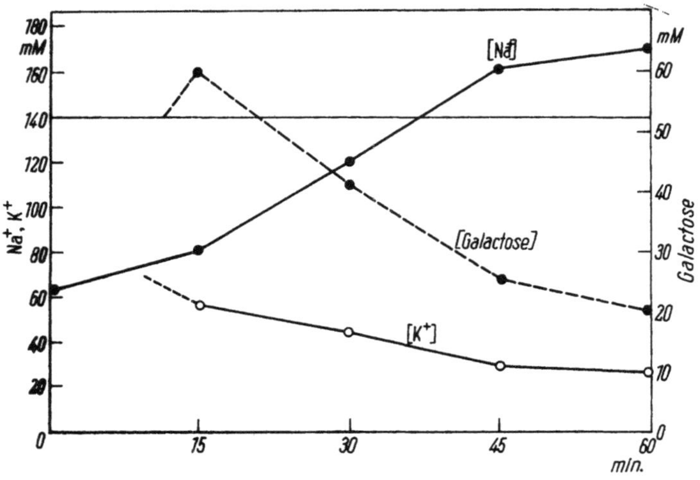

Fig. 6. Incubation of isolated mucosal sheets: Time-dependence of the intracellular galactose-, Na^+- and K^+-concentrations. Incubation medium: Krebs-Henseleit-buffer, monosaccharide: 10 mM, Na^+: 145 mM, tracer: 22 Na^+

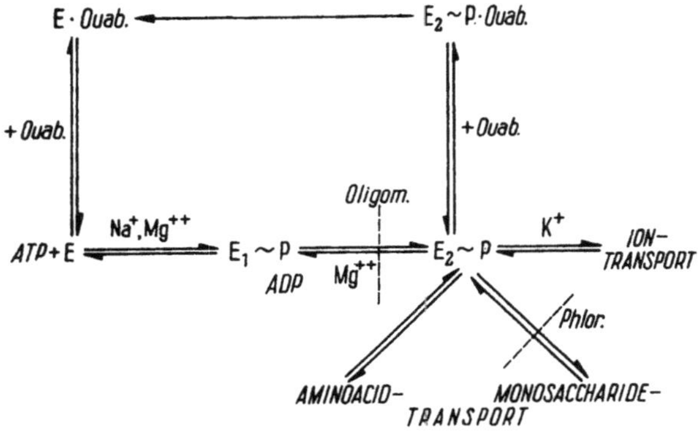

Fig. 7. Equilibrium reactions of the (Na, K)-dependent ATPase between ATP, the phosphorylated intermediats and their relations to monosaccharide transport (according to KIMMICH (6)).

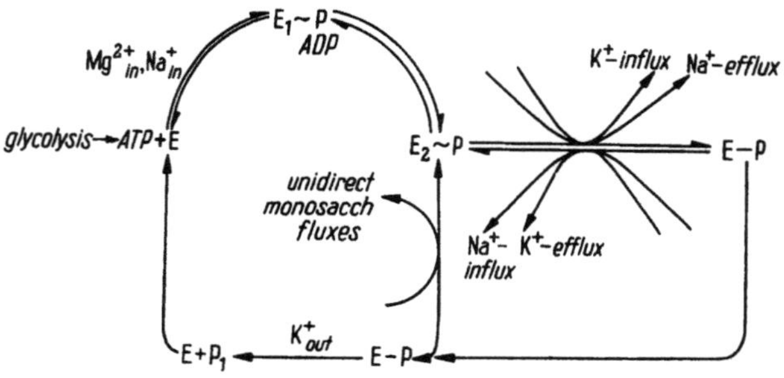

Fig. 8. **Modified scheme of the involvement of the (Na^+, K^+)-dependent ATPase in sugar and ion transport.**

3. MECHANISM OF ACTION AND CONTROL PROPERTIES OF SELECTED ENZYME SYSTEMS IN MAIN ROUTES OF METABOLISM

GENERAL PROBLEMS

Biochemisches Institut der Universität Freiburg im Breisgau and Gesellschaft für Strahlen- und Umweltforschung München

H. HOLZER

REGULATION OF ENZYMES BY PHOSPHORYLATION AND ADENYLYLATION

Introduction

Glycogen phosphorylase from muscle is the classical case of an enzyme which is regulated by phosphorylation/dephosphorylation. Since the discovery of this regulatory system by CORI et al. (1) and FISCHER and KREBS (2) some other enzymes have been found which are regulated by phosphorylation/dephosphorylation, and furthermore a distinct, but principally similar mechanism, namely adenylylation/deadenylylation has been established. Thus the importance of the regulation of enzymes by ATP-dependent chemical modification, i. e. phosphorylation and adenylylation, is becoming more and more generally recognized (for recent reviews see (3, 4)).

Table 1

ENZYMES REGULATED BY PHOSPHORYLATION/DEPHOSPHORYLATION AND ADENYLYLATION/DEADENYLYLATION

Phosphorylation/Dephosphorylation	Adenylylation/Deadenylylation
Glycogen phosphorylase	Glutamine synthetase
Phosphorylase b kinase	RNA Polymerase (?)
Phosphorylase a phosphatase (?)	
Glycogen synthetase	
Synthetase phosphatase (?)	
Fructose diphosphatase	
Pyruvate dehydrogenase complex	
Lipase (?)	
Palmityl CoA-Synthetase (?)	

Table 1 lists the enzymes known at present to be regulated by phosphorylation/dephosphorylation and adenylylation/deadenylylation. In some cases, as indicated by a question mark, there is only indirect evidence for such a mechanism. The first five enzymes of this list, all involved in the regulation of degradation and synthesis of glycogen, and the last, RNA polymerase, are treated in other parts of this symposium, and therefore will not be discussed here.

Fructose diphosphatase:

In 1966 MENDICINO et al. (5) demonstrated that purified fructose-1, 6-diphosphata from kidney is inactivated upon incubation with the crude kidney extract in the presence of ATP and Mg^{2+}. The inactivating system of the crude extract is tightly bound to mitochondria (6). In recent experiments MENDICINO (7) has shown that during incubation with γ-^{32}P-labeled ATP the in inactivation of the enzyme proceeds parallel to the incorporation of ^{32}P into trichloracetic acid-insoluble material. Results of a typical experiment are shown in Table 2. The results of MENDICINO et al. are summarized in Fig. 1. A stimulating effect of cyclic 3', 5'-AMP on inactivation of the enzyme, as described by MENDICINO

Table 2

TIME COURSE OF INACTIVATION OF LIVER-D-FRUCTOSE 1, 6-DIPHOSPHATASE AND TRANSFER OF PHOSPHATE FROM ATP^{32}

Time (min)	Radioactivity (cpm)	D-Fructose 1, 6-diphosphatase Activity (μmoles/min)
0	0	52, 8
5	6. 510	40. 5
10	12. 890	31. 7
15	18. 830	24. 6

J. MENDICINO: personal communication to H. HOLZER, 1970.

et al. (5) is included in this figure. It seems that this system is part of the general regulatory system which is controled by cyclic 3', 5'-AMP and includs several enzymes which participate as key enzymes in gluconeogenesis.

Pyruvate dehydrogenase complex:

Two years ago LINN et al. (8, 9) showed that the pyruvate dehydrogenase complexes (PDC) from beef kidney, beef heart, and pork liver mitochondria are inactivated by an enzyme-catalyzed, ATP-dependent phosphorylation and reactivated by enzyme-catalyzed dephosphorylation. Similar results have been obtained by WIELAND et al. (10, 11) working with PDC from pig heart mitochondria. Phosphorylation/dephosphorylation of PDC seems to play an important role in the hormonal regulation of this enzyme. Thus in fed rats about 70 % of the complex is in the active form, whereas in fasted rats less than 5 % is active WIELAND et al. (12) and treatment of alloxan-diabetic rats with insulin increases the ratio of active to inactive PDC.

Lipase:

As shown by CORBIN et al. (13) lipase from adipose tissue can be activated by a cyclic 3', 5'-AMP-dependent protein kinase from skeletal muscle. This finding makes it probable that this enzyme, like other enzymes participating in gluconeogenesis, is activated by cyclic AMP-dependent phosphorylation. HUTTUNEN et al. (14, 15) have now been able to demonstrate that phosphorylation of rat lipase by means of a protein kinase from rabbit muscle parallels its activation. This correlation may be seen clearly in Fig. 2, where both enzyme activity and incorporation of ^{32}P from γ-^{32}P-ATP into a trichloroacetic acid-insoluble fraction are plotted against time. Since there is at present only indirect evidence for a kinase in adipose tissue similar to that in muscle a question mark is suffixed to lipase (Table 1).

Palmityl CoA synthetase:

Microsomal palmityl CoA synthetase from rat liver can be activated by incubation with equimolar concentrations of ATP and Mg^{2+}, and inactivated by incubation with 5 mM Mg^{2+} in the absence of ATP (16, 17, 18). By analogy to results found with crude preparations of glycogen phosphorylase, which are explained by enzyme catalyzed phosphorylation/dephosphorylation, FARSTADT (19) suggest that palmityl CoA synthetase is also regulated by a phosphorylation/dephosphorylation mechanism.

Glutamine synthetase:
In 1966 an enzyme which inactivates purified glutamine synthetase from E. coli in the presence of ATP, Mg^{2+}, and glutamine was discovered and purified (20). Subsequent work by the group of EARL STADTMAN in Bethesda (21) and by the group in Freiburg (3, 4) has shown that specific enzymes catalyze effector-dependent adenylylation and deadenylylation of glutamine synthetase and thereby its inactivation and reactivation respectively. The current status of this regulatory system, based on the results of these two laboratories, is summarized in Fig. 3. Very recently ANDERSON and STADTMAN (22) have presented evidence, that the reactivation of glutamine synthetase b by deadenylylation is effected not by hydrolysis, but by phosphorolysis of the esterified AMP, forming ADP. Formation of the energy-rich pyrophosphate bond in ADP from adenylylated glutamine synthetase is reasonable from an energetic point of view, because MANTEL and HOLZER (23) have found that the adenylyl-O-tyrosin bond, linking AMP to the protein of glutamine synthetase, is an energy-rich bond as defined by LIPMAN (24). In a subsequent paper at this symposium Dr. WOHLHUETER will present some recent results on this new energy-rich bond.

Summary:
A list of the enzymes regulated by enzyme-catalyzed phosphorylation/dephosphorylation or adenylylation/deadenylylation is given. Some recent observations on the enzymes for which such a regulatory mechanism has been demonstrated or postulated are presented and discussed.

REFERENCES

(1) CORI, G., A.A. GREEN, J. Biol. Chem. 151, (1943), 31

(2) FISCHER, E.H., E.G. KREBS, J. Biol. Chem. 216, (1955), 121

(3) HOLZER, H., Advanc. in Enzymology 32, (1969), 297

(4) HOLZER, H., W. DUNTZE, Ann. Rev. Biochemistry 40, (1971), in press

(5) MENDICINO, J., C. BEAUDREAU, R.N. BHATTACHARYYA, Arch. Biochim. Biophys. 116, (1966), 436

(6) MENDICINO, J., H.S. PRIHAR, F.M. SALMA, J. Biol. Chem. 243, (1968), 2710

(7) MENDICINO, J., personal communication to H. HOLZER 1970

(8) LINN, T.C., F.H. PETTIT, L.J. REED, Proc. Natl. Acad. Sci. U.S.A. 62, (1969), 234

(9) LINN, T.C., F.H. PETTIT, F. HUCHO, L.J. REED, Proc. Natl. Acad. Sci. U.S.A. 64, (1969), 227

(10) WIELAND, O., E. SIESS, Hoppe-Seyler's Z. Physiol. Chem. 350, (1970), 1160

(11) WIELAND, O., B.v. JAGOW-WESTERMANN FEBS-Letters 3, (1969), 271

(12) WIELAND, O., E. SIESS, F.H. SCHULZE-WETHMAR, 1st Int. Symp. on Metabolic Interconversion of Enzymes, St. Margherita (1970), 52

(13) CORBIN, J.D., E.M. REIMANN, D.A. WALSH, E.G. KREBS, J. Biol. Chem. 245, (1970), 4849

(14) HUTTUNEN, J.K., D. STEINBERG, St.E. MAYER, Proc. Natl. Acad. Sci. U.S.A. 67, (1970), 290

(15) HUTTUNEN, J.K., D. STEINBERG, St.E. MAYER, Biochem. Biophys. Res. Commun. 41, (1970), 1350

(16) FARSTADT, M., Biochim. Biophys. Acta 146, (1967), 272

(17) AAS, M., Biochim. Biophys. Acta 202, (1970), 250

(18) FARSTADT, M., Acta Physiol. Scand. 74, (1968), 568

(19) FARSTADT, M., 1st Int. Symp. on Metabolic Interconversion of Enzymes, St. Margherita (1970), 70

(20) MECKE, D., K. WULFF, K. LIESS, H. HOLZER, Biochem. Biophys. Res. Commun. 24, (1966), 452

(21) STADTMAN, E.R., B.M. SHAPIRO, A. GINSBURG, H.S. KINGDON, M.D. DENTON, Brookhaven Symp. Biol. 21, (1968), 378

(22) ANDERSON, W.B., E.R. STADTMAN, Biochem. Biophys. Res. Commun. 41, (1970), 704

(23) MANTEL, M., H. HOLZER, Proc. Natl. Acad. Sci. U.S.A. 65, (1970), 660

(24) LIPMANN, F., Advances in Enzymol. 1, (1941), 99

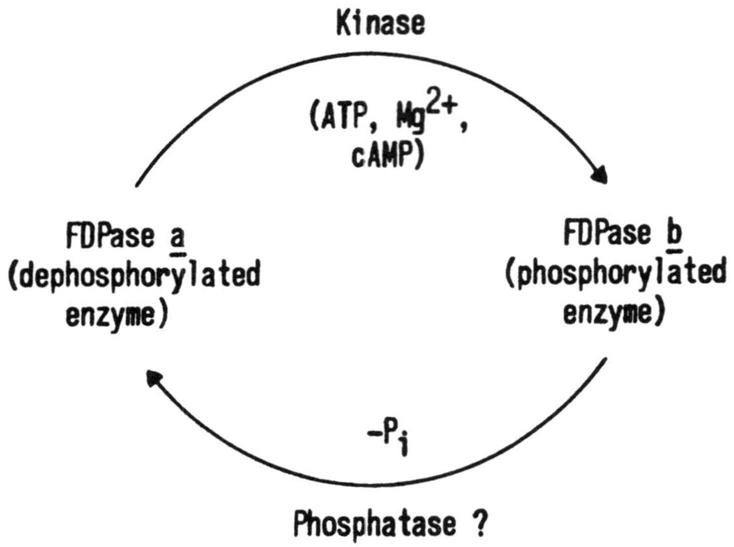

Fig. 1. Regulation of kidney fructose-1, 6-diphosphatase according to MENDICINO et al. (5, 6, 7).

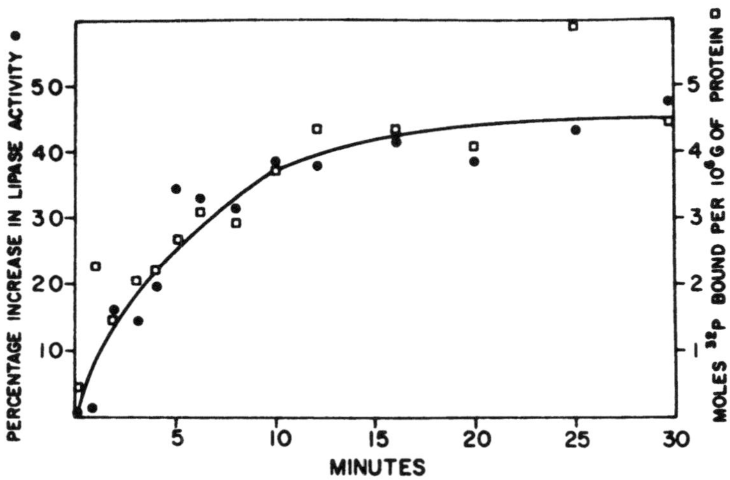

Fig. 2. Rat adipose tissue lipase and rabbit muscle protein kinase with γ-^{32}P-ATP (15).

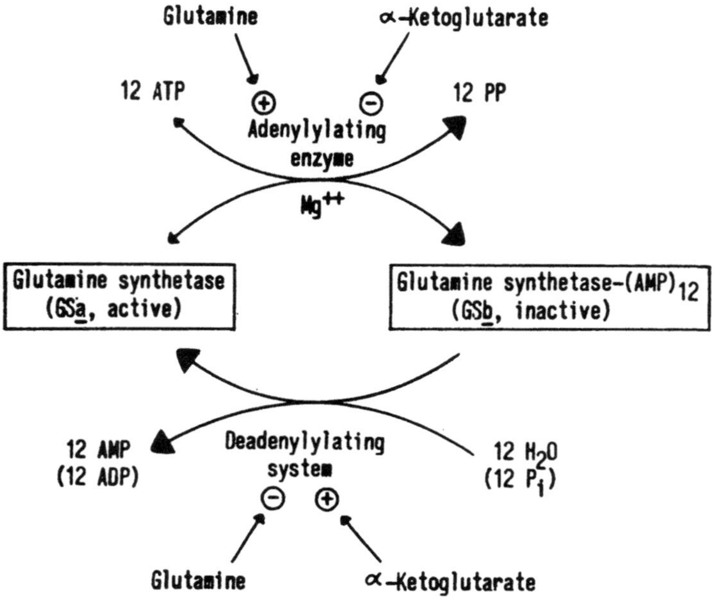

Fig. 3. Regulation of glutamine synthetase from E. coli by enzyme-catalyzed chemical modification.
⊕ = positive effector (stimulation),
⊖ = negative effector (inhibition).

Instituto de Enzimologia, Centro de Investigaciones Biologicas,
Consejo Superior de Investigaciones Cientificas, Madrid, Spain

ALBERTO SOLS

ALLOSTERIC EFFECTS BY PRODUCTS AND SUBSTRATES INVOLVING SPECIFICALLY REGULATORY SITES

In the three-dimensional world of regulation of enzyme activity, openend by UMBARGER and PARDEE fifteen years ago (1, 2) and generalized by MONOD, CHANGEUX and JACOB (3) in the early sixties, almost anything convenient is possible. Virtually any compound can be a specific effector of an enzyme, if such an effect is useful for the overall economy of a cell, no matter how chemically unrelated the compound is to the substrate(s) and product(s). Or viceversa: even for substrates and products there can be specific regulatory sites. Any "strange" effect by a substrate or product should be considered from the point of view of this possibility.

The term allosteric, introduced in metabolic regulation by the Pasteur group has been a success, but, although initially it filled an important role in the developing field of specific regulatory mechanisms, it has been mixed blessing. The initial physiological emphasis tended to be shifted to a certain mechanistic model (4). Unfortunately the two largely independent concepts have been widely used in parallel, and frequently in a rather loose way. The term allosteric in current enzymology and metabolic regulation has just become a fashionable term that tends uncosciously to be used in connection with almost of inhibitory or activation effects.

The famous nonbiochemist scientist popularly known as Lewis Carrol, in his logic-filled tale "Through the looking glass" quoted Humpty Dumpty as saying: "When I use a word, it means just what I choose it to mean -neither more nor less." Now, using the freedom allowed by the lack of formal agreement and the notorious ambiguity in current use, I will adopt for the term allosteric in this particular discussion the meaning of chemical transducer involving more than one kind of specific binding site in an enzyme, i.e., having "another specificity" in addition to that for substrate(s) and product(s). The aim of this discussion is precisely to present a few well defined cases of allosteric effects by products and substrates in the field of phosphotransferases and to discuss general criteria to distinguish effects by substrates and products involving different, specifically regulatory sites, from homotropic cooperative effects, positive or negative, in multimeric proteins as well as from competitive isosteric inhibition by products acting at the active site.

Four kinases are involved in the classical glycolytic pathway. In most organisms they are, in the order of their involvement in the pathway, hexokinase, phosphofructokinase, phosphoglycerate kinase and pyruvate kinase. Phosphofructokinase has long been recognized as a major pacemaker. Hexokinase was also a good candidate for metabolic regulation and this has been shown to be the case in most animal tissues. Less predictable, was the recent identification of regulatory in certain pyruvate kinases, in relation with potential gluconeogenesis in most eucaryotic non-photosynthetic organisms. For good reasons, although casually it look, like a mere chance, a common feature of these three kinases is that each of them is affected by a product or substrate in a "strange" way that can be shown to be genuinely allosteric, as defined above for the present discussion. Within the glycolytic pathway only phosphoglycerate kinase has never appeared as a likely candidate for metabolic regulation, ... at least until the identification of a polyphosphate dependent enzyme in certain primitive microorganisms described by KULAEV (5).

Inhibition of animal hexokinases by glucose 6-P

The hexokinases from animal tissues (but not yeast hexokinase nor liver glucokinase) are strongly inhibited by glucose 6-P at concentrations within the physiological range. Specificity studies indicate that there is an inhibitory site in the enzyme markedly different from that of the sugar subsite of the active site, as illustrated in Fig. 1 and Table 1. On this basis Crane and Sols postulated as early as 1954 the occurrence of a binding site specific for inhibition by glucose 6-P and presumably of a regulatory character (6). This strong inhibition by glucose 6-P cannot be reversed by an increase in the concentration of the sugar substrate (6), and acts on the enzyme by decreasing the affinity for the phosphoryl donor (8). In the reverse reaction, glucose 6-P (but not 2-deoxyglucose 6-P) gives a dramatically biphasic curve which reaches very strong inhibition at moderately high concentrations of glucose 6-P, as illustrated in Fig. 2 (9).

A summary of independent criteria, which taken together constitute overwhelming evidence of the allosteric character of the marked sensitivity of animal hexokinases

Table 1

DIFFERENTIAL SPECIFICITY OF THE SUGAR SUBSTRATE SITE AND THE GLUCOSE 6-P INHIBITORY SITE (6, 7)

Substrate site		Inhibitory site	
Glucose	+++	Glucose 6-P	+++
L-Sorbose	-	L-Sorbose 1-P	+++
α-Glucose 1-P	-	α-Glucose 1, 6-diP	+++
Mannose	+++	Mannose 6-P	-
2-Doxyglucose	+++	2-Deoxyglucose 6-P	-

Km 1, 5-anhydroglucitol ≫ Ki 1, 5-anhydroglucitol 6-P

to inhibition by glucose 6-P, is outlined in Table 2. The contrast between this inhibition by glucose 6-P involving a particular modifier site and the isosteric inhibition of hexokinases by ADP at its corresponding product site has been recently emphasized by KOSOW and ROSE (10).

It appears that at a certain level in evolution, hexokinase acquiredallosteric sensitivity to feedback inhibition by glucose 6-P, which usually is a primary product and is always a major metabolic crossroads. This allosteric feedback inhibition can efficiently control the glucose phosphorylation pathway in animal tissues, whenever the first step of this pathway, the reversible transport across the cell membrane, is not limiting (11). And since hexokinase is at the same time the irreversible step and the last step of a well defined metabolic pathway, it seems that evolution has favored the selection of an inhibitory site for the endproduct of the path way regardless of the fact that this endproduct is a primary product of the hexokinase reaction.

Inhibition of the liver isoenzyme of pyruvate kinase by ATP

An adaptive isoenzyme of pyruvate kinase was identified in liver by TANAKA et al. (12), who designated it as PK_L. The regulation of the activity of this isoenzyme has been characterized by LLORENTE et al. (13) as an interplay of at least four physiological factors: sigmoidal kinetics respect to the substrate

Table 2

ALLOSTERIC INHIBITION OF ANIMAL HEXOKINASES BY GLUCOSE 6-P

1) Not reverted by excess glucose

2) Differential specificity:
   a) 2-Deoxyglucose-6-P and several other primary products corresponding to high affinity substrates are not inhibitory
   b) L-Sorbose-1-P is a strong inhibitor although L-sorbose is not a substrate nor competitive inhibitor

3) Absence of significant inhibition in enzymes from other sources:
   a) yeast hexokinase
   b) liver glucokinase

4) In the reverse reaction glucose-6-P gives inhibition by "excess substrate"

phosphoenolpyruvate, activation by fructosediphosphate and inhibition by alanine and by ATP. Although the latter is a primary product of the pyruvate kinase reaction, and as such an isosteric inhibitor of the isoenzyme typical of the muscle (a tissue unable to carry out gluconeogenesis where phosphoenolpyruvate is not a major metabolic crossroads) as clearly shown by REYNARD et al. (14), ATP can inhibit PK_L by an additional mechanism. The inhibition of PK_L independent of the concentration of ADP is counteracted by the allosteric activator fructose-diphosphate, and the enzyme can be reversibly desensitized by cold treatment (13). We have also found (15) that the high specificity for ATP as inhibor is very different from the wide specificity of the enzyme respect to nucleotide diphosphate as phosphoryl acceptor substrate (Fig. 3). Moreover, specific removal of ATP with an hexokinase - 2-deoxyglucose trap leads to immediate disappearance of the inhibition (13). Table 3 summarizes these and certain other additional criteria (16, 17) that converge to substantiate the conclusion that the liver isoenzyme of pyruvate kinase has an allosteric site for inhibition by ATP, independently of the fact that it is a primary product of the reaction catalyzed by the enzyme. This situation parallels that of the inhibition of animal hexokinases by glucose-6-P. Although there are some qualitative and quantitative differences between the criteria summarized in Tables 2 and 3, (the combined weight of the evidence) amply suppost the conclusion that these two peculiar inhibitions by a primary product involve allosteric specifically regulatory sites.

Table 3

ALLOSTERIC INHIBITION OF THE LIVER ISOENZYME OF PYRUVATE KINASE (PK_L) BY ATP

1) Independent of the concentration of ADP

2) Differential specificity:
GDP, UDP and IDP are good substrates, while GTP, UTP and ITP are not good inhibitors

3) Reversible desensitization by standing in the cold

4) Counteracted by the allosteric effector FDP

5) Absence in enzyme from other sources:
   a) PK_M (muscle) and PK_A (adipose tissue) isoenzymes
   b) yeasts

Inhibition of yeast phosphofructokinase by ATP

Scarcely ten years ago a race on sensitivity of phosphofructokinases to a variety of factors was started by PASSONNEAU and LOWRY's report (18) of a somewhat bewildering array of metabolites able to affect the activity of phosphofructokinase in muscle extracts[1]. One of these effects was inhibition by excess of the substrate ATP. I obtained evidence, working with yeast preparations, that the inhibition of phosphofructokinase by ATP was not linked to the fact of its being a substrate, since other good phosphoryl donor substrates were not inhibitory (19), as illustrated in Fig. 4 and Table 4. This specific inhibition of yeast phosphofructokinase by ATP can be efficiently counteracted by AMP (20, 21) and by NH_4^+

1) In a visit to St. Louis, Mo., at about this time I gave a seminar on feedback controls in yeast glycolysis. The formal discussion was followed by an informal one between myself and Prof. Carl F. CORI, a distinguished pioneer in pointing to phosphofructokinase as probably critical for the control of glycolysis. At this time Prof. CORI muttered that phosphofructokinase looked like a "neurotic" enzyme. In tune with this light mood I offered the alternative that phosphofructokinase was perhaps an "allohysteric" enzyme.

ions (21), and reinforced by citrate (21, 22). The enzyme can be reversibly desensitized to a relatively stable insensitive form (23), by what seems to involve a complex metabolite induced change relatively stable (24). Irreversible desensitization can be achieved by treatment with trypsin (25, 26). The effects of various ligands, individually and in mixtures on the susceptibility of this desensitization by trypsin, gave additional evidence for the occurrence in native

Table 4

ALLOSTERIC INHIBITION OF YEAST PHOSPHOFRUCTOKINASE BY ATP

1) Differential specificity:

GTP, ITP, CTP and UTP are good substrates, but not inhibitors

2) Desensitization:

a) reversible change of form

b) trypsin treatment

3) Relationship to other effectors:

a) counteracted by AMP

b) " by NH_4^+

c) synergistic with inhibition by citrate (?)

4) Absence in enzyme from other sources

Dictyostelium discoideum (28)

Arthrobacter crystallopoietes (29)

Flavobacterium thermophilum (30)

yeast phosphofructokinase of an ATP regulatory site, and further indicated that the counteraction of the ATP inhibition by AMP had to involve an additional different site, as illustrated in the model represented in Fig. 6 (25, 27). Table 4 summarizes these observations and other indirect evidence that support the conclusion that the inhibition of yeast phosphofructokinase by ATP is indeed allosteric; the convergence of criteria constitutes overwhelming evidence for this conclusion. This conclusion seems to apply also for the phosphofructokinases from higher animals (31) and plants (32).

The ATP inhibitable phosphofructokinases are no "hara-kiri" enzymes. Simply,

although phosphofructokinase uses ATP as substrate, in most organisms the common glycolytic pathway, of which phosphofructokinase is the first physiologically irreversible enzyme, gives more ATP than that spent on this priming reaction. Hence ATP is a typical endproduct (11, 19) and as such can be a very convenient physiological inhibitor of phosphofructokinase, independently of the fact that it happens to be a substrate for the enzyme. The relative affinities for ATP of the corresponding substrate and allosteric site are adjusted so as to tend to make phosphofructokinase activity in situ proportional to the requirements of the cell for ATP, as can be appreciated from Fig. 4. Indeed, in the few cases of organisms with phosphofructokinase not inhibited by ATP, the enzyme seems to serve purposes other than that of energy producing glycolysis (28, 29) or use another energy-rich feedback inhibitor, phosphoenolpyruvate (30).
A considerable number of enzymes are somewhat inhibited by excess of a substrate. In the majority of the reported cases these inhibitions are moderate, usually up to about 50 % or so, require concentrations of the inhibitory substaates well above its physiological range, or some otherwise markedly unphysiological condition. No regulatory sites should be postulated on the basis of such flimsy evidence. Typical examples within the area of the phosphotransferases are fructosediphosphatase, the type III of animal hexokinase, and glyceraldehyde 3-phosphate dehydrogenase. Outside this area, the H isoenzyme of animal lactate dehydrogenase is a most conspicous case of premature hopes dissipated by hard facts.

Activation of certain phosphofructokinases by one of their products
Phosphofructokinase from E. coli can be markedly activated by ADP. Activation by fructosediphosphate has been reported for enzymes from several origins (33). From the results of ATKINSON and coworkers (see 33) and MONOD and collaborators (34), a number of criteria collected in Table 5 converge to amply justify the conclusion that ADP is an allosteric activator for the phosphofructokinase from E. coli, where it obviously constitutes the physiological equivalent to the AMP activation of the enzyme in most other organisms. The rationale is of course based on the fact well documented by Atkinson that either AMP or ATP can serve as metabolic signals of low energy charge in the cell (35). What is interesting from an evolutionary point of view and in relation with the possible functional interplays that could be significant for metabolic regulation suggested from some observations in yeasts and animal tissues, is the fact that the

Table 5

ACTIVATION OF CERTAIN PHOSPHOFRUCTOKINASES BY ONE OF THEIR PRODUCTS

1) E. coli: activation by ADP
   a) increases affinity for F6P, shifting the kinetics from sigmoid to hyperbolic
   b) counteracts allosteric inhibition by ATP
   c) GDP is also a good activator although GTP is only a weak substrate
   d) AMP does not activate

   Very likely allosteric

2) Yeast: activation by FDP
   a) dependent on low concentration of homologous substrate, with sigmoid kinetics
   b) 2, 5-anhydroglucitol-6-P, a non-phosphorylable analogue of F6P, behaves similarly
   c) high concentrations of either compound are inhibitory

   Conclusion: doubtful significance

fructosediphosphatase of E. coli is inhibited by AMP, just like in the other organisms mentioned. An additional contrast lies in the fact that the catabolic threonine deaminase is activated by AMP, while in Clostridium both enzymes are activated by ADP.
In contrast, the available information on the activation of yeast PFK by FDP (21) as summarized in Table 5, makes of doubtful significance the mechanism of this observation. More information on characteristics of the activation by fructosediphosphate reported for the phosphofructokinases of other tissues would be desirable.

Summing up
An outline of the authenticated cases of specific regulatory sites for products or substrates in glycolytic phosphotransferases, is summarized in Table 6. The present discussion should have substantiated the possibility that certain key enzymes could have allosteric, specifically regulatory sites for a substrate

Table 6

ALLOSTERIC EFFECTS BY PRODUCTS AND SUBSTRATES INVOLVING SPECIFICALLY REGULATORY SITES IN GLYCOLYTIC PHOSPHOTRANSFERASES

	Hexokinase, animal tissues	Pyruvate kinase L, liver	Phosphofructokinase, yeast	Phosphofructokinase E. coli
Regulatory effect	inhibition by product G6P	inhibition by product ATP	inhibition by substrate ATP	activation by product ADP
Non reversal by excess homologous substrate	+++	+	-	
Differential specificity	+++	++	++	++
Desensitization		+	++	
Relationship to other effectors		++	+++	+
Absence in enzyme from other sources	+++	++	+	+++
Reverse reaction	inhibition by substrate G6P		(inhibition by substrate ADP)	

or a primary product is quite real and not very infrequent. Nevertheless, it should be emphasized that it meets caution before reaching such conclusion. The criteria described in this discussion that directly or indirectly tend to support or discard such an interpretation are a sort of diagnostic tests, which although individually none of them is unequivocal (although some are much less equivocal than others), the coincidence of several of them can greatly help to validate the conclusion that a genuine specifically regulatory site for a substrate or product, but different from the corresponding active site does occur in an enzyme.
The examples analyzed in this discussion of allosteric effects by substrates and products include cases of inhibition by product, inhibition by substrate and activation by product. Also possible, but more difficult to analyze and outside my personal experience, is that some of the activation by substrate now so frequently recognized as sigmoidal kinetics could not be merely a case of interactions between identical sites in multimeric proteins but involve a specific regulatory site. If any such were demonstrated to occur it would complete the possibilities of allosteric effects by products and substrates in the restricted sense adopted for this discussion.

REFERENCES

(1) UMBARGER, H. E., Science 123, (1956), 848

(2) YATES, R. A., A. B. PARDEE, J. Biol. Chem. 221, (1956), 757

(3) MONOD, J., J.-P. CHANGEUX, F. JACOB, J. Mol. Biol. 6, (1963), 306

(4) MONOD, J., WYMAN, J.-P. CHANGEUX, J. Mol. Biol. 12, (1965), 88

(5) KULAEV, I. S. this volume

(6) CRANE, R. K., A. SOLS, J. Biol. Chem. 210, (1954), 597

(7) SOLS, A., R. K. CRANE, J. Biol. Chem. 210, (1954), 581

(8) FROMM, H. J., V. ZEWE, J. Biol. Chem. 237, (1962), 1661

(9) SOLS, A., G. DELA FUENTE, in preparation

(10) KOSOW, D. P., I. A. ROSE, J. Biol. Chem. 245, (1970), 198

(11) SOLS, A., In Carbohydrate Metabolism and its Disorders, (F. DICKENS, P. J. RANDLE, W. J. WHELAN, eds.), Academic Press, London, vol. 1, (1968), 53

(12) TANAKA, T., Y. HARANO, H. MORIMURA, R. MORI, Biochem. Biophys. Res. Commun. 21, (1965), 55

(13) LLORENTE, P., R. MARCO, A. SOLS, Eur. J. Biochem. 13, (1970), 45

(14) REYNARD, A. M., L. F. HASS, D. D. JACOBSEN and P. D. BOYER, J. Biol. Chem. 236, (1961), 2277

(15) CARBONELL, J., R. MARCO, A. SOLS, in preparation

(16) GANCEDO, J. M., C. GANCEDO, A. SOLS, Biochem. J. 102, (1967), 23

(17) MARCO, R., J. CARBONELL, P. LLORENTE, Biochem. Biophys. Res. Commun. 43, (1971), 126

(18) PASSONNEAU, J. V., O. H. LOWRY, Biochem. Biophys. Res. Commun. 7, (1962), 10

(19) VINUELA, E., M. L. SALAS, A. SOLS, Biochem. Biophys. Res. Commun. 12, (1963), 140

(20) RAMAIAH, A., J. A. HATHAWAY, D. E. ATKINSON, J. Biol. Chem. 239, (1964), 3619

(21) SOLS, A., M. L. SALAS, In Methods in Enzymology, (S. P. COLOWICK and N. O. KAPLAN, eds.), Academic Press, New York, vol. 9, (1966), 436

(22) SALAS, M. L., E. VINUELA, M. SALAS, A. SOLS, Biochem. Biophys. Res. Commun. 19, (1965), 371

(23) VINUELA, E., M. L. SALAS, M. SALAS, A. SOLS, Biochem. Biophys. Res. Commun. 15, (1964), 243

(24) AFTING, E. G., D. RUPPERT, H. HOLZER, this volume

(25) SALAS, M. L., J. SALAS, A. SOLS, Biochem. Biophys. Res. Commun. 31, (1968), 461

(26) FREYER, R., S. LIEBE, G. KOPPERSCHLÄGER, E. HOFMANN, Eur. J. Biochem. 17, (1970), 386

(27) FREYER, R., M. KUBEL, E. HOFMANN, Eur. J. Biochem. 17, (1970), 378

(28) BAUMANN, P., B. E. WRIGHT, Biochemistry 7, (1968), 3653

(29) FERDINANDUS, J., J. B. CLARK, Biochem. J. 113, (1969), 753

(30) YOSHIDA, M., T. OSHIMA, K. IMAHORI, Biochem. Biophys. Res. Commun. 43, (1971), 36

(31) LORENSON, M. Y., T. E. MANSOUR, J. Biol. Chem. 244, (1969), 6420

(32) PREISS, J., T. KASUGA, Ann. Rev. Plant Physiol. 21, (1970), 433

(33) ATKINSON, D. E., Ann. Rev. Biochem. 35, (1966), 85

(34) BLANGY, D., H. BUC, J. MONOD, J. Mol. Biol. 31, (1968), 13

(35) ATKINSON, D. E., In The Enzymes, P. D. BOYER, ed., Academic Press, New York 1, (1970), 461

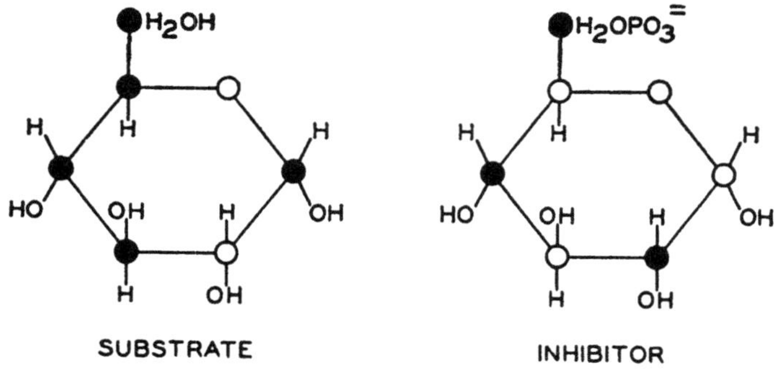

Fig. 1. Qualitative comparison of the main requirements of brain hexokinase for hexose substrate and inhibitor hexosephosphate (6).

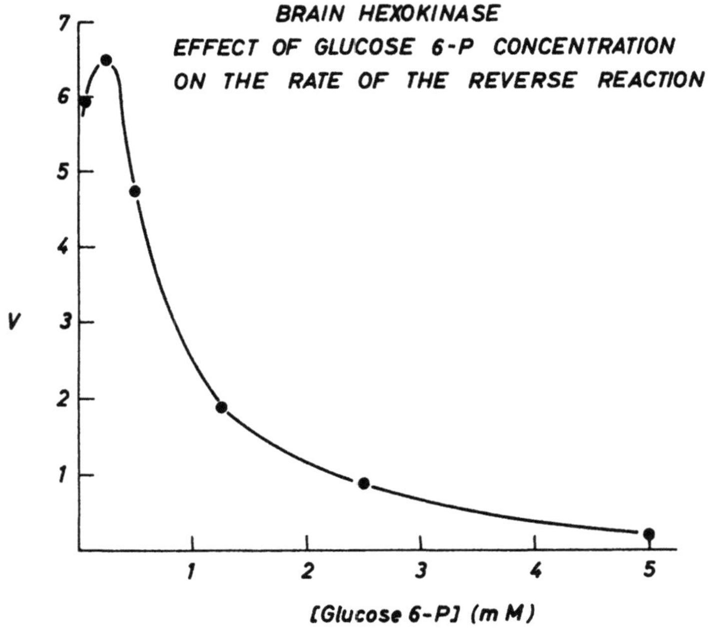

Fig. 2. Inhibition of brain hexokinase in the reverse reaction by "excess substrate".

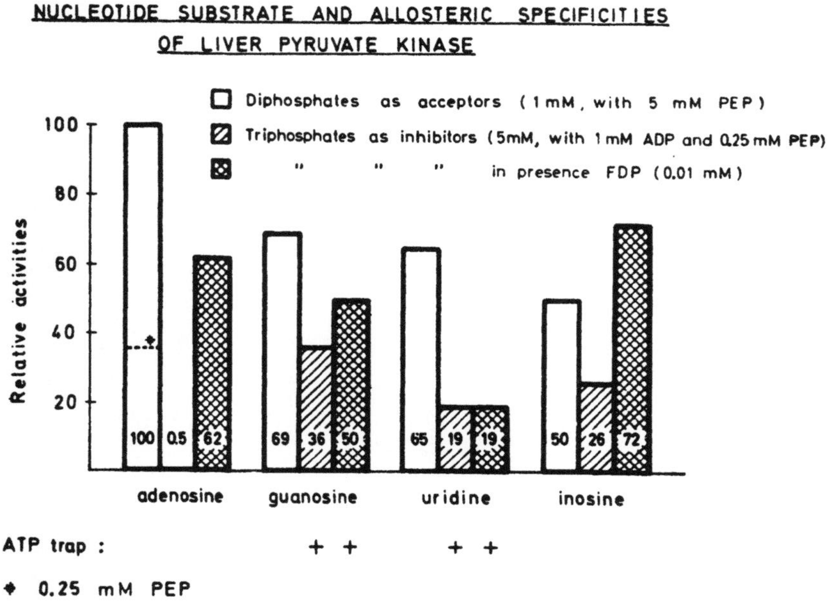

Fig. 3. Differential specificity for nucleotide phosphates as substrates and inhibition of liver pyruvate kinase.

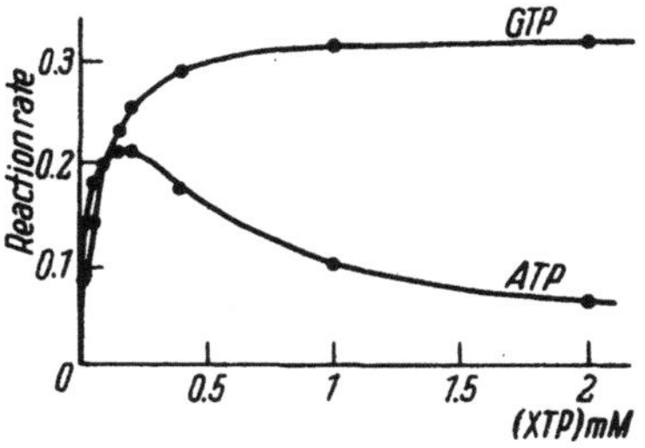

Fig. 4. Specific inhibition of yeast phosphofructokinase by ATP.

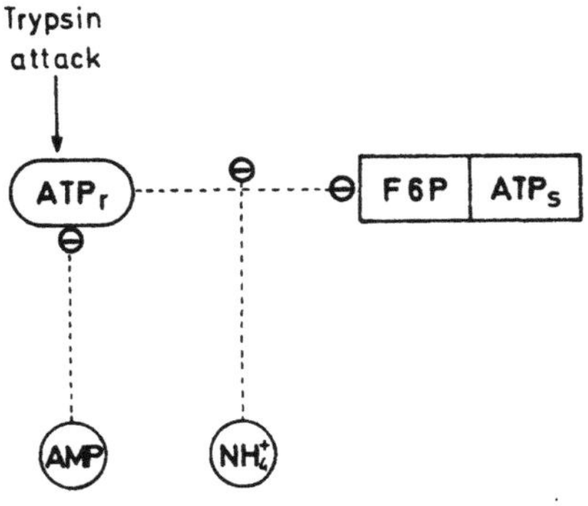

Fig. 5. Model of the different adenylic nucleotide binding sites in yeast phosphofructokinase and their functional significance.

Enzymological Department, Institute of Biochemistry, Hungarian Academy of Sciences, Budapest

T. KELETI

REGULATION OF ENZYMIC ACTIVITY BY INTRA- AND INTERSUBUNIT INTERACTIONS

Despite that the molecular mechanism of enzyme regulation was discovered on microbial enzymes and most of the known examples of regulatory enzymes are of microbial origin, glyceraldehyde-3-phosphate dehydrogenase from mammalian muscle seems to be the "Noe's ark" as far as this enzyme is endowed with almost all of the possible mechanisms of enzyme regulation.

The regulation of enzymic activity manifests itself in two different levels; these are 1. intrasubunit and 2. intersubunit interactions.

1. Intrasubunit interactions operate between the active center and some physiological modifier bound either to the regulatory center or to an aspecific site of the same subunit.

The physiological modifier may be the substrate or product or any metabolite in the cell. In the case of glyceraldehyde-3-phosphate dehydrogenase two of the substrates inhibit (1-3), one of the substrates and a substrate analogue (4) activate the enzyme. The kinetic analysis of enzymological data shows that there are distinct substrate- and product-binding sites on the enzyme, furthermore, the substrate and product may be bound to the binding site of each other. The NAD bound to the NADH-binding site inhibits the rate of breakdown of substrate (Fig. 1).

The excess of phosphate ion also inhibits the oxidative phosphorolysis of glyceraldehyde-3-phosphate. This inhibition is kinetically similar to the inhibition by excess NAD if the excess of phosphate is added in the assay mixture (3), however, the inhibition may be complete if the enzyme is preincubated with excess phosphate before the activity measurement (2, 5) (Fig. 2). The excess of both glyceraldehyde-3-phosphate (1) and glyceraldehyde (4) activate the reaction catalyzed by glyceraldehyde-3-phosphate dehydrogenase (Fig. 3). These data have recently been confirmed by KOSHLAND (6, 7).

NADH inhibits glyceraldehyde-3-phosphate oxidation in a parabolic competitive manner with respect to NAD (Fig. 4). Glyceric acid-3-phosphate, the analogue of the end-product glyceric acid-1, 3-diphosphate, also inhibits glyceraldehyde-3-phosphate dehydrogenase and this inhibition is of the parabolic mixed type with respect to glyceraldehyde-3-phosphate (Fig. 5).

These data can be interpreted as a consequence either of the existence of distinct substrate- and product-binding sites, where the substrate and product may be bound to either site and the substrate bound to the product-binding site serves as a modifier of the enzyme, or as a consequence of the existence of a regulatory center (3), which may be located outside the active center. In the former case the product-binding site plays the role of the regulatory center, whereas in the latter the regulatory center is topographycally independent of the active center. Although we cannot yet decide which of the above explanations holds, it has already been demonstrated that each glyceraldehyde-3-phosphate dehydrogenase subunit binds two moles of NAD (7-9).

The enzyme is inhibited by physiological modifiers such as adenine nucleotides: ATP, AMP, and adenine. As an example I mention the irreversible inhibition of glyceraldehyde-3-phosphate dehydrogenase by ATP in the absence of mercapto-ethanol, which is obviously formally non-competitive with NAD.

However, the reversible inhibition in the presence of mercaptoethanol is of mixed type (10) (Fig. 6). This experiment suggests that ATP cannot be bound to the active center of the enzyme. It is assumed that the adenine nucleotides are bound to the regulatory center of the enzyme; indeed, this is confirmed by the suspension by these modifiers of the inhibition caused by both excess NAD and phosphate (3). These data will be presented in detail in this symposium by Mrs. OVADI.

In our opinion a simple kinetic analysis is not sufficient to prove the existence of two types of binding sites on the same subunit. The method of double inhibitions provides valuable information on this problem.

This method has been succesfully applied in case of double pure competitive inhibition with alcohol dehydrogenase (35) and some other enzymes. Since not in all cases can be found purely competitive inhibitors we have elaborated the generalized theory of double inhibitions considering all cases, when each of the two inhibitors is purely or partially competitive, non-competitive or uncompetitive with the substrate and does or does not interact with the other (36) (Fig. 7). The value of the interaction constant of the two inhibitors can be determined numerically when both substances give complete inhibition (pure inhibitions). If, however, one or both inhibitors inhibit partially, only an approximate value of the interaction constant can be determined. In this case too, the method gives information not only on the binding-sites of substrate and the two inhibitors, but on the possible steric hindrance and steric changes in the enzyme induced by the substrate or inhibitors. We use this method, i. e. the double inhibition of glyceraldehyde-3-phosphate dehydrogenase by excess NAD and ATP, where ATP is not a competitive inhibitor of the enzyme. Preliminary results indicate that each subunit of the enzyme contains two binding sites to which the coenzyme can be bound.

2. Intersubunit interactions. a. The "classic" allosterism. According to the symmetry postulate, a change in the steric structure of one subunit induces an analogous change in the other subunits of the oligomer. The two different forms of the oligomer can bind either the substrate or the modifier, but not both at the same time (11, 12). b. The changes in the steric structure of one subunit caused by the substrate (induced fit (13, 14)) or by an inhibitor (induced misfit (15)) cause a similar change in the structure of another subunit in the same oligomer. c. Some enzymes contain two different types of subunit: one can bind and transform the substrate, whereas the other binds only the modifier and thus regulates the activity of the oligomer (16). d. The flip-flap mechanism (17), which may be called an asymmetry postulate.

Data characteristic of interactions between subunits may also be obtained without any interactions or even with monomeric enzymes. a. if a fluctuation fit is assumed, according to which enzyme molecules of different conformation are in equilibrium with each other and the binding of substrate or modifier shifts the equilibrium towards one or the other form stabilized by the ligand (18). b. if the enzyme can isomerize, follows steady-state kinetics and the substrate can bind to both forms of the enzyme (19); c. in the case of certain aggregates or populations of enzyme molecules containing fully active and partially damaged molecules.

The differentiation of real and kinetically apparent interactions is very difficult. It is well known that the results of kinetic analyses are often difficult to interpret and alone they give only limited information. Unambiguous information can be obtained only by the exclusion of certain mechaisms. In spite of this, we can find in the literature such daring attempts, where a complete mechanism is suggested for an enzyme from one single type of experiment (e. g. a sigmoid saturation curve).

A new approach to distinguish between allosteric interactions and pure kinetic causes of deviations from MICHAELIS-MENTEN kinetics was presented by BATKE (32).

The method is based on the analysis of specific activities of isoenzymes. Measuring the specific activity, one can avoid the uncertainties originating from the interpretation of sigmoidal saturation curves, since these are neither necessary nor an adequate criterion of allosteric interactions. In most cases it has been found that the number of the tetrameric isoenzymes is five, which is readily conceivable if two different subunits exist and their asymmetry is not detectable by the method used (see later). Since the interactions between subunits a and a, a and b, b and b may be different, the analysis of the specific activities of isoenzymes provides information on the allosteric interaction constants. When dissociation occurs, some of the interactions between the subunits are cancelled. Therefore we may expect that the changes of interactions between subunits are reflected in the specific activity of the enzyme molecule.

This theoretical approach to determine allosteric interactions by the analysis of the specific activity of isoenzyme systems can also be applied to oligomeric enzymes having no isoenzymes, if their subunits can be modified separately, specifically and irreversibly and if the modification results in a significant change in the specific activity of the enzyme molecule (32). By separating the oligomers modified to different extents and measuring their specific activities one can estimate the type and degree of interactions, since the oligomers containing modified and unmodified subunits in different ratios may be considered as isoenzymes.

Moreover the kinetic analysis of the irreversible modification itself may provide useful information about the presence or absence of interactions between subunits (33) (Fig. 8).

The kinetics of the irreversible modification of homotetrameric proteins show simple first order kinetics in the case of all-in-one mechanism. In this case there cannot be considerable interaction between the subunits. However, there must be a strong interaction if the irreversible modification of the protein follows the one intermediate mechanism. In this case the kinetics of the irreversible modification is identical with that of two consecutive first order reactions. If some modification of a protein follows the one-by-one mechanism we can determine whether the subunits are independent of each other or not. In this case it is possible to determine the fractional amounts of the different enzyme forms originally present in the protein preparation.

It is possible to calculate interaction constants even if tetramer $\rightleftharpoons$ dimer or dimer $\rightleftharpoons$ monomer dissociation occurs in the assay mixture. If dissociation occurs, the analysis of activity or of any other parameter depending on the interaction, as a function of the concentration of tetramer form actually present in the solution, shows directly whether only the tetrameric or the dimeric form or both are active. The experiments performed with glyceraldehyde-3-phosphate dehydrogenase confirm that:

1. there is a tetramer $\rightleftharpoons$ dimer equilibrium in solution.
2. the enzyme is active both in the tetrameric and dimeric forms.
3. the subunits interact with each other (10, 39).

The interaction between subunits depends also on the structure of the oligomer. In tetrameric molecules, theoretically, there are six possible arrangements of subunits: the tetrahedral, the plane, the cyclic, the stirrup, the triangel and the linear structure (37). As an example we present the possible isoenzymes in these six arrangements for a tetramer with two different subunits (Fig. 9).

Due to theoretical considerations on symmetry and thermodynamics the most probable structure is the tetrahedral one. In this case each subunit must have three surfaces contacting the other three subunits in the tetramer. These contacting surfaces may be functionally identical, i. e. they contain approximately the same number of hydrophobic and or polar amino acids in similar arrangement or some or all of these are different.

The subunits can be arbitrarily regarded as approximately symmetric or asymmetric. In fact, it is not reasonable to assume the existence of protein molecules which are strictly symmetric (geometrically, in charge distribution, etc.). However, it is possible to attribute an approximate symmetry to some protein molecules, the asymmetry of which cannot be detected, by the method used in the given experiment. Obviously, the approximate symmetry of a molecule is a relative term which changes with the development of methods. In other words we recognize isoenzymes if the asymmetry of subunits reaches the level of detection. If the asymmetry of only one of the two different subunits a and b is detectable we hay have 28 isoenzymes. If b subunit is asymmetric we have one a_4, one a_3b but six a_2b_2 (2 individual isoenzymes and 2 stereoisomeric pairs, see Fig. 10), 11 ab_3 (3 individual isoenzymes and 4 stereoisomeric pairs, see Fig. 11) and 9 b_4 (3 individual isoenzymes and 3 stereoisomeric pairs, see Fig. 12). We obtain the same (9) isoenzymes in a_4 and b_4 as in the last case (Fig. 12) if the asymmetry of both a and b subunits is detectable. However, in the case of a_3b and ab_3 we obtain 27 isoenzymes (3 individual ones and 12 stereoisomeric

pairs, see Fig. 13). In the case of a_2b_2 we have a camping site with 45 tents (3 individual isoenzymes and 21 stereoisomeric pairs, see Fig. 14), i. e. in this case the number of possible isoenzymes can be as high as 117. Five isoenzymes can be isolated if two different protomers form tetrahedral tetramers and the asymmetry of subunits is not detectable by the method used. Only in this latter case can we apply the method of the analysis of specific activities of isoenzymes to detect allosteric interactions between subunits. Previous data show the existence of five isoenzymes in glyceraldehyde-3-phosphate dehydrogenase. The analysis of interaction constants is in progress.
There are certain findings which suggest that either allosteric interactions (20, 21) or induced fit (8, 22) occurs in glyceraldehyde-3-phosphate dehydrogenase. The existence of induced fit in the case of this enzyme is very probable since the binding of one substrate influences the binding of the others, i. e. the binding of substrate changes the structure of at least the active center (Table 1).
Another evidence of structural changes is the stabilization of protein molecule by different substrates (24-88). This is also confirmed by thermodynamic data. The binding of NAD to the enzyme decreases the entropy of the complex as compared to the free enzyme (29). The stability of the enzyme-NAD complex is also reflected in the high activation energy of heat inactivation of tetrameric enzyme-coenzyme complex at the pH minimum of heat denaturation (30) (Fig. 15). However, these data merely show the changes in a subunit effected by the binding of coenzyme or substrate, but not the interaction between subunits. The phenomenon of induced misfit caused by ATP-inhibition or by maleylation provides evidence for the functional non-identity and interaction of subunits in glyceraldehyde-3-phosphate dehydrogenase.
Binding of 2 moles of ATP to the NAD-free tetrameric enzyme results in its inactivation. The same time course of inhibition was observed in the presence of 4 mole equivalents of ATP (10) (Fig. 16). The fact that no difference was found in the time course of inactivation with 2 or 4 ATP' s indicates that the subunits interact in the tetrameric molecule as will be presented in details by Mrs. OVADI. Similarly, maleylation of 1 side chain of the NAD-free enzyme, or enzyme $(NAD)_2$ complex completely inhibits the oxidation of glyceraldehyde-3-phosphate. The presence of the third and fourth, relatively loosely bound NAD' s in the complex enzyme-$(NAD)_4$ protects the enzyme against inactivation by maleylation (Fig. 17). The NAD-free maleylated tetrameric enzyme can bind only two moles of NAD, although the untreated enzyme binds all the four (10). These data also suggest that

Table 1

INTERACTIONS OF SUBSTRATES ON GLYCERALDEHYDE-3-PHOSPHATE DEHYDROGENASE

a. COENZYMES

determination	NAD		NADH	
	K_M	K_S	K_M	K_S
kinetically with GAP and arsenate or GA and arsenate	$1\text{-}6\times10^{-5}$	-	-	-
kinetically with GA and phosphate	-	2×10^{-5}	-	-
fluorimetrically, photometrically or with ultracentrifuge	-	$0.6\text{-}4\times10^{-7}$	-	2×10^{-7}
kinetically with GAP and phosphate or GA without anions	$1\text{-}2\times10^{-4}$	3×10^{-8}	-	-
kinetically with GSPP	-	-	3×10^{6}	

b. SUBSTRATES

determination	GAP		GA		PO_4^{3-}	
	K_M	K_S	K_M	K_S	K_M	K_S
kinetically with arsenate	$2\text{-}9\times10^{-5}$	-	1×10^{-1}	-	-	-
kinetically with phosphate	5×10^{-4}	7×10^{-8}	5×10^{-2}	2×10^{-2}	-	-
kinetically with GAP	-	-	-	-	2×10^{-4}	-
kinetically with GA	-	-	-	-	1×10^{-5}	7×10^{-5}

there are interactions between the subunits and the subunit pair binding the relatively loosely bound NAD's plays an important role in these interactions. These data along with those showing the instantaneous firm binding of only two mole equivalents of ATP to the tetrameric enzyme (10), the reducibility of only two moles of firmly bound NAD at high enzyme concentrations (10, 31, 32), all seem to indicate the functional dissimilarity of subunits in glyceraldehyde-3-phosphate dehydrogenase.

Since the thermodynamic parameters of enzymes in which subunit interactions occur may be different from those where such interactions are absent (34) it was of interest to analyze the thermodynamics of the reaction with various substrates. (38) From the Arrhenius plot of glyceraldehyde-3-phosphate oxidation under optimal conditions the activation energy of this reaction is about 12 kcal./mole. We have shown that this value is a real activation energy since both the pH optimum of the reaction and the MICHAELIS constants of NAD, glyceraldehyde-3-phosphate and phosphate ion remain unchanged if analyzed in the temperature range examined.
The activation energy of glyceraldehyde oxidation is about 8 kcal./mole. Since the chemical reaction is basically the same in both cases it may be assumed that the difference is due to changes in protein structure or to subunit interactions. The activation energy of the latter might be 1 kcal. per mole subunit. Inhibition by excess NAD or by ATP increases the activation energy of glyceraldehyde-3-phosphate oxidation with about 4 kcal./mole, whereas inhibition by excess phosphate, where not interaction between subunits was demonstrated does not change the activation energy of the reaction. The analysis of enthropy changes during the formation of activated complexes with NAD, glyceraldehyde-3-phosphate and phosphate ion gave a value of about -22 cal./degree/mole. This is in good agreement with the microcalorimetric measurements of the formation of enzyme-NAD complex (29).
As you can see we have made several attempts to sort out the regulation mechanisms from the many possible ones, which operate in glyceraldehyde-3-phosphate dehydrogenase. In spite of our different theoretical and experimental approaches we cannot yet give a clear-cut picture of the regulatory aspects of this enzyme. Our efforts remind me of the man who asked the sage cadi how to cram his geese. The cadi said: "Feed them on ground crane feather." The man came next day: "Cadi, 20 geese died." "Well, said the cadi, then give them minced asbestos slate." Again the man came next day: "Cadi, 40 geese died." The cadi blinked: "Well, then give them chopped lime stone." Next day the man rushed in upset: "Cadi, all geese died." "Oh", said the cadi, "stroking his beard, what a pity, when I have still so many good ideas."
Well, I also have a lot of good ideas how to study the mechanism of the regulation of glyceraldehyde-3-phosphate dehydrogenase.

REFERENCES

(1) BATKE, J., T. KELETI, Acta Biochim. Biophys. Acad. Sci. Hung. 3, (1968), 385

(2) KELETI, T., M. TELEGDI, Acta Physiol. Acad. Sci. Hung. 16, (1959), 235

(3) OVADI, J., M. NURIDSANY, T. KELETI, in preparation

(4) KELETI, T., Acta Physiol. Acad. Sci. Hung. 29, (1966), 101

(5) SEVERIN, S. E., N. K. NAGRADOVA, Dokl. Akad. Nauk SSSR 121, (1958), 519

(6) CORNISH-BOWDEN, R. J. D. E. KOSHLAND, Biochemistry 9, (1970), 3325

(7) SAPAG-HAGAR, M., Rev. Esp. Fisiol. 25, (1969), 201

(8) KOSHLAND, D. E., R. A. COOK, A. CORNISH-BOWDEN, In H. SUND (ed.) Pyridine Nucleotide Dependent Dehydrogenases. Springer Verlag, Berlin, (1970), 199

(9) CONWAY, A., D. E. KOSHLAND, Biochemistry 7, (1968), 4011

(10) OVADI, J., M. TELEGDI, J. BATKE, T. KELETI, in preparation

(11) MONOD, J., J. P. CHANGEUX, F. JACOB, J. Mol. Biol. 6, (1963), 306

(12) MONOD, J., J. WYMAN, J. P. CHANGEUX, J. Mol. Biol. 12, (1965), 88

(13) KOSHLAND, D. E., Proc. Natl. Acad. Sci. U. S. 44, (1958), 98

(14) KOSHLAND, D. E., Adv. Enzymol. 22, (1960), 45

(15) GERHART, J. C., A. B. PARDEE, J. Biol. Chem. 237, (1962), 891

(16) GERHART, J. C., H. K. SCHACHMAN, Biochemistry 4, (1965), 1054

(17) LAZDUNSKI, M., Battelle Symp. on Mol. Basis of Biol. Spec. and Control Mech. Seattle, 1970 (and private commun.)

(18) STRAUB, F. B., G. SZABOLCSI, In Molekul. Biol. Izd. Nauka, Moskva (1964), 182

(19) KELETI, T., Acta Biochim. Biophys. Acad. Sci. Hung. 3, (1968), 247

(20) KIRSCHNER, K., M. EIGEN, R. BITTMAN, B. VOIGT, Proc. Natl. Acad. Sci. U. S. 56, (1966), 1661

(21) KIRSCHNER, K., I. SCHUSTER, In H. SUND (ed.) Pyridine Nucleotide Dependent Dehydrogenases. Springer Verlag, Berlin. (1970), 217

(22) LEVITZKI, A., D. E. KOSHLAND, Proc. Natl. Acad. Sci. U. S. 62, (1969), 1121

(23) KELETI, T., In DEVENYI, T., P. ELÖDI, T. KELETI, G. SZABOLCSI, Strukturelle Grundlagen der biologischen Funktion der Proteine. Akadémiai Kiadó, Budapest (1969), 317

(24) ELÖDI, P., G. SZABOLCSI, Nature 184, (1959), 56

(25) LISTOWSKY, I., C. S. FURFINE, J. J. BETEIL, S. ENGLARD, J. Biol. Chem. 240, (1965), 4253

(26) KELETI, T., J. BATKE, Acta Physiol. Acad. Sci. Hung. 28, (1965), 195

(27) DEVENYI, T., P. ELÖDI, T. KELETI, G. SZABOLCSI, Biol. Közl. 8, (1960), 3

(28) VAS, M., L. BOROSS, private communication

(29) VELICK, S. F., J. P. BAGGOTT, J. M. STURTEVANT, Biochemistry 10, (1971), 779

(30) KELETI, T., M. SZEGVARI, In preparation

(31) BATKE, J., FEBS Letters 2, (1968), 81

(32) BATKE, J., In preparation

(33) KELETI, T., J. Theoret. Biol. 30, (1971), 545

(34) NOBLE, R. W., J. Mol. Biol. 39, (1969), 479

(35) YONETANI, T., H. THEORELL, Arch. Biochem. Biophys. 106, (1964), 243

(36) KELETI, T., Cs. FAJSZI, Mathemat. Biosci. in press (1971)

(37) FAJSZI, Cs., T. KELETI, In preparation

(38) KELETI, T., J. FÖLDI, S. ERDEI, T. TRO, In preparation

(39) BATKE, J., Studia Biophysica 25, (1970), 187

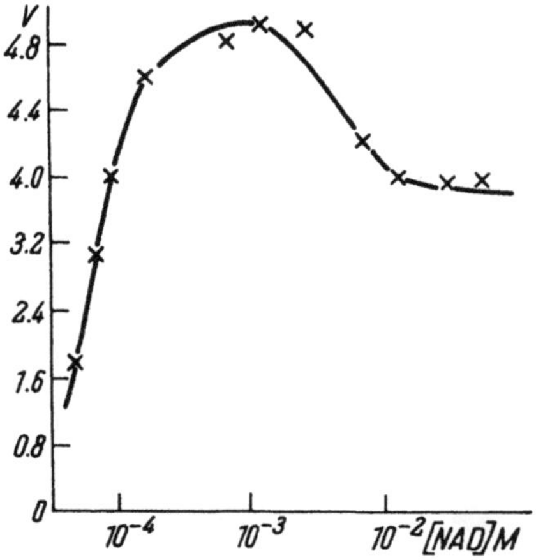

Fig. 1. The NAD-saturation curve of glyceraldehyde-3-phosphate dehydrogenase. Experimental conditions see ref. 1.

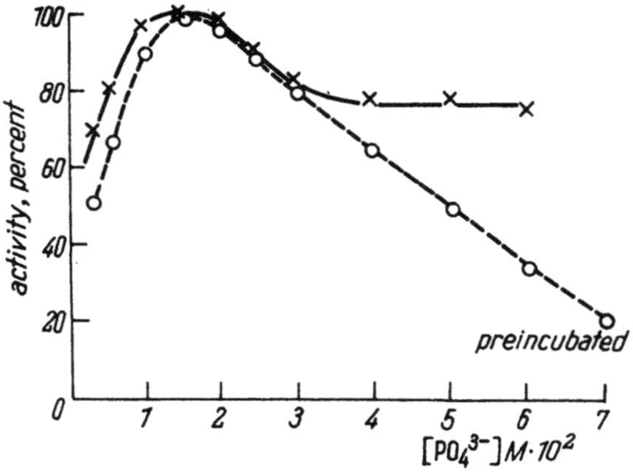

Fig. 2. The phosphate-saturation curve of glyceraldehyde-3-phosphate dehydrogenase. Experimental conditions see references 2 and 3.

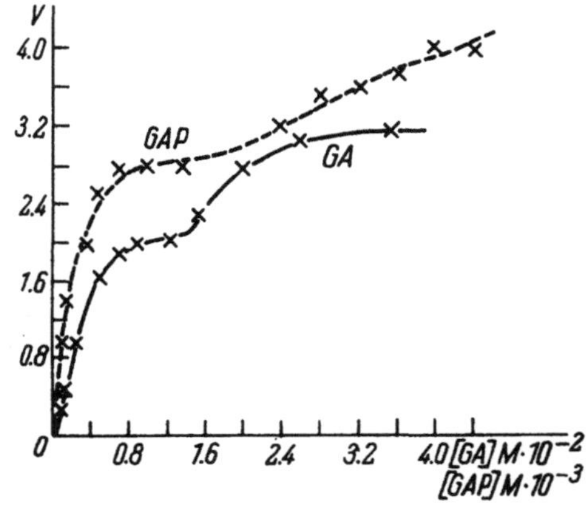

Fig. 3. The substrate-saturation curves of glyceraldehyde-3-phosphate dehydrogenase. Experimental conditions see references 1 and 4.

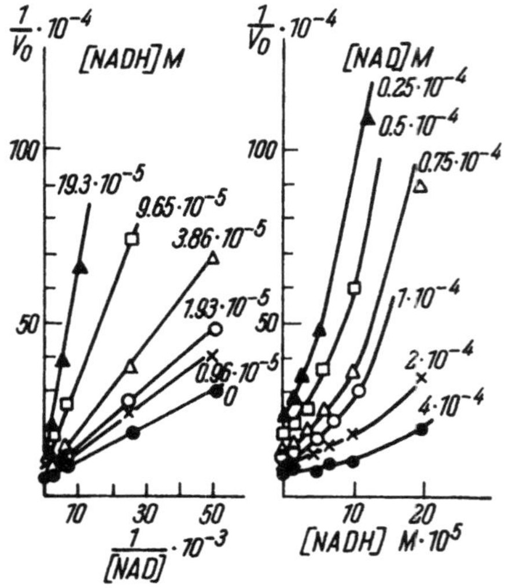

Fig. 4. The inhibition of glyceraldehyde-3-phosphate dehydrogenase by NADH. Experimental conditions see ref. 1.

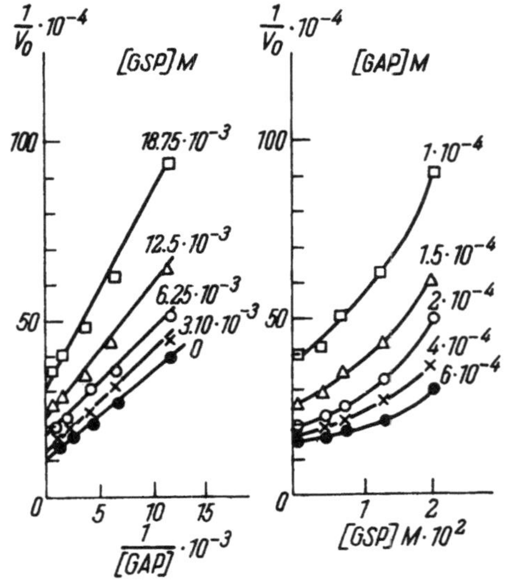

Fig. 5. The inhibition of glyceraldehyde-3-phosphate dehydrogenase by glyceric acid-3-phosphate. Experimental conditions see ref. 1.

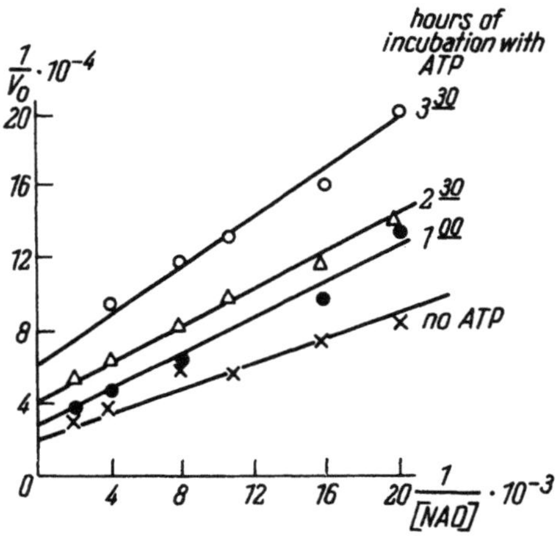

Fig. 6. LINEWEAVER-BURK plot of the inhibition of glyceraldehyde-3-phosphate dehydrogenase by ATP. Incubation was performed with 4 mole equivalents of ATP/mole of protein.

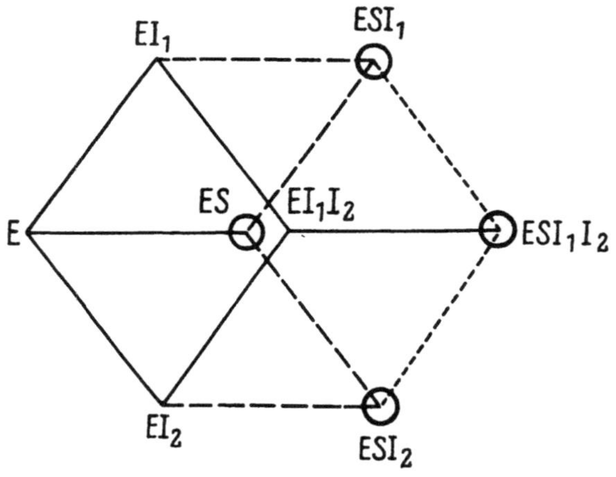

Fig. 7. Scheme of the general mechanism of double inhibitions. Broken lines show the elementary steps where the substrate and one of the inhibitors may interact; double lines, where the two inhibitors interact on the free enzyme; dotted lines, where the two inhibitors interact on the enzyme-substrate complex. The encircled symbols indicate complexes from which the product is formed.

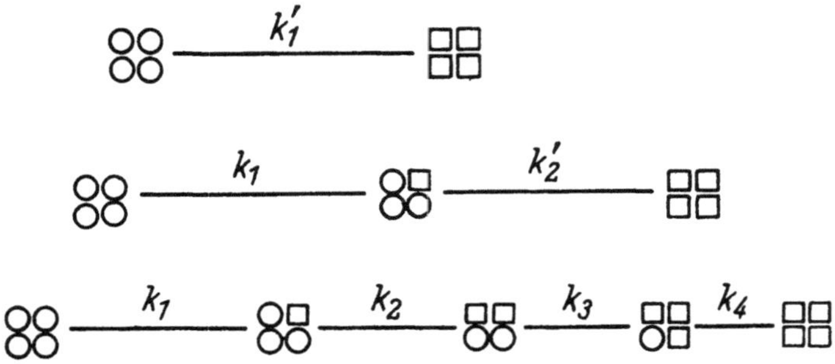

Fig. 8. All-in-one, one intermediate and one-by-one mechanism of irreversible modification of tetrameric proteins.
o: native subunit; □ : modified subunit.

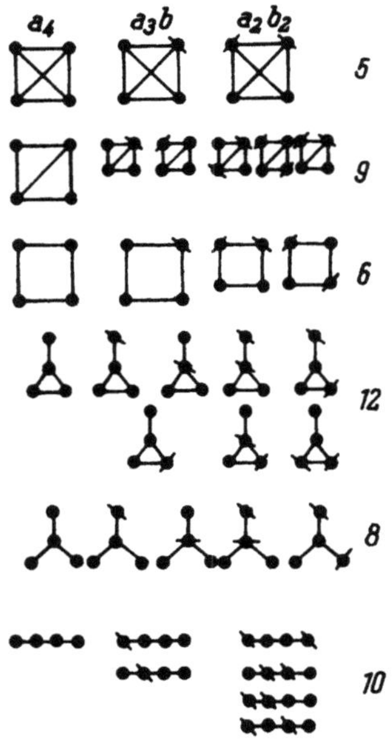

Fig. 9. Possible arrangements and number of isoenzymes of tetrameric molecules.

Fig. 10. Isoenzymes of a_2b_2 if b subunit is asymmetric.

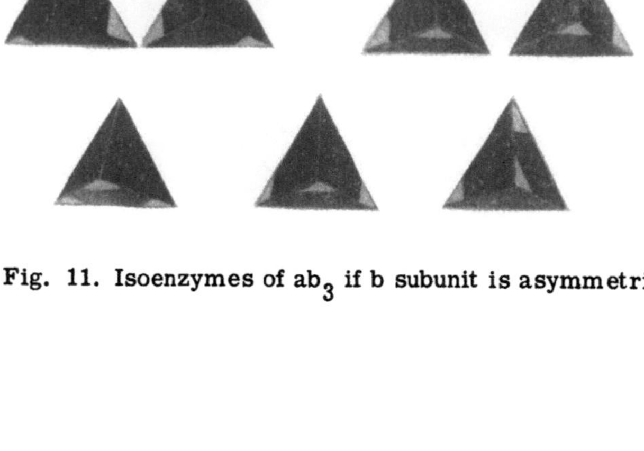

Fig. 11. Isoenzymes of ab_3 if b subunit is asymmetric.

Fig. 12. Isoenzymes of homotetrameric enzyme if the subunits are asymmetric.

Fig. 13. Isoenzymes of a_3b or ab_3 if both subunits are asymmetric.

Fig. 14. Isoenzymes of a_2b_2 if both subunits are asymmetric.

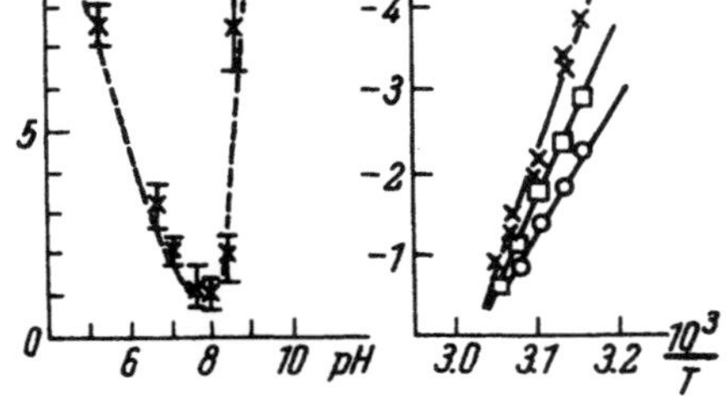

Fig. 15. Heat denaturation of glyceraldehyde-3-phosphate dehydrogenase. Left: pH dependence of the first order rate constant of inactivation at 50 °C. 1 mg enzyme/ml glycin buffer. Right: Arrhenius plot of heat inactivation at pH 6.1, 7.7 and 8.4, giving 112, 130 and 82 kcal./mole activation energy, resp.

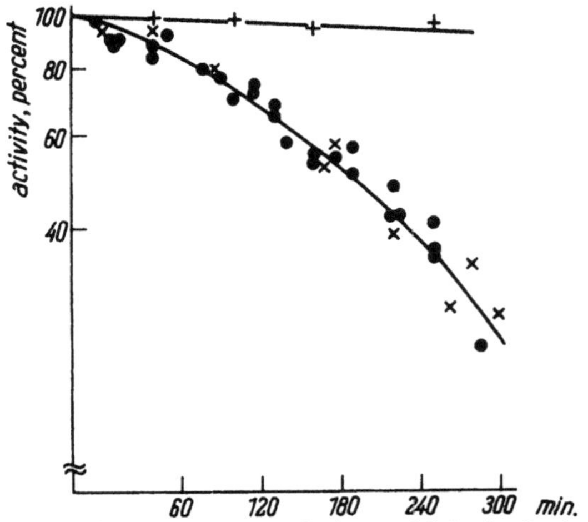

Fig. 16. Inactivation of glyceraldehyde-3-phosphate dehydrogenase by ATP. Experimental conditions see ref. 10. ×: 2 moles, ●:4 moles of ATP +: without ATP.

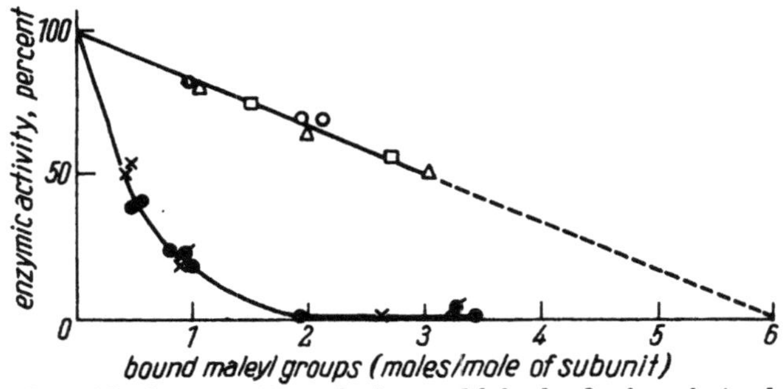

Fig. 17. Inactivation of glyceraldehyde-3-phosphate dehydrogenase by maleylation. Experimental conditions see ref. 10. o, □, Δ : enzyme-NAD_4; ●, ×, ● : enzyme-NAD_2 and free enzyme.

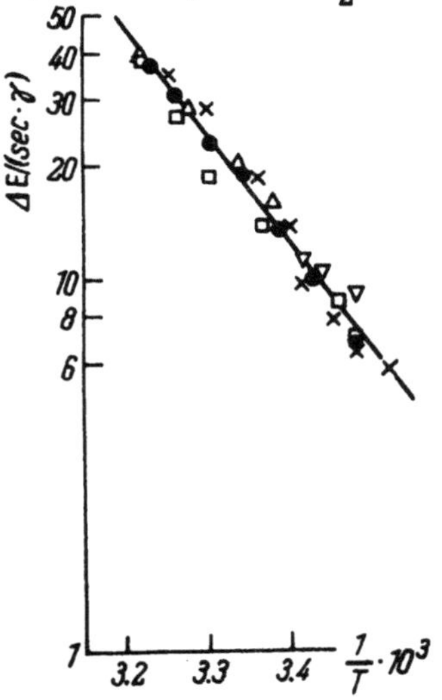

Fig. 18. Arrhenius plot of glyceraldehyde-3-phosphate oxidation.

SELECTED ENZYMES

The Department of Physiological Chemistry, The University of Würzburg

J. EHRLICH, K. FELDMANN, E. HELMREICH, T. PFEUFFER

THE FUNCTION OF PYRIDOXAL-5'-PHOSPHATE IN GLYCOGEN PHOSPHORYLASE

PLP is essential for the activity of all known α-glucan phosphorylases (FISCHER et al. (1)). The first phosphorylase enzyme shown to contain PLP was rabbit skeletal muscle phosphorylase a (BARANOWSKI et al. (2)). On removal of PLP - one mole per monomer of 100, 000 daltons (SEERY et al. (3) and Fischer, personal communication (1970)) - an inactive crystalline apoenzyme was obtained which regained its activity upon addition of PLP, ILLINGWORTH et al. (4). In the rabbit, there is more PLP bound to muscle phosphorylase than to all other PLP dependent enzymes taken together, FISCHER and KREBS (5). FISCHER et al. (6, 7) found that PLP bound to phosphorylase as an imine through its 4-formyl group to an ϵ-amino group of a lysyl residue can be reduced with $NaBH_4$ to a secondary amine, but only after the PLP binding region of the protein is reversibly "deformed" for example by acidification or high salt concentrations, GRAVES et al. (8). The reduced enzyme still retains about 60 % of its original activity. This rules out a participation of the 4-formyl group of PLP in phosphorylase catalysis. The 4-formyl group of PLP is the functional group in every other PLP dependent enzyme, SNELL (9). Ever since, the function of PLP in phosphorylase poses a challenging puzzle to enzymologists, because, if PLP should actually participate in the reaction catalyzed by the glycogen phosphorylases, it must do that through functional groups other than the 4-formyl group. Undoubtedly, PLP is a molecule tailored for a function as proton shuttle (JENCKS (10)). Thus it could substitute for functional groups of amino acid side chains and participate in general acid-base catalysis. Kinetic measurements with the temperatur jump method by AHRENS et al. (11) showed that substituted 3-hydroxy-pyridine-4-aldehydes lacking the 5'-phosphate group transfer protons in water between the ring nitrogen and the oxygen in position 3 mainly by an intermolecular route. On introduction of the phosphate group in 5'-position an additional intramolecular proton transfer occurs involving 3 rather than 2 proton-donor acceptor groups.

In order to see whether PLP could participate as proton shuttle in the phosphorylase reaction all 3 protonatable groups have been modified: 3' -O-Me PLP was synthesized and found to reactivate apophosphorylase to 20 to 25 % as compared with PLP, POCKER and FISCHER (12), SHALTIEL et al. (13). Thus the phenolic OH group is not essential for activity.

The nucleophilic pyridinium nitrogen of PLP (p_K 8.7) is another functional group. In the NO derivatives of PLP or 3' -O-MePLP, protonation of the nitrogen is blocked. The NO derivative of PLP was prepared by FUKUI et al. (14). In addition we have synthesized the N-oxide of 3' -O-Me PLP. The synthetic procedure is outlined in Fig. 1.

It is already known since 1958 from the work of CORI' s group (ILLINGWORTH et al. (15)) that the phosphate group of PLP does not exchange with Pi or glucose-1-P, the anionic substrates of phosphorylase. Recently we suggested the 5' -phosphate group of PLP (p_K 6.2) as a candidate for a catalytically functional group in phosphorylase, in analogy with the function of the carboxyl group of aspartate 52 in lysozyme (KASTENSCHMIDT et al. (16)). In order to block the phosphate group with p_K 6.2 of PLP the PL-5' -P-monomethylester was prepared. The preparative procedure is summarized in Fig. 2.

In order to prove unequivocally that one or the other protonatable group of PLP actually participates in the chemical reaction catalyzed by phosphorylase one must show that an inactive analogue can reconstitute a phosphorylase enzyme with the same or a very similar active site conformation as active PLP phosphorylase. This is necessary because in an enzyme, conformationally as flexible as phosphorylase, the proper fit of the prosthetic group to the enzyme and its conformational stabilization seem a prerequisite for its catalytic function.

Results

The role of the pyridinium nitrogen

BRESLER and FIRSOV (17) had assigned a role to the ring nitrogen of PLP in phosphorylase catalysis. Differential spectroscopy of phosphorylase b at high concentrations (≈ 20 mg/ml) in the presence of Pi or glucose-1-P indicated the appearance of a new maximum at 360 nm with a minimum at 330 nm. This was interpreted as indicating the formation of an ion pair between the pyridinium nitrogen of PLP and one of the negatively charged groups of Pi or glucose-1-P. This however, is very unlikely. FISCHER et al. (18) stated evidence which strongly argues against the proposal that the 360 nm absorbancy is indicating the formation of an enzyme substrate complex. Its appearance is more likely due to a conformational transition induced by the substrates which affects the interaction of PLP with neighbouring amino-acid side chains.

The N-Me derivative of PLP was prepared by POCKER and FISCHER. However, the N-Me PLP does not bind to apophosphorylase, presumably because of steric hindrance, electrostatic repulsion by the positively charged nitrogen or both (SHALTIEL et al. (13)). We have substantiated this conclusion by chemical analysis.

In contrast to N-Me PLP, NO-PLP binds to apophosphorylase b. Precautions must be taken however: PLP-NO is photoreactive. The reaction of PLP-N-O with apophosphorylase must be carried out in the dark and in the absence of oxygen. Reconstitution of apophosphorylase b with PLP-NO at 30°C at a molar ratio of 1.5 to 2:1 yielded an enzyme preparation with an activity of 20 to 25 % of that of the PLP reconstituted enzyme. This was also noted by FISCHER et al. (1, 18). We now could show that partial reactivation of apophosphorylase b is due to PLP formed by the reaction of apophosphorylase with PLP-NO. Use was made of the spectral differences between PLP-NO and PLP. (Fig. 3).

PLP-NO at neutral and alkaline pH has a strong absorbancy at 295 nm. At pH 7.0, ε = 9.800. PLP in this pH range lacks the 295 nm band. The disappearance of the 295 nm band provides therefore a convenient means to measure quantitatively the deoxygenation of PLP-NO. The deoxygenation of PLP-NO to PLP by apophosphorylase b is complete in 2 min at 30°C. It is rather specific: Neither lysozyme which has many lysyl groups or eggalbumin which has many SH groups did deoxygenate PLP-NO at comparable rates, nor did L-cysteine, 2-mercaptoethanol, lysine or ε-aminocaproic acid. This then poses the questions:

1. Which aminoacid side chain specifically reduces PLP-NO in apophosphorylase and

2. does this reaction take place at the PLP specific binding site?

PLP-NO phosphorylase b and PLP phosphorylase were reacted in 50 mM glycero-P-buffer pH 6.8 with dithio-bis nitrobenzoic acid ELLMAN (20), using conditions previously described by KASTENSCHMIDT et al. (16). Under these conditions only two exposed cysteinyl residues per monomer b form a mixed disulfide with DTNB in native phosphorylase b. Reaction of these two cysteinyl residues does neither affect catalytic activity or quarternary structure of phosphorylase b.

A comparison of the SH reactivity of PLP and PLP-NO phosphorylases revealed that for each mole of PLP formed a mole of SH becomes unreactive with DTNB (Fig. 4). This is a minimum estimate, because we would have missed SH groups which are reactive in native phosphorylase b only with alkylating agents or with DTNB at higher molar excess or which become reactive only on denaturation of the protein. Conceivably, additional SH groups not reactive with DTNB under the conditions of our experiments might take part in the deoxygenation of PLP-NO. Deoxygenation of PLP-NO by an SH group of the phosphorylase protein could lead to the formation of a disulfide via an intermediary sulfenic acid. Oxidation of a thiol to a sulfenic acid would be consistent with the 1:1 stoichiometry of loss of SH versus PLP formed. As is evident from Fig. 4 reduced PLP phosphorylase b with PLP irreversibly attached to the protein also reacts with PLP-NO. However, compared with apophosphorylase b reduced phosphorylase forms only half as much PLP and correspondingly loses half as many SH. This suggests that the deoxygenation reaction of PLP-NO$\rightarrow$PLP does not take place, at least not exclusively, at the PLP specific site. The cysteinyl residue that reacts with PLP-NO does not need to be involved directly in the binding of PLP and the assembly of the active site. It is more likely that the cysteinyl residue encounters the PLP-NO at some phase during the complicated structural transitions following the binding of PLP to apophosphorylase.

In view of the deoxygenation of PLP-NO to PLP by the phosphorylase protein, it was desirable to look for a stable analogue of PLP with an unprotonatable nitrogen in the pH range in which phosphorylase is active. For this purpose 3' -O-Me PLP-NO was synthesized. Although it is bound to apophosphorylase b, it is stable.

Experiments in Table 1 show that reconstitution of the apoprotein with 3' -O-Me PLP-NO yielded an inactive enzyme. The cofactor of which 0. 9 to 0. 95 moles per 100, 000 daltons of phosphorylase were bound to the protein was resolved and identified by paper electrophoresis in formic acid as the original 3' -O-Me PLP-NO. After removal of 3' -O-Me PLP-NO and a second cycle of reconstitution of the apoprotein - this time with PLP- a phosphorylase was obtained with 96 % of the specific activity of reconstituted active holophosphorylase b.

Table 1

REACTIVATION OF APOPHOSPHORYLASE b BY PLP ANALOGS

Analog	Moles of Analog/ Mole of apoenzyme	Specific Activity µmoles/mg/min	Extent of Reactivation %	Reactivation by PLP following Release of Analog or PLP	
				Specific Activity µmoles/mg/min	Extent of Reactivation %
PLP	2 : 1	65	100	62	96
3-O Me PLP NO	10 : 1	0	0	62	96
PLP-monomethylester	5 : 1	0	0	62	96

Apophosphorylase b (3×10^{-5}M) was reconstituted in glycero-P-buffer, pH 6.8 with analogs at 30°C at the molar ratios - analog: binding site per 100, 000 daltons - indicated. Time of reconstitution was 30 min in the case of PLP and PL-5' -P-monomethylester and 2 hrs. in the case of 3' -O-Me-PLP-NO. Enzymatic activity was determined with three different assay systems: In the direction of glycogen breakdown with the coupled enzyme assay of HELMREICH and CORI (21) or by arsenolysis (METZGER et al. (21)). In the direction of glycogen synthesis by measuring the release of P_i from glucose-1-P according to ILLINGWORTH and CORI (4) (see also: KASTENSCHMIDT et al. (16). Inorganic P was analyzed according to FISKE and SUBBAROW (22). Protein concentrations were determined, provided there was no interference by the analogs at 280 nm using an absorbancy index (1 % x cm) of 13.2 (KASTENSCHMIDT et al. (16)) otherwise the method of LOWRY (23) was used. Only the results obtained with the routine assay in the direction of glycogen synthesis are reported.

Modification of the phosphate of PLP

It is known that substitution of the 5' -phosphate group of PLP by a carboxyl or a sulfate group as in pyridoxal 5' -acetate or pyridoxal 5' -sulfate yields analogues which reconstitute inactive phosphorylases SHALTIEL et al. (13). The only active analogue is the 5' -methylene phosphate of pyridoxal, the PL-5' -phosphonate, which reactivates to about 20 % HULLAR (24), FISCHER and VIDGOFF (1).

The monomethylester of PLP was prepared by us because we wanted a PLP analogue with the phosphate group, p_K 6.2, blocked, but with the phosphate group $p_K < 2.5$ fully ionized at physiological pH. We hoped that the latter would suffice to properly position the pyridoxal phosphate to the apoprotein by ionic interaction with a positively charged amino-acid chain of the protein. As may be seen from Table 1 reconstitution of apophosphorylase b with PL-5' -monomethylester yielded a completely inactive phosphorylase.

Binding of PLP and analogues of PLP to apophosphorylase b

Using inactive analogues of a cofactor essential for enzymatic activity, one must prove that analogues which do not reactivate, still bind with sufficient strength to the protein. Table 2 indicates that this was the case for all the PLP-analogues under study.

But this is not enough: In order to prove that a functional group of PLP actually participates in catalysis, it must be convincingly shown that an analog with a particular functional group blocked still fits into its site exactly as the active cofactor. The steric and conformational requirements for the binding of a cofactor are set by the structure of the protein. These requirements must be met before a cofactor can fulfill its catalytic function.

SHALTIEL and CORTIJO (19) JOHNSON et al. (26), recently presented convincing evidence that at neutral pH PLP in native phosphorylase is buried in a hydrophobic pocket. This is shown schematically in Fig. 5, which was taken from the paper by SHALTIEL and CORTIJO (19).

PLP is therefore very strongly bound to phosphorylase. Thus, the binding of PLP (and PLP analogues) can not be treated on the basis of an equilibrium between free and bound ligand. From the data in Fig. 6 which serve as representative example, one would conclude that reactivation of apophosphorylase b with PLP or 3' -O-Me PLP is competitively inhibited by the inactive 3' -O-Me PLP-NO.

It is important to note that in all inhibition experiments the reactants including the inhibitor are added simultaneously at the start of reconstitution.

Table 2

STOICHIOMETRY OF BINDING OF PLP, 3-O Me PLPNO and PLP-MONOMETHYLESTER TO APOPHOSPHORYLASE b

Analog	Moles of Analog/ Mole of Apoenzyme	Type of Modification	Mole bound/ Mole apo-enzyme dimer
PLP	2 : 1	- -	1,85 - 2,2
3-O Me PLPNO	2 : 1	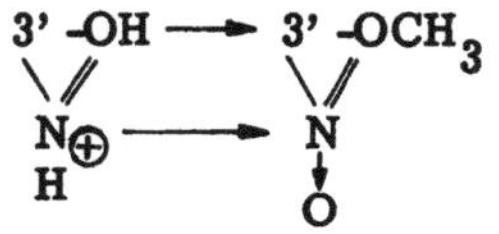	1,75 - 1,9
PLP-mono-monomethylester	2 : 1	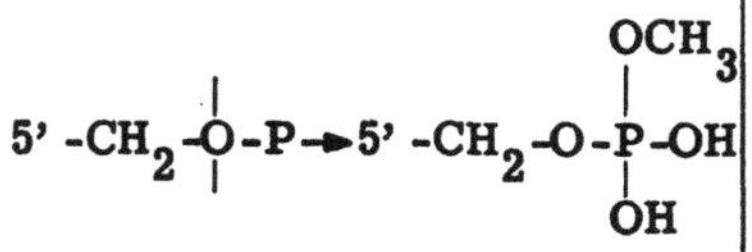	1,9 - 2,2

Reconstitution was carried out as described in the legend to Table 1. The reconstituted phosphorylases were passed over Sephadex G-25 and the bound cofactor was determined chemically by the method of WADA and SNELL (25). Moles of coenzyme bound are calculated per mole of dimer b of 200,000 daltons.

However, the experiment in Fig. 7 shows that inhibition by 3' -O-Me PLP-NO at a 1:1 molar ratio of 3' -O-Me PLP-NO : PLP is time dependent. If PLP is added to phosphorylase b already fully reconstituted with 3' -O-Me PLP-NO, the inactive analogue is very slowly replaced by active PLP. The half time of exchange is about seven hours.

On the basis of inhibition and exchange, we could divide the inactive PLP analogues we have studied, in two groups: Those which are and those which are not displaced by active PLP (or 3' -O-Me PLP). 3' -O-Me PLP-NO and PL belong to the former group. They are displaced, although very slowly, by PLP and 3' -O-Me PLP. To the latter group belong the inactive PL-5' -monomethylester and the active 3' -O-Me PLP. The latter analogues are not displaced by PLP. Thus, PLP and 3' -O-Me PLP bind for all practical purposes irreversibly to native phosphorylase b. The only inactive analogue which exhibits the same binding characteristics as the active cofactor, is the pyridoxal 5' -P-monomethylester.

On the other hand, 3' -O-Me PLP-NO appears to be bound differently to phosphorylase.
The imine bond linking PLP to an ε -aminolysyl group of the phosphorylase protein can not be reduced by $NaBH_4$ in native phosphorylase. SHALTIEL and CORTIJO (19) have assumed that the rate at which borohydride is decomposed in water is greater than the rate of reduction of the imine in the hydrophobic pocket, which shields the PLP from the aqueous environment. This offers an attractive handle to determine the position of a PLP analogue in the phosphorylase protein.
Native PLP phosphorylase b at pH 7 has major and minor absorption maxima at 333 nm and 425 nm respectively (Fig. 8). The ratio $\varepsilon_{333\ nm} : \varepsilon_{425\ nm}$ is 13. 3 : 1. For PLP-NO phosphorylase at pH 7 the ratio of $\varepsilon_{325\ nm} : \varepsilon_{425\ nm}$ is 1. 7 : 1 (Fig. 8). The spectral properties of PLP-NO phosphorylase are responsible for its yellow colour. PLP phosphorylase only turns yellow when the protein is reversibly denatured for example by acidification and denaturing agents. Under denaturing conditions and at acid pH, the PLP is turned from an apolar towards a more polar environment. This is responsible for the increase in absorbance at 425 nm and explains why PLP in the reversibly denatured protein is readily reduced by $NaBH_4$. From the spectral properties of PLP-NO phosphorylase we expected it to be reduced at neutral pH and under conditions where active PLP and 3' -O-Me PLP phosphorylase are not or only to a small extent reduced by $NaBH_4$. On the other hand, since phosphorylase reconstituted with PL-5' P-monomethylester has the same spectral (and fluorescence) properties as active PLP phosphorylase; it should not be reduced at neutral pH. These predictions could be verified experimentally, as shown in Table 3.
Thus, based on the data in Table 3, we again can divide the analogues with modified protonic groups into two categories: PLP-NO and 3' -O-Me PLP-NO phosphorylase are bound to the PLP site in apophosphorylase b, but they are turned towards a more polar environment. Hence, PLP-NO phosphorylase and 3' -O-Me PLP-NO phosphorylase are reduced with $NaBH_4$ at pH 6. 8 also several hours are required for complete reduction. This is very different from the behaviour of the PLP-holoenzyme, the active 3' -O-Me PLP phosphorylase and the inactive PL-5' -P-monomethylester phosphorylase. In these phosphorylases, the cofactors are bound to a more hydrophobic region. Hence, these phosphorylase enzymes are either not at all reduced by $NaBH_4$ at pH 6. 8 or to a lesser extent: 15 and 12. 5 % in the case of 3' -O-Me PLP and PLP ester phosphorylase, respectively. All phosphorylases are however rapidly reduced if the protein is reversibly "deformed".

Table 3

EXTENT OF REDUCTION WITH $NaBH_4$ AT pH 6.8, 0^oC AND IN THE ABSENCE OF DEFORMING AGENTS

Enzymes reduced	% Enzyme reduced in 3 hrs	Specific Activity before	after reduction
		μmoles/min/mg	
PLP-phosphorylase b	0	62	62
3' -O-Me PLP-phosphorylase b	15	16	12
PL-5' -P-Me-ester-phosphorylase	12.5	0	0
PLP-NO-phosphorylase b	65	14	14
3' -O-Me PLP-NO phosphorylase b	52	0	0

Apophosphorylase b (4×10^{-5}M) in 50 mM glycero-P, 25 mM 2-mercaptoethanol, 2 mM EDTA was reconstituted with a 3-fold molar excess over binding site per 100,000 daltons with PLP, PLP-NO and PL-5' -P-monomethylester for 20 min. In the case of 3' -O-Me PLP and 3' -O-Me PLP-NO a 6-fold molar excess was used and the time allowed for reconstitution was 3 hrs. Temp. was 30^oC in all instances. The reconstituted phosphorylases were reduced for 3 hrs. at 0^oC at pH 6.8. 6 portions - 50 μliters each - of a $NaBH_4$ solution containing 0.6 mg/ml were added stepwise during a 3 hr. period. After 3 hrs., the reduced enzymes were passed over a Sephadex G-25 column. The nonreduced coenzyme was determined with the method of WADA and SNELL (25). Enzymatic activity was determined before and after reduction in the routine assay in the direction of glycogen synthesis (see: Legend to Table 1). The activity of the PLP-NO phosphorylase b was the same before and after reduction, because it is due to the PLP formed from PLP-NO (see text). PLP bound to phosphorylase is not reduced under the conditions of these experiments.

To sum up: In the case of the analogues modified at the pyridinium nitrogen nothing conclusive can be said with respect to a possible participation of the pyridinium nitrogen - which is a strong nucleophile - in the phosphorylase reaction, because these analogues do not fit properly into the hydrophobic pocket. A protonatable pyridinium nitrogen is apparently necessary for holding PLP in the hydrophobic crevasse.

On the other hand, the inactive monomethylester of PLP and active PLP and 3' -O-Me PLP are all bound to an apolar region of the phosphorylase protein. We do not want to imply, of course, that the phosphate group like the pyridinium nitrogen is not also anchoring the PLP to its site, but one of the phosphates, the group with $p_K < 2.5$ may be sufficient to hold the PLP to its site. Only if one accepts that the inactive phosphorylase carrying the PL-5'-P-monomethylester and native PLP holophosphorylase have the same conformation, one can safely assume that the phosphorylase carrying the PLP-ester is inactive because the phosphate group, p_K 6.2, required for acid base catalysis, is blocked. However in order to prove or disprove this hypothesis, an entirely different experimental approach is required.

Fluorescence of PLP and PLP analogues bound to phosphorylase
(This part of the work was carried out in collaboration with Dr. H. Winkler, Max-Planck-Institute of Biophysical Chemistry, Göttingen).
Fluorescence spectroscopy is one of the most sensitive tools for studying structural transitions of proteins. SHALTIEL and FISCHER (27) found that pyridoxal-5'-phosphate in glycogen phosphorylase at neutral pH has a maximum fluorescence emission at 535 nm. The quantum yield is however very low: 0.012. The emission of pyridoxal-5'-phosphate at 535 nm can be elicited by excitation at 410 nm and at 333 nm. A fluorescence shift of such an order of magnitude has been interpreted by FÖRSTER et al. (28) and WELLER et al. (29, 30) as resulting from differences in the p_K values of protonatable groups in the ground and the excited state. The PLP fluorescence in phosphorylase has accordingly been interpreted by JOHNSON et al. (26). They have assumed that two tautometric forms of PLP bonded to a ε-aminolysyl group of the phosphorylase protein undergo intramolecular proton transfer in the excited state.
The pyridoxamine 5'-phosphate in reduced glycogen phosphorylase has a fluorescence excitation maximum at 330 nm and an emission maximum at 392 nm. Again the quantum yield is low but as CORTIJO and SHALTIEL (31) have shown the quantum yield increases about 10 fold with decreasing pH from 6.8 to 5.8. Use was made of this property of reduced phosphorylase b and a in the experiments in Figs. 9 and 10.
In Fig. 9 the pH dependence of the quantum yield of the fluorescence of pyridoxamine phosphate in active, reduced phosphorylase b and a, is compared with that of pyridoxamin 5'-phosphate monomethylester and pyridoxamin in inactive, reduced phosphorylase b.
It may be noted that the curves are rather similar for reduced phosphorylase b and a and for reduced phosphorylase b carrying the inactive PMP-5'-monomethylester. All three curves have a characteristic inflection between pH 6.0 and 6.3.

The curve for pyridoxamine phosphorylase b however, is shifted towards a higher pH range with an inflection around pH 7.2. In Fig. 10 the pH dependence of the fluorescence quantum yield for active reduced phosphorylase a and inactive reduced phosphorylase a carrying either the monomethylester of PMP or pyridoxamine are shown. The pH dependence of the fluorescence quantum yield is nearly the same for active reduced PMP phosphorylase a and inactive reduced PMP-5'-P-monomethylester phosphorylase a. PM-phosphorylase a as was the case with PM phosphorylase b behaves differently.

HELMREICH and CORI (21) could demonstrate large pH dependent changes in the affinity of phosphorylase a for the anionic substrate arsenate which replaced Pi in their experiments. The V_{max} values for arsenate from pH 6.0 to 7.5 were nearly the same with and without 5'-AMP, despite a five fold difference in the K_m values in the presence and absence of the cofactor. We have repeated these experiments with reduced phosphorylase a and have obtained similar results, i.e.: large increases in the K_m values for glucose-1-P in the pH range from 6.8 to 6.1 which are nearly completely reversed by 5'-AMP, whereas the pH dependent changes of V_{max} are less affected by 5'-AMP. Therefore, the enhancement of fluorescence of reduced phosphorylase with decreasing pH indicates a structural transition of the enzyme from a form with high affinity to one with low affinity for substrates. 5'-AMP shifts this allosteric equilibrium towards the high affinity form.

Fig. 11 shows as representative example experiments with reduced phosphorylase b and 5'-AMP. It may be seen that the fluorescence enhancement at low pH (i.e.: 6.5) can be completely quenched by 1×10^{-3} M 5'-AMP. 5'-AMP brings the quantum yield down to the level of 0.02. Such a low quantum yield is observed in the absence of 5'-AMP only at higher pH, i.e.: around 6.8. Similar results were obtained with reduced phosphorylase a with about $^1/100$ the amount of 5'-AMP. These observations allow us to correlate a functional, regulatory property of muscle phosphorylase, namely the effect of 5'-AMP on the affinity of the enzyme for its substrates, with a structural change of the protein, which is indicated by the reversal of the pH dependent fluorescence enhancement by 5'-AMP. Here it may be recalled, that the dependence of muscle phosphorylase on 5'-AMP for activity has been explained by an effect of 5'-AMP on the affinity of the enzyme for its substrates HELMREICH and CORI (21). Thus, the pH dependent enhancement of the quantum yield of PMP in reduced phosphorylase b and a and its reversal by 5'-AMP are suggestive of conformational transitions between inactive and active forms of the enzyme. It is therefore significant that the inactive phosphorylases b and a carrying the pyridoxamin phosphate-ester can undergo the same transitons as the active reduced holophosphorylases b and a. Precisely, what actual change in the microenvironment of the cofactor occurs and which amino acid residues are involved in the pH transitions remains to be seen.

Aggregation state of phosphorylases carrying PLP analogues

Polyacrylamide gel electrophoresis is widely used for probing into the association state of proteins.

The results presented in Figs. 12 and 13 show that the only inactive phosphorylase b which behaves on acrylamide gel electrophoresis like active PLP and 3' -O-Me PLP phosphorylases b is the PL-5' -monomethylester phosphorylase b. In contrast, 3' -O-Me PLP-NO- and PL-phosphorylases ressemble in their aggregation state more the heterogenous apophosphorylase b.

Summary

We have presented evidence that reconstitution of apophosphorylase with an analog of PLP whose 5' -phosphate group with p_K of 6.2 is esterified and thus not protonatable except at pH $<$ 2.5 yields an enzyme which by all criteria so far compared has a conformation which is very similar to that of the active PLP phosphorylase. However, the enzyme carrying the phosphate ester analogue of PLP is inactive. We would like to propose therefore that the phosphate group p_K 6.2 of PLP in glycogen phosphorylase participates in a general acid-base catalysis with nucleophilic substituion and retention of configuration.

REFERENCES

(1) FISCHER, E.H., J. VIDGOFF, unpublished results, personal communication (1970)

(2) BARANOWSKI, T., B. ILLINGWORTH, D.H. BROWN, C.F. CORI, Biochim. Biophys. Acta 25, (1957), 16

(3) SEERY, V.L., E.H. FISCHER, D.C. TELLER, Biochemistry 6, (1967), 3315

(4) ILLINGWORTH, B., G.T. CORI, Biochem. Prepn. 3, (1953), 1

(5) FISCHER, E.H., E.G. KREBS, Vitamins and Hormones 22, (1964), 399

(6) FISCHER, E.H., A.B. KENT, E.R. SNYDER, E.G. KREBS, J. Am. Chem. Soc. 80, (1958), 2906

(7) FISCHER, E.H., A.W. FORREY, J.L. HEDRICK, R.C. HUGHES, A.B. KENT, E.G. KREBS. Pyridoxal 5' -phosphate in the structure and function of phosphorylase. In: Chemical and Biological Aspects of Pyridoxal Catalysis, pp. 543-562. Ed. by SNELL, E.E., P.M. FASELLA, A. BRAUNSTEIN, A. ROSSI-FANELLI. Oxford: Pergamon Press, (1963), 543

(8) GRAVES, D.J., R.W. SEALOCK, J.H. WANG, Biochemistry 4, (1965), 290

(9) SNELL, E.E., Vitamins and Hormones 16, (1958), 77

(10) JENCKS, W.P., Catalysis in Chemistry and Enzymology, Mc Graw-Hill Book, (1969), 136

(11) AHRENS, M.L., G. MAASS, P. SCHUSTER, H. WINKLER, FEBS-Lett. 5 (1969), 327

(12) POCKER, A., E.H. FISCHER, Biochemistry 8, (1969), 5181

(13) SHALTIEL, S., J.L. HEDRICK, E.H. FISCHER, Biochemistry 8, (1969), 5189

(14) FUKUI, S., N. OHISHI, Y. NAKAI, S. SHIMIZU, Arch. Biochem. Biophys. 130, (1969), 584

(15) ILLINGWORTH, B., H.S. JANSZ, D.H. BROWN, C.F. CORI, Proc. Natl. Acad. Sci. U.S. 44, (1958), 1180

(16) KASTENSCHMIDT, L.L., J. KASTENSCHMIDT, E. HELMREICH, Biochemistry 7, (1968), 3590

(17) BRESLER, S., L. FIRSOV, J. Mol. Biol. 35, (1968), 131

(18) FISCHER, E.H., A. POCKER, J.G. SAARI, Essays in Biochemistry 6, (1970), 23

(19) SHALTIEL, S., M. CORTIJO, Biochem. Biophys. Res. Commun. 41, (1970), 594

(20) ELLMAN, G.L., Arch. Biochem. Biophys. 82, (1959), 70

(21) HELMREICH, E., C.F. CORI, Proc. Natl. Acad. Sci. U.S.A. 51, (1964), 131 and 52, (1964), 647

(22) FISKE, D.H., Y. SUBBAROW, J. Biol. Chem. 66, (1925), 375

(23) LOWRY, O.H., N.J. ROSEBROUGH, A.L. FARR, R.J. RANDALL, J. Biol. Chem. 193, (1951), 265

(24) HULLAR, T.L., Tetrahedron Lett. 49, (1967), 492

(25) WADA, H., E.E. SNELL, J. Biol. Chem. 236, (1961), 2089

(26) JOHNSON, G.F., J.I. TU, M.L.S. BARTLETT, D.J. GRAVES, J. Biol. Chem. 245, (1970), 5560

(27) SHALTIEL, S., E.H. FISCHER, Israel J. Chem. 5, (1967), 127

(28) FÖRSTER, Th., Z. Elektrochemie angew. physik. Chemie 54, (1950), 42

(29) WELLER, A.Z., Elektrochemie angew. physik. Chem 56, (1952), 662 and 58, (1954), 849 and 60, (1956), 1144

(30) WELLER, A., Naturwissenschaften 42, (1955), 175

(31) CORTIJO, M., S. SHALTIEL, Biochem. Biophys. Res. Commun. 39, (1970), 212

(32) SHALTIEL, S., J.L. HEDRICK, E.H. FISCHER, Biochemistry 5, (1966), 2108

(33) DIXON, M., E.C. WEBB, In "Enzymes", 2nd Ed. Longmans Ltd. London, (1964), 328

(34) HEDRICK, J.L., A.J. SMITH, Arch. Biochem. Biophys. 126, (1968), 155

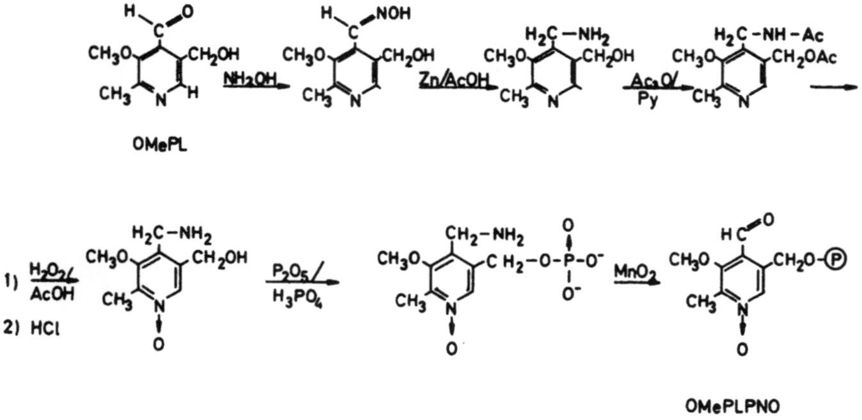

Fig. 1. Synthesis of 3′-O-Me-PLP-NO

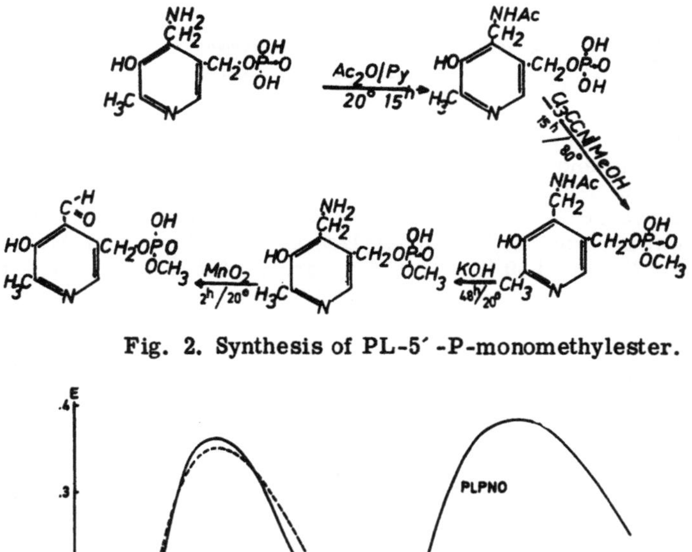

Fig. 2. Synthesis of PL-5′-P-monomethylester.

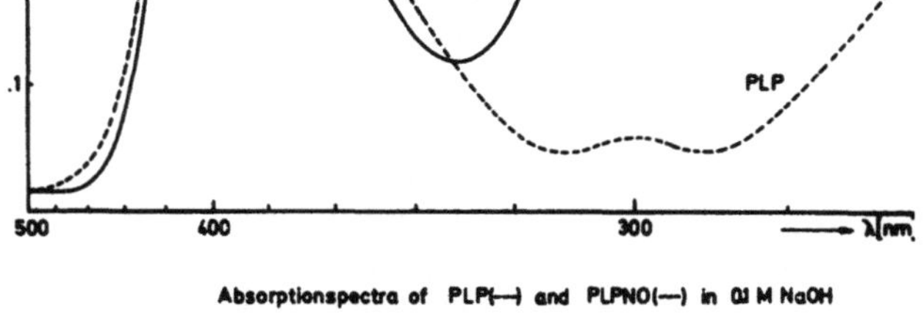

Fig. 3. The concentration of PLP was 5.2 x 10^{-5} M and that of PLP-NO 4.1 x 10^{-5} M. Temp.: 23 °C, Lightpath: 1 cm.

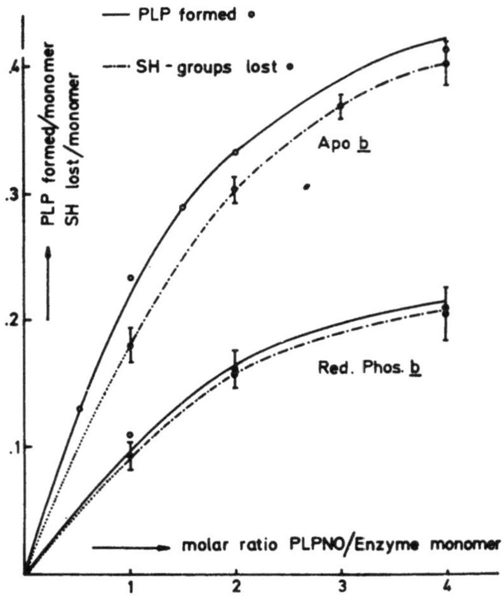

Fig. 4. Formation of PLP from PLP-NO by phosphorylase. Moles of PLP formed and moles of SH groups lost are calculated per monomer of 100, 00 daltons. Upper curves: Reaction of PLPNO with apophosphorylase b. Lower curves: Reaction of PLPNO with $NaBH_4$ reduced PLP-phosphorylase b. PLPNO was incubated with phosphorylase proteins (5×10^{-5} M) at the molar ratios indicated for 20 min. at 30 °C in 50 mM glycero-P, 2 mM EDTA buffer, pH 6. 8. Reaction was carried out in the dark and under nitrogen. For determination of PLP, and PLP-NO respectively, aliquots were removed from the reaction mixture after 20 min. and the cofactors released from the protein by precipitation with 0. 3 M $HClO_4$. To an aliquot of the supernatant fluid the same volume of 0, 5 M NaOH was added. The PLP formed from PLP-NO was determined from the decrease in the 295 nm absorbancy of the original PLP-NO solution (see Fig. 3 and text). The SH groups lost as given in this Fig., represent the difference between DTNB titratable SH groups of apophosphorylase b reconstituted with PLP and reconstituted with PLP-NO at the molar ratios indicated (upper curve). In the lower curve, the difference between DTNB titratable SH groups of reduced phosphorylase b and reduced phosphorylase b treated with PLP-NO at the molar ratios indicated, is shown. For SH-titration the enzymes ($7\text{-}9 \times 10^{-5}$ M) were freed from excess PLP-NO by gel filtration with Sephadex G-25, and then treated with a 20-fold molar excess of DTNB over phosphorylase dimer (200, 000 daltons) for 7 hrs at 23 °C in glycero-P buffer, pH 6. 8. The nitrophenylate anion was determined spectrophoto metrically at 412 nm KASTENSCHMIDT et al. (16). Apophosphorylase b was prepared according to SHALTIEL et al. (32). Reduced phosphorylase b was prepared as described by FISCHER et al. Crystalline rabbit skeletal muscle phosphorylase b was prepared according to Fischer and Krebs (1958). All the phosporylase preparations used in this work were more than three times recrystallized.

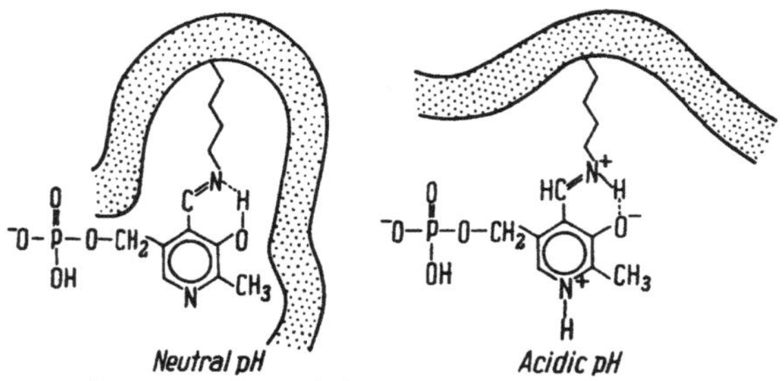

Fig. 5. Competitive inhibition of PLP- and 3-O-Me- PLP-phosphorylase b by the inactive analog 3-O Me PLP-NO. Apophosphorylase b was reconstituted for 30 min at 30 °C in 50 mM glycero-P-Na, 50 mM 2-mercaptoethanol buffer, pH 6. 8 with mixtures of the activating cofactors and the inactive 3-O-Me PLP-NO. PLP and 3-O-Me-PLP were added at the molar ratios indicated. The inactive analog 3-O Me PLP NO was added at varying ratios up to a 25-fold molar excess over the active analog. All reactants were added simultaneously at the start of the recombination experiment (see text). After reconstitution was completed, aliquots were withdrawn, diluted 150-fold with buffer and assayed for activity in the direction of glycogen synthesis. Protein concentration was measured spectrophotometrically (see legend to Table I). The data are plotted according to the method of Dixon and Webb (33). K_i is the apparent inhibitor constant for 3-O-Me-PLP-NO. The lines were drawn by the method of least squares.

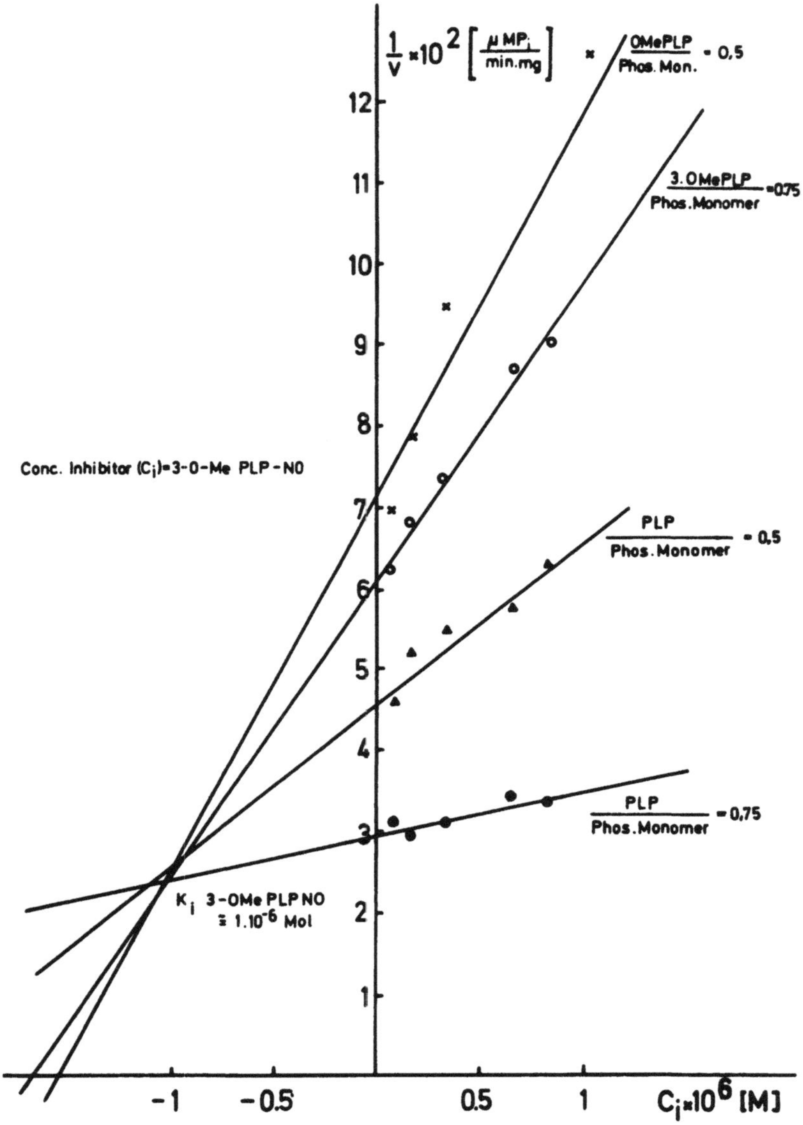

Fig. 6. Mode of attachment of PLP to phosphorylase.

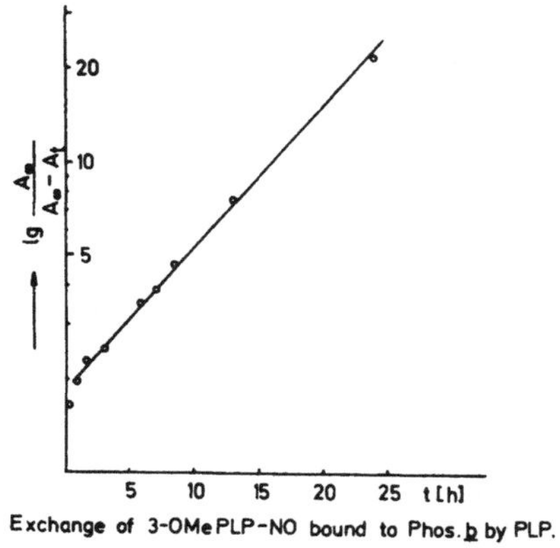

Fig. 7. Stoichiometrically reconstituted 3-O-Me-PLP-NO phosphorylase b (1.75×10^{-5} M) was incubated with an amount of PLP equimolar to 3-O-Me PLP-NO in 50 mM glycero-P-Na, 50 mM 2-mercaptoethanol buffer, pH 6.8 at 20 °C for the time indicated on the abscissa. Aliquots were withdrawn, diluted and the specific activity determined (see legend to Table I). A ∞ represents the specific activity of the enzyme extrapolated to infinite time when all of the inactive analog would have been replaced by PLP. Exchange is treated as a first order process.

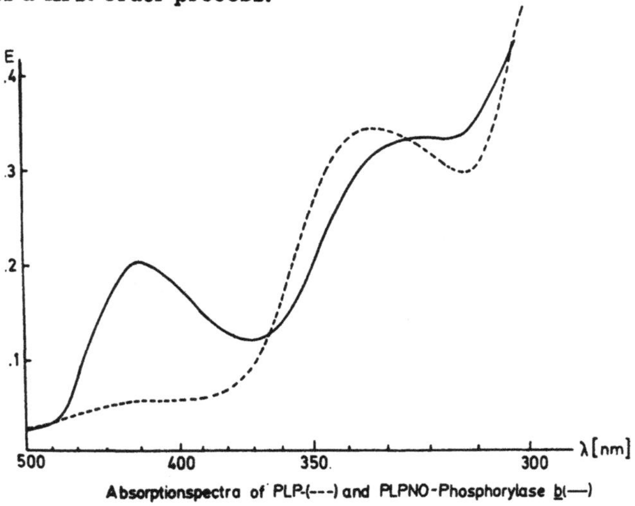

Fig. 8. PLP-phosphorylase b (7.4 mg/ml) and PLP-NO-phosphorylase b (5 mg/ml) in 50 mM glycero-P, 2 mM EDTA buffer, pH 6.8. Temp. 23 °C. Lightpath: 1 cm.

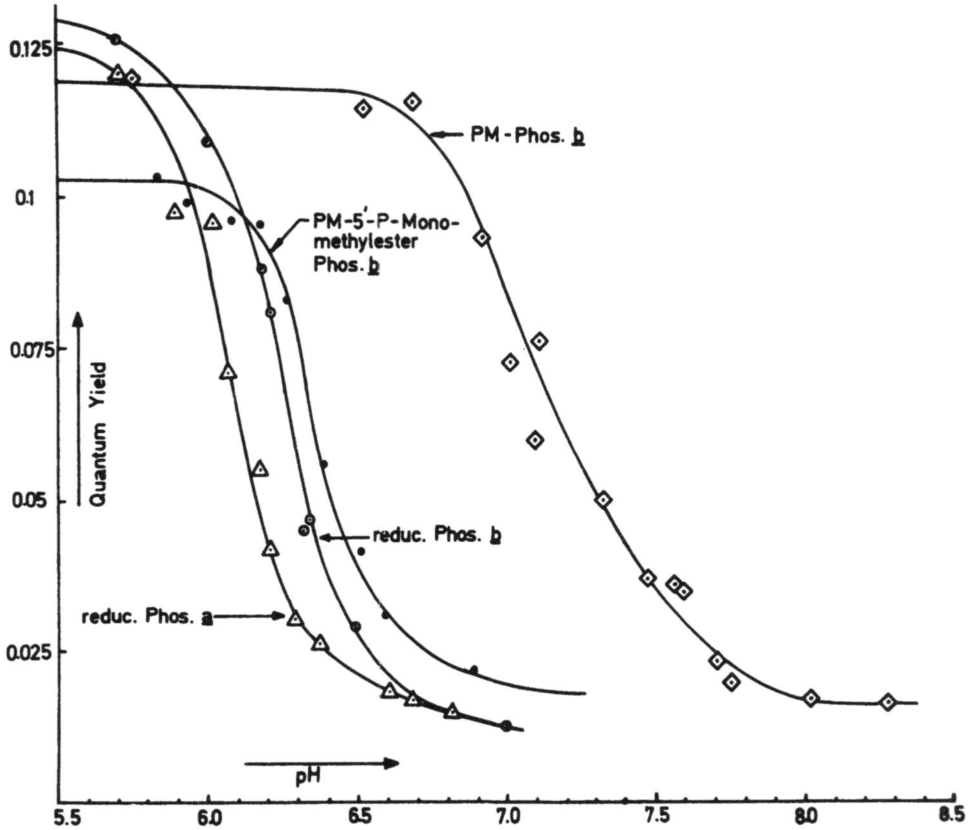

Fig. 9. pH dependence of the fluorescence quantum yield of $NaBH_4$ reduced phosphorylases b and a. The concentration of the enzymes ranged from 0. 3 to 3 mg/ml in 50 mM glycero-P, 50 mM 2-mercaptoethanol, 1 mM EDTA buffer. Incubation was at 25 °C. The pH was adjusted by addition of HCl. Excitation was at 334 nm. The analyzing wavelengths were in the range from 250 nm to 550 nm.

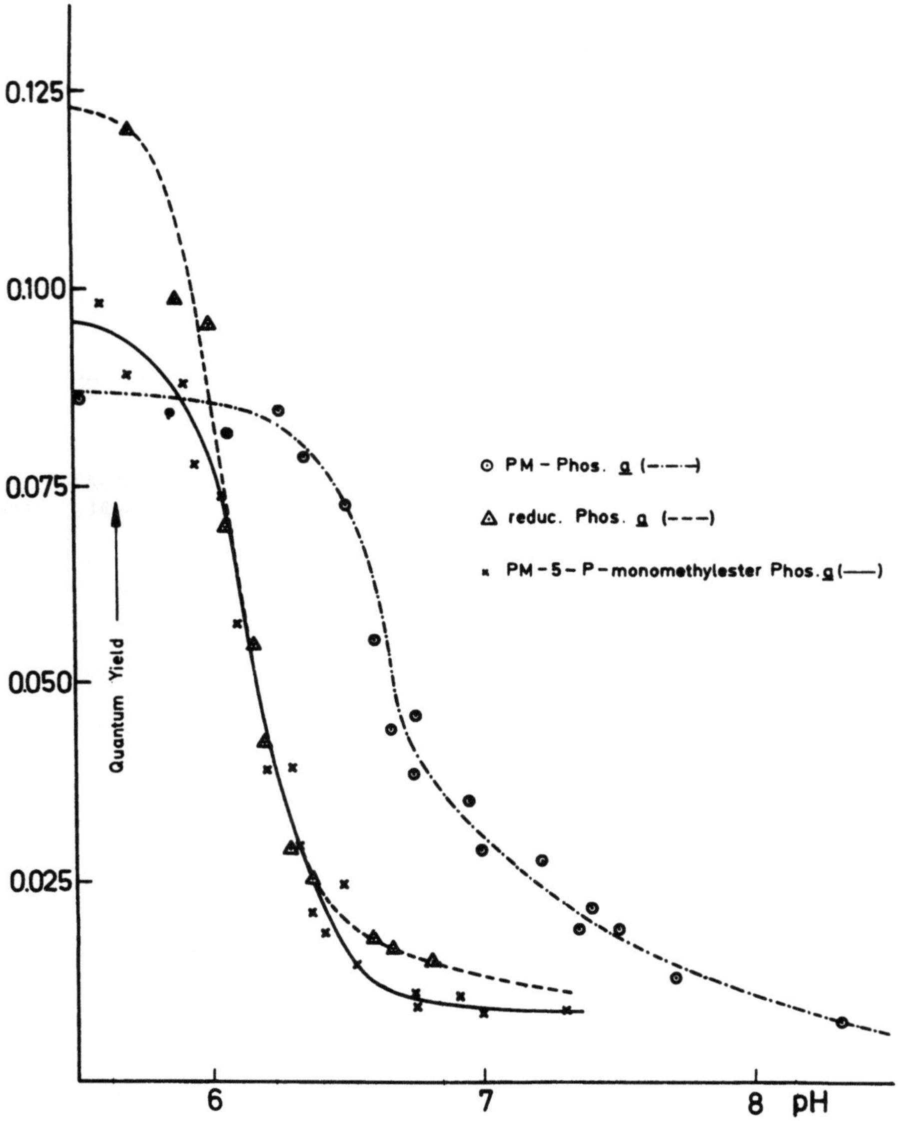

Fig. 10. pH dependece of the fluorescence quantum yield of reduced phosphorylases a. The experimental conditions were those described in the legend to Fig. 9.

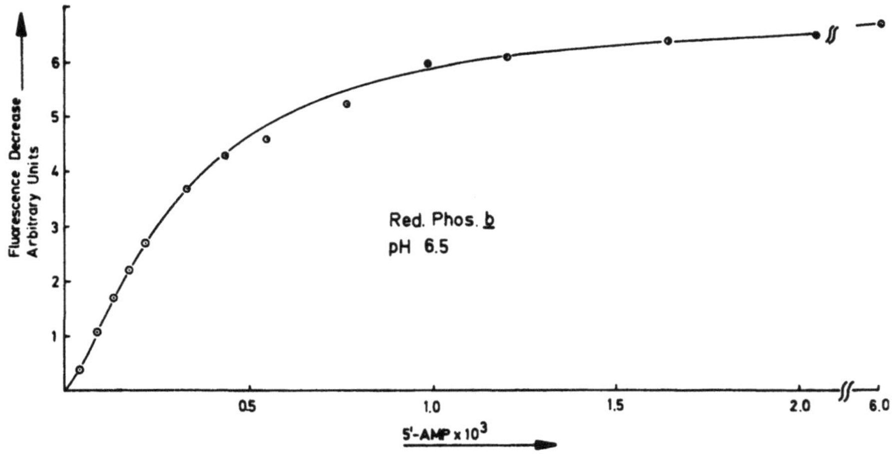

Fig. 11. Fluorescence quenching of reduced phosphorylase b by 5′-AMP. The concentration of the reduced enzyme was 2.87 mg/ml of 50 mM glycera-P, 50 mM 2-mercaptoethanol, 1 mM EDTA buffer, pH 6.5 at 25 °C. Excitation was at 334 nm and emission wavelength was 392 nm. From the fluorescence titration curve one would calculate a K_{diss} for 5′-AMP and reduced phosphorylase b at pH 6.5 and 25 °C of about 2.7×10^{-4} M if one were to neglect cooperation among binding sites.

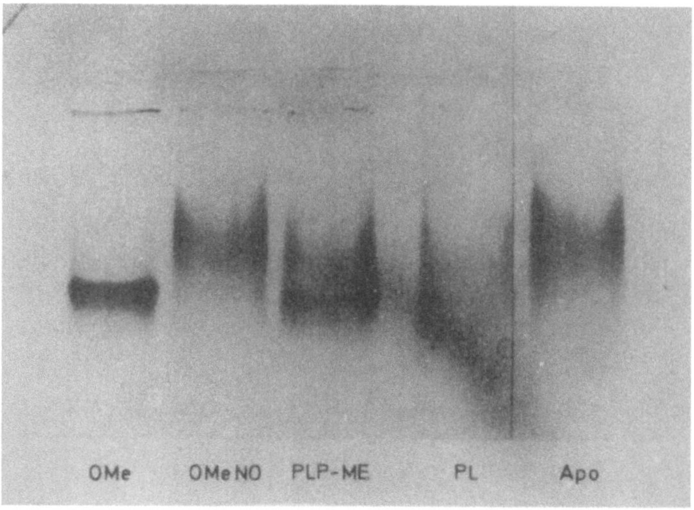

Fig. 12. Polyacrylamide gel electrophoresis of analog reconstituted- and apophosphorylase b. Electrophoresis was carried out at pH 8. 5. 0. 034 M Asparagine - Tris, pH 7. 3, was used as reservoir buffer. 30 μgs of the phosphorylase proteins in 30 μ liters of buffer were applied to the gel (6 %). 150 volts were applied to each gel block. Time was 3 hrs. Temperature was kept constant at 20 °C. The direction of migration is downward from the top (-) to the bottom (+). For further details, see HEDRICK and SMITH (34). O Me is active 3-O-Me-PLP phosphorylase b; O Me NO is inactive 3-O-Me-PLP-NO- phosphorylase b. PLP-Me is inactive PL-5′ -P-monomethylester-phosphorylase b; PL is inactive pyridoxal-phosphorylase b and Apo is inactive apophosphorylase b.

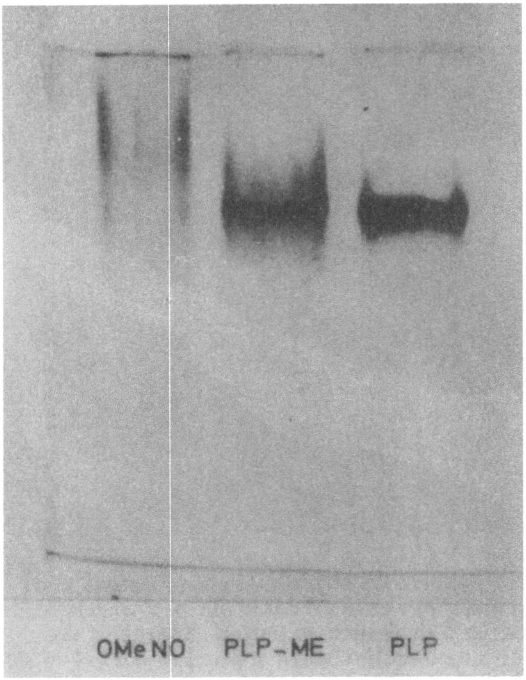

Fig. 13. Polyacrylamide gel electrophoresis of analog reconstituted- and PLP-holophosphorylase b. For conditions see legend to Fig. 13. PLP is native, active holophosphorylase b.

The Department of Physiological Chemistry, The University of Würzburg

D. PALM and H. D. WEISSHAAR

CONFORMATIONAL CHANGES UPON BINDING OF PLP, PLP-ANALOGUES AND 5'-AMP TO THE APOENZYME OF RABBIT SKELETAL MUSCLE PHOSPHORYLASE b

Phosphorylase b can be resolved from pyridoxal-5'-phosphat (PLP) in "deforming buffers" such as imidazol-citrate with L-cysteine as carbonyl reagent (1). The apoenzyme, which is crystallizable, can be fully reactivated on reconstitution with PLP (2). 5'-AMP, the allosteric modifier of phosphorylase, blocks resolution of phosphorylase b from PLP. In the preceding paper by HELMREICH (3) some of the conditions and structural prerequisites of PLP and PLP-analogues necessary for reconstitution of an enzymatically active phosphorylase have already been discussed. PLP has a double function: It participates in the formation of the active site conformation and in addition it may have an as yet undefined function in the chemical reaction catalyzed by phosphorylase (4). We therefore were interested in a method which would enable us to observe the complex conformational transitions accompanying the binding of PLP or PLP-analogues to apophosphorylase. Refinements of the hydrogen protein exchange method proved to be a valuable tool. This method (5) as originally conceived by LINDERSTROM-LANG was intended for estimation of the extent of α-helical structures in proteins. However, later studies made it clear that hydrogen exchange merely reflects the inaccessibility of the interior of native proteins. Further improvements of the method by ENGLANDER (6) made it possible for the first time to measure rapidly exchanging protons as well. Finally ANFINSEN (7) has quite recently introduced an additional important improvement by showing, that originally fast exchanging hydrogens exchange much more slowly or are even excluded from back-exchange, if a ligand shields hydrogens from the solvent in the protein-ligand complex. To describe this effect he has used the terms "suppression of back-exchange" or "trapping" of hydrogens.

We show, that conformational changes induced by a small ligand can also bring about this exclusion of hydrogens of the polypeptide backbone from the solvent.

Specifically, we will present evidence that back-exchange of hydrogens in phosphorylase is retarded to a large extent by the binding of PLP or PLP-analogues and 5' -AMP to the apophosphorylase protein.

We have measured the rate and extent of hydrogen back-exchange from the phosphorylase protein after it was equilibrated to a known extent in tritiated water. An outline of the experimental procedure is given in Fig. 1. Exchange-in with tritium labeled water results in an equilibrium between the peptide bond hydrogens and the aqueous solvent. The contribution of faster exchanging protons originating from other heteroatoms of amino acid residues may be neglected. The excess tritiated water was rapidly separated by passage of the protein through short Sephadex G-25 columns. Back-exchange starts during this transfer to unlabeled solvent (8). At the times stated (t_n) the amount of hydrogen, which the protein still retained, was determined by again passing the protein through Sephadex. In the effluent at each time intervall the protein concentration and radioactivity were measured. The protein concentration could easily be determined by using ^{14}C labeled acetylated polyethylene glycol as a marker. Polyethylene glycol moves exactly together with the protein through the column. The $^{14}C/^{3}H$ ratio is a direct indicator of the degree of back-exchange.

Hydrogen exchange equilibration between solvent and protein might not be obtained even after a 24 - 48 hour period. This was the case for phosphorylase. Thus, after short periods of exchange-in (5 min) only fast exchanging protons, which are readily accessible to the solvent are equilibrated. "Buried" parts of the polypeptide backbone will not be labeled extensively. If, at this early stage of exchange-in, a ligand is added, which induces changes in conformation or motility of the protein, some formerly exposed peptide bond hydrogens exchange very slowly. The effect exceeds direct shielding from the solvent by the ligand. All hydrogens with half times considerably longer than exchange-in time must have been excluded from exchange by the conformational alteration of the protein. This is shown in Fig. 1 by the change from circles to squares. Enough time was allowed in the experiments to bring the reconstitution of the apophosphorylase to completion. As controls, the experiments were performed exactly in the same manner, but PLP or other ligands were added prior to exchange-in.

The kinetics of exchange-out may be represented as a series of simultaneous first order processes: $H_t = \sum_i^n c_i \cdot e^{-k_i \cdot t}$ where H_t is the number or percentage of hydrogens not exchanged at time t, k_i is the rate constant, c_i is the number of hydrogens in the i-th kinetic class and n is the number of classes. Theoretically, n approaches the number of potentially exchangeable peptide bonds. Actually the problem can be simplified because the number of rate constants can be reduced to 2 to 4 average rate constants (9). The calculations were carried out with a Wang 720 computer. Average rate constants were determined by the method of successive substraction. Experimental and calculated data were fitted by the least square method to a multiple linear regression.

Results and discussion

Exchange-out of fully reconstituted enzymatically active phosphorylase b is shown in Fig. 2. The already completely reconstituted enzyme was incubated with HOT for 25 min. prior to exchange-out. Only fast exchanging protons with an average half time of 14 min. are labeled. If now reconstitution is initiated by adding PLP at 1 : 1 molar ratios calculated per monomer of phosphorylase b of 100 000 daltons to apophosphorylase b 5 min. after exchange-in is started and exchange-in is continued for another 20 min., the percentage of hydrogens per mole of protein, which are not exchangeable under the conditions of the experiment, rises by 23.3 $\pm$ 1.2 % of the potentially exchangeable hydrogens. This experiment quite clearly demonstrates that a total of 23 % of the peptide backbone, i. e. about 230 amino acid residues of 100 000 daltons, must have changed its conformation in such a manner that the peptide bond hydrogens became less accessible to the solvent. For the sake of discussion we can divide the hydrogen atoms excluded from back-exchange into fast, medium fast, slow, and extremely slow exchanging hydrogens. The respective half-times in min. are: 14, 125, 1 000, > 1 000.

Approximately 11 % of all potentially exchangeable hydrogens have an average half-time of about 1 000 min. This suggests that binding of PLP to apophosphorylase triggers a conformational transition which removes a rather large part of the protein from contact with the solvent. Since the molecular weight of the phosphorylase monomer is 100.000 daltons (10) about 100 amino acids seem to participate in this conformational transition. HELMREICH (3) has already presented other evidence indicating that PLP is bound and immobilized in a hydrophobic pocket of the phosphorylase protein (11). He has also presented data showing that reconstitution of apophosphorylase with PLP causes dramatic changes in the quaternary structure of the protein. Active reconstituted phosphorylase b is predominantly dimeric and apophosphorylase b predominantly monomeric at least at higher temperatures (12). We would like therefore to suggest that the fast and medium fast exchanging protons originating from approximately 12 % of the potentially exchangeable hydrogens, or 120 peptide bonds of the protein, are retarded from back-exchange because of the monomer-dimer transition. It should be recalled that the average exchange rates of these hydrogens differ from those involved in the formation of the hydrophobic pocket by a factor of more than 8 fold.

A promising way to assess the contribution of the PLP site to the overall conformational change of apophosphorylase was opened to us by comparing the PLP binding to crystalline D-serine dehydratase from E. coli. This is the only monomeric PLP-dependent enzyme so far described (13). Thus reconstitution of the apoenzyme can be studied without interference of a monomer-dimer transition.

The percentage of slow exchanging hydrogens, trapped by PLP on reconstitution of serine dehydratase was 10; there was very little trapping of fast exchanging hydrogens, i. e. 3 %, as was to be expected. Corresponding to the molecular weight of serine dehydratase, which is about one half of the phosphorylase subunit, approximately 45 amino acids are involved in the conformational change necessary for specific PLP binding in serine dehydratase. This may be compared to 100 amino acids in phosphorylase b. In the former enzyme, we could neither observe a significant influence on fast exchanging hydrogens by PLP nor by 5' -AMP.

For a more detailed study of the slow exchange, which is related to the hydrophobic inclusion of PLP, we used several analogues of PLP (3) for reconstitution. All of them stabilize a predominantly dimeric composition of the reconstituted phosphorylase protein at the protein concentrations used.

Pyridoxal (PL), although it is bound to apophosphorylase at pH 6. 8 (3), causes very little trapping. (Fig. 3) This is even true using 100 fold excess of PL. Thus PLP on removal of the 5' -phosphate group seems to loose its capacity to elicit a protein conformational change. PL is no longer capable of inducing the formation of a hydrophobic pocket.

The likewise enzymatically inactive analogue pyridoxal-5' -phosphat-monomethylester, however, shows remarkable trapping (Fig. 3). The number of fast exchanging hydrogens is 20 % less as compared to PLP-phosphorylase. The portion of the slowly exchanging hydrogens was still two thirds of those trapped with PLP. The slowly exchanging hydrogens are those which are involved in the formation of the hydrophobic binding region for PLP. In this respect it may be unexpected, that the partially active analogue, 3-0-methyl-PLP, entrapps even slightly less hydrogens than PLP-monomethylester. It resembles more the behavior of PLP as far the fast exchanging hydrogens are concerned (Fig. 3).

Both analogues, the active 3-0-methyl-PLP and the inactive PLP-monomethylester are capable in forming a cofactor binding site from which the solvent is largely excluded. These were also essential features of PLP binding. When we compare the structural features of the analogue-reconstituted enzymes, we find that the quaternary structure of the PLP-monomethylester enzyme is somewhat less stable, but its hydrophobic binding site is even more stable than that of the 3-0-methyl-PLP enzyme.

We found good agreement between the PLP binding stoichiometry and the extent of the trapping, assuming that 1 mole of PLP is bound per monomer of phosphorylase b. If during reconstitution hybrids were formed, one would expect more than proportional trapping with less than the stoichiometric amount of PLP bound. This however was not the case. The same proportionality was obtained with PLP-monomethylester (assay of the analogue according to the method of WADA and SNELL (14)). Although the active 3-0-methyl-PLP showed proportionality of trapping hydrogens to the binding of the analogue, the enzymatic activity was restored in this experiment only to about 1/2 the maximal reactivation after 20 min. (100 % after prolonged time). This suggests that the complex reconstitution sequence can be divided into an initial binding, coinciding with a change in the protein conformation; the last phase, probably rate limiting, is then a slow reshuffling process, ultimately leading to an enzymatically active and stable conformation.

For studying the fast exchange, which might be related to a change in the quaternary structure, 5' -AMP seemed to offer an attractive handle. 5' -AMP is the allosteric modifier of phosphorylase. There is an absolute requirement of 5' -AMP for phosphorylase b but only a relative requirement for phosphorylase a. As discussed earlier it influences the association of apophosphorylase b.

With fully reconstituted phosphorylase b no difference in exchange-out is observed wether or not 5' -AMP is added. If 5' -AMP is added to apophosphorylase in the absence of PLP, only fast and medium fast exchanging hydrogens are trapped (Fig. 4). This allows us to assign the fast and medium fast exchanging hydrogens to the association reaction of the protein. These hydrogens are prevented from exchange because access of the solvent is more restricted when the monomers make contact with each other through their contact surfaces and form dimeric and higher structures.

If apophosphorylase is preincubated with 5' -AMP, less hydrogens are trapped on subsequent reconstitution with PLP. 5' -AMP can cooperate with PLP in bringing about dimerisation. PLP has an effect on dimerisation and an effect on a rather restricted region of the protein, which is considered to be the PLP site. The former effect is similar to the effect of 5' -AMP. In the presence of 5' -AMP the PLP-specific effect on the slowly exchanging hydrogens is even somewhat higher and increases from 11.6 to 13 or 14 % of all potentially exchangeable hydrogens. If one compares the data of PLP and 3-0-methyl PLP in the presence of 5' -AMP one finds that the specific trapping effect in the case of 3-0-methyl PLP is only 5 % and only slightly influenced by AMP, distinguishable by a shift of fast exchanging hydrogens to slow exchanging hydrogens.

Our data may best be interpreted by assuming that PLP is enclosed in a hydrophobic region which is formed during reconstitution. PLP is firmly bound to this hydrophobic area. In the presence of 5' -AMP access of the solvent to this apolar region is even more restricted (15). The 3-0-methyl PLP seems to be more mobile and less restricted in the corresponding region which also shows less restricted motility. Moreover 5' -AMP has only minor effects on the 3-0-methyl PLP binding site. The PLP-monomethylester prevents access of solvent to the hydrophobic binding region to even a greater extent than the active 3-0-methyl PLP. On removal of the 5' -phosphate group no conformational changes are observable by hydrogen exchange. If we now look at the dimerisation reaction, we can say, that the apophosphorylase dimer formed by 5' -AMP is a rather loose structure. Once PLP is bound, more hydrogens are trapped, indicating that PLP considerably tightens the dimeric structure of phosphorylase b.

The synthetically prepared PLP-analogues were kindly provided by Dr. Th. PFEUFFER.

REFERENCES

(1) SHALTIEL, S., J. L. HEDRICK, E. H. FISCHER, Biochemistry 5, (1966), 2108

(2) SHALTIEL, S., J. L. HEDRICK, A. POCKER, E. H. FISCHER, Biochemistry 8, (1969), 5189

(3) HELMREICH, E., preceding paper of this series

(4) FISCHER, E. H., A. POCKER, J. C. SAARI, In "Essays in Biochemistry", ed. by P. N. CAMPELL and F. DICKENS, Academic Press, London and New York, 6, (1970), 23

(5) HVIDT, A., S. O. NIELSEN, Advances in protein chemistry, 21, (1966), 287

(6) ENGLANDER, S. W., Biochemistry 2, (1963), 798

(7) SCHLECHTER, A. N., L. MORAVEK, C. B. ANFINSEN, J. Biol. Chemistry 244, (1969), 4981

(8) ENGLANDER, S. W., Methods of Enzymology 12, (1968), 379

(9) LAIKEN, S. L., M. P. PRINTZ, Biochemistry 9, (1970), 1547

(10) FISCHER, E. H., personal communication

(11) SHALTIEL, S., M. CORTIJO, Biochem. Biophys. Res. Comun. 41, (1970), 594

(12) HEDRICK, J. L., S. SHALTIEL, E. H. FISCHER, Biochemistry 5, (1966), 2117

(13) DOWHAN jr., W., E. E. SNELL, J. Biol. Chemistry 245, (1970), 4618

(14) WADA, H., E. E. SNELL, J. Biol. Chemistry 236, (1961), 2089

(15) ULLMANN, A., P. R. VAGELOS, J. MONOD, Biochem. Biophys. Res. Comun. 17, (1964), 86

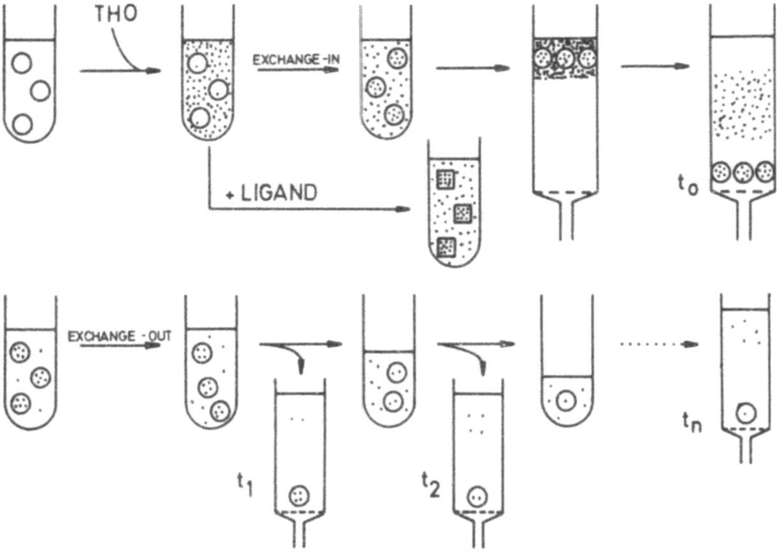

Fig. 1. Schematic presentation of hydrogen - tritium exchange of proteins. See text for further details.

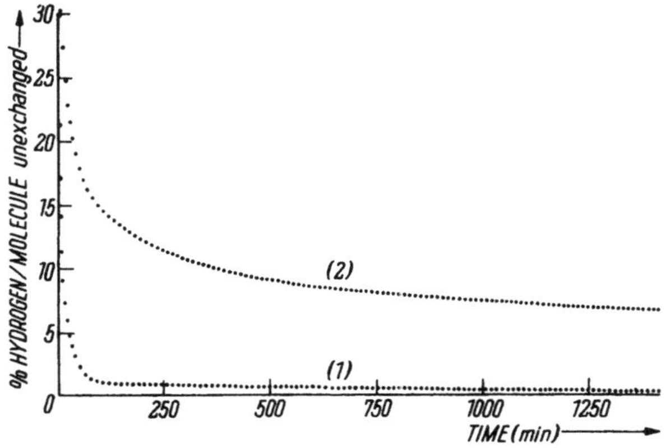

Fig. 2. Hydrogen - tritium exchange-out of phosphorylase b at 23 °. Glycerophosphat 50 mM, mercaptoethanol 50 mM, pH 7.0
(1) Apophosphorylase b, AMP free, (5mg/ml), reconstituted with PLP 5 min after addition of HOT. Total exchange-in time 25 min.
(2) Apophosphorylase b, reconstituted with PLP 20 min prior to addition of HOT. Total exchange-in time 25 min.

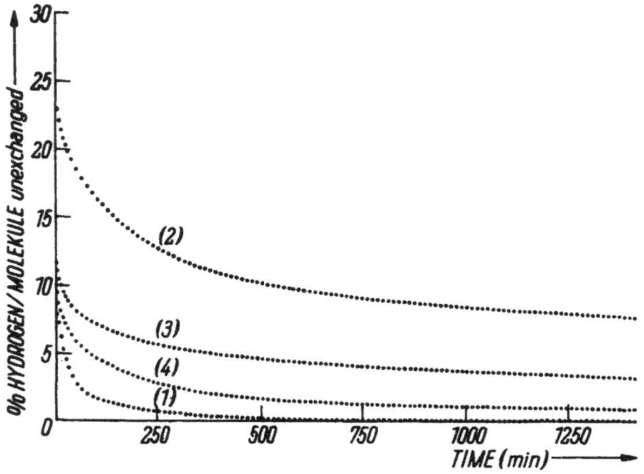

Fig. 3. Retarded exchange-out of labeled peptide bonds upon addition of PLP-analogues to apophosphorylase b. Experimental details see Fig. 2. (1) PLP added in a molar ratio of 2:1 per mole of subunit 5 min. after addition of HOT. (2) PLP-monomethylester, (2:1). (3) 3-0-methyl-PLP, (2:1). (4) PL, (100:1). The values given are the differences with controls, where the analogue was added prior to addition of HOT.

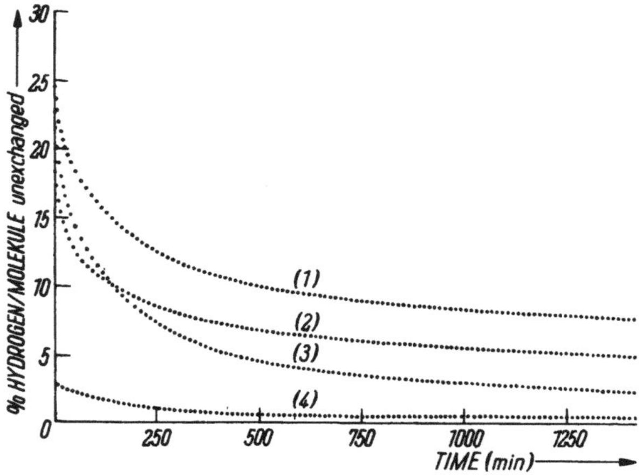

Fig. 4. Trapping of hydrogens upon addition of 5'-AMP or 5'-AMP + PLP to apophosphorylase b. Experimental details see preceding figures. (1) AMP added 5 min after addition of HOT. (2) PLP added 5 min after addition of HOT. (3) AMP added prior to addition of HOT; PLP added 5 min after addition of HOT. (4) AMP added prior to addition of HOT; 3-0-methyl-PLP added 5 min after addition of HOT.

A. N. Bach Institute of Biochemistry, Academy of Sciences and Institute for Vitamin Research, USSR, Moscow

N. B. LIVANOVA, B. I. KURGANOV, N. P. LISSOVSKAJA

ABOUT ALLOSTERIC INTERACTIONS ON PHOSPHORYLASE b FROM RABBIT MUSCLE

The allosteric and structural properties of phosphorylase b from muscle (α - 1,4 - glucan: orthophosphate glucosyl transferase, EC. 2.4.1.1.) have been the subject of much extensive investigation in recent years. Phosphorylase b has a subunit structure and belongs to the class of allosteric enzymes. Muscle glycogen phosphorylase b has an absolute requirement for the allosteric activator AMP and is inhibited by ATP, ADP, adenine, adenosine, glucose, glucose - 6 phosphate and UDPG.

Relatively little is known about the localization of binding sites for these effectors.

The term "allosteric regulation" is based on the assumption that active and allosteric sites are spatially distinct and that interaction between these two sites is mediated by conformational changes in protein molecule. The effects of synergism and antagonism in the combined action of two effectors on the activity of allosteric enzyme give prove to spatial arrangement of these effectors binding sites. For a quantitative evaluation of synergism or antagonism, we used in previous work (1) parameter j which for two negative effectors has the following form:

$$j = \frac{i_{1,2}}{1 - (1 - i_1)\ (1 - i_2)} ,$$

where i_1 and i_2 are individual degrees of inhibition ($i = 1 - V_i/V_o$; V_i and V_o are velocities of enzymatic reaction in the absence and in the presence of the inhibitor, respectively) and $i_{1,2}$ is the degree of inhibition after combined action of two inhibitors. The synergistic effects result in the values of j which are essentially greater than unit, whereas the antagonistic effects result in the values of j which are essentially less than unit. It is obvious that effects of

synergism and antagonism in the action of two effectors on the enzyme activity may be applied only for analysis of the spatial arrangement of two inhibitors or two activators binding sites. This approach was used in our provious work (1) for analysis of combined action of two inhibitors on phosphorylase b activity.
The study of two ligands combined binding at enzyme molecule is more general approach which allows to study the relative spatial arrangement of active and allosteric sites, and of positive and negative effectors binding sites as well.
It has been shown recently (2, 3) that glucose, glucose-6-phosphate, ATP and AMP decrease the rate of tryptic digestion of phosphorylase b. MUSZBEK and SÜMEGI (3) observed the decrease of fluorescence intensity during the phosphorylase b tryptic digestion.
In present work the mutual influence of the ligands on their binding with phosphorylase b has been studied by measuring the rate of phosphorylase b fluorescence intensity decrease during the tryptic digestion of the enzyme in the presence of pair of ligands to receive the data on spacial arrangement of binding sites for these ligands.
Glycogen phosphorylase b was prepared by the method of Fischer and Krebs and was recrystallized four times. The crystallized enzyme was treated with Norit before use to free it from nucleotides. Tryptic digestion was carried out at 43°C, buffered in 25 mM β-glycerophosphate - 1 mM EDTA - 0, 5 mM mercapto-ethanol at pH 7, 2. The final concentration of phosphorylase b was 0, 1 mg/ml and trypsin - 0, 067 mg/ml. The tryptic digestion was followed by the decrease of phosphorylase b fluorescence intensity (the excitation at 290 nm). The digestion was carried out in the constant temperature compartment of an Hitachi spectro fluorometer MRF-2A. In one minute after trypsin addition the plot of fluorescence intensity versus time is linear and its linearity maintains for at least 7 min. The initial velocity of fluorescence intensity decrease was calculated from the slop of the linear part of the kinetic curve. The results of the effect of phosphorylase b inhibitors - glucose and glucose-6-phosphate on the rate of fluorescence intensity decrease during tryptic digestion of the enzyme are summarized in Fig. 1.
Glucose and glucose-6-phosphate diminish the rate of phosphorylase b fluorescence intensity. Since the protective effect of these ligands at the same concentration is equal it is clear that the combined action of glucose + glucose-6-phosphate can be determined by taking of these ligands in the equal molar concentration.
Fig. 1 shows that the combined protective action of glucose+glucose-6-phosphate

is more pronounced than the individual action of each ligand. The synergism in the combined action of glucose and glucose-6-phosphate indicates that these two ligands may react with separate sites on the enzyme. This result is in agreement with our data on combined inhibitory effect of glucose and glucose-6-phosphate on the activity of phosphorylase b (1).

To decide whether the inhibitor - ATP binds at the same binding site as the activator - AMP or it has a separate binding site, the combined effect of AMP and ATP on the rate of phosphorylase b fluorescence intensity decrease during the tryptic digestion has to be determined. Both AMP and ATP have protective effect at the tryptic digestion of phosphorylase b but ATP is less effective; the effect of ATP is equal to that of AMP only if ATP concentration is 32 times more than AMP concentration. Coefficient 32 was obtained by shift along the abscissa axis and superposition of the curves of dependence on initial velocity of fluorescence intensity decrease upon logarithm of AMP and ATP concentration (Fig. 2).

Fig. 2 shows that combined action of AMP and ATP at molar ratio of their concentrations 1:32 practically coincides with the individual action of these ligands at their proper concentration. This indicates that AMP and ATP act at the same site on the enzyme, and is in accordance with the competitive inhibition of phosphorylase b activity by ATP with respect to AMP (4).

The results on combined action of activator and inhibitors (AMP + glucose and AMP+glucose-6-phosphate) on the rate of fluorescence intensity decrease during tryptic digestion of phosphorylase b are summarised in Figures 3 and 4. There is synergism in combined action of AMP and glucose (Fig. 3) and antagonism in combined action of AMP and glucose-6-phosphate (Fig. 4). This result suggests that AMP, glucose and glucose-6-phosphate have separate binding sites on enzyme.

The problem of spatial arrangement of binding sites of substrate - glucose -1-phosphate and of inhibitors - glucose and glucose-6-phosphate has a special interest. In contrast to other studied ligands glucose-1-phosphate accelerates the process of fluorescence intensity decrease during the tryptic digestion of phosphorylase b (Fig. 5).

The accelerating action of glucose-1-phosphate has been studied in the presence of glucose (10^{-2} M) and glucose-6-phosphate (10^{-2} M). If glucose and glucose-6-phosphate compete with glucose-1-phosphate for the same binding site on the enzyme, these inhibitors might be expected to lower the accelerating action of glucose-1-phosphate. But from the data in Fig. 5 ist is evident that the accelerating action of glucose-1-phosphate on the decrease of fluorescence intensity during the tryptic digestion of phosphorylase b does not diminish in the presence of glucose and glucose-6-phosphate at high concentrations. So, glucose-1-phosphate at the concentration 2.10^{-2} M without any additions accelerates the decrease of fluorescence intensity in 1.5 times and in the presence of glucose (10^{-2} M) or glucose-6-phosphate (10^{-2} M) in 2.1 times and 1.8 times, respectively. This result indicates that glucose-1-phosphate, glucose and glucose-6-phosphate bind at separate sites on the enzyme.

The data obtained by GRAVES et al. (5) are compatible with the conclusion that the binding site of glucose-6-phosphate overlaps neither the binding site of AMP nor the binding site of glucose-1-phosphate. The work by GRAVES et al. showed that phosphorylase b derived from phosphorylase a by partial tryptic hydrolysis is activated by AMP, but is not inhibited by glucose-6-phosphate. It must be noted, however, that the problem of spatial arrangement of binding sites for glucose, glucose-1-phosphate and glucose-6-phosphate awaits further investigation because it is known that glucose is competitive with respect to glucose-1-phosphate (6).

Thus, the data presented suggest that phosphorylase b possesses separate binding sites for AMP, glucose, glucose-6-phosphate and glucose-1-phosphate and the same binding site for AMP and ATP.

REFERENCES

(1) SILONOVA, G.V., N.B. LIVANOVA, B.I. KURGANOV, Molekularnaja Biologia, 3, (1969), 768, (in Russian)

(2) MUSZBEK, L., S. DAMJANOVICH, B. CSABA, Biochem. biophys. Acta, 167, (1968), 464

(3) MUSZBEK, L., J. SÜMEGI, Experientia, 26, (1970), 25

(4) MORGAN, H.E., A. PARMEGGIANI, J. biol. Chem., 239, (1964), 2440

(5) GRAVES, D.J., S.A. MANN, G. PHILIP, R. OLIVERA, J. biol. Chem., 243, (1968), 6090

(6) CORI, G.T., A.A. GREEN, J. biol. Chem., 151, (1943), 31

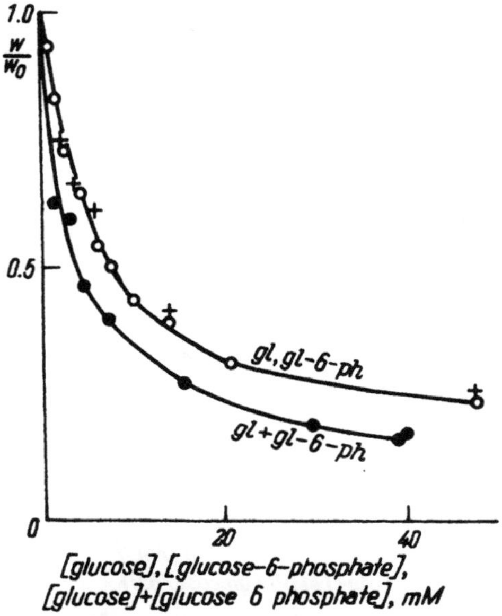

Fig. 1. A plot of the relative velocity of fluorescence intensity decrease during the tryptic digestion of phosphorylase b (w/w_0) vs. concentration of glucose (o), glucose-6-phosphate (+) and glucose + glucose -6- phosphate in molar ratio 1:1 (●).

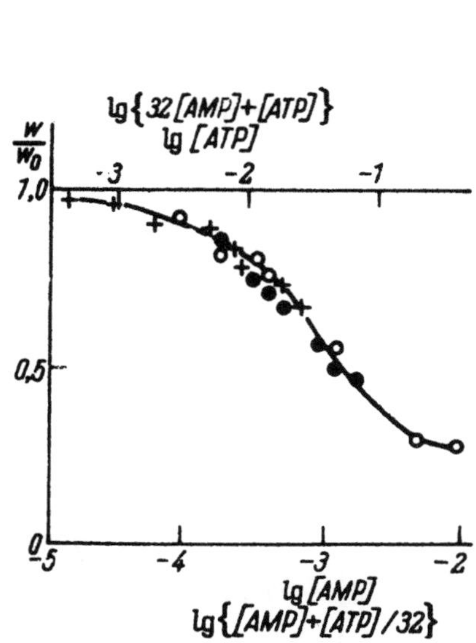

Fig. 2. A plot of the relative velocity of fluorescence intensity decrease during the tryptic dygestion of phosphorylase b vs. logarithm of the molar concentration of AMP (o), ATP (+) and AMP+ATP in molar ratio 1:32 (●).

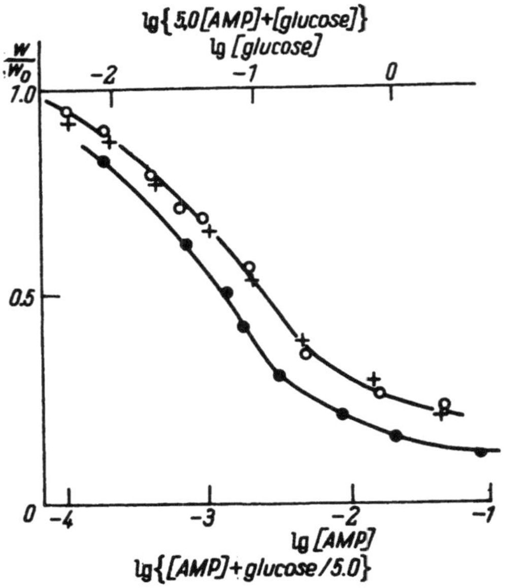

Fig. 3. A plot of the relative velocity of fluorescence intensity decrease during the tryptic digestion of phosphorylase b vs. logarithm of the molar concentration of AMP (o), glucose (+) and AMP+glucose in molar ration 1:5 (●).

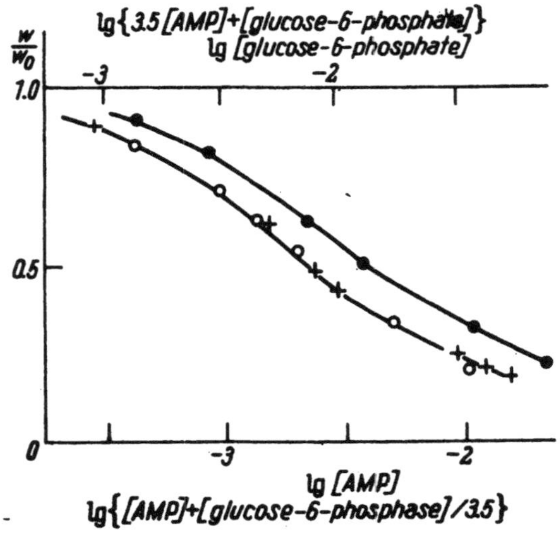

Fig. 4. A plot of the relative velocity of fluorescence intensity decrease during the tryptic digestion of phosphorylase b vs. logarithm of the molar concentration of AMP (o), glucose-6-phosphate (+), and AMP + glucose -6- phosphate in molar ratio 1:3, 5 (●).

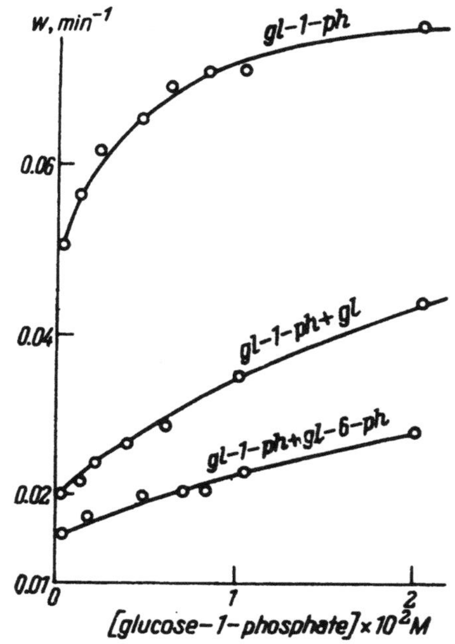

Fig. 5. A plot of the relative velocity of fluorescence intensity decrease during the tryptic digestion of phosphorylase b vs. concentration of glucose-1-phosphate in the absence of any additions and in the presence of 10^{-2} M glucose and of 10^{-2} M glucose-6-phosphate.

Forschungszentrum für Molekularbiologie und Medizin der Deutschen Akademie der Wissenschaften zu Berlin, Berlin-Buch

H. WILL, J.U. REICH, A. WOLLENBERGER

ALLOSTERISCHE EIGENSCHAFTEN DER GLYKOGENPHOSPHORYLASE b AUS HERZ- UND SKELETTMUSKEL DES SCHWEINS - EINES GEMISCHTEN V, K-ENZYMSYSTEMS

Für die Beschreibung der komplexen kinetischen Eigenschaften der Glykogenphosphorylase b der Kaninchenskelettmuskulatur hat sich das von MONOD, WYMAN und CHANGEUX vorgeschlagene Modell der allosterischen Transformation (1) sehr nützlich erwiesen. Substrat (Orthophosphat (OP), Glukose-1-phosphat (G1P)) und Aktivator (AMP) erhöhen einer des anderen Affinität für das Enzym und heben in genügend hohen Konzentrationen gegenseitig ihre Kooperativität auf. Daneben bewirkt AMP eine Stimulierung der katalytischen Effektivität des Enzyms. Tatsächlich wurde von einer Reihe von Autoren über eine starke Abhängigkeit der maximalen Geschwindigkeit (V_{max}) von der AMP Konzentration sowohl für die Reaktion der Phosphorylase des Glykogens als auch für die Reaktion der Glykogensynthese berichtet (2). Mit G1P als Substrat wird in Abwesenheit von AMP nur eine sehr geringe Aktivität gemessen. Erst bei hohen Aktivatorkonzentrationen wird V_{max} unabhängig von AMP. Demgegenüber ist nach BUC (3) in Richtung des Glykogenabbaus die maximale Aktivität auch in Abwesenheit von AMP bei genügend hohen Orthophosphatkonzentrationen erreichbar. Diese Aktivierung scheint aber mit einer unspezifischen Anionenaktivierung zusammenzuhängen, welche auch durch andere Anionen, z.B. Sulfatanionen, bewirkt werden kann (4).

Für die Isoenzyme der Glykogenphosphorylase b aus Herz- und Skelettmuskel des Schweins wurde analog der Skelettmuskelphosphorylase des Kaninchens ein komplexer Einfluß des AMP auf die Enzymkatalyse gefunden. Die Isolierung der beiden Enzyme erfolgte nach Methoden der Literatur; die gereinigten Enzyme erwiesen sich in der Disk-Elektrophorese als weitgehend einheitliche Protein- und scharf voneinander getrennte Aktivitätsbanden.

Abb. 1 zeigt die Substratsättigungskurven für das herzspezifische Isoenzym und die Skelettmuskelphosphorylase in Abhängigkeit von den angegebenen AMP-Konzentrationen. AMP verändert sowohl die Kooperativität als auch die maximale

Geschwindigkeit der kinetischen Kurven. Das läßt sich aus den durch einen Rechner ermittelten Parametern der HILL-Gleichung ersehen. In Abb. 2 ist die Computeranpassung entsprechend der HILL-Gleichung an die experimentellen Werte in einem halblogarithmischen Koordinatensystem dargestellt. Für die Skelettmuskelphosphorylase sind die HILL-Parameter in der Tabelle 1 zusammengefaßt. Für beide Enzyme werden mit fallender AMP-Konzentration sowohl K als auch der HILL-Koeffizient $\bar{n}$ größer, während V_{max} abnimmt. Nach MONOD et al. (1) können allosterische Enzyme als K oder V-Systeme eingeordnet werden.

Tabelle 1

SKELETAL MUSCLE ISOENZYM

(AMP) mM	V_{max} μmoles OP(min mg)$^{-1}$	K mM	$\bar{n}$
1, 0	12, 2 $\pm$ 0, 3	5, 21 $\pm$ 0, 4	1, 17 $\pm$ 0, 07
0, 1	9, 65 $\pm$ 0, 2	9, 0 $\pm$ 0, 6	1, 39 $\pm$ 0, 07
0, 05	7, 8 $\pm$ 0, 4	14, 5 $\pm$ 1, 7	1, 29 $\pm$ 0, 1
0, 01	1, 72 $\pm$ 0, 2	25, 2 $\pm$ 4, 5	1, 63 $\pm$ 0, 2

In einem K-System wirkt der Aktivator (oder Inhibitor) durch Änderung der Bindungsaffinität des Enzyms für das Substrat, in einem V-System dagegen nur durch Modifikation der katalytischen Effektivität des Enzyms. Diese Systeme sind eher als Grenzzustände eines allgemeineren Mechanismus, des gemischten K, V-Enzymsystems aufzufassen, wie es sich hier für die Isoenzyme der Phosphorylase b abzeichnet. Bei der mathematischen Modellierung dieses Systems wurde das MONOD-Modell der allosterischen Transformation zugrunde gelegt. Abb. 3 zeigt alle sich im Rahmen dieses Modells zwischen dimerem Enzym, Substrat und Aktivator bildenden Komplexe. Die kinetischen Eigenschaften dieses Systems hängen 1. von den Konstanten der hier formulierten Gleichgewichte und 2. von den Reaktionskonstanten für den Zerfall der Enzym-Substrat-Komplexe zu den entsprechenden Endprodukten ab.
Eine allgemeine Lösung dieses Systems ist komplex, die eindeutige Bestimmung aller auftretenden Parameter aus kinetischen Messungen allein nicht möglich.

Es wurde deshalb zusätzlich zu den MONOD' schen Postulaten (welche sich streng genommen nur auf die Bindung beziehen) angenommen, daß 1. die Gleichgewichtseinstellung zwischen allen Enzymkomplexen schnell im Vergleich zur Geschwindigkeit der Bildung der Reaktionsprodukte erfolgt, 2. die Geschwindigkeitskonstanten für den Zerfall aller Substrat enthaltender Spezies mit dem Enzym in der R-Konformation gleich sind, 3. die Enzym-Substrat-Komplexe mit dem Enzym in der T-Konformation katalytisch inaktiv sind.
Für die Abhängigkeit der Geschwindigkeit von den Substrat- und Aktivatorkonzentrationen ergibt sich dann:

$$\frac{v}{V_{max}} = \frac{\alpha(1+\alpha)}{(1+\alpha)^2 + L(1+c\alpha)^2\left(\frac{1+d\beta}{1+\beta}\right)^2} \qquad (1)$$

Hier bedeuten L die allosterische Konstante, und β die normierten Substrat- und Aktivatorkonzentrationen,

$$c = \frac{K_R^{G1P}}{K_T^{G1P}} \quad \text{und} \quad d = \frac{K_R^{AMP}}{K_T^{AMP}} \cdot \quad K_R^{G1P},\ K_T^{G1P},\ K_R^{AMP},\ K_T^{AMP}$$

sind die Dissoziationskonstanten der Enzym-Liganden-Komplexe für die beiden Konformationen R und T.
Abb. 4 zeigt die in Übereinstimmung mit Gleichung (1) durchgeführte Computeranpassung an die experimentellen Werte. Sie fällt für das herzspezifische Enzym wegen der im Modell nicht berücksichtigten Substrathemmung bei hohen G1P-Konzentrationen schlechter aus als für die Skelettmuskelphosphorylase. Die numerische Auswertung erlaubt alle in Gleichung (1) auftretenden Parameter zu bestimmen. In Tabelle 2 sind diese für die Skelettmuskelphosphorylase wiedergegeben. Der hohe Wert für L bedeutet, daß in Abwesenheit von Liganden die Phosphorylase b fast ausschließlich in der inaktiven T-Konformation vorliegt. Diese Form hat auch eine geringere Affinität für die Liganden G1P und AMP als die R-Konformation. Mit steigender Ligandenkonzentration wird das Gleichgewicht deshalb in Richtung der aktiven R-Konformation verschoben.
Die ermittelten Parameter gestatten spezifische Eigenschaften des allosterischen Enzymsystems, wie z. B. die Bindung des Aktivators AMP an die Phosphorylase b in Abwesenheit von Substrat, vorauszusagen. Tatsächlich gilt für das

Tabelle 2

SKELETAL MUSCLE ISOENZYM

V_{max}, μmoles OP $(\text{min mg})^{-1}$	12 ± 0,49
L	10600 ± 2400
K_R^{G1P}, mM	1,85 ± 0,4
K_T^{G1P}, mM	20 ± 1,8
K_R^{AMP}, mM	0,00258 ± 0,00035
K_T^{AMP}, mM	0,154 ± 0,012

einfache System: dimeres Enzym-Ligand, bei nichtexclusiver Bindung des letzteren nach RUBIN und CHANGEUX (5):

$$\bar{y} = \frac{\beta(1+\beta) + Ld\beta\,(1+d\beta)}{(1+\beta)^2 + L(1+d\beta)^2} \qquad (2)$$

In Abb. 5 ist die unter Verwendung der Konstanten aus der Tabelle 2 und der Gleichung (2) berechnete Abhängigkeit $\bar{Y}$ vs (AMP) für die Skelettmuskelphosphorylase dargestellt. Die eingezeichneten Punkte dagegen geben die experimentellen Werte der Bindungsuntersuchung wider. Die Bindung des AMP wurde nach einer von Colowick und Womack vorgeschlagenen Dialysemethode (6) untersucht, die auf der Messung der Dialysegeschwindigkeit von radioaktiv markiertem Liganden aus einer Gleichgewichtsmischung von Enzym und Ligand beruht.
Der Vergleich der experimentellen und berechneten Abhängigkeit sowohl der kinetischen als auch der Bindungsuntersuchungen zeigt die Möglichkeit der Simulation von Enzym-Ligand-Beziehungen durch das vorgeschlagene Modell. Die erzielte Anpassung schließt aber andererseits weitere mögliche Modelle nicht aus.

LITERATUR

(1) MONOD, J., J. WYMAN, J.-P. CHANGEUX, J. Mol. Biol., 12, (1965), 88

(2) BLACK, W.J., J.H. WANG, J. Biol. Chem., 243, (1968), 5892

(3) BUC, H., Biochem. Biophys. Res. Commun., 28, (1967), 59

(4) ENGERS, H.D., N.B. MADSEN, Biochem. Biophys. Res. Commun., 33, (1968), 49

(5) RUBIN, M.M., J.-P. CHANGEUX, J. Mol. Biol., 21, (1966), 465

(6) COLOWICK, S.P., F.C. WOMACK, J. Biol. Chem., 244, (1968), 774

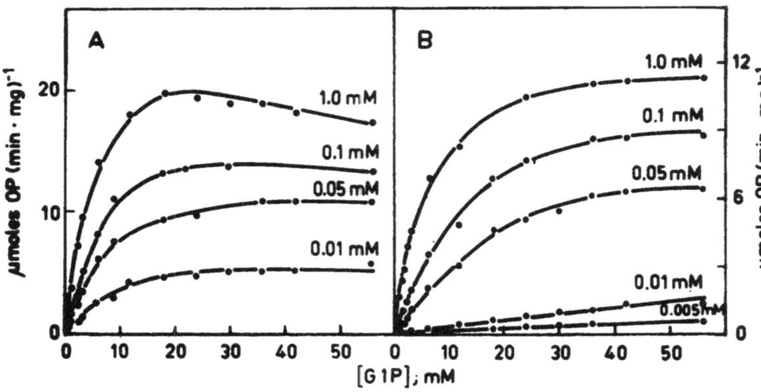

Fig. 1. Abhängigkeit der Phosphorylaseaktivität von der G1P- und AMP-Konzentration. A: Herzspezifisches Isoenzym. B: Skelettmuskelphosphorylase. Versuchsbedingungen: Puffer: 5 mM-Na-Glycerophosphat-2 mM EDTA-1 mM Mercaptoäthanol, pH 6, 8. Die Konzentration der Phosphorylasen betrug 0, 2-1· 10^{-7} M, des Glykogens 0, 5 %. μ = 0. 12 (KCl), t = 20 °C.

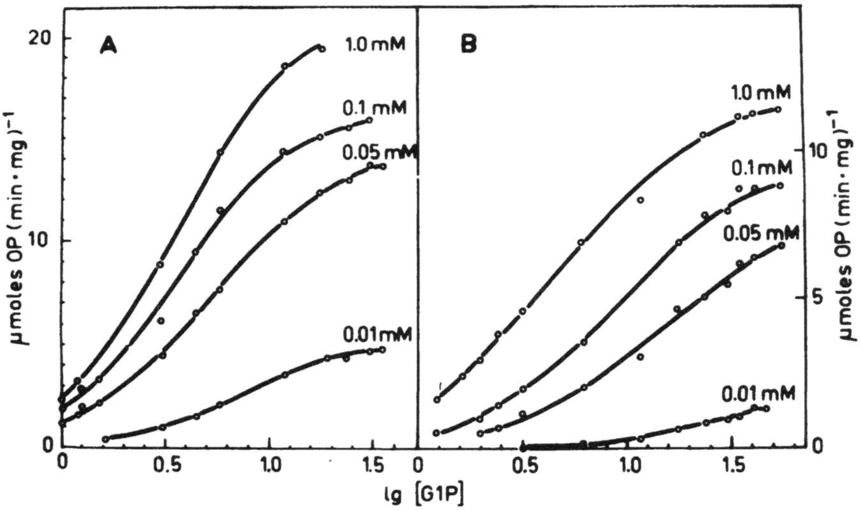

Fig. 2. HILL-Simulation der experimentellen Werte aus Abb. 1. Die durchgezogenen Linien repräsentieren die errechneten Kurven. Die Punkte sind die experimentellen Werte aus Abb. 1.

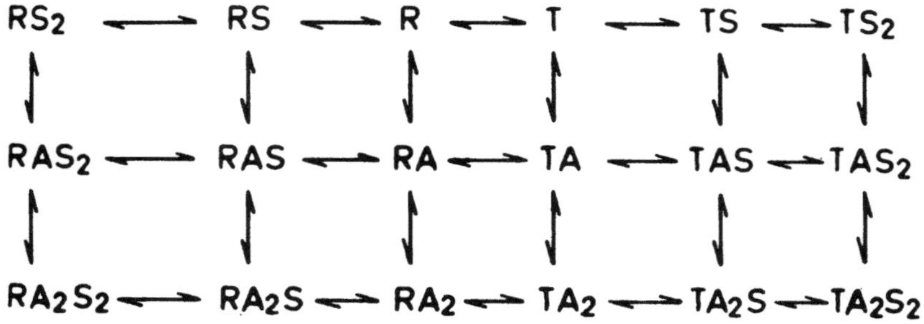

Fig. 3.

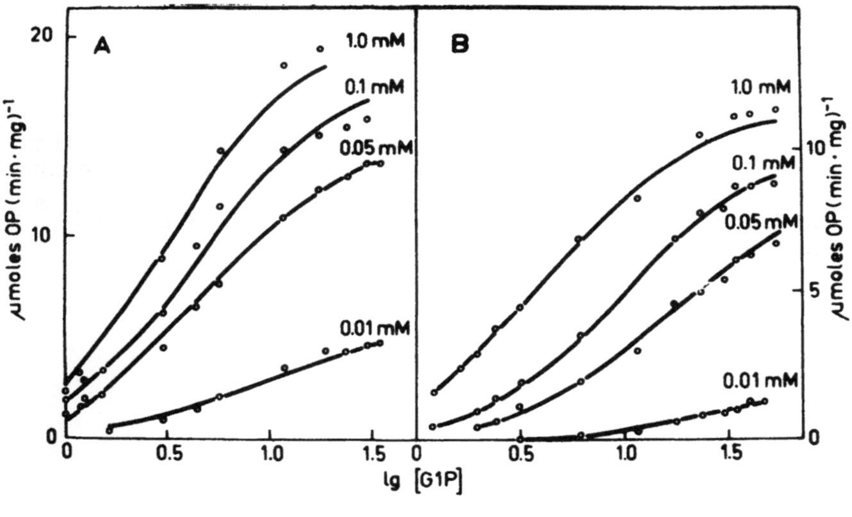

Fig. 4. Simulation der experimentellen Werte entsprechend Gleichung (1). Die durchgezogenen Linien sind die errechneten Kurven. Die Punkte sind die experimentellen Werte aus Abb. 1.

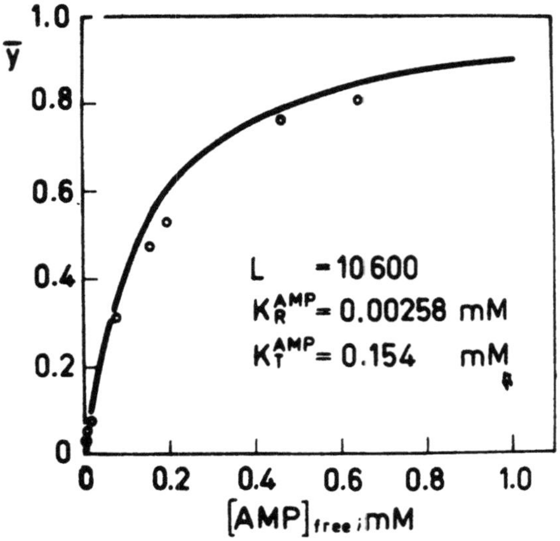

Fig. 5. Simulation der AMP-Bindung entsprechend Gleichung (2) und experimentelle Werte der Bindungsuntersuchung. Die durchgezogene Linie gibt die errechnete Kurve wider.
Die Konzentration der Skelettmuskelphosphorylase b betrug $5,5 \cdot 10^{-5}$ M. Sonstige Bedingungen siehe Abb. 1.

Forschungszentrum für Molekularbiologie und Medizin der Deutschen Akademie der Wissenschaften zu Berlin, Berlin-Buch

E.-G. KRAUSE, H. WILL, A. WOLLENBERGER

ISOENZYME DER PHOSPHORYLASE b IM HERZMUSKEL DES SCHWEINES: EINIGE ASPEKTE ZUR REGULATION DER GLYKOGENOLYSE

Die den phosphorolytischen Abbau des Glykogens katalysierende Phosphorylase (EC 2.4.1.1) zeichnet sich durch vielfältige Beziehungen der katalytischen Aktivität und der molekularen Struktur aus. Diese beinhalten neben der allosterischen Regulation durch Substrate, AMP und andere Effektoren, insbesondere Glukose-6-Phosphat (G6P) und ATP, die enzymatische Modifizierung der Glykogenphosphorylase durch die Phosphorylase Kinase und Phosphorylase Phosphatase (1). Neben diesen bereits seit längerer Zeit im Muskelgewebe bekannten Regulationsmechanismen der Glykogenolyse kommt einer weiteren Komponente Aufmerksamkeit zu: der Existenz von Phosphorylase-Isoenzyme im Herzmuskel verschiedener Säugetiere (2, 3, 4). Im Herzmuskel können 3 verschiedene Isophosphorylasen vorkommen, die in der Reihenfolge ihrer elektrophoretischen Beweglichkeit in Richtung der Anode als Isoenzym 1, 2 und 3 bezeichnet werden. Das Isoenzym 3 ist identisch mit der Glykogenphosphophosphorylase aus Skelettmuskel (5), wogegen das Isoenzym 1 im Herzmuskel, in der glatten Muskulatur und anderen Geweben vorkommt.

Da über die Regulationscharakteristik des herztypischen Isoenzyms bisher nur wenig bekannt ist, sein Anteil und seine Bedeutung für die myokardiale Glykogenolyse deshalb nur abgeschätzt werden kann, wurden von uns eine Reihe von Effektoren hinsichtlich ihrer regulatorischen Potenz für die Aktivität der beiden Isoenzyme getestet. Die vorgelegten Befunde geben einen Vergleich zwischen einigen kinetischen Parametern der Isophosphorylase 1 und 3 aus Schweinemuskel, wobei neben gleichartigem Verhalten deutlich differierende Eigenschaften der 2 Phosphorylase-Spezies zu erkennen sind.

Die Isolierung der beiden Enzyme aus Herz- bzw. Skelettmuskel und einige Reinheitskriterien der benutzten Enzympräparate wurden bereits andernorts dargestellt (4). In der Tabelle 1 sind die für beide Isophosphorylasen erhaltenen

K_m-Werte für die Substrate Glykogen, Glukose-1-Phosphat (G1P) und Orthophosphat (OP) angegeben, die unter den benutzten Testbedingungen nur gering voneinander abweichen.

Table 1

MICHAELIS CONSTANTS OF PHOSPHORYLASE b FROM HEART AND SKELETAL MUSCLE OF THE PIG

	K_m (glycogen)[a)] %	K_m (G1P)[a)] mM	K_m (OP)[b)] mM
Isophosphorylase 1 from heart muscle	0.060	3.5	7.6
Isophosphorylase 3 from skeletal muscle	0.039	6.0	4.6

a) Determined according to ref. 8;
b) Determined according to ref. 9.

Auf die Unterschiede in der Affinität der beiden Isoenzyme für den Aktivator AMP und die Beziehungen zwischen AMP und V_{max} wurde in der vorhergehenden Arbeit (WILL, REICH und WOLLENBERGER) ausführlich eingegangen. Die aus dem LINEWEAVER-BURK-Diagramm erhaltenen K_m-Werte der Isophosphorylasen (R-Zustand entsprechend dem MONOD-Modell) für AMP betragen für das Isoenzym 1 $0,9 \times 10^{-5}$ M und für das Isoenzym 3 $1,8 \times 10^{-5}$ M.
Zur weiteren Diskriminierung der zwei Isoenzyme wurde der Einfluß von ATP und anderen Effektoren auf die Enzymaktivität geprüft. Wie die Abb. 1 zeigt, führt die Anwesenheit von ATP in Gegenwart von 1 mM AMP zu einer signifikanten Enzymhemmung und die Sigmoidität der Kurven wird verstärkt. Die erforderliche Konzentration an G1P für $V_{max\ 1/2}$ erhöht sich unter diesen Bedingungen für das Isoenzym 1 von 3,9 auf 7,6 und für das Isoenzym 3 von 4,6 auf 19,9 mM. Die eintretenden Veränderungen des HILL-Koeffizienten (n) können der Abb. 1 entnommen werden.
Gegenüber den nur graduellen Unterschieden bezüglich der Abhängigkeit der

Enzymaktivität von den Substraten, AMP und ATP ergeben sich für G6P unter den benutzten Testbedingungen ausgeprägte qualitative Differenzen. Während die Aktivität des herztypischen Enzyms durch 0,63 mM G6P stark gehemmt wird und sich in Gegenwart von 1 mM AMP $\underline{n}$ von 1,4 auf 1,81 erhöht, erweist sich die Skelettmuskelphosphorylase als nicht hemmbar (Abb. 2).
Die unterschiedliche Affinität der beiden Isoenzyme aus Schweinemuskel für G6P steht im Gegensatz zu Befunden an Skelettmuskelphosphorylase des Kaninchens (6) und gereinigten Isophosphorylasen aus Kaninchenherz (5), wonach der Inhibitor G6P beide Enzyme im gleichen Ausmaß beeinflußt.
Die für die zwei Phosphorylasen aus Schweinegewebe nachgewiesenen qualitativen Unterschiede im Verhalten gegenüber G6P sind bemerkenswert, da i) die selektive Inhibition des herztypischen Enzyms eine Analyse der Isoenzymverteilung in Gewebsextrakten und Gewebsschnitten gestattet und ii) die G6P-Kontrolle des herztypischen Enzyms, das im Schweinemyokard etwa 2/3 der Gesamtphosphorylase $\underline{b}$ ausmacht (4), für die Regulation der Glykogenolyse von Bedeutung sein kann.
Außer den bereits beschriebenen Metaboliten wurde noch eine Vielzahl weitere getestet, u.a. Aminosäuren, Zwischenprodukte des Zitronensäurezyklus und der Glykose, deren Effekte auf beide Enzyme aber nicht mehr als $\pm$ 15 % der Kontrollaktivität ausmachten. Die Hemmung der Aktivitäten durch Glukose, UDPG, Adenosin und einigen Fettsäuren und ihren CoA-Derivaten zeigte in keinem Falle Differenzen im Ausmaß der gefundenen Hemmung durch G6P.
Den Phosphorylasen $\underline{a}$ und $\underline{b}$, als Schlüsselenzyme der Glykogenolyse in Skelett- und Herzmuskel, kommen insbesondere im Myokard dann eine entscheidende Rolle zu, wenn das Organ unzureichend oder nicht mehr mit Sauerstoff versorgt wird. Bekanntermaßen kommt es im Herzmuskel unter hypoxischen oder anoxischen Bedingungen zu einer raschen Erniedrigung des Phosphatpotentials, die in den ersten Minuten nach Unterbrechung der Sauerstoffversorgung insbesondere durch die rasche Vermehrung an Orthophosphat zustande kommt (7). Das myokardiale Orthophosphat erhöht sich innerhalb von 2 min nach Einsetzen der Anoxie von 2 mM auf Werte zwischen 15 und 25 mM, wogegen sich AMP als Aktivator der Phosphorylase $\underline{b}$ nur verdoppelt und das inhibierende ATP nur gering erniedrigt wird (7). Dieser Gesichtspunkt sowie Untersuchungen von KASTENSCHMIDT u. Mitarb. (8), wonach mit Erhöhung der Temperatur die Kooperativität der Skelettmuskelphosphorylase zunahm, veranlaßte uns, die Phosphorylasegeschwindigkeit des Glykogens in Abhängigkeit von Orthophosphat bei 35° C zu untersuchen (Abb. 3).

Die in Abb. 3 demonstrierte Abhängigkeit der Reaktionsgeschwindigkeit von der OP-Konzentration in Gegenwart von 0,1 und 1 mM AMP zeigt eine erhöhte Kooperativität bei 35° C als bei einer Temperatur von 20° C (nicht demonstriert). Die Affinität der beiden Isophosphorylasen für OP ist unter diesen Bedingungen deutlich unterschieden, so daß das herztypische Isoenzym in Gegenwart von 0,1 mM AMP und 5 mM OP bereits mehr als 40 % der Maximalaktivität erreicht, gegen nur 10 % für das Skelettmuskelenzym. Letzteres bleibt jedoch bis zu einem Konzentrationsbereich von 10 - 20 mM OP substratabhängig.
Da im Herzmuskel des Schweines und auch des Menschen beide Enzymspezies vorliegen, sollten die nachgewiesenen Affinitätsunterschiede der Isophosphorylasen für OP, die z.T. durch Differenzen in der AMP-Affinität bedingt sein können, zu einer fein abgestuften Substratkontrolle der Phosphorylase des Glykogens führen.
Die bisherigen Untersuchungen zum kinetischen Verhalten der zwei Phosphorylasetypen aus Muskelgewebe führten somit zu zwei bemerkenswerten Unterschieden: a. in der Affinität für G6P und b. bezüglich der Abhängigkeit der Reaktionsgeschwindigkeit der Phosphorylase vom OP. Inwieweit diese Differenzen in den kinetischen Parametern der zwei Enzymspezies für die Kontrolle und Steuerung des myokardialen Glykogenabbaus von Bedeutung sind, wird in weiteren Untersuchungen abzuklären sein.

LITERATUR

(1) HELMREICH, E., Compr. Biochem. 17, (1969), 17

(2) YUNIS, A.A., E.H. FISCHER, E.G. KREBS, J. Biol. Chem. 237, (1963), 2809

(3) HANABUSA, K., H. KOHNO, J. Biochem. 66, (1969), 69

(4) WILL, H., E.-G. KRAUSE, A. WOLLENBERGER, Biochem. Biophys. Res. Commun. 40, (1970), 7

(5) SCHLISELFELD, L.H., C.H. DAVIS, E.G. KREBS, Biochem. 9, (1970), 4959

(6) MORGAN, H.E., A. PARMEGGIANI, J. Biol. Chem. 239, (1964), 2240

(7) WOLLENBERGER, A., E.-G. KRAUSE, Amer. J. Cardiol. 22, (1968), 349

(8) KASTENSCHMIDT, L.L., J. KASTENSCHMIDT, E. HELMREICH, Biochem. 7, (1968), 4543

(9) HELMREICH, E., C. F. CORI, Proc. Nat. Acad. Sci. 51, (1964), 131

(10) WILL, H., U. J. REICH, E. -G. KRAUSE, A. WOLLENBERGER, in Vorbereitung

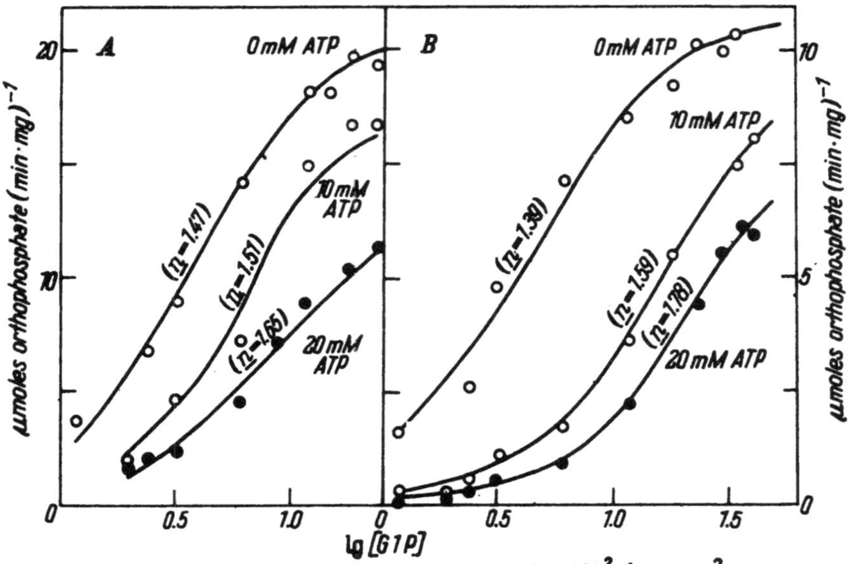

$$\frac{v}{V_{max}} = \frac{\alpha(1+\alpha)}{(1+\alpha)^2 + L'(1+c\cdot\alpha)^2}\ ; \qquad L' = L\cdot\left(\frac{1+d\beta}{1+\beta}\right)^2\cdot\left(\frac{1+e\gamma}{1+\gamma}\right)^2$$

Fig. 1. Activity of phosphorylase b as a function of the ATP concentration. The circle represent the experimental data; the curves are obtained by computer simulations according to the MONOD-equation given under the diagram.
A: Isophosphorylase 1 from heart muscle. B: Phosphorylase from skeletal muscle. (From data of WILL et al. (10.)

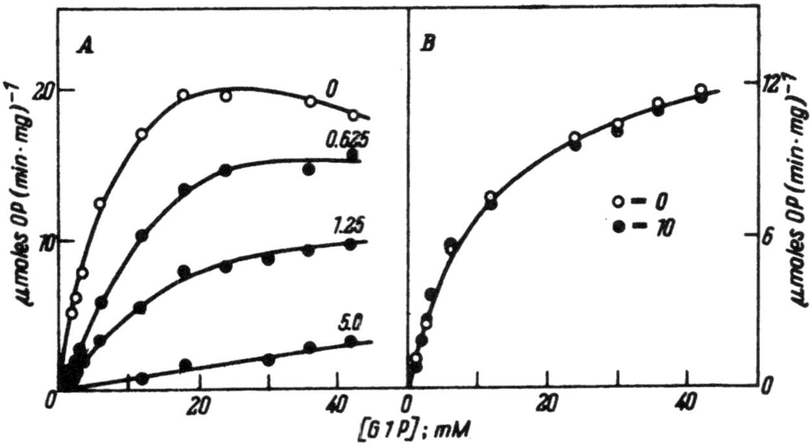

Fig. 2. Dependence of phosphorylase activity upon the concentration of G6P. The figures on the top of the curves denote the concentration of G6P. A: Isophosphorylase 1; B: Phosphorylase from skeletal muscle. (From WILL et al. (4).)

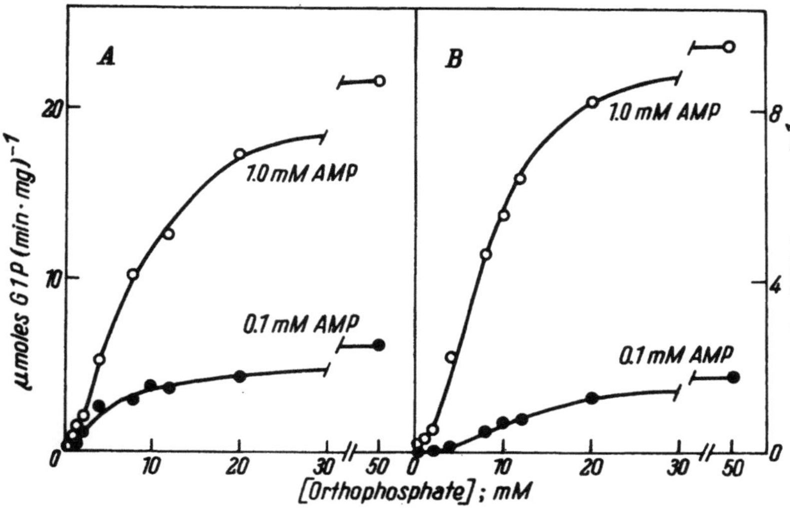

Fig. 3. Activity of isophosphorylase as a function of the OP concentration in presence of 0.1 and 1 mM AMP. Experimental conditions were those specified in ref. 9; temperature of the test at 35 °C. A: Isophosphorylase 1 from heart muscle. B: Phosphorylase from skeletal muscle.

Laboratoire de Physiologie Cellulaire Vègètale, associè au C. N. R. S.,
Centre Universitaire de Marseille-Luminy - France

J. RICHARD, G. NOAT, J. C. MEUNIER

STRUCTURE AND CATALYTIC MECHANISMS OF SOME HEXOKINASES

The main objective of this lecture is to discuss some recent results concerned with the kinetics of the transfer reaction catalyzed by hexokinases. However, one must start this discussion with a brief survey of our knowledge pertaining to the structure, and particularly the quaternary structure, of these enzymes.

Some problems related to the structure of hexokinases

It seems now accepted that Yeast hexokinases consist of the association of two subunits (1, 2, 3) each with a molecular weight close to 50, 000 (1, 2, 3). From the recent results of PRINGLE (4), each subunit is made up of one polypepteide chain (MW $\sim$ 50, 000). In various experimental conditions (glucose, phosphate, dilution of the enzyme, alkaline pH), Yeast hexokinases dissociate into their subunits (1). This dissociation is reversible and the isolated subunits can associate if those experimental conditions are reversed (no glucose, no phosphate, increase in the enzyme concentration, acidic pH). This situation can be pictured by the scheme of Fig. 1. Two isohexokinases, called P_I and P_{II} exist in Yeast cells (1, 2).

It seems that a close similarity exists in the structure of hexokinases from various origins. Thus, for example, four isohexokinases have been isolated from Wheat germ and obtained in a highly purified state (5). Two of them, called L_I ("light" I) and L_{II} ("light" II) have a molecular weight close to 50, 000. The two others, called H_I ("heavy" I) and H_{II} ("heavy" II) have a molecular weight close to 110, 000.

As can be seen from the results of Table 1, a close similarity exists between the amino acid compositions of L hecokinases from Wheat, and P hexokinases from Yeast. With the technique (6) used by PRINGLE (polyacrylamide gel electrophoreses in sodium dodecyl sulfate or in 6 M guanidine-HCl), we have obtained evidence that L isohexokinases also consist of one polypeptide chain (MW $\sim$ 50, 000) and that

H isohexokinases consist of two polypeptide chains (MW ~ 55,000). However, at least one major difference occurs between Wheat germ and Yeast hexokinases. The dissociation of "heavy" forms, which can be obtained with sodium dodecyl sulfate, guanidine-HCl or urea, does not occur on diluting the enzyme, or in the presence of glucose, of phosphate, or in alkaline media. On the other hand, aggregation of the "light" forms is not obtained if those experimental conditions are reversed. It thus appears that the reversible interconversion,

L ("light" form) $\rightleftharpoons$ H ("heavy" form)

probably does not occur with Wheat enzymes.

The mechanism of the transfer reaction catalyzed by Yeast hexokinase

In the author's opinion three main questions pertain to this problem:

1 is the binding of substrates ordered or random?

2 is the release of products ordered or random?

3 what is the mechanism of the inhibitory effect of free ionic species, ATP^{4-} and Mg^{2+}, on the transfer reaction?

This discussion will now be an attempt to answer these questions.

Ordered or random binding of substrates

Until now two types of models have been presented to explain the binding of substrates on Yeast hexokinase. The first mechanism, proposed by HAMMES and KOCHAVI (7, 8, 9) and by NOAT and RICARD (10, 11, 12, 13) implies that the binding is ordered, glucose being the first substrate. In addition, the system would not be in a pseudo-equilibrium state, but in steady-state. The second mechanism, proposed by FROMM and coworkers (14, 15, 16, 17) implies, on the contrary, a random binding of substrates, and the existence of a quasi-equilibrium between all the enzyme forms. These two alternatives are depicted in Fig. 2. Both mechanisms give rise to the same rate equation, which can be written with the DALZIEL (18) symbolism,

$$\frac{[E]_0}{v} = \Phi_0 + \frac{\Phi_1}{[G]} + \frac{\Phi_2}{[MA]} + \frac{\Phi_{12}}{[G][MA]} \qquad (1)$$

where $[E]_0$, $[G]$, $[MA]$ are, respectively, the total enzyme, glucose, and Mg ATP^{2-} concentrations, and Φ the DALZIEL coefficients.

It is possible, at least in theory, to distinguish the two mechanisms of Fig. 3

Table 1

COMPARATIVE AMINO ACID COMPOSITION (MOLES PER MOLE PROTEIN, MW 48.000) OF WHEAT GERM ISOHEXOKINASES L_I AND L_{II} and YEAST HEXOKINASES P_I AND P_{II}

	Wheat germ hexokinases		Yeast hexokinases	
	L_I	L_{II}	P_I	P_{II}
Lysine	31	32	33	32
Histidine	8	7	8	5
Arginine	17	17	15	17
Aspartic acid	48	49	49	49
Threonine	22	20	26	26
Serine	28	26	22	21
Glutamic acid	55	55	45	48
Proline	19	26	21	24
Glycine	46	45	37	36
Alanine	43	34	30	30
1/2 Cystine	4	3	4	3
Valine	36	36	26	21
Methionine	7	6	10	10
Isoleucine	22	21	28	32
Leucine	39	40	39	32
Tyrosine	13	11	15	14
Phenylalanine	16	16	16	21

either by directly proving the binding of only one substrate (or the binding of both substrates) on the enzyme (equilibrium dialysis technique, for instance), or by a comprehensive kinetic analysis of the reaction using either alternate substrates and product inhibition. This last way will be examined later.

Recently, by using different dialysis techniques, COLOWICK and WOMACK (19) and our group (20) showed the binding of glucose on Yeast hexokinase, but were unable to observe the binding of the chelate $MgATP^{2-}$ in the absence of the sugar. Obviously, it is difficult to reconcile this experimental fact with a rapid equilibrium random mechanism. The results obtained with glucose and magnesium-ATP are

presented in Tables 2 and 3. Due to the precision of the technique used, the affinity constant of the hexokinase for $MgATP^{2-}$ has to be lower than 800 M^{-1}. The affinity constant of the enzyme for glucose, obtained from equilibrium dialysis measurements, is very close to that obtained from kinetic analysis (Table 2). With the dialysis technique (20), it is also possible to obtain evidence for the binding of one molecule of glucose per monomer (MW ~ 50,000).
The existence of low ATPase activity of hexokinase (21) has been used by FROMM (22) as an argument against the ordered model. This argument seems to us of little value. In the random mechanism, the hexokinase activity of the enzyme is reaction (I) and ATPase activity by reaction (II). If the above mechanism is in a pseudo-equilibrium state, the apparent rate constants k and k' must be very low as compared to all the other rate constants. Thus, the Michaelis constant

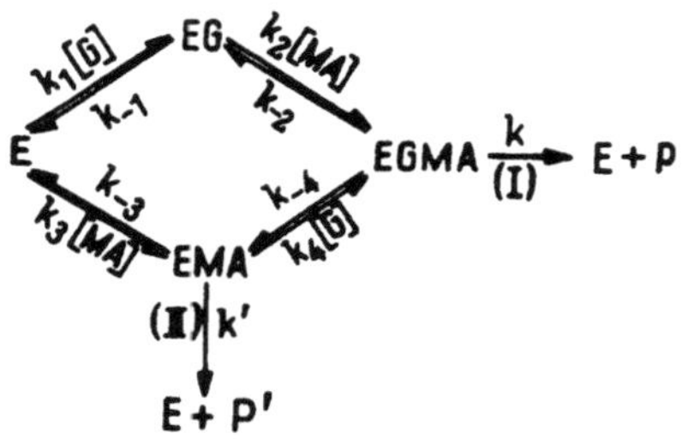

for ATPase activity ($K_m = k' + k_{-3} / k_3$) must be practically equal to the dissociation constant ($K_{MA} = k_{-3} / k_3$) of the complex enzyme-magnesium-ATP. $1/K_m$ is thus a measure of the affinity of hexokinase for $MgATP^{2-}$. The values $1/K_m$ have been determined in various experimental conditions by COLOWICK and coworkers (21) and by our group (20). They are very low and thus incompatible with a random model in a pseudo-equilibrium state.
Recently, taking into account these results, FROMM (23) tried to reconcile the low ATPase activity of hexokinase with the random model. He suggested that the discrepancy between the K_m for ATPase and hexokinase reactions is due to a difference in the aggregation or conformation of the enzyme in the experimental conditions used to determine each activity. This argument also seems of little value. It is well known (see above) that Yeast hexokinase exists only as a monomer or as a dimer. In both the equilibrium dialysis and kinetic measurements, the enzyme was in a monomeric state. As a matter of fact, the experimental conditions (enzyme concentration, ionic strength, pH...) used to determine, by equilibrium

Table 2

AFFINITY CONSTANTS OF YEAST HEXOKIN SE FOR [^{14}C] GLUCOSE

Buffers	[^{14}C] glucose (-)Hexokinase (A) μM	[^{14}C]glucose (+)Hexokinase (B) μM	(B-A) μM	K M^{-1}	$1/\overline{K}_m$ (kinetics) M^{-1}
Glycyl-glycine	212	223	11	9450	8000
Tris	212	222	10	7100	8000
Triethanolamine	212	224	12	11500	10000

Experimental conditions are given in ref. (20). [^{14}C] glucose concentration is determined "outside" the gel in the presence (+) or in the absence (-) of hexokinase.

Table 3

AFFINITY CONSTANTS OF YEAST HEXOKINASE FOR [^{14}C] $MgATP^{2-}$

Buffers	[^{14}C] $MgATP^{2-}$ concentration (-)Hexokinase A μM	[^{14}C] $MgATP^{2-}$ concentration (+)Hexokinase B μM	B - A μM	ATP_h μM min^{-1}	$1/\overline{K}_m$ (ATPase) M^{-1}
Glycyl-glycine	207	207	-	2, 5	100
Tris	205	205	-	2, 8	100
Triethanolamine	207	207	-	4, 2	250

Experimental conditions are given in ref. (20). [^{14}C] $MgATP^{-2}$ concentration is determined "outside" the gel in the presence (+) or in the absence (-) of hexokinase. $[ATP]_h$ is the amount of ATP hydrolyzed during the experiment.

dialysis, the affinity constants of enzyme for glucose and magnesium-ATP (Tables 2 and 3) are identical. Thus, one cannot postulate different aggregation or conformation states of the same enzyme in the same experimental conditions.

Moreover, the agreement between the glucose affinity constants obtained on dialysis and kinetic measurements (Table 2) clearly rule out FROMM' s assumption (23).

A second powerful tool to distinguish between the ordered and the random mechanisms is to perform a kinetic study of the transfer reaction in the presence of various substrates or mixtures of substrates (alternate substrates). This has been done both by FROMM (24) and by NOAT and RICARD (11).

The abscissa of the intersection point of the primary Lineweaver-Burk plots (when $1/[MA]$ is the variable) is given by

$$\frac{1}{[MA]} = - \frac{\Phi\ 1}{\Phi\ 12} \quad (3)$$

which is equivalent to

$$\frac{1}{[MA]} = - \frac{k_2 k_3}{k_{-1}(k_{-2}+k_3)} \quad (4)$$

in the case of the ordered model and to

$$\frac{1}{[MA]} = - K_{MA} \quad (5)$$

for the random model in a quasi-equilibrium state. In the last case, one should obtain the same value of the abscissa with different hexoses. Obviously, this is not the case (Fig. 3).

If two hexoses G and G' are simultaneously phosphorylated, one has one of the situations described by the following models (Fig. 4).

If the rate measurements correspond to the sum of the two reaction velocities ($v = k\,[EGMA] + k'[EG'MA]$), as is the case for pH-stat measurements, one will have for the first model

$$\frac{[E]_o}{v} = \frac{N}{D} \quad (6)$$

with

$$\begin{aligned} N = {} & \alpha\beta\left\{k_{-1}k'_{-1} + k_1k'_{-1}\,[G] + k_{-1}k'_1\,[G']\right\} + \\ & + [MA]\left\{k'_{-1}k_2k\alpha + k_{-1}k'_2k'\beta + (k_1k'_2k'\beta + k_1k'_{-1}k_2\alpha)\,[G] + \right. \\ & \left. + (k'_1k_2k\alpha + k_{-1}k'_1k'_2\beta)\,[G']\right\} + \\ & + [MA]^2\left\{k_2k'_2kk' + k_1k_2k'_2k'[G] + k'_1k_2k'_2k\,[G']\right\} \end{aligned} \quad (7)$$

and

$$D = [MA]\left\{k_1 k_2 k \,(k'_{-1}\alpha + k' k'_2\,[MA])\,[G] + k'_1 k'_2 k'\,(k_{-1}\beta + \right.$$
$$\left. + k_2 k\,[MA]\,)\,[G']\right\} \tag{8}$$

where

$$\alpha = k'_{-2} + k' \qquad \text{and} \qquad \beta = k_{-2} + k \tag{9}$$

Thus the plots 1/v versus 1/[MA] could be curvilinear. If the rate measurements correspond to one reaction velocity (v = k [EGMA]), as is the case for spectrophotometric measurements, then the previous equation becomes

$$\frac{[E]_o}{v} = \Phi_o + \frac{\Phi_1}{[G]}\left\{1 + X + Y\right\} + \frac{\Phi_2}{[MA]} + \frac{\Phi_{12}}{[G]\,[MA]}\left\{1 + X + Y\right\} \tag{10}$$

where the Φ are the usual DALZIEL coefficients and

$$X = \frac{k'_1\,[G']\quad(k'_{-2}+k')}{k'_{-1}(k'_{-2}+k') + k'_2 k'\,[MA]} \qquad Y = \frac{k'_2\,[MA]\,X}{k'_{-2} + k'} \tag{11}$$

Equation 10 is that recently given by RUDOLPH and FROMM (24). However, it is possible to rewrite this expression in a more symmetrical form using only the DALZIEL coefficients pertaining to the substrate (Φ) and to the alternate substrate (Φ'). One gets

$$\frac{[E]_o}{v} = \Phi_o + \frac{\Phi_1}{[G]} + \frac{\Phi_2}{[MA]} + \frac{\Phi_{12}}{[G][MA]} + \frac{[G']}{[G]}\left\{\Phi_1 + \frac{\Phi_{12}}{[MA]}\right\}\frac{\Phi'_2\,/\,[MA] + \Phi'_o}{\Phi'_{12}\,/\,[MA] + \Phi'_1} \tag{11}$$

In the case of the random model, and if the rate measurements correspond to the sum of two reaction velocities (v = k [EGMA] + k' [EG' MA]), then the rate equation is

$$\frac{[E]_o}{v} = \frac{1 + K_{MA}[MA] + K_G\,[G] + K_{G'}[G'] + K_{MA}K'_{G'}[MA][G'] + K'_G K_{MA}[G][MA]}{K_{MA}\,[MA]\{kK'_G\,[G] + k'\,K'_{G'}\,[G']\}} \tag{12}$$

If the rate measurements correspond to one reaction velocity, the previous equation can be written

$$\frac{[E]_o}{v} = \Phi_o + \frac{\Phi_1}{[G]}\left\{1 + X\right\} + \frac{\Phi_2}{[MA]} + \frac{\Phi_{12}}{[G][MA]}\left\{1 + Y\right\} \tag{13}$$

with

$$X = K'_{G'}\,[G'] \qquad \text{and} \qquad Y = K_{G'}\,[G'] \tag{14}$$

As before, this equation can be rewritten using the Φ pertaining not only to the substrate, but also to the alternate substrate, Φ'. One gets

$$\frac{[E]_o}{v} = \Phi_o + \frac{\Phi_1}{[G]} + \frac{\Phi_2}{[MA]} + \frac{\Phi_{12}}{[G][MA]} + \frac{[G']}{[G]}\left\{\frac{\Phi_1 - \Phi'_o}{\Phi'_1} + \frac{\Phi_{12}\,\Phi'_2}{\Phi'_{12}\,[MA]}\right\} \tag{15}$$

In a case of a random model in a quasi-equilibrium state, both LINEWEAVER-BURK plots (reciprocalof v versus reciprocal of glucose and $MgATP^{2-}$ concentrations) are, of necessity, linear (Equation 15). In the case of the ordered model, the plots 1/v versus reciprocal of glucose are also linear, but the plots are not necessarily linear when the reciprocal of magnesium-ATP is taken as a variable. In the last case, the plots can be linear, but it must be stressed that it is not obligatory. Thus, for example, if, in the equation 11, the Φ are equal, or almost equal, to the Φ', then the equation becomes

$$\frac{[E]_o}{v} = \Phi_o + \frac{\Phi_1}{[G]} + \frac{\Phi_2}{[MA]} + \frac{\Phi_{12}}{[G][MA]} + \frac{[G']}{[G]}\left\{\frac{\Phi_2}{[MA]} + \Phi_o\right\} \tag{16}$$

In that case, it is clear that double reciprocal plots, using either glucose or magnesium-ATP, are both linear. In the same way, it can be shown that if

$$\frac{\Phi'_2}{\Phi'_{12}} = \frac{\Phi'_o}{\Phi_1} \tag{17}$$

then the ratio

$$\frac{\Phi'_2 / [MA] + \Phi'_o}{\Phi'_{12} / [MA] + \Phi'_1} \tag{18}$$

will be independant of [MA] and thus equal to a constant, Φ^A. In that case, equation 11 becomes

$$\frac{[E]_o}{v} = \Phi_o + \frac{\Phi_1}{[G]} + \frac{\Phi_2}{[MA]} + \frac{\Phi_{12}}{[G][MA]} + \frac{[G']}{[G]}\left\{\Phi_1 + \frac{\Phi_{12}}{[MA]}\right\}\Phi^A \tag{19}$$

Again, both plots (with respect to glucose and magnesium-ATP) will be linear. Recently, FROMM (24) proposed using, instead of an alternate substrate, a labelled substrate ("isotope competition" in FROMM's terminology), for instance

glucose and [^{14}C] glucose. This idea was presented as a general tool to distinguish between multisubstrate mechanisms (24). FROMM (24) applied this reasoning to the case of Yeast hexokinase, and found the mechanism random in pseudo-equilibrium state.
In fact, this reasoning is erroneous. Since the labelled substrate must possess the same kinetic parameters as the unlabelled substrate, the Φ and Φ' are identical and the rate equation is thus equation 16. As a consequence, it is impossible to use radioactive isotopic substrates as a tool to distinguish between multi-substrate mechanisms. The systematic use of "isotope competition" would have invariably given a random equilibrium state for every mechanism... It must be also emphasized that obtaining linear Lineweaver-Burk plots with regard to both substrates is not a proof per se of the existence of a random equilibrium mechanism, since equations 16 and 19, which correspond to an ordered model, also give straight lines. On the other hand, curvilinear plots in the presence of an alternate substrate, are incompatible with the random equilibrium model. These curvilinear plots have been obtained (11) with fructose as an alternate substrate (Fig. 5).
The existence of an obligatory order in the binding of substrates implies the existence of a conformational change induced or stabilized by the first substrate. This conformational change would allow the binding of the second substrate.
Very recently, DE LA FUENTE and SOLS (25) demonstrated the existence of an induced fit of Yeast hexokinase in the presence of non-phosphorylable analogues of the sugar substrate. This induced fit much increases the ATPase activity of the enzyme. These results are clearly in agreement with the ordered model.

Ordered or random release of products

Probably, the best way to obtain information on the mechanism of the release of glucose-6-phosphate and $MgADP^-$, is to study product inhibition of the reaction. This has been done recently by various groups (11, 26). One should first note that if the binding of substrates is really ordered, the release of products may be either ordered or random. However, in either case, the system would not be in a pseudo-equilibrium state but only in steady-state.
The results of Fig. 6 and 7 show that $MgADP^-$ behaves as a non-competitive inhibitor with regard to both glucose and magnesium-ATP. On the other hand, it seems that glucose-6-phosphate is a competitive inhibitor with regard to glucose, and a non-competitive inhibitor with regard to magnesium-ATP (Fig. 7). Moreover, the variation of the intercepts of the primary plots 1/v

versus 1/ [G] are curvilinear (Fig. 8). The only way to explain these results is to consider that the transfer reaction follows an ordered Bi Bi mechanism with a dead-end complex enzyme-glucose-magnesium-ADP. Such a situation, and the corresponding rate equations, are depicted in Fig. 9.

If there is little doubt that $MgADP^-$ is a non-competitive inhibitor with regard to $MgATP^{2-}$, since this result has been obtained in different laboratories (11, 26), there seem to be some inconsistencies as to the behavior of glucose-6-phosphate. Recently (27), DE LA FUENTE and SOLS obtained evidence for a non-competitive inhibition of glucose-6-phosphate with regard to glucose. We recently also noticed, that depending in the origin of the hexokinase and the method used to prepare the enzyme, glucose-6-phosphate is either a competitive or a non-competitive inhibitor with regard to glucose. If, for instance, bentonite is used (28) to purify the enzyme, glucose-6-phosphate appears to be competitive to glucose. Perhaps the best way to explain these results, is to suppose that the release of products follows a partly random model in steady-state in which the mechanism tends to favor the formation of $MgADP^-$ as the first product (Fig. 9). If this tendency is very marked, glucose-6-phosphate will be a competitive inhibitor with regard to glucose. If not, glucose-6-phosphate will be a non-competitive inhibitor. Recently KOSOW and ROSE (26) explained their results of isotope exchange with this hypothesis.

Inhibition of the transfer reaction by the free ionic species

We recently noticed (13) that free ATP (ionic species ATP^{4-}) behaves non-competitively with regard to $MgATP^{2-}$ at low glucose concentration, and competitively at high glucose concentration (Fig. 10).

Moreover, free magnesium (Mg^{2+}) behaves as a competitive inhibitor with regard to total ATP, and as a non-competitive inhibitor with regard to glucose (Fig. 11). Of all the a priori possible explanations of the inhibition due to the ionic species, it has been recently shown (13) that only one is compatible with the results of Fig. 10 and 11. The data of Figure 10 are in agreement only with a binding of free ionic species ATP^{4-} on the free enzyme and on the enzyme-glucose complex (Fig. 12). On the other hand, the results of Fig. 11 are not in agreement with a binding of free metal on an enzyme form, but rather, with the formation of a ternary chelate Mg_2ATP, also unable to bind any enzyme form. The analytical formulation of this situation is equation

$$\frac{[E]_o}{v} = \Phi_o + \frac{\Phi_1}{[G]} + \frac{\Phi_2}{[A]_o} + \frac{\Phi_{12}}{[G][A]_o} + \frac{K_1\Phi_2[M]}{[A]_o} + \frac{K_1\Phi_{12}[M]}{[G][A]_o} \tag{20}$$

where M is the free ionic species Mg^{2+}, $[A]_o$ the total concentration of ATP and K_1 the equilibrium constant of the postulated reaction

$$MgATP^{2-} + Mg^{2+} \rightleftharpoons Mg_2ATP \tag{21}$$

The determination of the DALZIEL coefficients and the value of the thermodynamic equilibrium constant of the system, allow one, using the HALDANE relationships, to calculate most of the apparent velocity constants of the reaction. The values obtained are given in Fig. 13. Moreover, the stability constant K_1 of the ternary chelate has been estimated, from our results, to be 40 M^{-1}.

Knowing the values of the apparent velocity constants, and writing down the differential equations for all the enzyme forms, one can simulate with an analog computer, the kinetics of the appearance, with time, of any product. The results of Fig. 14, which correspond to the formation of glucose-6-phosphate, clearly show an excellent agreement between the experimental results and the computer trace, at least for one hour.

It thus appears that the proposed model, and the values of the apparent rate constants are in accord with the experimental results. In the author's opinion, the phosphotransferase reaction catalyzed by Yeast hexokinase follows an ordered Bi mechanism, or, more probably, an ordered Bi random Bi mechanism, the first model being a limiting case of the second.

It is obvious that this kinetic behavior is linked to the structure of the enzyme, and to the fact that Yeast hexokinase is monomeric in the presence of glucose. It would be thus particularly tempting to perform kinetic studies on similar enzymes which do not dissociate into subunits after the binding of their sugar substrate, as is the case for H forms of Wheat germ hexokinases.

REFERENCES

(1) SCHULZE, I. T., S. P. COLOWICK, Journ. Biol. Chem., 224, (1969), 2306

(2) GAZITH, J., I. T. SCHULZE, R. H. GOODING, F. C. WOMACK, S. P. COLOWICK, An. New York Acad. Sci., 151, (1968), 1

(3) LAZARUS, N. R., A. H. RAMEL, Y. M. RUSTUM, E. A. BARNARD, Biochemistry, 5, (1966), 4003

(4) PRINGLE, J. R., Biochem. Biophys. Res. Com., 39, (1970), 46

(5) MEUNIER, J. C., J. BUC, J. RICARD, F. E. B. S. Letters, (1971), (in the press)

(6) SHAPIRO, A. L., E. VINUELA, J. V. MAIZEL, Biochem. Biophys. Res. Com., 28, (1967), 815

(7) HAMMES, G. G., D. KOCHAVI, Journ. Am. Chem. Soc., 84, (1962), 2069

(8) HAMMES, G. G., D. KOCHAVI, Journ. Am. Chem. Soc., 84, (1962), 2073

(9) HAMMES, G. G., D. KOCHAVI, Journ. Am. Chem. Soc., 84, (1962), 2076

(10) NOAT, G., J. RICARD, M. BOREL, C. GOT, Eur. Journ. Biochem., 5, (1968), 55

(11) NOAT, G., J. RICARD, Eur. Journ. Biochem., 5, (1968), 71

(12) NOAT, G., J. RICARD, J. BUC, C. R. Acad. Sci., 266, (1968), 1891

(13) NOAT, G., J. RICARD, M. BOREL, C. GOT, Eur. Journ. Biochem., 13, (1970), 347

(14) FROMM, H. J., V. ZEWE, Journ. Biol. Chem., 237, (1962), 1661

(15) HANSON, T. L., H. J. FROMM, Journ. Biol. Chem., 240, (1965), 4133

(16) FROMM, H. J., V. ZEWE, Journ. Biol. Chem., 237, (1962), 3027

(17) FROMM, H. J., Eur. Journ. Biochem., 7, (1969), 385

(18) DALZIEL, K., Acta Chem. Scand., 11, (1957), 1706

(19) COLOWICK, S. P., F. C. WOMACK, Journ. Biol. Chem., 244, (1969), 774

(20) NOAT, G., J. RICARD, M. BOREL, C. GOT, Eur. Journ. Biochem., 11, (1969), 106

(21) KAJI, A., S. P. COLOWICK, Journ. Biol. Chem., 240, (1965), 4454

(22) ZEWE, V., H. J. FROMM, R. FABIANO, Journ. Biol. Chem., 239, (1964), 1625

(23) RUDOLPH, F. B., H. J. FROMM, Journ. Biol. Chem., 245, (1970), 4047

(24) RUDOLPH, F. B., H. J. FROMM, Biochemistry, 9, (1970), 4660

(25) FUENTE, G. DE LA, A. SOLS, Eur. Journ. Biochem., 16, (1970), 226

(26) KOSOW, D. P., I. A. ROSE, Journ. Biol. Chem., 245, (1970), 198

(27) FUENTE, G. DE LA, A. SOLS, Eur. Journ. Biochem., 16, (1970), 234

(28) DARROW, R. A., S. P. COLOWICK, In Methods in Enzymology (edited by S. P. COLOWICK and N. O. KAPLAN) Academic Press, New York 5, (1961), 226

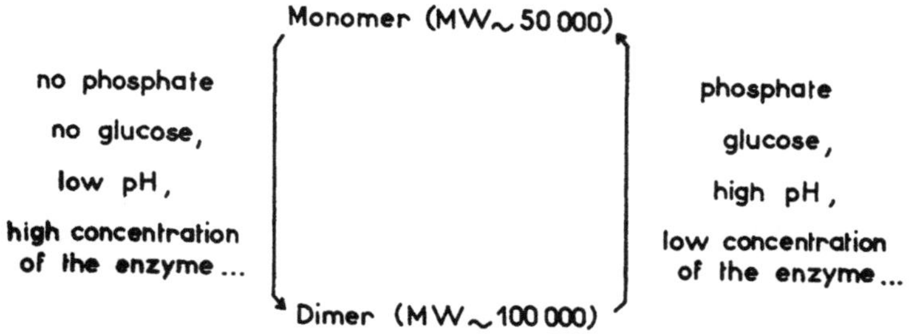

Fig. 1. Reversible dissociation of Yeast hexokinases.

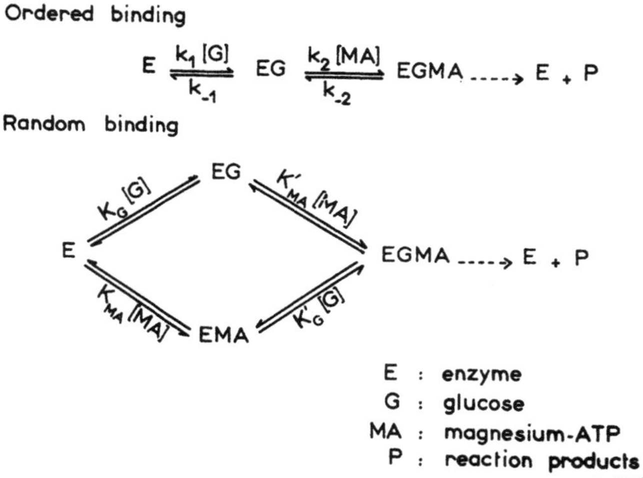

Fig. 2. Ordered and Random binding of substrates (G and MA).

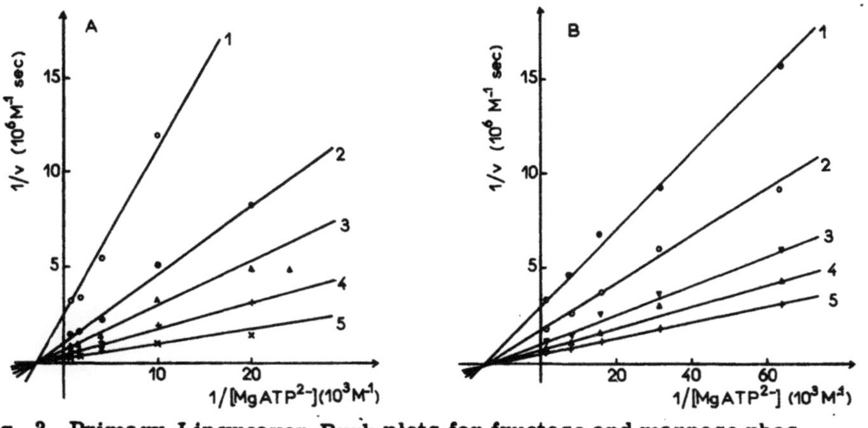

Fig. 3. Primary Lineweaver-Burk plots for fructose and mannose phosphorylation.
The fructose concentrations are respectively: 0.1 mM, 0,25 mM, 0.5 mM, 1 mM and 2.5 mM. The mannose concentrations are respectively: 0.025 mM, 0.05 mM, 0.1 mM, 0.25 mM and 0.5 mM. More detailed experimental conditions are given in ref. (10).

Ordered model with alternate substrate

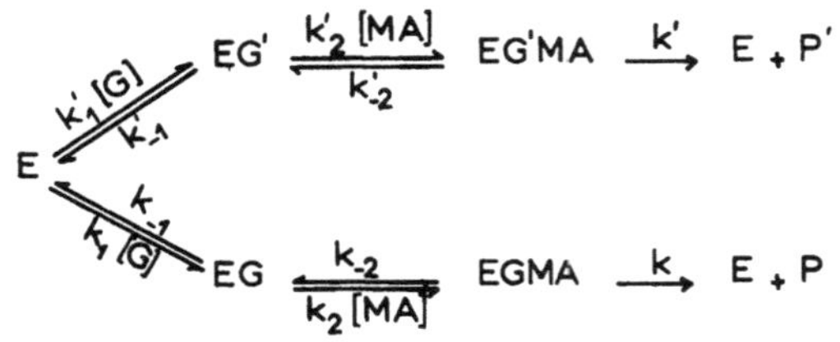

Random model with alternate substrate

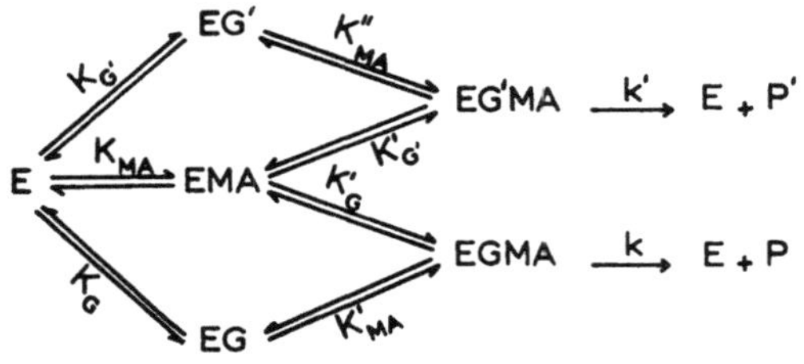

Fig. 4. Ordered and Random binding of substrates (G and MA) in the presence of an alternate substrate (G').

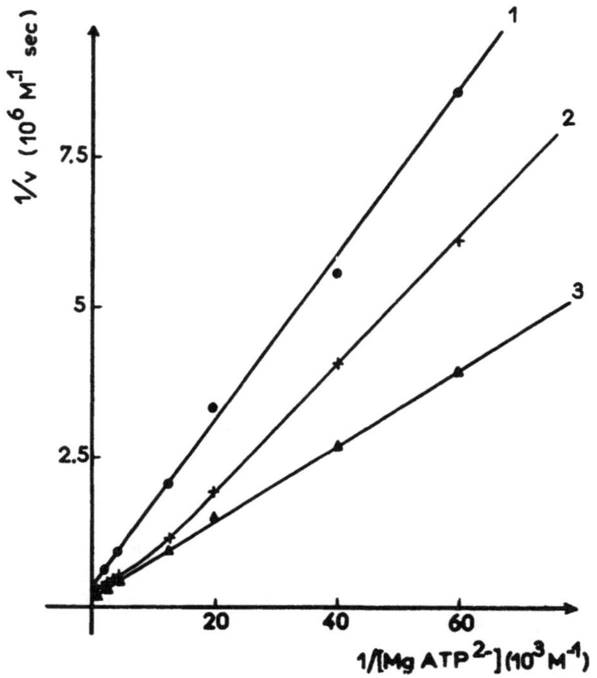

Fig. 5. Simultaneous phosphorylation of glucose and fructose.
Upper plot: glucose phosphorylation (0.5 mM), lower plot: fructose phosphorylation (2.5 mM), middle curve: glucose plus fructose phosphorylation. More detailed experimental conditions are given in ref. (10).

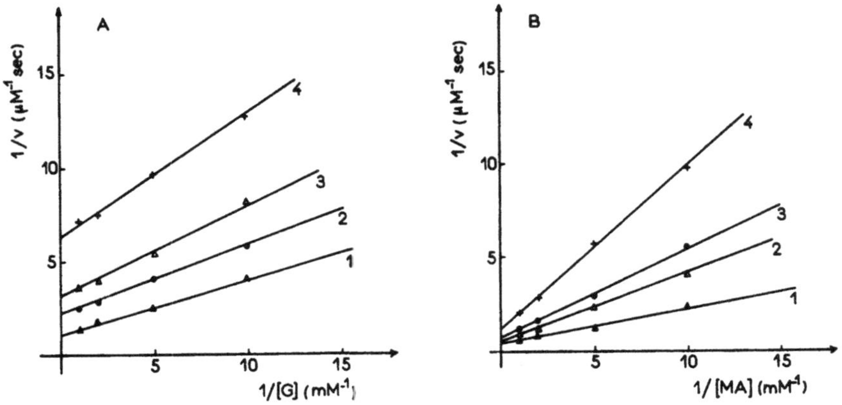

Fig. 6. Magnesium-ADP inhibition of the transfer reaction.
A. $MgADP^-$ as a non-competitive inhibitor with regard to glucose.
The $MgADP^-$ concentrations are respectively: 0, 2.25 mM, 4.65 mM, 9.5 mM.
The $MgATP^{2-}$ concentration is 0.1 mM.
B. $MgADP^-$ as a non-competitive inhibitor with regard to $MgATP^{2-}$.
The $MgADP^-$ concentrations are respectively: 0, 2.25 mM, 4.65 mM and 9.5 mM. The glucose concentration is 0.2 mM. More detailed experimental conditions are given in ref. (10).

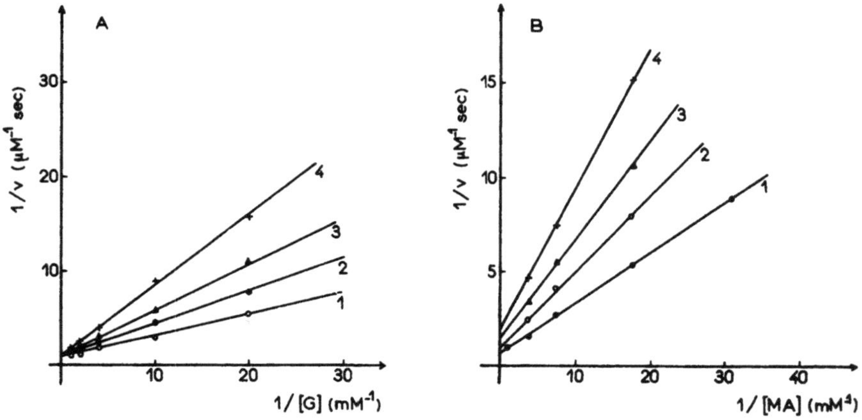

Fig. 7. Glucose-6-phosphate inhibition of the transfer reaction.
A. Glucose-6-phosphate as a competitive inhibitor with regard to glucose.
The glucose-6-phosphate concentrations are: 0, 10 mM, 20 mM, 40 mM.
The $MgATP^{2-}$ concentration is 0.125 mM.
B. Glucose-6-phosphate as a non-competitive inhibitor with regard to $MgATP^{2-}$. The glucose concentrations are those of Fig. 7A. The glucose concentration is 0.125 mM.

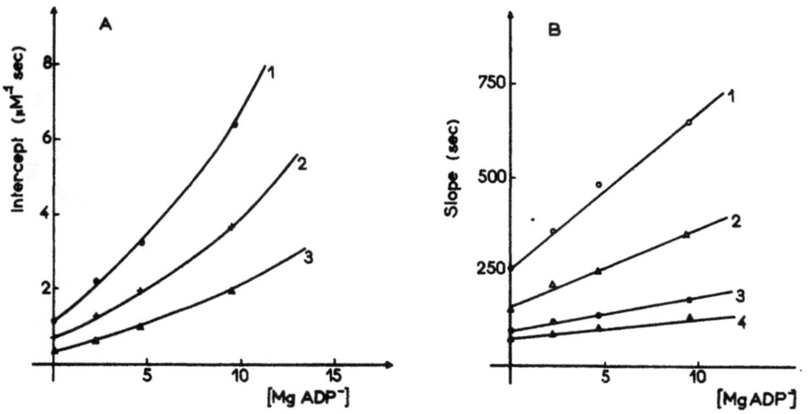

Fig. 8. Secondary plots for the transfer reaction.
A. Intercepts of the primary plots (reciprocal of rate versus reciprocal of glucose concentration) versus $MgADP^-$ concentration. The $MgATP^{2-}$ concentrations are respectively: 0.1 mM, 0.2 mM and 0.5 mM.
B. Slopes of the primary plots (reciprocal of rate versus reciprocal of glucose concentration) versus $MgADP^-$ concentration. The $MgATP^{2-}$ concentrations are respectively: 0.1 mM, 0.2 mM, 0.5 mM, 1 mM. More detailed experimental conditions are given in ref. (10).

Ordered Bi Random Bi model

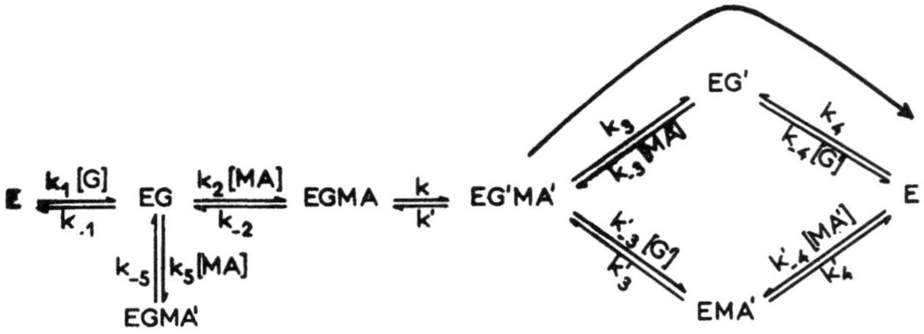

Ordered Bi Bi model

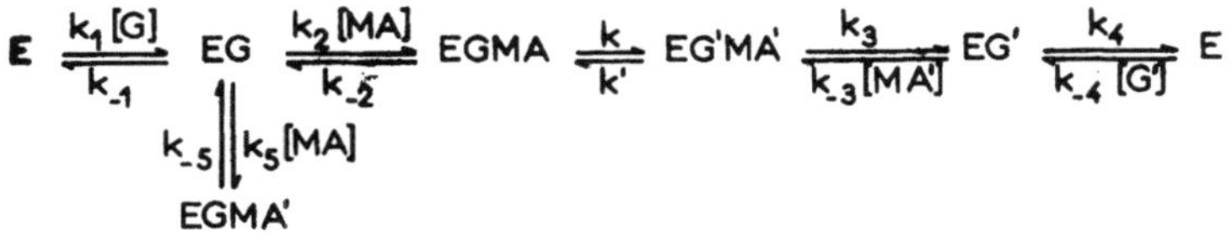

with MgADP⁻ as inhibitor

$$\frac{[E]_0}{v} = \Phi_0 + \frac{\Phi_1}{[G]} + \frac{\Phi_2}{[MA]} + \frac{\Phi_{12}}{[G][MA]} + \Phi_0^I[MA'] + \frac{\Phi_2^I[MA']}{[MA]} + \frac{\Phi_{2'}^I[MA']^2}{[MA]} + \frac{\Phi_{12}^I[MA']}{[G][MA]}$$

with glucose-6-phosphate as inhibitor

$$\frac{[E]_0}{v} = \Phi_0 + \frac{\Phi_1}{[G]} + \frac{\Phi_2}{[MA]} + \frac{\Phi_{12}}{[G][MA]} + \frac{\Phi_1^I[G']}{[G]} + \frac{\Phi_{12}^I[G']}{[G][MA]}$$

Fig. 9. Models of the transfer reaction catalyzed by Yeast hexokinase.

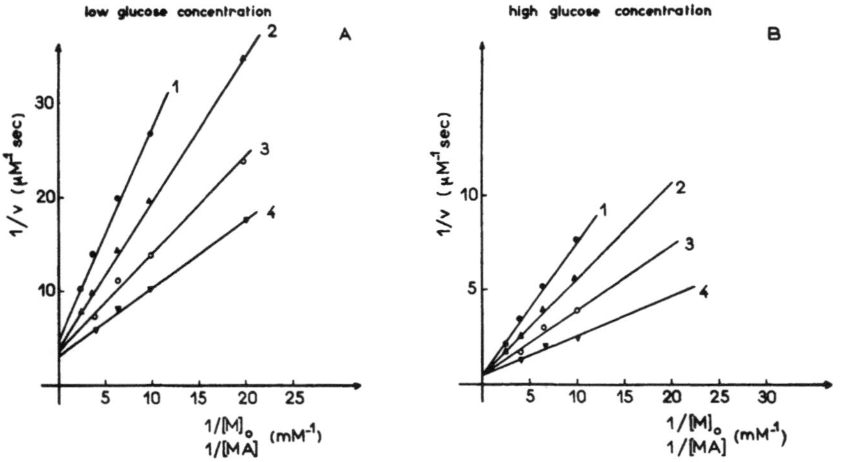

Fig. 10. Free ATP (ATP^{4-}) inhibition of the transfer reaction.
A. "Low" glucose concentration (0. 5 mM).
B. "High" glucose concentration (5 mM).
In both cases, the total ATP concentrations are respectively (decreasing slopes): 12. 5 mM, 8 mM, 5 mM, 2. 5 mM. More detailed experimental conditions are given in ref. (13).

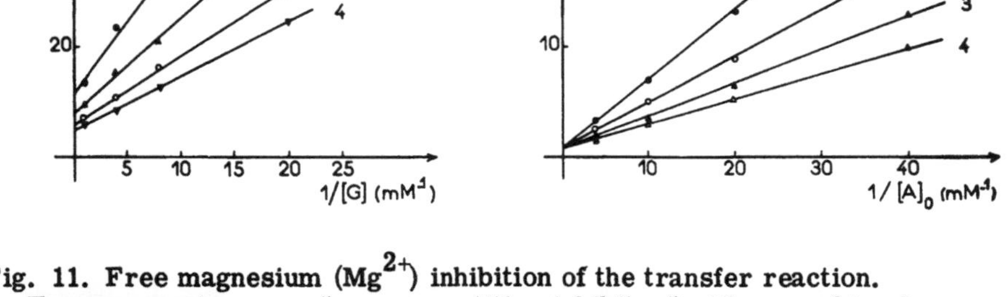

Fig. 11. Free magnesium (Mg^{2+}) inhibition of the transfer reaction.
A. Free magnesium as a "non-competitive inhibitor" with regard to glucose. The total magnesium concentrations are respectively (decreasing slopes): 50 mM, 25 mM, 10 mM and 2. 5 mM. The "total" ATP concentration is 0. 025 mM.
B. Free magnesium as a "competitive inhibitor" with regard to "total" ATP. The magnesium concentrations are respectively (decreasing slopes): 50 mM, 25 mM, 10 mM and 2. 5 mM. The glucose concentration is 0. 25 mM. More detailed experimental conditions are given in ref. (13).

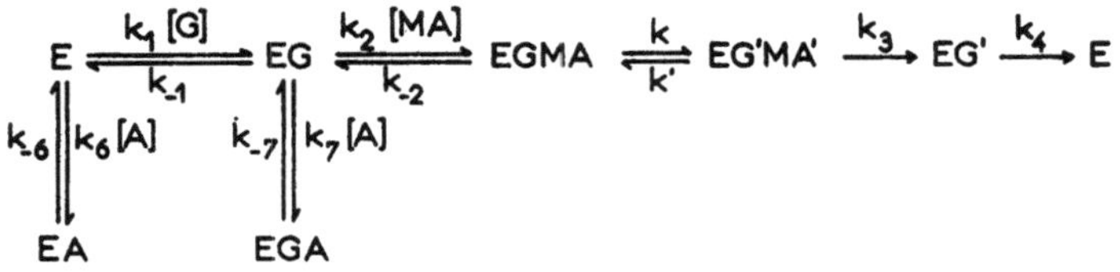

$$\frac{[E]_0}{v} = \Phi_0 + \frac{\Phi_1}{[G]} + \frac{\Phi_2}{[MA]} + \frac{\Phi_{12}}{[G][MA]} + \frac{\Phi_1^I [A]}{[G]} + \frac{\Phi_3^I [A]}{[MA]} + \frac{\Phi_{12}^I [A]}{[G][MA]}$$

Fig. 12. Model for free ATP (ATP^{4-}) inhibition of the transfer reaction.

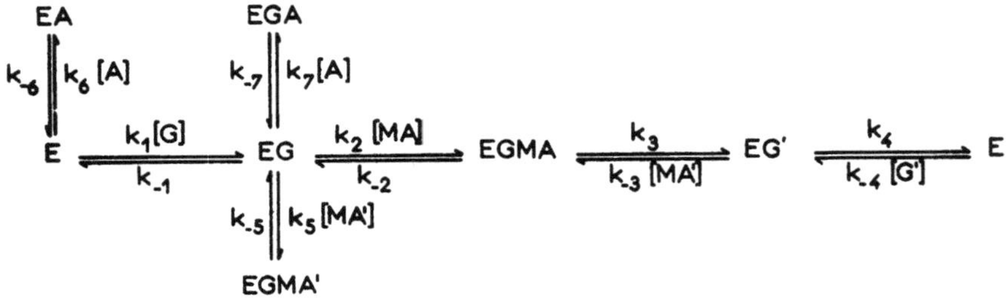

Apparent rate constants

$k_1 = 8.3 \times 10^6 M^{-1} sec^{-1}$	$k_{-1} = 1.9 \times 10^3 sec^{-1}$	$\frac{k_5}{k_{-5}} = 110 M^{-1}$
$k_2 = 6.5 \times 10^6 M^{-1} sec^{-1}$	$k_{-2} = 1.0 \times 10^3 sec^{-1}$	
$k_3 = 4.3 \times 10^3 M^{-1} sec^{-1}$	$k_{-3} = 5.6 \times 10^6 M^{-1} sec^{-1}$	$\frac{k_6}{k_{-6}} = 200 M^{-1}$
$k_4 = 5.9 \times 10^3 M^{-1} sec^{-1}$	$k_{-4} = 4.0 \times 10^5 M^{-1} sec^{-1}$	$\frac{k_7}{k_{-7}} = 600 M^{-1}$

Fig. 13. Apparent rate constants of the transfer reaction.

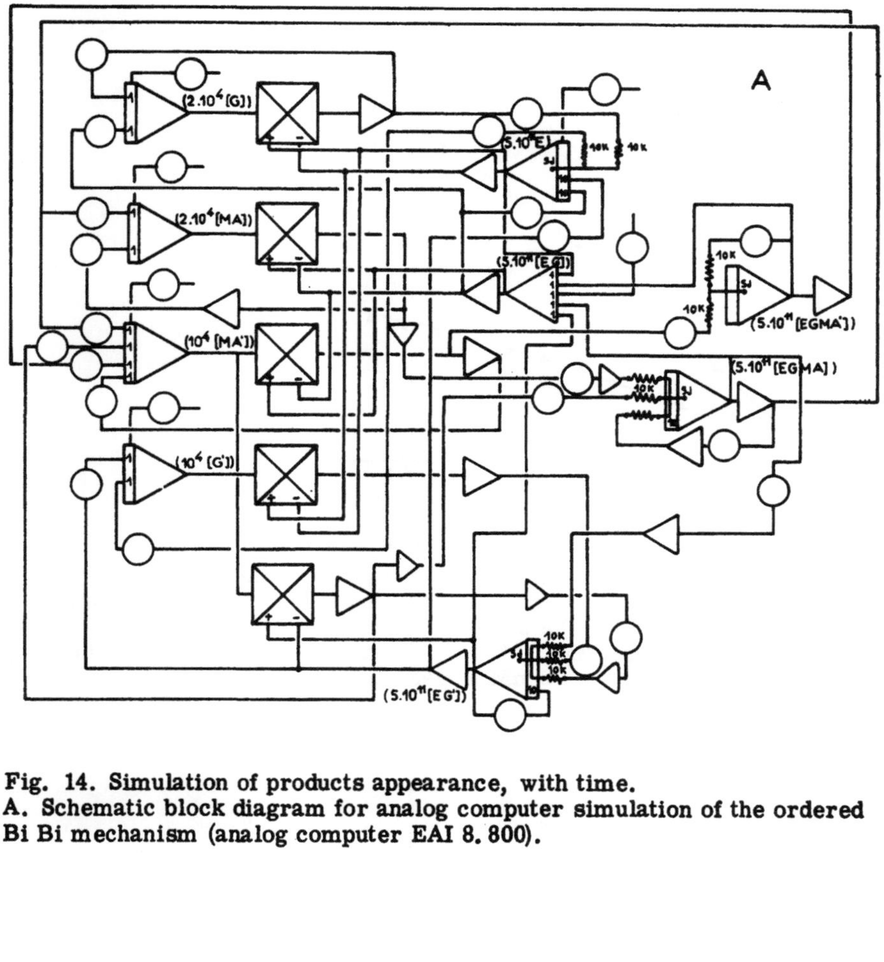

Fig. 14. Simulation of products appearance, with time.
A. Schematic block diagram for analog computer simulation of the ordered Bi Bi mechanism (analog computer EAI 8.800).

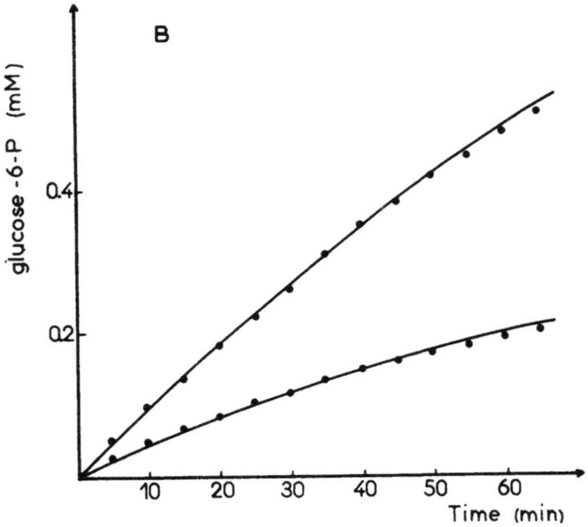

Fig. 14. Simulation of products appearance, with time.

B. Kinetics of the release of glucose-6-phosphate in the absence of products at the beginning of the reaction. Points are experimental results. Curves are the computer traces. The initial substrate concentrations are respectively: [glucose] = 0.5 mM, [$MgATP^{2-}$] = 0.5 mM (lower curve) and [glucose] = 5 mM and [$MgATP^{2-}$] = 2 mM (upper curve).

Institut für Pflanzenphysiologie der Akademie der Wissenschaften der UdSSR, Moskau

S. V. SOKOLOVA

LOKALISIERUNG UND EIGENSCHAFTEN DER HEXOKINASE PFLANZLICHER GEWEBE

Die Hexokinase der tierischen Gewebe ist bisher recht intensiv erforscht worden. Dabei wurde gezeigt, daß dieses Enzym in verschiedenen Formen auftreten kann: a) als unspezifische Hexokinase (E. C. 2. 7. 1. 1.), die Glukose, Fructose und Mannose phosphoryliert, b) als spezifische Glukokinase (E. C. 2. 7. 1. 2.), die nur Glukose phosphoryliert und die eine hohe Michaelis-Konstante besitzt und c) als spezifische Fruktokinase (E. C. 2. 7. 1. 3.), die Ketozucker phosphoryliert. Außer in ihrer Substrat-Spezifität und der Affinität unterscheiden sich diese Formen durch ihr Verhalten bei der Polyacrylamid- oder Stärkegelelektrophorese, durch unterschiedliche Hemmbarkeit durch die Endprodukte der Reaktion und durch andere Eigenschaften.

In pflanzlichen Geweben wurde die Hexokinase viel weniger untersucht. Eine der ersten größeren Untersuchungen wurde von SALTMAN durchgeführt (1), KURSANOV, PAVLINOVA und AFANASIEVA konnten eine Hexokinase in den Leitbündeln der Zuckerrübe nachweisen (2). Die vorliegende Arbeit ist der weiteren Untersuchung dieses Enzyms gewidmet.

Zur Durchführung der ersten Versuche wurde ein Enzympräparat verwendet, welches durch Ammoniumsulfat-Fraktionierung aus dem Rohextrakt von isolierten Zuckerrübenleitbündeln gewonnen wurde.

Die Hexokinaseaktivität konzentriert sich dabei auf die Fraktion, die einer $(NH_4)_2SO_4$-Sättigung von 0 bis 60 % entspricht. Durch Bestimmung der Enzymaktivität in der mitochondrialen, in der mikrosomalen und in der löslichen Fraktion bekamen wir eine Vorstellung über die Lokalisierung der Hexokinase in der Zelle. Die Bestimmung der Fermentaktitvität erfolgte über die Messung der Zuckerabnahme im Inkubationsmedium oder durch spektrophotometrische Messung der NADP-Reduktion bei 340 nm in einem System mit Glukose-6-P-dehydrogenase. Zur Untersuchung der Wechselbeziehungen zwischen Glukose und Fruktose wurden radioaktive Zucker verwendet.

Die Eigenschaften der Enzym-Rohpräparate

Wie aus der ersten Tabelle ersichtlich ist, kann dieses Enzympräparat sowohl Glukose als auch Fruktose phosphorylieren (Tab. 1).

Tabelle 1

DIE HEXOKINASE-AKTIVITÄT DES ENZYM-ROHPRÄPARATES
(in μMol phosphorylierten C^{14}-Zucker/10g. Gewebe . 30 min.)

Substrat	Aktivität	Hemmung in %
2,5 mM C^{14}-Glukose	12,2	-
2,5 mM C^{14}-Glukose + 10 mM Fruktose	10,3	15,6
Fruktose-C^{14} 2,5 mM	19,3	-
2,5 mM C^{14}-Fruktose + 10 mM Glukose	16,65	14,5

Bei einer Substratkonzentration von 2,5 mM ist die Reaktionsgeschwindigkeit von Fruktose etwas größer als die der Glukose. Bei hohen Substratkonzentrationen zeigt sich bei beiden Zuckern eine deutliche Abnahme der Hexokinaseaktivität, was offensichtlich auf eine Hemmung durch überschüssiges Substrat zurückzuführen ist. Das Vorhandensein von Glukose hat praktisch keinen Einfluß auf die Fruktose-Phosphorylierung. Ein 4-facher Überschuß an Glukose führt hier nur zu einer Hemmung von 14,5 %. Ein ähnliches Verhalten konnten wir für Fructose im Falle der Glukose-Phosphorylierung beobachten. Die zweite Tabelle zeigt die Verteilung der Hexokinase auf einzelne Zellfraktionen. Der größte Teil der Aktivität konzentriert sich hierbei auf die lösliche (cytoplasmatische) Fraktion (Tab. 2). 20-30 % der Gesamtaktivität sind an die mitochondriale Fraktion gebunden.

Die Eigenschaften der mitochondrialen Hexokinase

Bekanntlich kann man die mitochondriale Hexokinase durch Behandlung mit verschiedenen Agenzien in Lösung bringen. Wir verwendeten dazu 0,9 nNaCl, 0,33 % DOC oder 1 % Triton X-100. 0,9 n Na Cl-Lösung, die sich bei tierischen Geweben recht gut zur Lösung der mitochondrialen Hexokinase eignet, erwies

Tabelle 2

DIE VERTEILUNG DER HEXOKINASE-AKTIVITÄT AUF VERSCHIEDENE ZELLFRAKTIONEN
(in μmoles phosphorylierten Zucker und % der Gesamtaktivität, 30 min, 30 °C)

Die Zellfraktionen	pro 10 g der Gewebe				pro 10 mg Protein der Fraktion	
	Glukose		Fruktose		Glukose	Fruktose
	μmoles	%	μmoles	%	μmoles	μmoles
Zellwände	0, 13	1, 0				
Mitochondrien	2, 6	20, 1	2, 1	12, 3	2, 2	1, 8
Mikrosomen	1, 5	12, 0	1, 8	10, 7	2, 3	2, 8
Lösliche Fraktion	8, 3	66, 9	13, 9	77, 0	4, 1	6, 9
Gesamtaktivität	12, 53	100	17, 8	100, 0		

sich in unserem Falle als ungeeignet. DOC löst zwar ebenfalls die mitochondriale Hexokinase, es kommt hierbei allerdings rasch zum Absinken der Enzymaktivität. Die besten Ergebnisse konnten wir mit einer 1%-igen Lösung von Triton X-100 erzielen. (Tab. 3)

Tabelle 3

SOLUBILISATION DER MITOCHONDRIALEN HEXOKINASE DURCH BEHANDLUNG MIT 1% TRITON X-100

Enzym-Präparat	Hexokinase-Aktivität (μ Mol)	Protein (mg)	Spezifische Aktivität
mitochondriale Suspension[1]	2,5	11,6	0,210
solubilisierte mitochondriale Hexokinase[2]	4,8	11,5	0,416

1 Bestimmung der Hexokinase-Aktivität erfolgte über die Messung der Zuckerabnahme im Inkubationsmedium.
Inkubationsmedium: 2,5 mM Glukose, 10 mM Mg-ATP, 50 mM NaF, 0,1 M Tris-HCl-Puffer (pH 7,4)

2 Bestimmung der Hexokinase-Aktivität erfolgte über spektrophotometrische Messung der NADP-Reduktion bei 340 nm.
Inkubationsmedium: 2,5 mM Glukose, 2,5 mM Mg-ATP, 50 mM NaF, 0,05 M Tris-HCl-Puffer (pH 7,4), 0,31 mM NADP, Glukose-6-P-dehydrogenase, 1 I. E.

Wie aus der Tabelle 3 zu entnehmen ist, bringt die letztgenannte Behandlung das gesamte mitochondriale Eiweiß in Lösung und die Hexokinaseaktivität erhöht sich nahezu auf das Doppelte. Möglicherweise aktiviert Triton X-100 die Hexokinase, wie das bereits im Falle der tierischen Hexokinase nachgewiesen werden konnte (3). Eine andere Möglichkeit kann jedoch auch nicht ausgeschlossen werden. Im Unterschied zu den Versuchen mit Mitochondriensuspensionen wurde die Messung der Enzymaktivität beim löslichen Enzym spektrophotometrisch in einem System mit Glukose-6-P-dehydrogenase vorgenommen. Unter diesen Bedingungen kommt es zu keiner Glukose-6-P-Anreicherung.

Das Optimum der Hexokinaseaktivität wurde bei einem pH-Wert von 8, 0 gefunden.

Das Enzym besitzt eine wesentlich größere Affinität zu Glukose als zu Fruktose (Abb. 1). In der Abbildung 1 ist die Abhängigkeit der Reaktionsgeschwindigkeit von der Substratkonzentration dargestellt (nach LINEWEAVER-BURK).

Auf der Grundlage dieser Ergebnisse berechneten wir die Michaelis-Konstanten für Glukose mit $1,25 \times 10^{-4}$ M und für Fruktose mit $4,16 \times 10^{-3}$ M. Diese Werte stimmen recht gut mit den Angaben überein, die man in der Literatur für die unspezifische Hexokinase findet.

Tabelle 4

WECHSELBEZIEHUNGEN ZWISCHEN GLUKOSE UND FRUKTOSE BEI DER PHOSPHORYLIERUNG DURCH HEXOKINASE DER VERSCHIEDENEN ZELLFRAKTIONEN
(in μmoles phosphorylierten C^{14}-Zucker/10 g Gewebe, 30 min., 30 °C)

Substrat	Mitochondrien	Mikrosomen	Lösliche Fraktion
Glukose C^{14}	3, 4	1, 0	3, 9
Glukose-C^{14} + Fruktose	3, 0	0, 9	3, 9
Hemmung m %	11, 7	10, 0	0
Fruktose-C^{14}	3, 4	1, 4	4, 9
Fruktose-C^{14} + Glukose	0, 1	0, 0	4, 3
Hemmung in %	97, 4	100, 0	12, 2

Wie aus Tabelle 4 zu ersehen ist, hemmt Glukose die Phosphorylierung der Fruktose durch mitochondriale Hexokinase fast völlig. Alle diese Angaben sprechen nach unserer Meinung für die Annahme, daß die mitochondriale Hexokinase der Leitbündel mit der unspezifischen Hexokinase der tierischen Hexokinase vergleichbar ist.

Die Eigenschaften der löslichen (cytoplasmatischen) Hexokinase

Wie bereits in der Tabelle 4 gezeigt wurde, ist die Geschwindigkeit der Fruktose-Phosphorylierung durch die lösliche Fraktion der Zelle etwas höher, als die der Glukose-Phosphorylierung. Glukose hat hier nur geringe hemmende Wirkung.

Die Phosphorylierung der Fruktose wird durch 1 M KCl fast verdoppelt. (Tab. 5)

Tabelle 5

BEEINFLUSSUNG DER HEXOKINASE-AKTIVITÄT DURCH 1 M KCl
(in μmoles phosphorylierten Zucker/10 g Gewebe, 30 min)

Substrat	Mitochondrien	Lösliche Fraktion
Glukose	3,9	4,7
Glukose + K^+	1,1	3,4
Aktivierung (+) oder Hemmung (-) in %	-71,8	-27,3
Fruktose	3,3	5,8
Fruktose + K^+	0,9	10,3
Aktivierung (+) oder Hemmung (-) in %	-72,7	+77,8

Bekanntlich wirken Kaliumionen als spezifischer Aktivator der Fruktokinase. Aus diesen Ergebnissen läßt sich auf das Vorhandensein einer spezifischen Fruktokinase in der löslichen Fraktion schließen. In der mitochondrialen Fraktion wird sowohl die Phosphorylierung der Glukose als auch der Fruktose sehr stark durch Kalium gehemmt. Es kann angenommen werden, daß die geringe Hemmung der Phosphorylierung von Glukose durch Kaliumionen in der cytoplasmatischen Fraktion auf die Anwesenheit einer unspezifischen Hexokinase hinweist, die mit dem mitochondrialen Enzym identisch ist.
Im Zusammenhang mit der möglichen Rolle dieser Enzyme ist das Vorhandensein der verschiedenen Hexokinasen im Zuckerrübenleitbündel von besonderem Interesse. Vor kurzem konnte bei Aerobacter aerogenes gezeigt werden, daß die Fruktokinase-Aktivität durch Sacharose induziert werden kann und daß dieses Enzym Bestandteil des Hauptweges beim Stoffwechsel der Sacharose ist. (4) Sacharose ist praktisch der einzige Zucker, der im Phloemexudat nachgewiesen werden kann. Deshalb ist nicht auszuschließen, daß der Metabolismus bei pflanzlichen Leitgeweben durch den Stoffwechsel dieses Zuckers gewährleistet wird. Der erste Schritt wäre dann: Sacharose + UDP $\rightleftharpoons$ UDPGl + Fruktose.
Die freie Fruktose würde dann durch eine spezifische Fruktokinase phosphoryliert.

Auf der Grundlage dieser Ergebnisse läßt sich die überwiegende Lokalisierung der Hexokinaseaktivität in der löslichen Fraktion, die im wesentlichen durch die Fruktokinase bedingt wird, verstehen.
Die unspezifische Hexokinase ist an Membranstrukturen gebunden. Schon früher haben GREEN u.a. das Vorliegen aller Glykolyse-Enzyme in den Erythrocyten-Membranen gezeigt (5). Es ist möglich, daß die Aufgaben der unspezifischen Hexokinase in der Phosphorylierung der Hexosen während des Transports durch die Membran besteht.
Damit sind Transport und Stoffwechsel als eine Einheit zu betrachten, deren detailliertere Untersuchung jedoch weiteren Arbeiten vorbehalten bleiben muß.

LITERATUR

(1) SALTMAN, P. J., biol. Chem., 200, (1953), 145

(2) KURSANOV, A. L., O. A. PAVLINOVA, T. P. AFANASIEVA, Fiziologiya Rast, 6, (1959), 286

(3) ETINHOFF, P. N., D. BIESOLD, P. TEICHGREBER, L. MÜLLER, Biochimija, 34, (1969), 1034

(4) KELKER, N. E., T. E. HANSON, R. L. ANDERSON, J. biol. Chem., 245, (1970), 2060

(5) GREEN, D. E., E. MURER, H. O. HULTIN, S. H. RICHARDSON, B. SALMON, G. P. BRIERLEY, H. BAUM, Arch. Biochem. Biophys., 112, (1965), 635

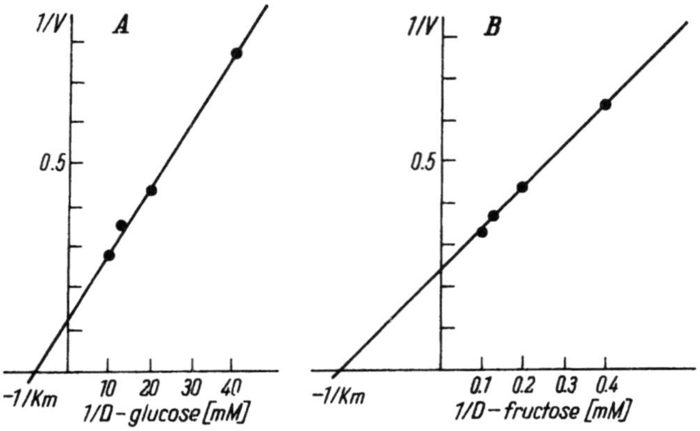

Fig. 1. Abhängigkeit der Reaktionsgeschwindigkeit von der Substratkonzentration.

Physiologisch-chemisches Institut der Karl-Marx-Universität Leipzig

E. HOFMANN

UNSOLVED PROBLEMS OF PHOSPHOFRUCTOKINASE

In the last years considerable progress was acchieved in respect to understanding of the kinetics and cooperative interactions of substrates and effectors on phosphofructokinases from different sources as well as of the general regulatory properties of this enzyme including its subunit structure in a variety of organs and organisms.
Several models and computer applications have been developed. Further, sedimentation constants and molecular weights have been measured as well as subunit structure and interconvertible forms studied.
Various papers demonstrated, that not all phosphofructokinases isolated from different species revealed allosteric properties. For example according to BAUMANN and WRIGHT phosphofructokinase from the slime mold Dictyostelium discoideum shows neither cooperative interactions between F-6-P-binding sites nor ATP-inhibition nor activation by AMP. It is further evident, that phosphofructokinase from different sources may differ in their kinetic properties, molecular weights and subunit structure. Moreover, the enzyme from one and the same organism may show different properties depending on isolation procedures, time of storage of the enzyme and so on. I think we will discuss this in detail in this afternoon, so that I do not intend to give a survey about all these problems, but I wish to focuss your attention to some striking problems and results, not regarding problems, which probably will be dealt with in the forthcoming papers.
In this introduction, I have predominantly in mind phosphofructokinase from yeast, but in some respects this enzyme shall be compared with phosphofructokinase from other sources.

Problems concerned with substrates

It is probable, although not definitely proved that the furanoside form of fructose-6-phosphate binds to the enzyme. To our knowledge, data are lacking for the necessary stereospecific prerequisites of the sugar phosphate for acting as phosphate acceptor in the phosphofructokinase-reaction.

In comparison, much more is known for instance about the hexokinase reaction. According to SOLS and CRANE this enzyme binds β-D-glucopyranose in the chair-equatorial-configuration and changes it to the chair-axial-form, which then is phosphorylated.

The phosphate donor of phosphofructokinase is the MgATP-chelate. However, we don't know, if Mg binds also directly and independently to the enzyme protein. The role of K^+ or NH_4^+ in the catalytic process in addition is quite obscure. NH_4^+ and K^+ increases the affinity of fructose-6-phosphate to the enzyme. It seems, that, K^+ or NH_4^+ bring the enzyme into an active conformational state, ready for binding of fructose-6-phosphate. Further studies have to be performed to clear up these questions.

Binding sites and cooperative interactions

Yeast-phosphofructokinase possesses several binding sites for substrates and effectors. The cooperative interactions between fructose-6-phosphate-binding sites are independent of ATP concentration. Lowering the pH in the range from 7, 5 to 6, 1 a weakening of cooperative interactions and a change of the HILL-coefficient from nearly 2 to one occur. It seems worthwile to mention, that mammalian and yeast phosphofructokinase show just opposite behaviour in changing cooperative interactions and ATP inhibtion with changing pH-value.

It must be pointed out, that at least with yeast phosphofructokinase and probably also with muscle phosphofructokinase, the inhibitor ATP seems not to be responsible for the homotropic cooperative interactions between fructose-6-phosphate binding sites. With ITP as noninhibiting phosphate donor for instance, which is bound only to the catalytic site of the enzyme without occupying the ATP regulatory site, also cooperative interactions with fructose-6-phosphate occur.

Order of reaction

According to SOLS and coworkers and of UYEDA, as well as of LAZER, ROWLAND and BANK muscle and erythrocyte phosphofructokinase at pH 8 and with noninhibiting ATP-concentrations show a ping-pong-behaviour, indicating that the first product, probably ADP, is released before fructose-6-phosphate is bound. In this case a phosphorylated enzyme intermediate might be postulated, but is should be underlined, that there is no direct evidence to support this

speculation. In addition, UYEDA found that sceletal muscle phosphofructokinase reveals ATPase activity.
In extension of these studies by HULME and TIPTON using the isotope exchange method gave evidence that ox heart phosphofructokinase reacts according the following scheme

$$E \rightleftharpoons EATP \rightleftharpoons E^{P}_{ADP} \rightleftharpoons E^{P}_{F\text{-}6\text{-}P}ADP \rightleftharpoons E^{P}_{F\text{-}6\text{-}P} \rightleftharpoons EFDP \rightleftharpoons E$$

Molecular weight of the yeast-enzyme and its subunit structure
From sedimentation velocity experiments in the analytical ultracentrifuge a molecular weight of 560000 ($M_{S^o/\bar{D}app}$) for the yeast enzyme has been determined in our institute. By high speed sedimentation equilibrium we determined a molecular weight of 550000. The method for preparation of homogenous yeast phosphofructokinase has been worked out by NISSLER, EINATZ and FRIEDRICH; the molecular parameters were evaluated by KOPPENSCHLÄGER, DIEZEL and LIEBE.
The apparent diffusioncoefficient $D_{app;20^o}$ is equal to $2,867 \cdot 10^{-7} \frac{cm^2}{sec}$.
The frictional coefficient f/fo is 1, 23.
The sediment coefficient $S^o_{20;w}$ = 17, 85 S.
In comparison, centrifugation in a sucrose density gradient at pH 7, 0 gave a s-value of 18, 5 and a calculated molecular weight of 565000.
At pH 7, 0 up to 8 mg protein/ml there is a normal dependence of the measured S-values from the protein concentrations.
Neither at high nor at low protein concentrations an aggregation or dissociation behaviour of the yeast enzyme could be detected. In comparison, muscle phosphofructokinase shows, as the LARDY-group pointed out, an association-dissociation behaviour depending on the protein concentration. At pH 6 however at protein concentrations of less then about 2 mg/ml a dissociation of the 18 S-molecule occurs.
A significant difference can be observed between freshly prepared enzyme and stored enzyme from a commercial preparation, which has been purified to homogenity, using affinity chromatography with the chromophore of Blue dextran covalently bound to Sephadex and eluting the enzyme with increasing ionic strength. The stored enzyme dissociates already at higher protein concentrations

than the fresh one, although s-values of both enzymes are quite similar at pH 7 and pH 6.
In the following report from LIEBE, KOPPERSCHLÄGER and DIEZEL these problems will be further evaluated especially in respect to ligand and pH stabilization of various interconvertible molecular forms of yeast phosphofructokinase.
Some words about the subunit structure of yeast phosphofructokinase; after dialysis in alkaline media (pH 10) most of the enzyme splits into fragments having 60 000 daltons, as has been measured in a graded polyacrylamide gel electrophoresis system. From results, obtained by different methods like electrophoresis, density gradient and analytical ultracentrifugation, a model of yeast phosphofructokinase subunit structure has been developed.
Three subunits with sixty thousand daltons combine to a unit of 180 000, detectable by polyacrylamide gel electrophoresis and density gradient centrifugation. Evidently, either this is the smalles enzymatically active unit of yeast phosphofructokinase per se or, being itself inactive, it can easily be transformed by substrates and effectors into an active form, perhaps by dimerization.
The basic unit (monomer of 180 000 daltons) aggregates to oligomeric forms such as dimers (360 000) and trimers (550 -560 000). The dimer and trimer are interconvertible forms and can be stabilized under a variety of conditions. This will be shown in the paper of LIEBE. These forms can be found by analytical ultracentrifugation, as well as in sucrose density centrifugation and PAA-electrophoresis.
However, some questions are still open in respect to the smallest subunit. The 60 000 entity can be obtained until now only after alkaline splitting of the 550 000 molecule.
Fragmentation of the enzyme by SDS-treatment, followed by SDS-gel electrophoresis, results predominantly in the occurence of fragments with 90 000, only sometimes contaminated with a 60 000 component, however, in a comparable small amount. Interestingly in these cases hitherto no 30 000 fragments can be detected. The isolated 90 000 fragment cannot be splitt neither in presence of high concentrations of mercaptoethanol nor with guanidine-HCl, urea and SDS.
Until now, we have no good explanation for this curious situation and further experiments using carboxymethylation and/or maleylylation are necessary for completing the subunit structure of yeast phosphofructokinase.

As it will be shown in this afternoon by ZIMMERMANN, WENZEL and GAUER, also crystalline erythrocyte phosphofructokinase occur in several oligomeric enzymatic active forms, which can be separated by sepharose chromatography. Among them multiple equilibria exist.

Fachbereich Biologie der Universität Konstanz

H.W. HOFER

RELATIONS BETWEEN STRUCTURE AND CATALYTIC ACTIVITY OF RABBIT MUSCLE PHOSPHOFRUCTOKINASE

In discussions dealing with the regulation of the glycolytic pathway, phosphofructokinase plays an important role. The enzyme was purified and studied from a great number of species and organs and a lot of data are available. Yet the mechanisms of action of this enzyme are so complex that our knowledge on its catalytic mechanism and the factors regulating it in vivo is very limited.
In this study, some experiments are reported which were performed on crystalline rabbit skeletal muscle phosphofructokinase (1). A model is also presented, which explains some aspects of the kinetic behaviour of the enzyme by means of its structural properties.

The Molecular Structure of Rabbit Muscle Phosphofructokinase

Some charateristics of the molecular structure of rabbit muscle phosphofructokinase are well established. The molecular weights for several states of the enzyme are summarized in the Table. The data were taken from a publication of PAETKAU, YOUNATHAN and LARDY (2) and are in good agreement with our own measurements.
The molecular weight of the smallest unit of phosphofructokinase which exerts catalytic activity, is 374,000. This molecule is able to dissociate into inactive subunits of half the molecular weight. Smaller particles, e.g. subunits of 93,200 daltons, were obtained only in the presence of denaturing agents. There is some evidence that these particles are the smallest identical subunits of phosphofructokinase, the protomers. This was suggested primarily by the characterization of certain amino acid residues in the intact enzyme. For example, the 93,200 molecular weight subunit contain 18 cysteines (2, 3). KEMP (4) demonstrated that the cysteine residues can be classified with respect to their reactivity towards ELLMAN's reagent. There was only a single sulfhydrylgroup which reacted with maximum velocity. We confirmed these results, and demon-

Table

MOLECULAR WEIGHTS OF THE SUBUNITS AND OLIGOMERIC FORMS OF RABBIT MUSCLE PHOSPHOFRUCTOKINASE (from: PAETKAU, YOUNATHAN and LARDY (2).

Condition	Mol. weight, experimental	N	N· 93,200
0.1 M Tris-phosphate pH 8.0	$1.6 \cdot 10^6$	16	$1.49 \cdot 10^6$
	$7.7 \cdot 10^5$	8	$7.46 \cdot 10^5$
	$3.6 \cdot 10^5$	4	$3.74 \cdot 10^5$
0.8 M urea, 0.8 mM FDP, pH 5.8	$1.9 \cdot 10^5$	2	$1.86 \cdot 10^5$
4 mM sodium dodecyl sulfate, pH 11	$9.3 \cdot 10^4$	1	$9.32 \cdot 10^4$
5.5 M guanidinium chloride, 0.01 M DTT, pH 8	$5.1 \cdot 10^4$	0.5	$4.66 \cdot 10^4$
6.5 M guanidinium chloride, 0.1 M DTT, pH 8	$2.5 \cdot 10^4$	0.25	$2.36 \cdot 10^4$

Abbreviations: FDP, fructose 1,6-bisphosphate; DTT, dithiothreitol

strated that the catalytic activity of the enzyme was abolished when the fast reacting sulfhydryl group was modified by N-ethyl maleimide or p-chloromercuri benzoate (5). The modification of other sulfhydryl groups had almost no effect on catalytic activity, when the fast reacting sulfhydryl group was protected from the reaction with N-ethyl maleimide. The significance of the fast reacting sulfhydryl group for the catalytic function of the enzyme was underlined by the fact that its reaction with N-ethyl maleimide was prevented by the presence of ATP and Mg^{2+}.

Under strongly denaturing conditions, still smaller particles of 51,000 and 25,000 molecular weight were found, which may represent isolated peptide chains (Table). The amino acid modification experiments would suggest a heterogeneity of these chains.

It is evident from the Table that there are also states of phosphofructokinase with molecular weights higher than 374,000. The values of 770,000 and 1.6 million, as reported by PAETKAU, YOUNATHAN and LARDY suggest the existence of dimers and trimers. These molecular weight determinations were performed in nonhomogeneous systems, in which several states of phosphofructokinase were present. The problems of such experiments are well known. Some of these difficulties may be by-passed using the method of sucrose density gradient centrifugation according to MARTIN and AMES (6). Yet in such experiments, which we performed in collaboration with Dr. J. NOLTE, separation into several distinct peaks, which would represent different states of aggregation of the enzyme, was not achieved. In the case of yeast phosphofructokinase much better separation of aggregated enzyme species was reported by LIEBE et al. (7).

In our experiments the homogeneous reference enzyme (Aldolase) was distributed over five fractions. The phosphofructokinase was present in about 25 fractions of the gradient tube, whose content was divided into 40 fractions. For evaluation of the experiments we calculated a numerical mean value of molecular weight, which showed a marked dependency on the amount of enzyme applied to the gradient. The mean molecular weight increased from 355,000 when 0.15 mg phosphofructokinase were added to the gradient tube, to 940,000 in the presence of 4 mg enzyme (8). In collaboration with Dr. H. MARKAU we determined mean molecular weights by light scattering. Here also we found a distinct dependency of the molecular weight on the concentrations of phosphofructokinase. It was suggested by these experiments that a concentration-dependent association - dissociation equilibrium exists. Our measurements can be sufficiently described by a model as indicated in Fig. 1.

The model shows the following structural charateristics: the phosphofructokinase molecule of 374,000 daltons is able to dissociate reversibly into subunits of half molecular weight and to associate into oligomers of double, threefold, fourfold etc. molecular weight. The grade of association is not limited. K_1, K_2, K_2' etc. refer to equilibrium constants. In all calculations reported in this study, we took $K_2 = K_2' = K_2''$ etc.

Some evidence for this model can be drawn from electronmicrographs published by PARMEGGIANI et al. (3). Dr. D. SCHWAB has repeated electronmicroscopy on our phosphofructokinase preparations and obtained the same results. He used ammonium molybdate, which has only slight effects on enzyme activity, for negative staining. The electronmicroscopic pictures showed some phosphofructokinase molecules isolated and some aggregated to chains of different lengths.

Concentration - Dependent Changes in Phosphofructokinase Activity

Since there is much evidence for concentration-dependent changes in the molecular structure of rabbit muscle phosphofructokinase, the question arises, whether there are changes in the kinetic behaviour, too.

The results of some earlier investigations (9) are demonstrated in Fig. 2. The specific activities of phosphofructokinase under standardized conditions were determined with different concentrations of the enzyme. In the figure, the specific activities were plotted against the enzyme concentrations in the assay. Curve A represents experiments in which the enzyme was diluted in phosphate buffer; curve B shows the specific activities after dilution in glycylglycine buffer. The pH of dilution buffers and assays was 7. 1.

Both curves show a marked decrease in specific activity after dilution. The decrease is more pronounced in glycylglycine buffer, whereas potassium phosphate tended to stabilize the enzyme. The same effect can be obtained by the addition of AMP or fo the substrate to the dilution system.

In the assays whose results were plotted in Fig. 2, the concentrations of substrates remained constant. Fig. 3 shows another set of experiments, in which the concentrations of ATP were varied. The specific activities of phosphofructokinase have been plotted versus ATP concentrations in the assays. The tests wer performed at pH 7. 0 in the presence of 1. 0 mM fructose 6-phosphate and 4. 0 mM $MgCl_2$. ATP acts as a strong inhibitor under these conditions. The curves were measured at different concentrations of the enzyme. There was a marked dependence of optimal specific activity on the enzyme concentration present in the assay. This effect was already demonstrated by the preceding figure. Moreover, with decreasing enzyme concentrations the ATP inhibition was enhanced. The ATP concentrations leading to 50 % inhibition were markedly lower at lower enzyme concentrations. When 0. 66 mg phosphofructokinase were present in the solution, 50 % inhibition was reached at 2. 15 mM ATP. When the enzyme was diluted to 0. 04 mg/ml only 0. 7 mM ATP was required for the same inhibition.

The concentration of phosphofructokinase affects not only the inhibition by ATP, but also the affinity for fructose 6-phosphate. Fig. 4 shows saturation curves for fructose 6-phosphate at different concentrations of phosphofructokinase. The specific activity of the enzyme, as obtained in the assays, was plotted versus fructose 6-phosphate concentrations. Again, the specific activity of the enzyme has its optimum at high phosphofructokinase concentrations. Evidently, it is impossible to reach the same specific activities at low phosphofructokinase concentrations as at higher enzyme levels, even at high concentrations of fructose 6-phosphate.

Moreover, fructose 6-phosphate concentrations required for half maximum reaction rate are lower at increasing enzyme concentrations. When 2.42 mg phosphofructokinase were present in 1 ml of stock solution, halb-saturation ocurred at 1.2 mM fructose 6-phosphate. In the presence of 0.06 mg enzyme/ml, 2.9 mM fructose 6-phosphate was required for half-saturation. In addition, the steepness of the curves is altered. At lower enzyme concentrations the curves are flattened. On the other hand, the ratio of substrate concentrations required for 90 % saturation to substrate concentration required for 10 % saturation, decreases systematically from 7 at 2.42 mg phosphofructokinase/ml to 2.35 at 0.06 mg/ml.

It is evident that these results cannot be explained by common kinetic laws, which are based on a linear relationship between enzyme activity and enzyme concentration, and which consider the amount of enzyme to be very low compared with those of substrates. In order to analyze the influence of phosphofructokinase on substrate affinity, we attempted at first to study the binding of fructose 6-phosphate and ATP to the enzyme by means of equilibrium dialysis and gel filtration. Using the gel filtration technique with very low fructose 6-phosphate concentrations, KEMP and KREBS (10) postulated the binding of 1 mole of fructose 6-phosphate and 3 moles of ATP to one protomer of rabbit skeletal muscle phosphofructokinase; in the case of sheep heart phosphofructokinase, LORENSON and MANSOUR (11) reported the binding of 3.75 moles of ATP and upto 11.8 moles of fructose 6-phosphate per 100,000 g of enzyme; the binding of fructose 6-phosphate was reduced to 1.75 moles in the presence of phosphate.

For binding studies we preferred the technique of equilibrium dialysis which provided constant concentrations of the enzyme. There was a marked dependency of substrate binding on the enzyme concentration. This fact has already been pointed out by LORENSON and MANSOUR. The number of bound fructose 6-phosphate molecules increased, and the binding constants decreased, with lower phosphofructokinase concentrations. However, it was impossible to discern between "specific" substrate binding sites (which are involved in substrate turnover) and other, perhaps "nonspecific" sites, even in the presence of high concentrations of potassium phosphate and potassium chloride. At pH 7.6, and in the presence of 0.1 M potassium phosphate and 0.2 M potassium chloride, up to 200 moles of fructose 6-phosphate were bound to one protomer. This amount could be reduced to almost zero by varying the conditions. For that reason the direct measurement of the influence of enzyme concentration on substrate binding was not effective in solution of the problem discussed in this study.

Application of the Structure Model to Phosphofructokinase Kinetics

For sufficient explanation of the dependence of phosphofructokinase activity on enzyme concentration, we tried to relate the kinetic analyses to structural transitions, which accompany the changes in phosphofructokinase concentration as described by the association dissociation model.

It was confirmed by various measurements that the smallest active unit of phosphofructokinase has a molecular weight of about 370 000. Therefore, a loss of activity may be due to dissociation into inactive subunits of half molecular weight, which is induced by low enzyme concentrations. At high phosphofructokinase concentrations and in the presence of one of the substrates or of effectors, e.g. phosphate, fructose bisphosphate, AMP, the specific activity of phosphofructokinase at saturating substrate levels is almost independent of enzyme concentration (Fig. 2). In the absence of these conditions, significant amounts of deactivated phosphofructokinase occur. This may be due to the presence of inactive subunits in the association-dissociation equilibrium of the enzyme.

The specific activity at substrate saturation, which is in the following designated as V_{max}, should depend on the relation between active and inactive species of phosphofructokinase, which is a function of total phosphofructokinase concentration according to the model presented by Fig. 1.

Using suitable values for K_1 and $K_2 = K_2' = K_2''$ etc., it is possible to calculate a theoretical curve and to compare it with the experimental data. In Fig. 5, the maximal velocities obtained from the curves in Fig. 4, which were extrapolated to infinite substrate concentrations, were plotted versus phosphofructokinase concentration. An absolute scale on the ordinate was established by plotting the values for optimal specific velocities divided by the maximal specific rate, which is designated as V_{max} . This rate is equivalent to the phosphofructokinase specific activity at infinite substrate concentration, when the fraction of inactive enzyme species approaches zero.

The curve connecting the points was computerized for the following values: $K_1 = 0.5 \cdot 10^{6}$ mole^{-1}, and $K_2 = K_2' = K_2''$ etc. $= 2 \cdot 10^{6}$ mole^{-1}. Obviously the association-dissociation model is able to explain not only the data obtained from the determination of the mean molecular weight, but also independent kinetic measurements.

An explanation for differences in substrate (and inhibitor) saturation curves with respect to enzyme concentration is more complicated. To solve this problem, we tried to combine the association model for the heterogeneous phosphofructokinase with established theories of allosteric enzyme action, which are based on a homogeneous enzyme composed of subunits.

EIGEN (12) has proposed a general plan, which takes account of all possible interactions of subunits, provided that each subunit can exist in only two states. This scheme would lead to very complex mathematical expressions with a great number of constants. Yet it was shown by various authors that some special cases of this general scheme, above all the model of MONOD, WYMAN and CHANGEUX (13) and the induced fit models of KOSHLAND et al. (14) are good descriptions for the binding functions of allosteric enzymes of homogeneous molecular weights. These models involve at least three constants: in the case of the "MONOD model" they are the binding constants for the substrate in the states of high and low affinity and the equilibrium constant between the different states of the allosteric enzyme. Moreover, the number of interacting sites must be considered. Alteration of one of these terms as a function of enzyme concentration could result in a shift of saturation functions as measured by steady - state kinetics. From the studies reported in the literature and from our own measurements, it is not clear, however, whether the binding constants for fructose 6-phosphate and ATP are dependent on phosphofructokinase concentration. The data obtained from equilibrium dialysis and gel filtration experiments suggested a shift in the binding constants for fructose 6-phosphate and ATP related to enzyme concentration and a reduction in the number of binding sites at high phosphofructokinase concentrations. Yet the significance of these results remains uncertain, because it was impossible to distinguish between specific binding sites, which are involved directly in substrate turnover, and nonspecific binding sites. Another method, which would be independent of the models based on steady state kinetics, would be the kinetic measurement of substrate binding and allosteric transitions, for example by means of relaxation methods. Unfortunately, we have until now no indicator for substrate binding which could be applied to rapid reaction techniques. We have performed some orienting assays with a rapid mixing technique. In the assay system which was used for these experiments, the proton formed by the phosphofructokinase reaction was used as an indicator of substrate turnover. As the assay system was only weakly buffered, an acidification took place during the reaction. The actual pH value and proton release was evaluated from the optical densitiy of the assay solution, which contained 4-nitrophenol as pH indicator. In summary, the assay reaction can be described by the slow proton formation, as catalyzed by phosphofructokinase:

$$\text{Fru 6-P} + \text{ATP} \xrightarrow{Mg^{2+}} \text{Fru 1,6-P}_2 + \text{ADP} + H^+,$$

and the fast reaction of the proton with the yellow nitrophenolate ion to the colourless acid. We evaluated only those data which were calculated from deviations of less than 0. 2 pH units from the starting pH. For this reason, the data available until now are restricted to the reaction period in the pre-steady state.

Phosphofructokinase which was preincubated with fructose 6-phosphate, exerted its optimal activity within less than 5 msec after addition of ATP. The same was true when the reaction was started with both substrates, ATP and fructose 6-phosphate. If phosphofructokinase was preincubated with ATP and Mg^{2+}, the enzyme showed a different behaviour after addition of fructose 6-phosphate. After mixing in the stopped flow apparatus, there was a lag phase, in which only little substrate turnover was observed. This phase was followed by a period of increasing enzymic activity, which is apparently due to a fructose 6-phosphate induced activation process. The time requirement suggests that a conformation change is the basis of this activation. When the enzyme is present in an active state from the beginning, substrate binding is a much faster process.

The actual substrate turnover, as evaluated from stopped flow experiments, was plotted against time in Fig. 6. It shows the dependence of the activation process on the phosphofructokinase concentration present in the assay. The actual specific activities of phosphofructokinase in terms of μmole substrate turnover/ min. per mg protein was plotted versus time. The origin of the time axis is equivalent to the moment of mixing the fructose 6-phosphate solution with the phosphofructokinase, which was preincubated with ATP and Mg^{2+}. The time requirement of activation is evidently increased at low enzyme concentrations. Although these experiments are purely qualitative, a decrease in the time required for activation, and a promotion of the active state by high enzyme concentrations, is suggested. Experiments which will allow quantitative statements are in progress.

In the following we want to suggest a model which gives a fairly good fit to the data from kinetic measurements. It is based on the structural association model. The changes in the saturation curves for fructose 6-phosphate at varying enzyme concentration may be due to structural transitions as association and dissociation. The individual amounts of phosphofructokinase species which are present at different enzyme concentrations, are demonstrated in Fig. 7 as they are calculated from the model. Oligomers are taken into account only up to the tetramer. At high phosphofructokinase concentrations a deviation of these curves from the model must be expected.

At low concentrations of phosphofructokinase, the inactive subunit is favoured; the active phosphofructokinase is primarily represented by the monomer. At higher concentrations the amounts of dimer, trimer etc. increase. The distribution of phosphofructokinase species in Fig. 7 was calculated with the same values for K_1 and K_2 as in Fig. 5. Whether the saturation functions of the associated phosphofructokinases are different or not, the saturation of the total enzyme solution is given by the sum of the saturation functions of the monomer and the oligomers, weighted by their concentrations. We took into account only the oligomers up to the tetramer and calculated the curves for fructose 6-phosphate saturation of the monomeric, dimeric, trimeric, and tetrameric state of phosphofructokinase from the data shown in Fig. 4. They are represented by the lines in Fig. 8. Obviously, the shapes of the curves are not symmetrical, thus indicating that neither the "KOSHLAND" (except the "linear" case), nor the "MONOD" model, can fit exactly. This may be due rather to experimental data and inexact constants underlying our calculations and does not necessarily implicate a failure of the models. The broken lines represent curves which were calculated from the "MONOD" model for an optimal fit. It was implied in the calculations that the oligomers consist of identical monomers and that the subunits of phosphofructokinase are symmetrical even in the oligomers. Moreover, it was suggested by the shape of the saturation curves that only slight binding of fructose 6-phosphate to one state of the allosteric enzyme occurs. Obeying these restrictions, the curves were fitted by variation only of the transformation constants K_t, which

describes the equilibrium between the allosteric states (Fig. 9). The fit for all curves was optimal when the binding constant for fructose 6-phosphate was taken as $1.5 \cdot 10^{-4}$ M. The fitting values for the transformation constant K_t decreased from $3.5 \cdot 10^5$, for the monomer, to $1.5 \cdot 10^3$ for the tetramer. A schematic representation of the transitions of "active" into "inactive" states of monomeric and aggregated phosphofructokinase is shown in Fig. 9 k_a and k_b refer to equilibrium constants of association of the allosteric states of the enzyme.
The variation of the transformation constants K_t with respect to the association of phosphofructokinase is demonstrated in Fig. 10. The logarithm of K_t shows an almost linear decrease with the grade of oligomerisation. In the approximation used above, this constant is the single parameter which is variable depending on the oligomerisation grade. It is responsible for the differences in the binding functions for fructose 6-phosphate, and any difference in free energies of fructose 6-phosphate binding between oligomers are therefore expressed by K_t.
Fig. 10 suggests equal differences between the transformation free energies in the individual steps of oligomerisation. It may be noted that K_t can also be regarded as a measure of the differences in interaction energies between the enzyme subunits in the allosteric states.
The significance of Fig. 10 becomes more lucid by the following reaction scheme:

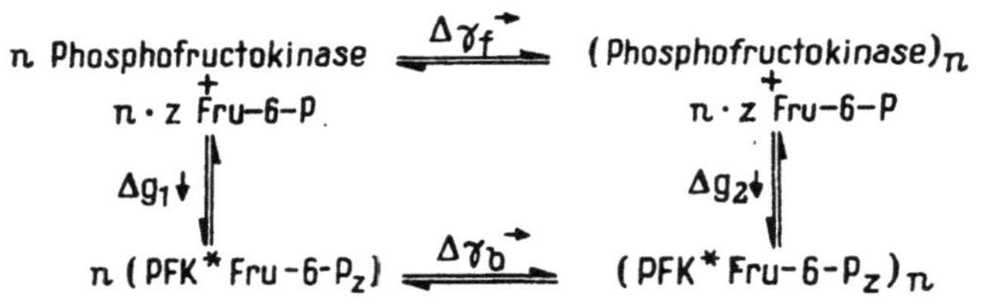

The scheme separates the processes of stepwise aggregation, which are characterized by the free energy changes $\Delta\gamma_f$ and $\Delta\gamma_b$, and the process of allosteric state transition, which is accompanied by fructose 6-phosphate binding and is described by the free energy changes Δg_1 and Δg_2. Around the circle the sum of reaction free energies must be equal to zero, because the free energy change accompanying the transition from one state to another is independent of path.
It follows that

$$\Delta g_1 - \Delta g_2 = \Delta\gamma_f - \Delta\gamma_b$$

This means that the difference in the free energies of fructose 6-phosphate binding (accompanied by state transition) between two differently associated phosphofructokinases must be equal to the difference in the association energies of fructose 6-phosphate free and fructose 6-phosphate binding phosphofructokinase.

Fig. 10 showed equal differences in the free energies (as expressed by log K_t) of fructose 6-phosphate binding between the associated phosphofructokinases; now, it is obvious that this is trivial, because the values of the association constants were the same for K_2, K_2' etc. (cf. Fig. 9: $k_a = k_a' = \ldots$, and $k_b = k_b' = \ldots$). Fig. 10 demonstrates only that the predictions of our model and the experimental data are in agreement.

Final conclusions

In summary, the essential results of this study are:

It is a well-known fact that phosphofructokinase exist in different molecular species. Experimental data are in the case of rabbit muscle phosphofructokinase described by a reversible association and dissociation. Moreover, the enzyme shows concentration dependent changes in its maximal specific activity, as well as in its saturation functions for fructose 6-phosphate. Parallel to these changes are alterations in the inhibition by ATP. The steady state kinetics can also be explained by the association dissociation model, with the additional assumption that the association in the presence and in the absence of the substrate is different. It may be noted that this assumption was experimentally confirmed. The difference in the free energies of association is reflected in the different binding energies of the substrates. This statement is independent of any kinetic model. If a kinetic model is taken into account which explains allosteric substrate binding by means of subunit interaction, the differences in association energies are also reflected by the differences in subunit interaction energies of substrate-free and substrate-binding phosphofructokinase.

The considerations on the concentration dependence of phosphofructokinase specific activity are valid not only in vitro but also in the muscle. SIGEL and PETTE (15) confirmed the localisation of phosphofructokinase in the I - band of cross striated muscle. This space amounts to about 15 % of the total muscle and contains the phosphofructokinase as well as most of the other glycolytic enzymes, in high concentrations. In the white muscle of the rabbit the concentration is 3.7 mg/ml (about 10^{-5} M in terms of the monomer molecular weight); in the red

muscle the mean concentration is 0.35 mg/ml or 10^{-6} M.
At these concentrations high association of the enzyme takes place. Although it is not clear whether different states of aggregation play a role in regulation of the enzyme under physiological conditions, it is evident that the aggregated enzyme reveals altered properties. This should be taken into account, when in vitro data of phosphofructokinase are compared with in vivo measurements. For example, phosphofructokinase at concentrations which are normally used in the absorption assays, is completely inhibited in the presence of substrate concentrations as they were reported by several authors in muscle, even in the presence of positive effectors like fructose bisphosphate, glucose 1,6-bisphosphate (16), or AMP. At the in vivo concentrations of the enzyme, the ATP inhibition is much decreased and the values required for half maximum activity are close to the over-all substrate and effector concentrations as reported by HOHORST (17) and others.

REFERENCES

(1) HOFER, H.W., D. PETTE, Hoppe Seyler's Z. physiol. Chem. 349, (1968), 995

(2) PAETKAU, V.H., E.S. YOUNATHAN, H.A. LARDY, J. molecular Biol. 33, (1968), 721

(3) PARMEGGIANI, A., J.H. LUFT, D.S. LOVE, E.G. KREBS, J. Biol. Chem. 241, (1966), 4625

(4) KEMP, R.G., Biochemistry (Washington) 8, (1969), 3162

(5) HOFER, H.W., Hoppe Seyler's Z. physiol. Chem. 351, (1970), 649

(6) MARTIN, R.G., B.N. AMES, J. Biol. Chem. 236, (1961), 1372

(7) LIEBE, ST., G. KOPPERSCHLÄGER, W. DIEZEL, K. NISSLER, J. WOLFF, E. HOFMANN, FEBS Letters 8, (1970), 20

(8) HOFER, H.W., H. MARKAU, J. NOLTE, in preparation

(9) HOFER, H.W., D. PETTE, Hoppe Seyler's Z. physiol. Chem. 349, (1968), 1105

(10) KEMP, R.G., E.G. KREBS, Biochemistry (Washington) 6, (1967), 423

(11) LORENSON, M.Y., T.E. MANSOUR, J. Biol. Chem. 244, (1969), 6420

(12) EIGEN, M., in: Fast Reactions and Primary Processes in Chemical Kinetics (ed. S. Claesson), Nobel Symp. 5, Stockholm, (1967), 333

(13) MONOD, J., J. WYMAN, J.P. CHANGEUX, J. molecular Biol. 12, (1965), 88

(14) KOSHLAND, D. E., G. NEMETHY, D. FILMER, Biochemistry (Washington) 5, (1966), 365

(15) SIGEL, P., D. PETTE, J. Histochem. Cytochem. 17, (1969), 225

(16) HOFER, H. W., D. PETTE, Hoppe Seyler's Z. physiol. Chem. 349, (1968), 1378

(17) HOHORST, H. J., M. REIM, H. BARTELS, Biochem. biophys. Res. Commun. 7, (1962), 137

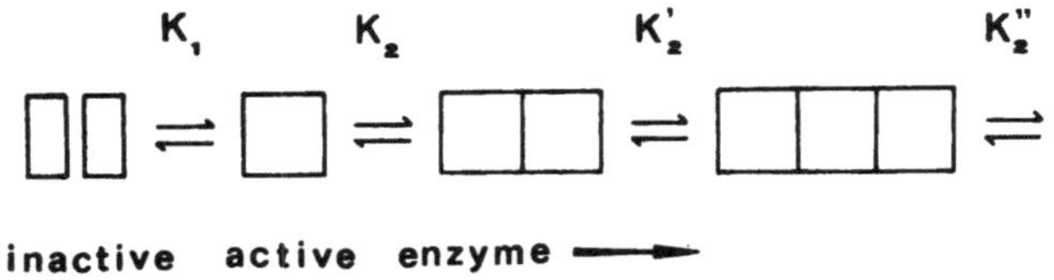

Fig. 1. A model for association and dissociation reactions of rabbit muscle phosphofructokinase.

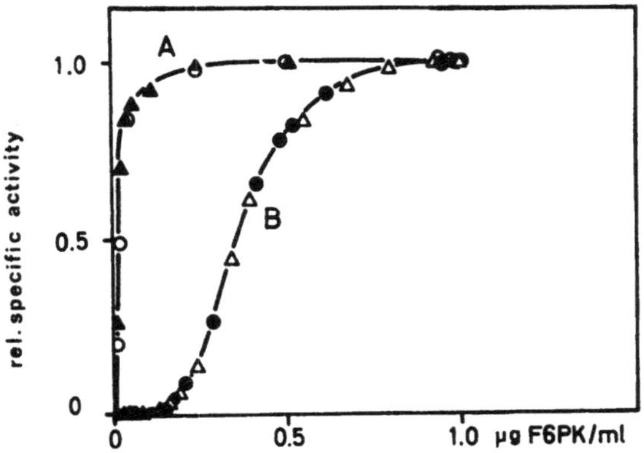

Fig. 2. Dependence of the specific activity of phosphofructokinase on enzyme concentration. The abscissa refers the final phosphofructokinase concentrations in the assays.

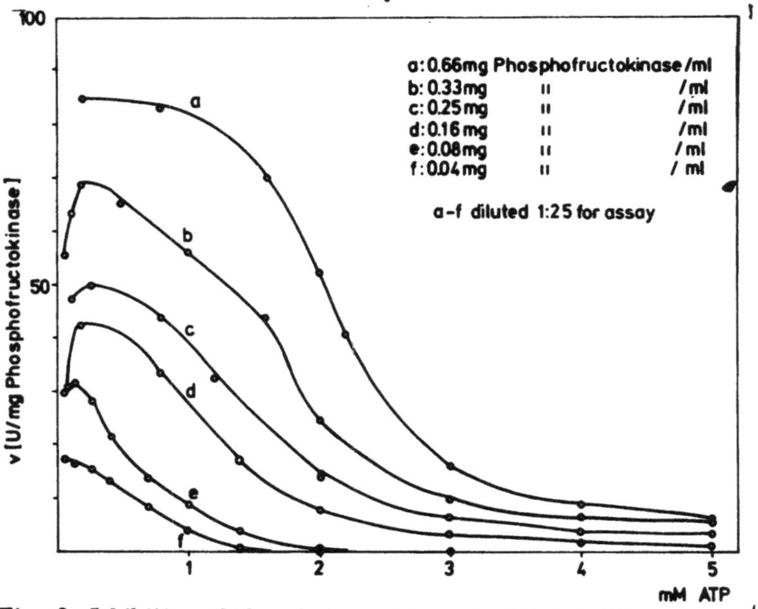

Fig. 3. Inhibition of phosphofructokinase by ATP at different concentrations of the enzyme.

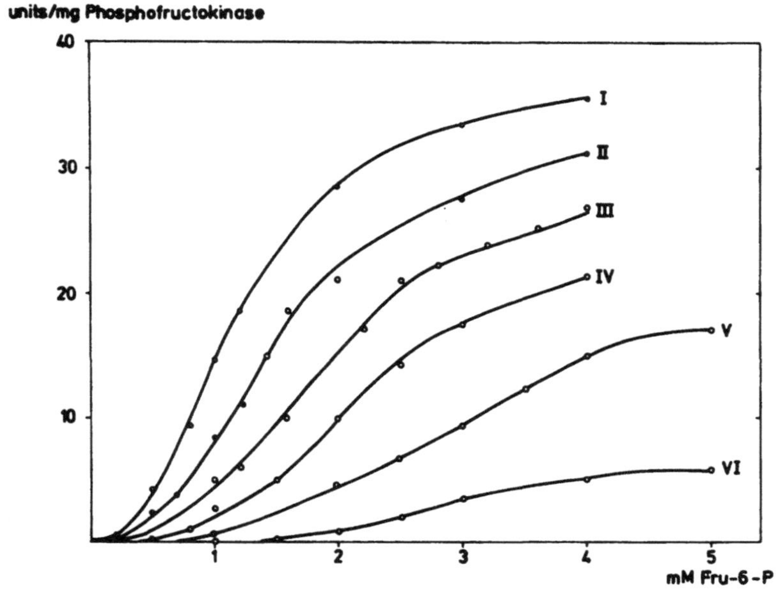

Fig. 4. "Saturation curves" of phospofructokinase for fructose 6-phosphate at different enzyme concentrations. The phosphofructokinase concentrations in the stock solutions were: (I) 2.42 mg/ml, (II) 1.21 mg/ml, (III) 0.60 mg/ml, (IV) 0.31 mg/ml, (V) 0.15 mg/ml, (VI) 0.06 mg/ml. The solutions were diluted 1 : 25 to reach the final enzyme concentration for assay.

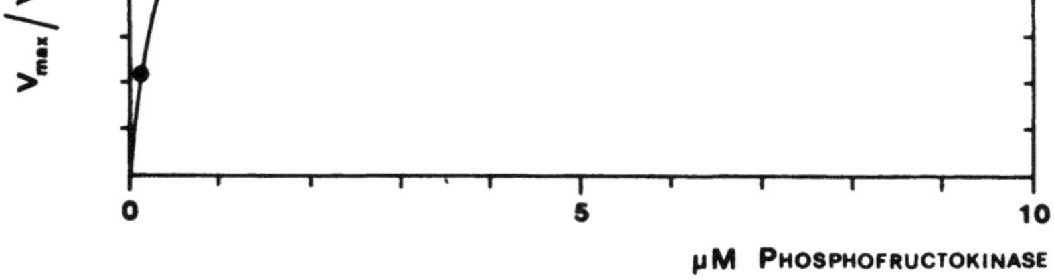

Fig. 5. Dependence of the specific activity of phosphofructokinase at saturating fructose 6-phosphate levels on enzyme concentration.

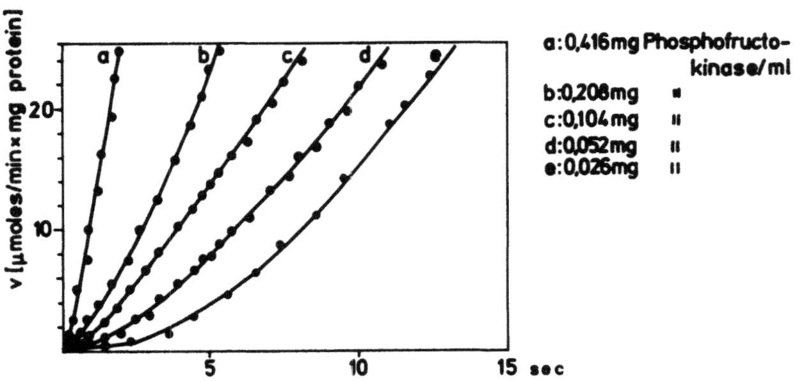

Fig. 6. Time requirement of fructose 6-phosphate-dependent activation of ATP-inhibited phosphofructokinase at different concentrations of the enzyme.

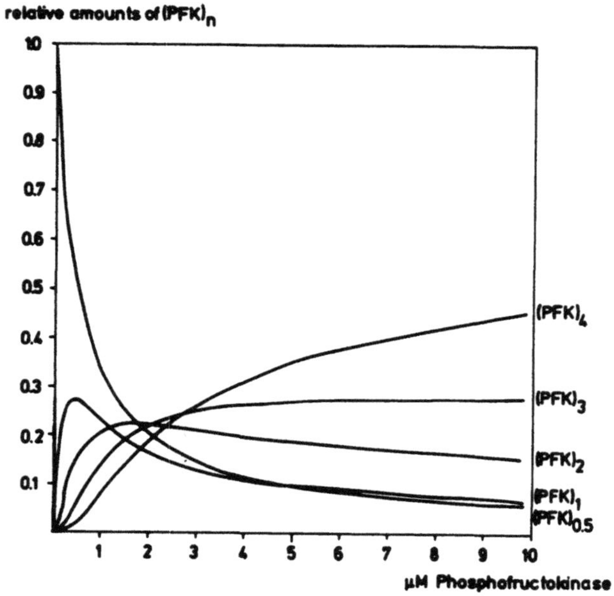

Fig. 7. Relative amounts of associated and dissociated phosphofructokinase species in dependence on the enzyme concentration. The abscissa refers to the enzyme concentration in terms of the monomer of 374, 000 molecular weight.

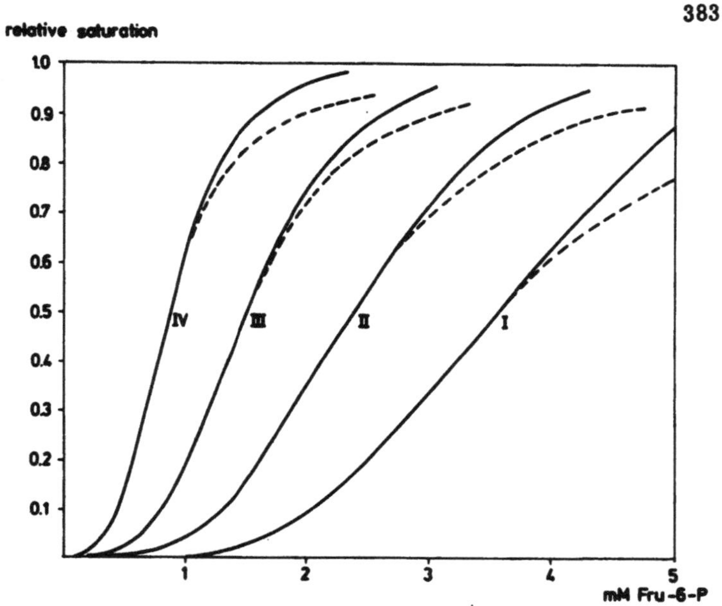

Fig. 8. Calculated saturation curves for fructose 6-phosphate of the monomer (I), dimer (II), trimer (III), and tetramer (IV) of phosphofructokinase.

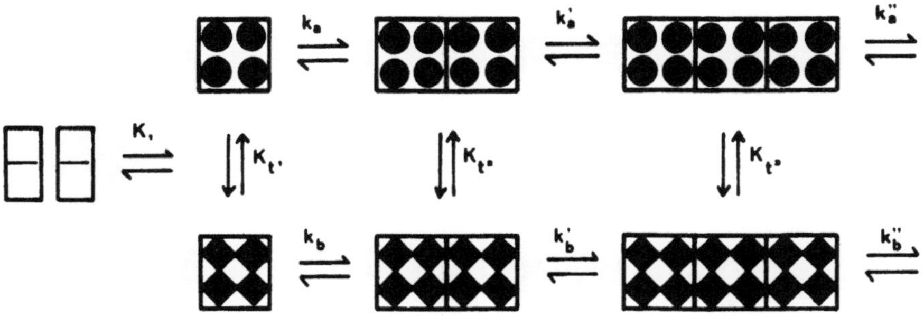

Fig. 9. Schematic representation of the "concerted" transitions of allosteric states of monomeric and aggregated phosphofructokinase.

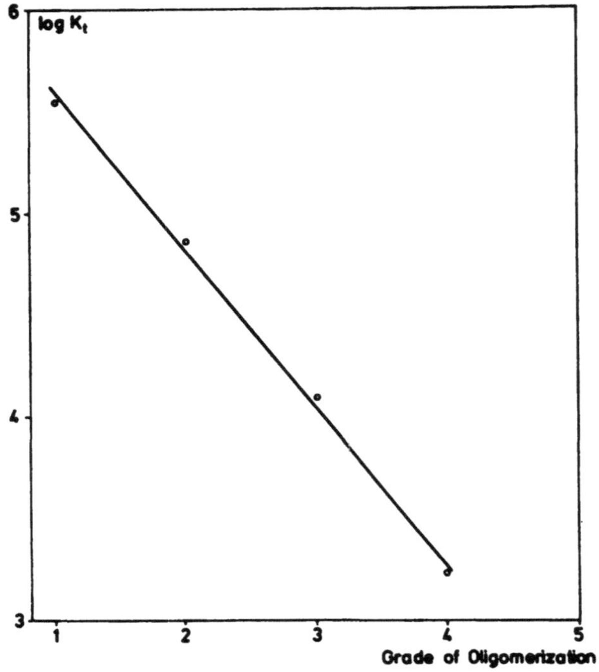

Fig. 10. Dependence of log K_t on the oligomerization grade of phosphofructokinase.

Biochemisches Institut der Universität Freiburg im Breisgau and Gesellschaft für Strahlen- und Umweltforschung München

E. -G. AFTING, D. RUPPERT and H. HOLZER

INTERCONVERSION OF TWO FORMS OF YEAST PHOSPHOFRUCTOKINASE

Control enzymes generally occur at branch points in a metabolic pathway. They are subject to several types of regulation. In pricipal, control results either from a change in the amount of enzyme, e. g. induction or repression, or from a change in activity of preexisting enzyme. According to HOLZER and DUNTZE (1) the latter mechanism may be divided into two main classes: first, enzyme regulation by "chemical modification", and second, enzyme regulation by "physical modification".

The term "chemical modification" indicates that specific groups are covalently attached to or removed from a protein by enzyme catalysis leading to a change in protein structure and activity. The classic example of an enzyme regulated by "chemical modification" was glycogen phosphorylase from mamalian tissues. (For a review see (2)). Nowadays several enzymes are known, which are regulated by "chemical modification". (For a review see (1)).

On the other hand "physical modification" implies that effectors are bound noncovalently to an enzyme (e. g. by electrostatic forces or hydrophobic or hydrogen bonds etc.). "Physical modification" encompasses all enzymes, which are regulated by allosteric control. Numerous enzymes are so regulated.

In yeast, phosphofructokinase plays an important role in carbohydrate metabolism. The reaction catalyzed by this enzyme is the first reaction commom to degradation of all sugars (glucose, mannose, fructose, galactose). If the catabolism of these sugars is the only available source of energy, the rate of glycolysis is determined by the flux through the phosphofructokinase reaction. In yeast phosphofructokinase is the rate limiting step of glycolysis and the gateway to all following catabolic and anabolic processes.

ATP, which may be considered to be the main product of glycolysis, provides a feedback signal to phosphofructokinase, the enzyme which catalyzes the first irreversible step in the glycolytic pathway (Fig. 1). This control by means of physical modification was first described by Vinuela et al. (3). As shown in Fig. 2, ATP concentrations higher than 0.1 mM inhibit phosphofructokinase (solid line). This form is therefore called ATP-sensitive (abbreviated PFK_s). In 1964 VINUELA et al. (4) identified a second form of phosphofructokinase, which is not inhibited by excess ATP (dashed line). This form has been designated as desensitized phosphofructokinase (abbreviated PFK_d). When a crude yeast extract is incubated with ATP, cyclic AMP and sodium fluoride its content of PFK_s is converted to PFK_d. This is true also for partially purified PFK_s. VINUELA et al. (4) interpreted these results as evidence for a "converting enzyme" which regulates yeast PFK by "chemical modification". ATZPODIEN et al. (5) succeeded in purifying this "converting or desensitizing" protein about 50-fold by means of protamine sulfate precipitation, ammoniumsulfate precipitation, acetone fractionation and a second ammoniumsulfate precipitation.

In addition to "desensitizing protein", purified PFK_s (6), ATP and sodium fluoride, a heat stable fraction of yeast extract was necessary for desensitization (7). This heat stable fraction could be replaced by a mixture of fructose-6-phosphate and AMP. Yet the effect remained a mystery, for the concentration of fructose-6-phosphate and AMP in the heated extracts were too low to explain the stimulatory effect on PFK conversion. Nevertheless, based on the foregoing observations it was postulated (4, 5, 7) that PFK is regulated by ATP-dependent, enzyme-catalyzed "chemical modification". We will show here that this hypothesis is not correct. PFK is not regulated by "chemical modification" but by an effector-stimulated "physical modification".

Table 1 shows the conversion of PFK_s to PFK_d in the complete mixture as described by ATZPODIEN (5) containing fructose-6-phosphate, ATP, AMP and sodium fluoride and "desensitizing protein". The conversion is followed by measuring the R_{ATP} value, defined as the ratio of the activities at 0.05 mM ATP, where neither of the two forms is inhibited, and at 1.0 mM ATP, where only PFK_s is inhibited. An R_{ATP} value of 3.0-5.0 is typical for ATP-sensitive PFK_s, whereas an R_{ATP} value of 0.4-0.6 is typical for desensitized PFK_d. During incubation of 20 minutes the R_{ATP} value changes from 3.5 to o.65; that means PFK_s is converted completely to PFK_d. If one now removes PFK_d and "desensitizing protein" by means of ultrafiltration and then adds fresh PFK_s, it too is converted to PFK_d in the absence of "desensitizing protein". "Desensitizing protein", therefore, does not alter PFK_s itself, but alters the

Table 1

DESENSITIZATION OF PFK_s WITHOUT "DESENSITIZING PROTEIN". THE "COMPLETE MIXTURE" CONTAINED "DESENSITIZING PROTEIN" AND PFK_s

Incubation mixture	$R_{ATP} = \frac{\text{activity at 0.05 mM ATP}}{\text{activity at 1.0 mM ATP}}$	
	Before incubation	After incubation
Complete mixture	3.5	0.65
Mixture from above after incubation following ultra-filtration, and readdition of PFK_s	-	0.54

Table 2

ADENOSINE NUCLEOTIDES AND DESENSITIZATION OF PFK_s

	Standard mixture before incubation	Standard mixture after incubation
ATP (mM)	0.1	0.0
ADP (mM)	0.0	0.09
AMP (mM)	1.0	1.0
R_{ATP} of PFK_s after incubation in the above mixtures of adenosine nucleotides (without "desensitizing protein" + NH_4^+)	4.8	0.8

medium in such a way, that desensitization can take place. Control experiments proved that PFK_d and "desensitizing protein" were removed by this ultrafiltration technique.

Table 2 gives the measured concentrations of adenosine nucleotides in the conversion mixture before and after incubation. It can be seen, that during incubation ATP completely disappears and is recovered to 90 % as ADP. The AMP concentration is not measurably altered during incubation. Direct measurement of the ATP-ase activity (8) of "desensitizing protein" suffices to account for the observed ATP hydrolysis. The lower part of this table shows that a mixture constituted in analogy to that found after incubation is fully effective without "desensitizing protein". On the contrary, the complete mixture containing ATP, AMP, fructose-6-phosphate and sodium fluoride without "desensitizing protein" is ineffective in the PFK_s to PFK_d conversion. The reason why in previous control experiments without "desensitizing protein" the conversion was not seen, is that ATP was always used in excess compared to fru-6-P. So fru-6-P was phosphorylated completely to fru-1, 6-P_2 by means of phosphofructokinase reaction and the essential effector for the conversion, fru-6-P, was not available.
Those effectors, which are essential to the conversion are summarized in table 3. Omission of fructose-6-phosphate, ammonium, magnesium, fluoride or ADP each prevents conversion. In contrast to the report of Vinuela et al. cyclic AMP could not substitute for ADP, nor work synergistically with ADP in the complete mixture.
These experiments show that phosphofructokinase of yeast is not altered by "chemical modification" as suggested by VINUELA and ATZPODIEN. The results reported here prove that enzyme conversion from an ATP-sensitive form to an ATP-desensitized form is achieved by "physical modification". PFK_d, converted as shown in table 3 shows the same ATP dependence as described by VINUELA (Fig. 2). On the other hand, the reaction kinetics with respect to fructose-6-phosphate are changed from sigmoidal for PFK_s to hyperbolic for PFK_d at 2.0 mM ATP (Fig. 3).
Fig. 4 summarizes the regulation of PFK with respect to PFK_s to PFK_d conversion. Estimation of ATP levels in yeast under growth conditions so far investigated indicates ATP concentrations higher than 10^{-4} moles per liter of cells. In this concentration range PFK_s is subjected to inhibition. It follows that conversion of PFK_s to PFK_d in intact cells augments the flux of fructose-6-phosphate to fructose-1, 6-diphosphate. Therefore we consider PFK_d to be the more active form of the enzyme. Stimulation of the conversion by fructose-6-phosphate is physiologically useful: namely, availibility of substrate promotes activation of PFK_s. The usefulness of the positive effect of ADP on the conversion is also

Table 3

EFFECTORS WHICH ARE NECESSARY FOR DESENSITIZATION IN THE ABSENCE OF "DESENSITIZING PROTEIN"

Incubation mixture	$R_{ATP} = \frac{\text{activity at 0.05 mM ATP}}{\text{activity at 1.0 mM ATP}}$	
	Before incubation	After incubation
Complete desensitizing mixture[1]	6.3	0.3
- fructose-6-phosphate	6.0	6.8
- NH_4^+ [2]	6.1	6.1
- Mg^{2+}	6.6	6.6
- F^-	3.5	3.6
- ADP	3.7	3.5
- ADP, + cycl. AMP	5.4	4.5
+ cycl. AMP	3.8	0.7

1 50 mM imidazol/HCl-buffer, pH 6.8, 0.07 mM fru-6-P, 0.1 mM ADP, 10 mM $(NH_4)_2SO_4$, 1 mM $MgCl_2$, 10 mM NaF, and 40-80 μg PFK_s

2 NH_4^+ is necessary for the conversion. In earlier experiments it was added with ammonium-precipitated "desensitizing protein".

clear, for as long as the cell's ADP is not fully phosphorylated to ATP the cell requires active PFK_d. The energy situation of the cell thus regulates the energy-yielding catabolism of carbohydrates.

The physiological purpose of ammonia-stimulated conversion to PFK_d may be rationalized as follows. In yeast, ammonium fixation occurs principally through amination of α-ketoglutarate and subsequent amidation to glutamine (9, 10). An abundance of ammonia therefore requires a supply of α-ketoglutarate for its assimilation, which in turn is provided by glycolysis in concert with citric acid cycle. It is thus an advantage to the cell that ammonium stimulates conversion to PFK_d and consequently glycolysis.

The physiological significance of the fluoride effect is not clear.

REFERENCES

(1) HOLZER, H., W. DUNTZE, Ann. Rev. Biochem. 40, (1971), in press

(2) HELMREICH, E., Comprehensive Biochemistry 17, (1969), 17

(3) VINUELA, E., M.L. SALAS, A. SOLS, Biochem. Biophys. Res. Commun., 12, (1963), 140

(4) VINUELA, E., M.L. SALAS, M. SALAS, A. SOLS, Biochem. Biophys. Res. Commun. 15, (1964), 243

(5) ATZPODIEN, W., J.M. GANCEDO, V. HAGMAIER, H. HOLZER, European J. Biochem. 12, (1970), 6

(6) ATZPODIEN, W., H. BODE, European J. Biochem. 12, (1970), 126

(7) GANCEDO, J.M., W. ATZPODIEN, H. HOLZER, FEBS-Letters 5, (1969), 199

(8) LAMPRECHT, W., J. TRAUTSCHOLD, in "Methoden der enzymatischen Analyse" (H.U. BERGMEYER, ed.), Verlag Chemie, Weinheim (1962), 543

(9) WITT, I., H. HOLZER, Biochem. Z. 339, (1964), 255

(10) HOLZER, H., G. HIERHOLZER, I. WITT, Coll. du C.N.R.S., No 124, (1965), 407

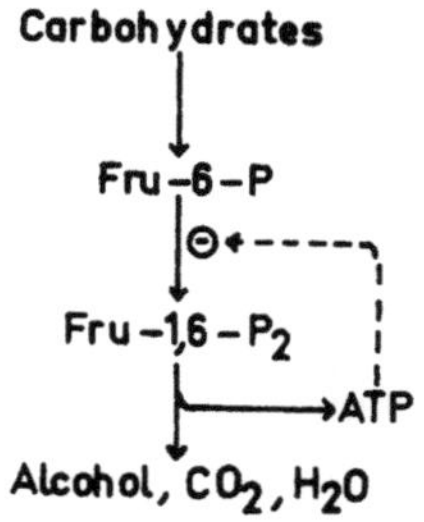

Fig. 1. PFK regulation by ATP (schematic).

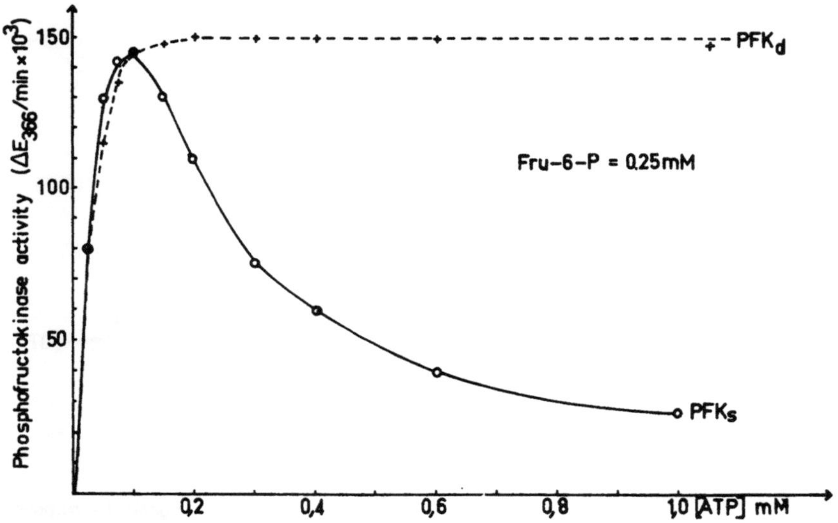

Fig. 2. Dependence of PFK activity on ATP.
+++ = PFK_d; ooo = PFK_s; (Fru-6-P = 0. 25 mM).

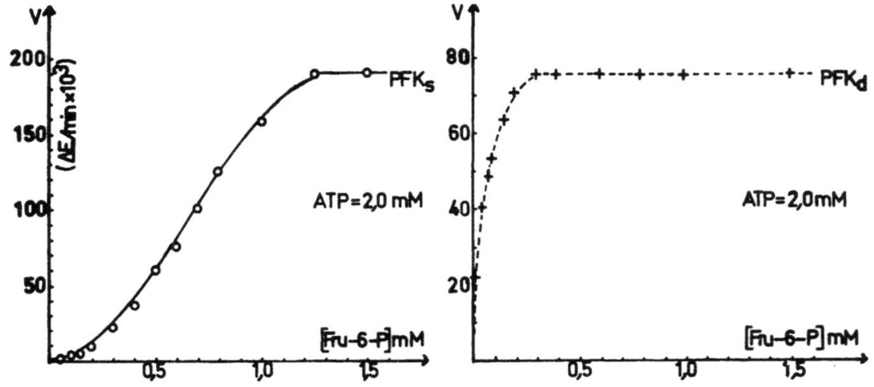

Fig. 3. Dependence of PFK activity on fru-6-P.
+++ = PFK_d; ooo = PFK_s; (ATP = 2.0 mM).

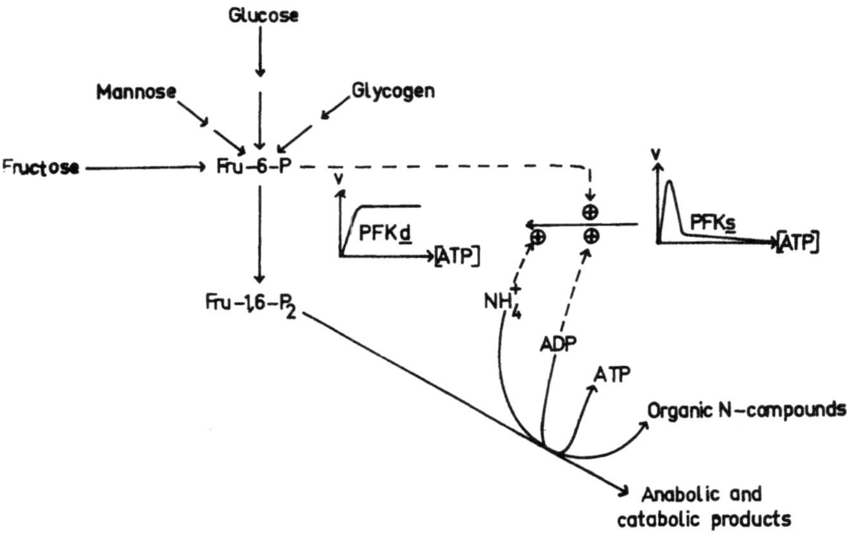

Fig. 4. Scheme of PFK regulation including conversion.

Physiologisch-chemisches Institut der Karl-Marx-Universität Leipzig

St. LIEBE, G. KOPPENSCHLÄGER and W. DIEZEL

INFLUENCE OF EFFECTORS ON THE MOLECULAR PROPERTIES AND ENZYMIC ACTIVITY OF YEAST-PHOSPHOFRUCTOKINASE

According to our results the kinetic behaviour and the molecular properties of yeast-phosphofructokinase depend on the source of the enzyme. For the investigations about the influence of effectors on the molecular properties and enzymic activity we used two preparations of yeast-phosphofructokinase:
1. Yeast-phosphofructokinase from BOEHRINGER, Mannheim (for this preparation we use the name stored enzyme). This commerciel enzyme was further purified using preparative ultracentrifugation.
2. In our laboratory freshly prepared yeast-phosphofructokinase. Both enzymes show no difference in sedimentation pattern at pH 7.0 (Fig. 1).
Stored yeast-phosphofructokinase can be converted into several molecular forms depending on pH and ligand concentrations (Tab. 1). The trimer with about 18 s (this is identical with the 17 s fraction of the first slide) having a molecular weight of about 540 000 is stable at pH 7.0 independent of substrates and effectors and at pH 6.0 with Fru-6-P or high ATP concentrations. The dimer with about 14 s and a molecular weight of about 360 000 was obtained at pH 6.0. This dimer tends to split further under these conditions. Stabilization of the dimer is achieved at pH 6.0 with 1 mM ATP and at pH 9.0 with 2 mM Fru-6-P.
The dimeric form of yeast phosphofructokinase can be detected, too, after incubation of the stored enzyme at pH 9.0 with 2 mM Fru-6-P, using analytical ultracentrifugation (Fig. 2). On the left you see the stored enzyme at pH 7.1, in the middle after incubation at pH 9.0 with 2 mM Fru-6-P for one day and right after 5 days. It seems that the dimer is an intermediate form during the fragmentation of the enzyme into smaller subunits. The formation of the dimer from the trimer is reversible.
At pH 9 in presence of 2 mM ATP we observed two peaks after density gradient centrifugation, having 17.4 s with 530 000 daltons and 10.6 s with 240 000 daltons (Tab. 1). Variation of pH between pH 8 and 9 as well as variation of ATP

Table 1

INFLUENCE OF pH AND SUBSTRATES ON SEDIMENTATION VELOCITY OF YEAST-PHOSPHOFRUCTOKINASE IN DENSITY GRADIENT CENTRIFUGATION

buffer medium		s-value of yeast-phosphofructokinase	
		stored enzyme	fresh prepared enzyme
pH 6.0		14.5 (Dimer)	21.5
	+ 1 mM ATP	14.5 (Dimer)	18.5 (Trimer)
	+ 6 mM ATP	18.5 (Trimer)	-
	+ 2 mM Fru-6-P	18.5 (Trimer)	21.5
pH 7.0	-	18.5 (Trimer)	18.5 (Trimer)
	+ 2 mM Fru-6-P	18.5 (Trimer)	18.5 (Trimer)
	+ 2 mM ATP	18.5 (Trimer)	18.5 (Trimer)
pH 9.0	-	17.4 (Trimer)	18.5 (Trimer)
	+ 2 mM Fru-6-P	13.5 (Dimer)	18.5 (Trimer)
	+ 2 mM ATP	17.4 (Trimer)	18.5 (Trimer)
		10.6	

concentration between 2 and 6 mM do not change this distribution. The position of the slower fraction depends on protein concentration. With higher protein concentrations a higher sedimentation coefficient has been observed. With polyacrylamide-gel-electrophoresis the 10 s fraction can be resolved into three bands having molecular weights of 540 000, 180 000 and 90 000 daltons (Fig. 3). The molecular weight of 180 000 is identical with the value for the monomer of phosphofructokinase. Therefore we believe that the 10 s component represents an equilibrium involving the monomer and trimer of posphofructokinase. The isolated 10 s fraction shows after a second centrifugation at pH 9 with ATP the same distribution as the native enzyme (Fig. 4 a), indicating an equilibrium between the 10 and 17 s component. The 10 s fraction shows a similar behaviour as the trimer by variation of pH and substrates (Fig. 4 a-c). At pH 9 with Fru-6-P the fraction is transformed into the dimer and at pH 7.1 with Fru-6-P into the trimer with a small amount of dimer. This behaviour corresponds to that of the native enzyme.

In contrast to the characteristics of stored enzyme the freshly prepared yeast-phosphofructokinase shows in some respects another behaviour. The enzyme is more stable upon changing pH and substrates (Tab. 1). Using density-gradient-centrifugation neither at pH 6 nor at pH 9 molecular forms with lower molecular weights can be observed.

Fig. 5 demonstrates the dependence of s-values on protein concentration at pH 7. 0 and pH 6. 0, using analytical ultracentrifugation. At low protein concentrations we found at pH 6. 0 a sharp decrease of s-values. The curve with the open circles is an experiment with an enzyme one week after preparation and the curve with the cross after four weeks, indicating a dependence of the dissociation on the time of storage. From the sedimentation and diffusion coefficients (S_w^o, D_w^o) at pH 6. 0 we calculated a molecular weight of about 560 000 daltons. Therefore we assume that the higher s-value at pH 6. 0 is caused by a conformational change of the enzyme between pH 6 and 7 and not by an association. Also with respect to the stability of enzymic activity there is a difference between freshly prepared and stored enzyme. In very dilute solutions stored enzyme is inactivated very rapidly (Fig. 6). In the double logarithmic plot we obtained two straight lines indicating different processes for the first (fast) and the second (slow) period of inactivation. If we add Fru-6-P to the dilution medium after 2 or 4 hours, we observe a reactivation of the enzyme, but the full activity ist not regained.

In contrast freshly prepared enzyme is inactivated more slowly in the first four hours (Fig. 7). The different inactivation rate can be observed at pH 7 as well as at pH 6 and 9. If the dilution medium contains Fru-6-P or ATP we observe an other behaviour. Both fresh and stored enzyme are not inactivated within the first four hours after dilution. Even at pH 7 and pH 9 an increase in enzymic activity within the first two to four hours could be detected.

In concentrated solutions, too, the stored enzyme is inactivated during dialysis, whereas the fresh enzyme is stable. As in dilute solutions the enzymic activity of the dialyzed stored sample is regained after redialysis against buffer with Fru-6-P. Although the enzymic activity of the samples with and without Fru-6-P differs of about 50 % the sedimentation patterns are equal in the analytical ultracentrifuge. Therefore the loss of activity is not accompanied by a desaggregation of the enzyme. Also in diluted solutions there is no close relationship between enzymic activity and desaggregation of the enzyme molecule, because at pH 9 with ATP or Fru-6-P we found forms with lower molecular weights in density gradient centrifugation and under the same conditions an increase in enzymic activity.

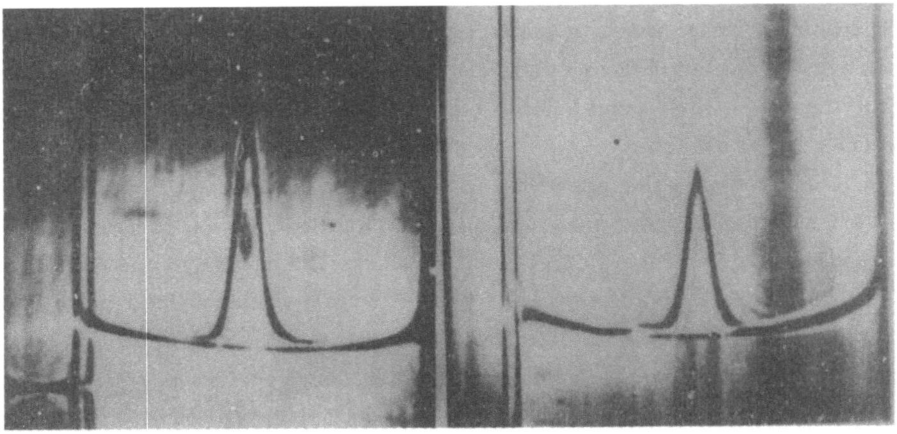

Fig. 1. Sedimentation pattern of freshly and stored Yeast-Phosphofructokinase.

Fig. 2. Sedimentation pattern of stored Yeast-Phosphofructokinase after incubation at pH 9.

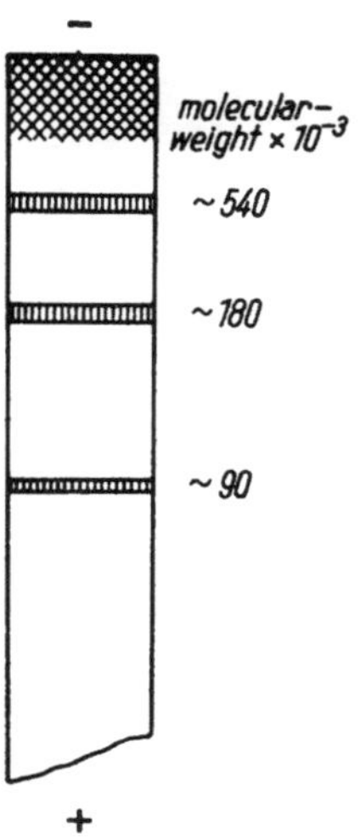

Fig. 3. Polyacrylamide-gel electrophoresis of Yeast-Phosphofructokinase.
The enzyme was incubated at pH 9.0 with 3 mM ATP and 3 mM 2-Mercaptoethanol for 12 hours.

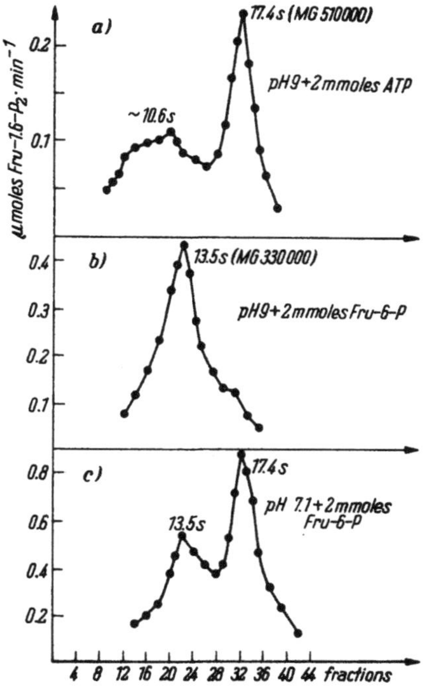

Fig. 4. Sedimentation pattern of the 10 s-fraction in density gradient centrifugation.
The 10 s - fraction was dialyzed against Kalium-phosphate-buffer with 0.5 mM EDTA and 3 mM 2-Mercaptoethanol for 17 hours. pH-value and ligand concentrations are on the fig.

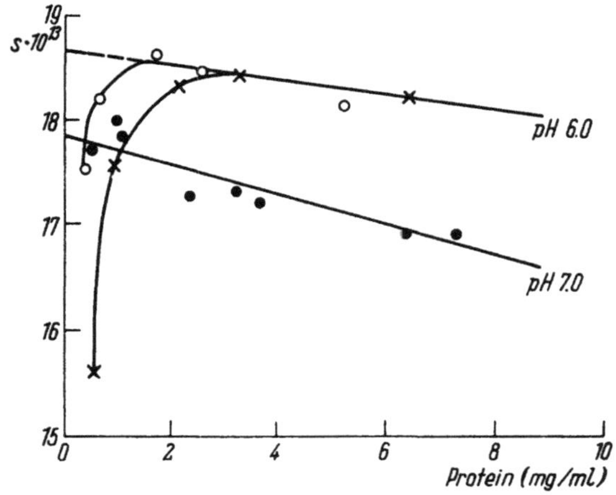

Fig. 5. Dependence of s-values on protein concentration.

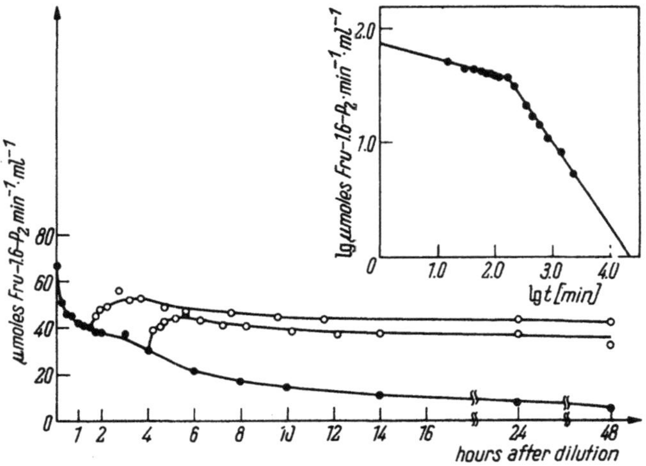

Fig. 6. Inactivation of stored Yeast-Phosphofructokinase after dilution. The final concentration of the enzyme was 50 µg/ml.
Dilution medium: ●——● 0.1 M Kalium-phosphate-buffer pH 7.1 with 0.5 mM EDTA and 3 mM 2-Mercaptoethanol.
o——o After addition of 2 mM Fru-6-P.

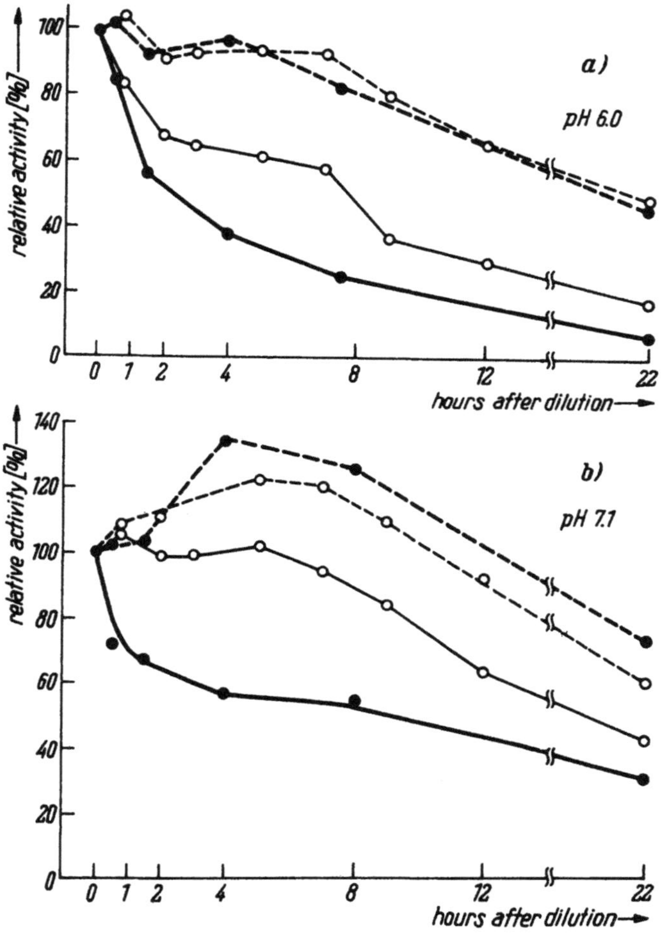

Fig. 7. Inactivation of stored (●——●) and fresh prepared (○——○) Yeast-Phosphofructokinase after dilution.
The final concentration of the enzyme was 50 μg/ml.
Dilution medium: ——— 0.1 M Kalium-phosphate-buffer with 0.5 mM EDTA and 3 mM 2-Mercaptoethanol.
- - - - - 0.1 M Kalium-phosphate-buffer with 0.5 mM EDTA. 3 mM 2-Mercaptoethanol, and 2 mM Fru-6-P.

Unitè de Biochimie, Universitè Catholique de Louvain (Belgium)

G. F. DOMAGK

ISOLATION AND CHARACTERIZATION OF FRUCTOSE DIPHOSPHATE ALDOLASE FROM BEEF MUSCLE

The experimental work to be reported here was begun 6 years ago when I had the opportunity of spending one year with B. L. HORECKER in the Department of Molecular Biology, ALBERT EINSTEIN College of Medicine, Bronx, N.Y., U.S.A. This study was continued later at the University of Göttingen, and is now receiving renewed attention at Louvain.

Aldolase had first been described by MEYERHOF and LOHMANN in 1934, when they had found that muscle extracts convert fructose diphosphate into triose phosphate. With a purified enzyme the stoichiometric and reversible reaction shown in Fig. 1 was demonstrated: dihydroxyacetone phosphate (DHAP) condenses with glyceraldehyde phosphate to form fructose diphosphate (FDP). Early studies had shown that DHAP is an essential reactant, whereas glyceraldehyde phosphate can be replaced by various other aldehydes. Of these the reactions with D-glyceraldehyde (yielding fructose-1-phosphate) and with erythrose-4-phosphate (to give sedoheptulose-1, 7-diphosphate) might be of physiological significance.

Aldolase has been found in a great variety of animal tissues, in plants, and in many microorganisms; it may be simpler to mention the organisms which do not contain aldolase, namely Leuconostoc mesenteroides and a few heterofermentative microorganisms, which lack completely the EMBDEN-MEYERHOF pathway.

Crystalline preparations of aldolase were obtained in 1939 by BARANOWSKI and in 1943 by WARBURG, who observed that the aldolase from rat muscle behaves different from the aldolase found in yeast by the fact, that the latter is inhibited by metal complexing agent, whereas these chelators do not affect the animal enzyme.

The mechanism of aldol condensation of trioses had been studied already by EMIL FISCHER in 1887, and his son, H.O.L. FISCHER, showed in 1936 that the aldolase catalyzed condensation of DHAP with D, L-glyceraldehyde yields only the ketohexoses fructose and sorbose (with a transconfiguration of the

hydroxyl groups at the carbon atoms C-3 and C-4), whereas the corresponding cis-derivatives (tagatose and psicose) are not formed.
From spectral studies performed in 1957 TOPPER had postulated the intermediate formation of a Schiff base complex between aldolase and its substrate, and in 1958 ROSE showed that a base-catalyzed aldol condensation with the formation of an DHAP carbanion is involved in this biocatalysis (23):
In 1961 HORECKER and his coworkers had applied successfully to aldolase EDMOND FISCHER' s method of stabilizing Schiff bases by reduction with boro-hydride; the radioactive reactant $DHA^{32}P$ could be fixed to the protein (Fig. 3). Acid hydrolysis of the modified enzyme protein yielded ε-glyceryl lysine (Fig. 4), which proved that the ε-amino group of an active lysine residue must be involved in the catalytic function of aldolase.
When I joined Professor HORECKER' s group in 1965 many comparative studies involving aldolases from rabbit muscle, rabbit liver (16), yeast (10), spinach (4), and even Boa constrictor (24) were on the way, and I chose the beef muscle enzyme as the new field for me. This enzyme could be purified by simple steps described in Table 1, by which large crystals (Fig. 5) can be obtained in good yield. The molecular weight was determined in the ultracentrifuge to be 156 000, and the aminoacid composition given in Table 2 was found to be very similar to that of the rabbit muscle enzyme. Only methionine contents seem to be significantly higher than the value reported for the rabbit muscle enzyme.

Table 1

PURIFICATION OF BEEF MUSCLE ALDOLASE

Fraction	Vol. (ml)	Units	Protein (g)	Spec. Activity	Yield (%)
Crude extract	530	17,000	13.35	1.3	100
1st Ammon. sulfate pptn s= 0.44-0.65	75	16,400	5.12	3.3	96
2nd Ammon. sulfate pptn s= 0.48-0.53	53	15,200	1.68	9	90
DEAE Cullulose filtrate	154	14,800	1.1	13.4	87
Crystals	180	13,500	1.1	12.2	77

Table 2

AMINOACID COMPOSITION OF BEEF MUSCLE ALDOLASE

Lysine	106	Alanine	153
Histidine	37	Cystein	31
Arginine	62	Valine	84
Aspartate	132	Methionine	19
Threonine	76	Isoleucine	80
Serine	76	Leucine	146
Glutamate	167	Tyrosin	48
Proline	95	Phenylalanine	35
Glycine	122	Tryptophan	12

In 1965, when careful aminoacid analyses of the β-glyceryl phosphate aldolase were performed by LAI in HORECKER' s group, it was reported that the rabbit muscle aldolase had not 1 but 2 active sites (13). I was not very popular when my work indicated the presence of 3 active sites in both beef muscle aldolase and rabbit muscle aldolase. However, I was rather sure of my results since the ^{14}C-DHAP, which was prepared from radioactive fructose by a 2 steps enzymatic reaction as indicated in Fig. 6, had been purified by high voltage eletrophoresis. The specific activity of the eluted ^{14}C-DHAP could be determined exactly by enzymatic assay and scintillation counting respectively.

A structural investigation of the active site of beef muscle aldolase became possible by following the techniques developed in HORNECKER' s group (14), the principles of which are listed in Table 3.

Table 3

STEPS USED FOR THE ISOLATION OF THE DHAP BINDING SITE

1) Schiff base formation with ^{14}C-DHAP
2) Schiff base stabilized by means of $NaBH_4$
3) Precipitation of enzyme-substrate complex by trichloroacetic acid
4) Reduction of the protein by mercaptoethanol; carboxymethylation
5) Tryptic digestion
6) Isolation of peptide (Gel filtration and ion exchange chromatography)

3 subsequent chromatographic steps with the reduced, carboxymethylated and tryptically cleaved Schiff base enzyme gave a radioactive peptide (Fig. 7), which in the fingerprint technique moved as a singel uniform spot. The aminoacid composition of this peptide was determined to be identical to that of the octaeicosa peptide which represents the substrate binding site of rabbit aldolase.

Table 4

28 AMINOACIDS FOUND IN THE DHAP SITE OF BEEF MUSCLE ALDOLASE

1	β-Glyceryl-lysine	2	Proline
1	Lysine	2	Glycine
3	Histidine	2	Alanine
1	Carboxymethyl-cystein	1	Valine
2	Aspartic acid	1	Methionine
3	Threonine	1	Isoleucine
1	Serine	4	Leucine
2	Glutamic acid	1	Tyrosine

The primary structure of the active site peptide of beef muscle aldolase has been elucidated partially: a kinetic test using carboxypeptidases A and B according to Fig. 8 gave evidence for its 3 C-terminal aminoacids, and after the peptide had been cleaved at its single methionine residue by means of cyanogen bromide, 5 further C-terminal residues could be determined by successive Edman degradation steps in combination with the aminoacid anayzer (Fig. 9).
Fig. 10 shows the primary structures of the substrate binding sites of rabbit muscle aldolase and rabbit liver aldolase as determined by HORECKER (17).
It can be seen that both these peptides are very similar in their structure: the liver enzyme shows only 4 aminoacid exchanges (underlined in Fig. 10) and one deletion as compared to muscle aldolase. The 4 exchanges can be explained by single base mutations in the DNA fragment coding for this peptide.
For some years there had been good evidence for aldolase to be composed of 3 subunits; Table 5 presents some of the arguments used to support this assumption. More recent data, however, given in Table 6, make it almost sure that there are 4 very similar subunits of a molecular weight of 40 000 each.

Table 5

EVIDENCE FOR 3 SUBUNITS IN RABBIT MUSCLE ALDOLASE

3 C-terminal tyrosine residues	ref. (3)
Dissociation by base yields 1/3 and 1/6 of molecular weight	ref. (6)
3 binding-sites for DHAP	ref. (2)
3 N-terminal proline residues	ref. (26)

Table 6

EVIDENCE FOR 4 SUBUNITS IN RABBIT MUSCLE ALDOLASE

Five-membered hybrids	ref. (20)
Ultracentrifugation after dissociation by urea	ref. (25)
Eletronmicroscopy of enzyme molecules	ref. (21)
Cleavage of CNBr and peptide analyses	ref. (12)

The fact that only 3 substrate binding sites were found in aldolase now thought to be composed of 4 subunits was explained by the fact, that earlier studies of KOBASHI et al. (8) had shown that there is one mole of organic phosphorus bound per mole of aldolase.

Following a technique developed for the rabbit muscle enzyme (18) we were able to isolate 2 species of protein subunits from our beef enzyme; (Fig. 13) shows a chromatography of carboxymethylated beef muscle aldolase performed in the presence of 8 M urea. For the rabbit muscle aldolase it has been shown (11) that the difference between α- and β-subunits is due to a single asparagine residue in α-chains, which in the β-subunits is replaced by an aspartic acid. The α_2 β_2 structure, which had been postulated for aldolase 3 years ago, has recently been questioned again, when it was found, that in young rabbits the $\alpha:\beta$ ratio is definitely age-dependent (9). Shortly after birth of the animal the aldolase of muscle is composed almost exclusively of α-subunits; during the following months of life the molar fraction of β-subunits increases steadily from 0.08 to 0.5, which value is reached after about 6 months (11). The continued replacement of only 2 β-chains of aldolase has been discussed as a normal turnover of this enzyme. The search for differences in the structures of muscle aldolases from beef and rabbit so far has not given definite results. We plan to use the high contents of methionine residues (20 as compared to 12 methionines found in rabbit muscle aldolase) in the beef enzyme to use for the production of peptides by protein cleavage by means of cyanogen bromide. The existence of structural differences in the proteins of rabbit and beef muscle aldolases, respectively, is very likely: ORD measurements performed by my former coworker W. DOMSCHKE at Freiburg University showed a higher helical content in the rabbit aldolase. With both enzymes a specific inhibition by homologous antibodies was observed; these data are shown in Table 7.

Table 7

SPECIFIC INHIBITION OF RABBIT MUSCLE ALDOLASE (RMA) AND BEEF MUSCLE ALDOLASE (BMA) BY HOMOLOGOUS ANTIBODIES

Enzyme + Additive in pre-incubation	Aldolase Activity (U/ml)	% Inhibition
RMA + normal rabbit serum	42	0
RMA + anti-RMA	22.4	47
RMA + anti-BMA	42.5	0
BMA + normal rabbit serum	42.3	0
BMA + anti-RMA	42.7	0
BMA + anti-BMA	19.5	54

As a last point in this review I want to bring up the question of the existence of various aldolases in the animal organism. About 30 years ago LEUTHARDT had postulated the occurence of an aldolase in liver, which was thought to be specific for the cleavage for fructose-1-phosphate (15). Comparative studies of recent years have shown that there is not such an enzyme. Some aldolases of tissues other than muscle have, however, a higher affinity for fructose-1-phosphate. The relative activities toward fructose diphosphate and fructose-1-phosphate as well as different mobilities have been used to classify animal aldolases into 3 groups named A, B and C (Fig. 14). Minor chemical modifications like acetylation of part of the tyrosine residues in muscle aldolase converts the A type enzyme into an aldolase of the B type (22; Fig. 15).

Fig. 16, taken from the review written by MORSE and HORNECKER (14), shows that the active site of muscle aldolase seems to be composed of a narrow hydrophobic pocket, in which several aminoacid residues establish contact with the substrate.

In the beginning of this review the stereospecificity for substrates with a trans-configuration of the hydroxygroups at carbon atoms C-3 and C-4 has been mentioned. In 1962 HEATH (4) has isolated an aldolase from fucose adapted E. coli, which showed a specificity for substrates with both OH-groups at C-3 and C-4 at the "right side" of the molecule; our own studies on degradation of L-rhamnose by Lactobacillus plantarum (1) led us to the isolation of an aldolase specific for a trans-configuration of the hydroxylgroups at C-3 and C-4, but opposite to that required by fructose diphosphate aldolase. At present we are studying the bacterial metabolism of D-tagatose, which hopefully will lead to the isolation of an aldolase not described so far and specific for a cis-configuration (Fig. 17).

REFERENCES

(1) DOMAGK, G. F., R. HEINRICH, Biochem. Zeitschr. 341, (1965), 420

(2) DOMAGK, G. F., unpublished experiments (1965)

(3) DRECHSLER, E. R., P. D. BOYER, A. G. KOWALSKY, J. Biol. Chem. 234, (1959), 2627

(4) FLURI, R., T. RAMASARMA, B. L. HORECKER, Europ. J. Biochem. 1, (1967), 117

(5) GHALAMBOR, M. A., E. C. HEATH, J. Biol. Chem. 237, (1962), 2427

(6) HASS, L. F., M. S. LEWIS, Biochemistry 2, (1963), 1368

(7) KOCHMAN, M., E. PENHOET, W. J. RUTTER, Fed. Proc. 27, (1968), 590

(8) KOBASHI, K., C. Y. LAI, B. L. HORECKER, Arch. Biochem. Biophys. 117, (1966), 437

(9) KOIDA, M., C. Y. LAI, B. L. HORECKER, Arch. Biochem. Biophys. 134, (1969), 623

(10) KOWAL, J., T. CREMONA, B. L. HORECKER, Arch. Biochem. 114, (1966), 13

(11) LAI, C. Y., B. L. HORECKER, J. Cell. Physiol. 76, (1970), 381

(12) LAI, C. Y., C. CHEN, Arch. Biochem. Biophys. 128, (1968), 212

(13) LAI, C. Y., O. TCHOLA, T. CHENG, B. L. HORECKER, J. Biol. Chem. 240, (1965), 1347

(14) LAI, C. Y., P. HOFFEE, B. L. HORECKER, Arch. Biochem. Biophys. 112, (1965), 567

(15) LEUTHARDT, F., E. TESTA, H. P. WOLF, Helv. Chim. Acta 36, (1953), 227

(16) MORSE, D. E., B. L. HORECKER, Arch. Biochem. Biophys. 125, (1968), 942

(17) MORSE, D. E., B. L. HORECKER, Adv. Enzymol. 31, (1969), 125

(18) MORSE, D. E., W. CHAN and B. L. HORECKER, Proc. Nat. Acad. Sci. U. S. 58, (1967), 628

(19) LEBHERZ, H. G., W. J. RUTTER, Biochemistry 8, (1969), 109

(20) PENHOET, E., T. RAJKUMAR, W. J. RUTTER, Proc. Nat. Acad. Sci. U. S. 56, (1966), 1275

(21) PENHOET, E., M. KOCHMAN, R. VALENTINE, W. J. RUTTER, Biochem. 6, (1967), 2940

(22) PUGH, E. L., B. L. HORECKER, Biochem. Biophys. Res. Comm., 26, (1967), 360

(23) ROSE, I. A., Proc. Nat. Acad. Sci. U. S. 44, (1958), 10

(24) SCHWARTZ, E., B. L. HORECKER, Arch. Biochem. Biophys. 115, (1966), 407

(25) SIA, C. L., B. L. HORECKER, Biochem. Biophys. Res. Comm. 31, (1968), 731

(26) SINE, H. E., L. F. HASS, J. Biol. Chem. 244, (1969), 430

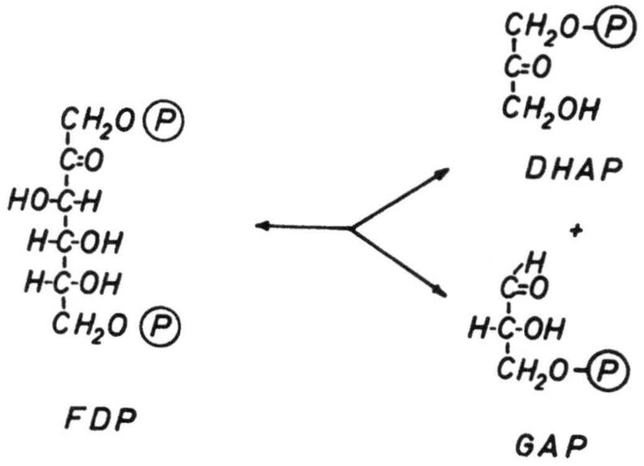

Fig. 1. The aldolase reaction.

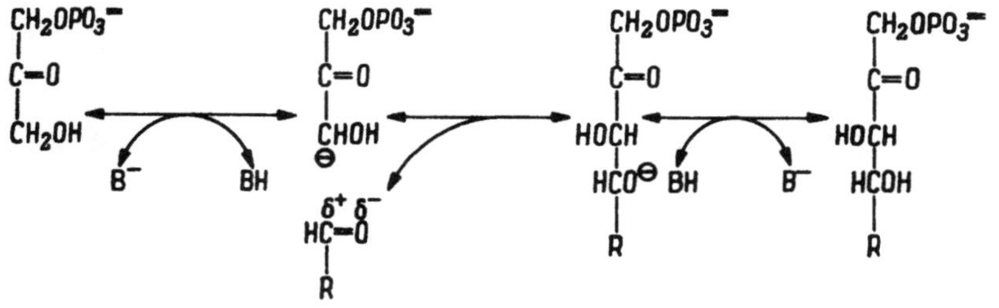

Fig. 2. Reaction mechanism of aldolase.

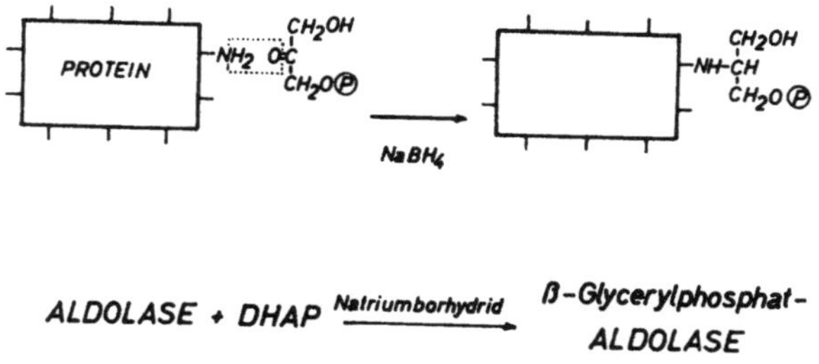

Fig. 3. Formation of β-glyceryl-phosphate aldolase.

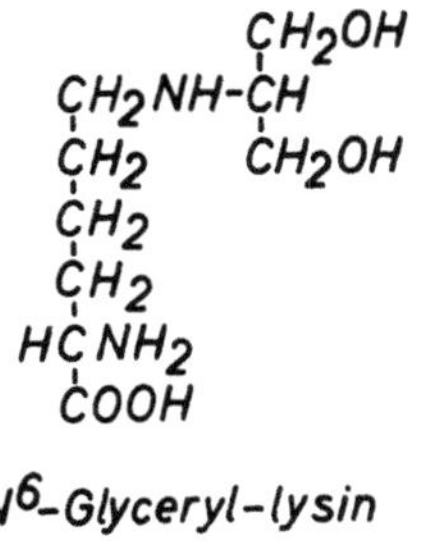

Fig. 4. N^6-β-glyceryl lysine.

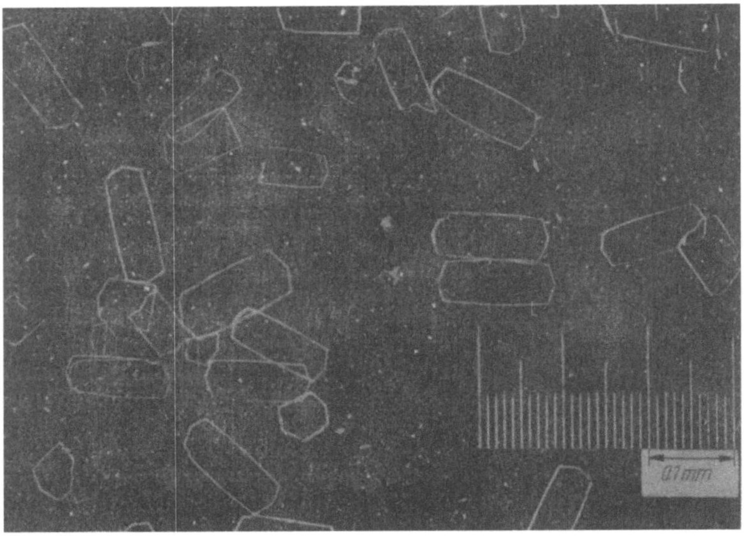

Fig. 5. Micrograph of crystalline beef muscle aldolase.

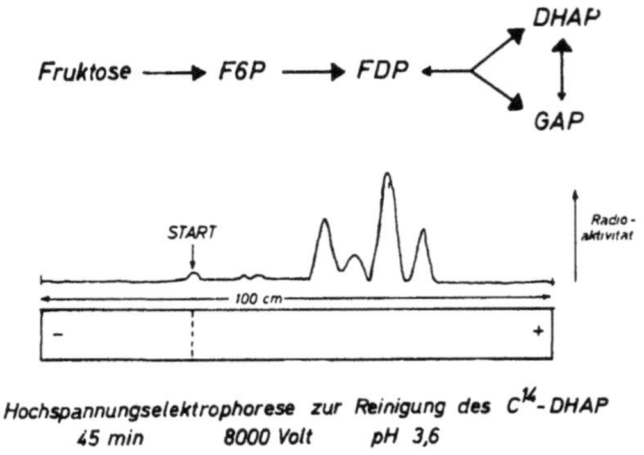

Fig. 6. Preparation and purification of radioactive DHAP (the 3rd peak was eluted from the paper).

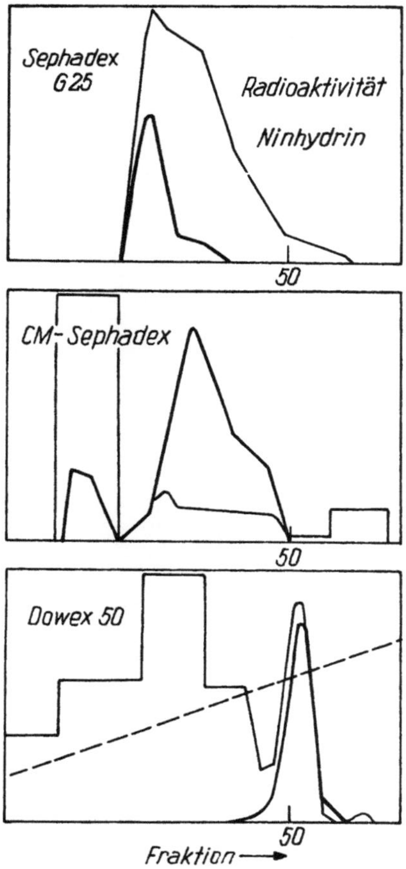

Fig. 7. Isolation of the DHAP binding site of Beef Muscle Aldolase.

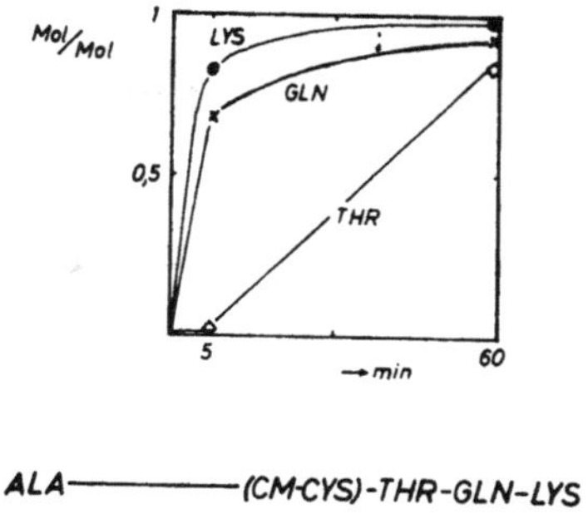

Fig. 8. C-terminal structure of the DHAP binding site.

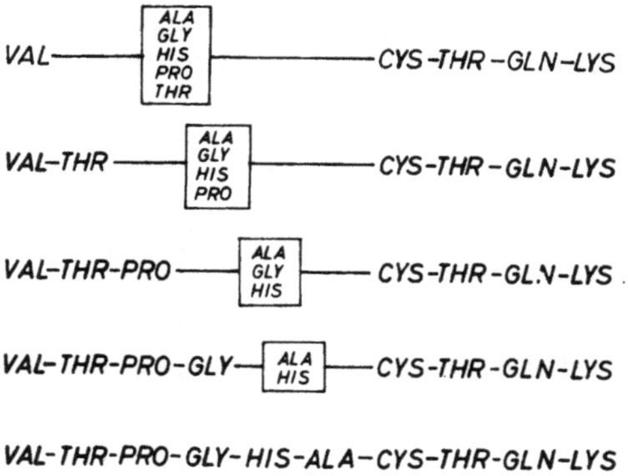

Fig. 9. Edman degradations after CNBr cleavage.

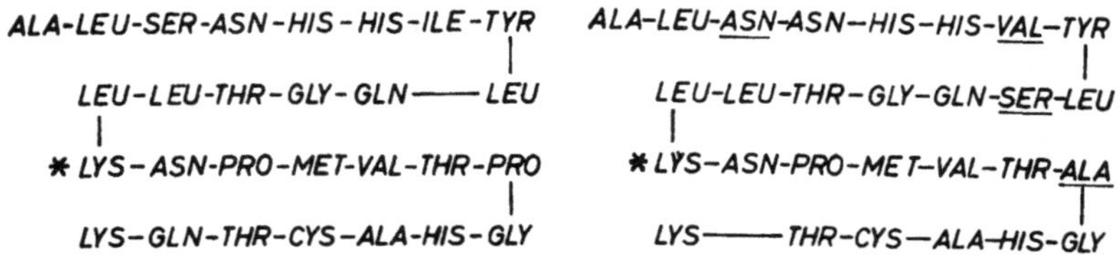

Fig. 10. Substrate binding sites of Muscle (left) and Liver (right) Aldolase (17).

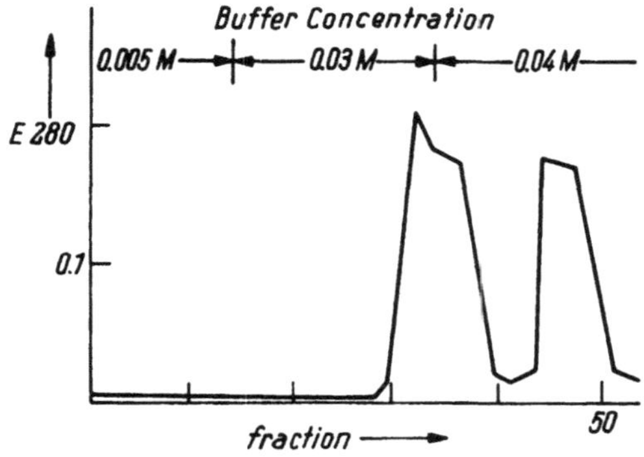

Fig. 11. Preparation of subunits of Beef Muscle Aldolase DEAE-cellulose; Tris pH 8; 8 M Urea.

Type	Organ	FDP/F1P	Electrophoresis
A	Muscle	40	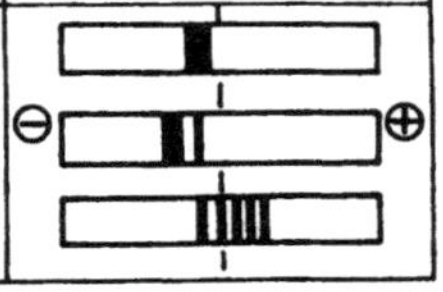
B	liver	1.2	
C	brain	3	

Fig. 12. Different types of class I aldolases from rat tissues (after (19)).

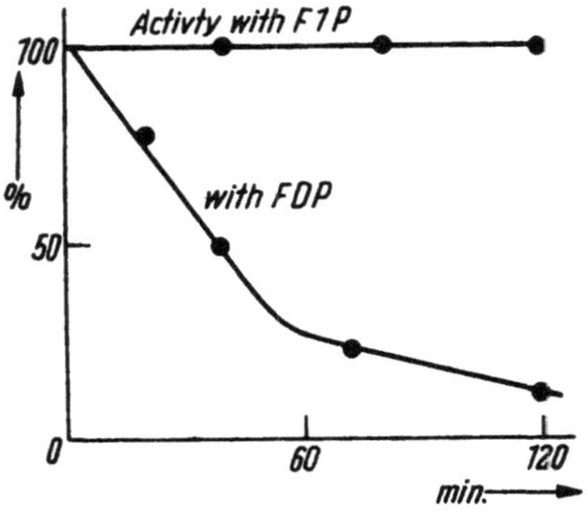

Fig. 13. Modification of beef muscle aldolase by incubation with acetyl-imidazole.

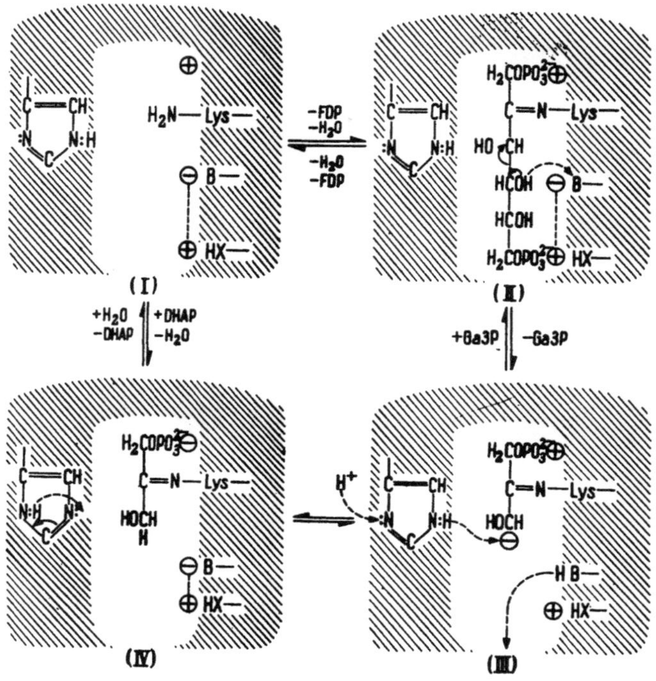

Fig. 14. Mechanism of aldolase.

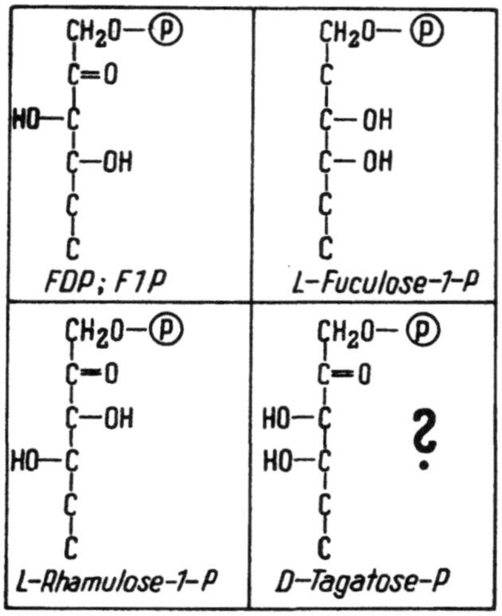

Fig. 15. Stereospecificity of various aldolases (their substrates are mentioned).

Physiologisch-chemisches Institut der Karl-Marx-Universität Leipzig

G. ZIMMERMANN, J. GAUER, W. DIEZEL, K. -W. WENZEL

STUDIES ON SUBUNIT STRUCTURE OF ERYTHROCYTE PHOSPHOFRUCTOKINASE

In this communication we will report about the behaviour of several oligomeric forms of phosphofructokinase from human red blood cells. Association-phenomena of erythrocyte phosphofructokinase has been detected both LAYZER and coworkers as well as in our laboratory.

As recently described, the molecular weight of the enzyme depends on the concentration of the enzyme protein. This has been found by sucrose density gradient centrifugation. The molecular weight was estimated to be 400 000. At higher protein concentrations an increase of the sedimentation coefficient has been found.

In eletrophoresis and gel filtration experiments we could observe various oligomeric forms of erythrocyte phosphofructokinase. Now, we are able to give a more precise characterization of the molecular weight distribution of erythrocyte phosphofructokinase.

In our experiments we used an enzyme prepared according to the procedure shown in Table 1.

The erythrocyte phosphofructokinase was purified using DEAE-cellulose, ammonium sulfate fractionation, heat treatment, Sephadex G 200 and Sepharose 4 B. The final specific activity of the purified enzyme was 136 U/mg.

This enzyme was dissolved in a stabilizing medium, such as 100 mM phosphate buffer at pH 7 and 2 mM fructose-6-phosphate.

The purified enzyme in a protein concentration of 7 mg/ml has been found to be suitable for crystallization. To the enzyme solution solid ammonium sulfate was stepwise added to a final saturation up to 45 %. In this suspension the phosphofructokinase crystallized by changing the temperature between -20° and $+5^{o}$. Crystallization occurs in presence of 2 mM fructose-6-phosphate. Figure 1 shows enzyme crystals.

This high purified enzyme shows only a single band in the sodium dodecylsulfate polyacrylamide gel electrophoresis (Fig. 2).

Table 1

Purification of Human Erythrocyte Phosphofructokinase

Step		Volume ml	Total Activity U	Total Protein mg	Specific Activity U/mg	Purification	Yield %
1.	Haemolysate	4800	1920	585000	0, 003	1	100
2.	Batch DEAE-Cellulose	1410	1560	5250	0, 3	100	81
3.	Ammonium Sulfate Fractionation	240	1460	1620	0, 9	300	76
4.	Heat treatment	230	1310	350	3, 7	1230	68
5.	Ammonium Sulfate Fractionation	5	1150	255	4, 5	1500	60
6.	Sephadex G 200	3, 5	1080	54	20	6700	56
7.	Sepharose 4 B	1, 5	970	7, 1	136	45000	50

Fig. 3 shows the molecular weight determination of the fragment produced by SDS. Trypsin, aldolase and catalase as reference proteins we used for calculation of molecular weight. By this method the molecular weight of this unitwas determined to be about 104 000 dalton. This is only an approximated value owing to a lack of suitable reference proteins in the range of this phosphofructokinase subunit.
For studying the oligomeric structure of phosphofructokinase we applied the native enzyme to gel filtration and disc eletrophoresis.
The polyacrylamide gel electrophoresis was carried out with a linear gel concentration gradient as described by DIEZEL and KOPPENSCHLÄGER. The separated protein bands were detected with coomassie brilliant blue.
The results are shown in figure 4.
The native enzyme reveals two distinct bands and a broad region of higher molecular forms. The molecular weights related to this bands were determined under the same conditions using a reference run with catalase and monomeric and dimeric serum albumin as standards. The molecular weights were found to be about 220 000, 450 000 and in the range between 600 000 and 900 000. Simultaneously the enzyme activity of these separated forms has been measured.
The 450 000 band and the higher molecular region shows phosphofructokinase activity. The fractions containing the 220 000 molecules are under the conditions of electrophoresis inactive.
In addition, the native enzyme was studied by means of gel filtration using Sepharose 4 B. The column had a length of 70 cm and a diameter of 1, 5 cm. This gel chromatography system was equilibrated with buffer containing 100 mM potassium phosphate at pH 7 and in presence of 2 mM fructose-6-phosphate. The resulte of these experiments are presented in figure 5.
Five enzymatic active peaks are obtained. Using reference proteins as standards for the molecular weight determination these five peaks represent oligomeric forms of the enzyme with the molecular weights of 210 000, 420 000, 630 000, 840 000 and 1 050 000. The higher molecular forms could only be roughly estimated, because unfortunately we did not get protein standards in that high molecular range. Urease and apoferritin available in our laboratory unexpectedly split into smaller entities under the conditions of chromatography. So they could not be used as standards.
From the linear relation between the eluting volume and the molecular weights of the proteins taken as references the calculation of the approximated values by extrapolation has been done.

It seems, that between these forms a multiple equilibrium consists. This can be concluded from the following experiments. After gel filtration on Sepharose 4 B the peak of 840 000 has been isolated. Upon rechromatography a redistribution under appearance of all five active forms of the erythrocyte phosphofructokinase could be observed.

In addition, the isolated 840 000 molecules come to a rearrangment during gel electrophoresis and are converted to forms with molecular weights of 220 000, 450 000 and 900 000.

Comparing both results from electrophoresis and gel chromatography experiments we can conclude, that indeed different oligomeric forms of erythrocyte phosphofructokinase occur. Despite the fact, that differences in the results between both methods exist, which could be explained methodically. In general it can be said, that in the polyacrylamide gel electrophoresis proteins tend to aggregate to higher molecular forms.

It should be pointed out, that the occurence of the five oligomeric forms of the erythrocyte phosphofructokinase fits well to the SDS fragment with 104 000. However, we do not know whether this form is really the smallest unit.

Owing to this fact, we conclude, that Sepharose 4 B chromatography is able to separate the dimer, tetramer, hexamer, octamer and decamer of the 104 000 unit.

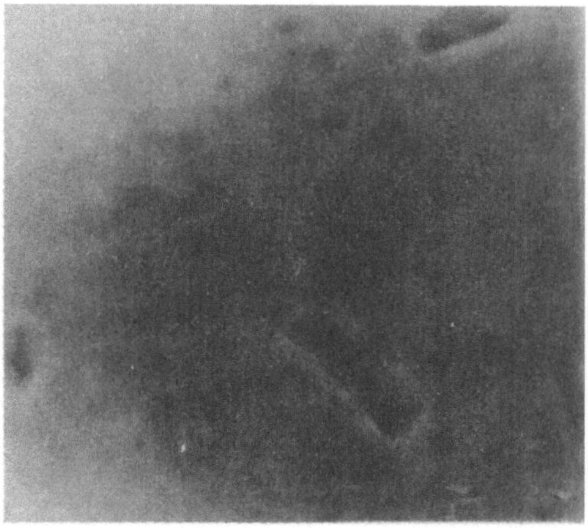

Fig. 1. Crystals of erythrocyte phosphofructokinase (600-fold).

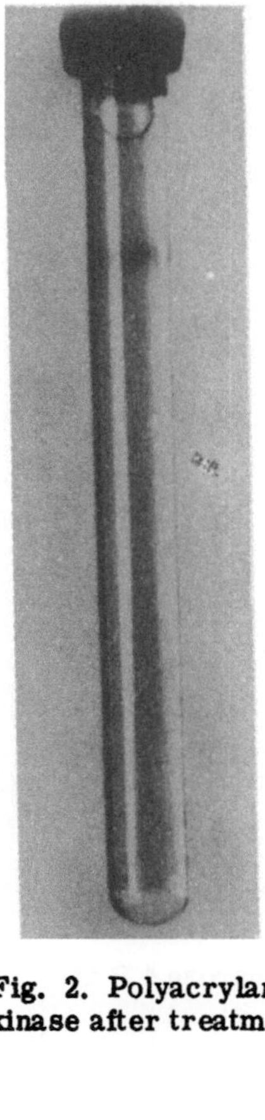

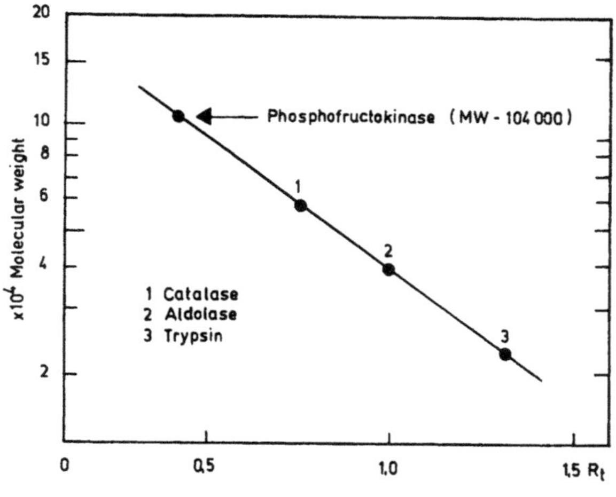

Fig. 3. Molecular weight determination of the SDS fragment by polyacrylamide gel electrophoresis.

Fig. 2. Polyacrylamide gel electrophoresis of erythrocyte phosphofructokinase after treatment with SDS.

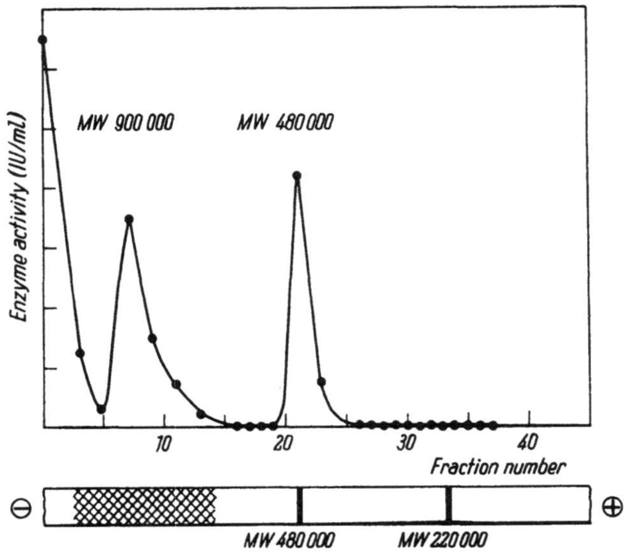

Fig. 4. Distribution of erythrocyte phosphofructokinase on polyacrylamide gel electrophoresis using a linear gel concentration gradient. Top: enzyme activity; bottom: protein bands.

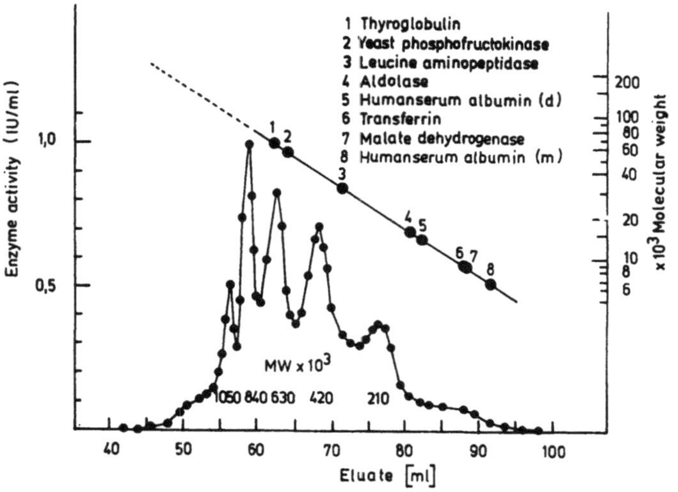

Fig. 5. Fractionation of the enzyme by gel filtration on Sepharose 4 B at pH 7.0 in presence of 2 mM F-6-P.

Institute of Physiological and Biological Chemistry, Humboldt-University Berlin

B. KÜHN, G. JACOBASCH, CH. GERTH, S. RAPOPORT

SIGNIFICANCE OF THE PFK-REACTION FOR THE REGULATION OF GLYCOLYSIS

Systematic investigations during the last years have shown that the regulation of glycolysis takes place at several steps. Important results were obtained by variation of pH, temperature and ionic conditions. From determinations on the profile of the erythrocyte metabolites an crossover-point between G6P/F6P and FDP was found (1, 2). It was therefore postulated PFK was the main controlling step in glycolysis.

Kinetic studies on the purified PFK from rat erythrocytes were carried out to elucidate the regulatory properties of the enzyme.

Kinetic measurements were carried out in ionic conditions corresponding to conditions within erythrocytes. Ammoniumions which have been used frequently influence the PFK in a specific manner as we have shown previously. Therefore ammonium sulfate was replaced by potassium chloride in the test system. The maximal activities in the potassium system are half as high as in the ammonium system. The differences of potassium and ammonium are not explicable by ionic strength alone.

Ammoniumions decrease the cooperativity of PFK and shift the F6P dependence curve to the left (Fig. 1).

In presence of 90 mM KCl the F6P-dependence curve is sigmoidal even at pH 8, 4. The HILL-coefficient is 1, 48. Replacement of potassium chloride by ammonium sulfate yields a Michaelis-Menten-curve with a HILL-coefficient of 1. The sigmoidicity is much more pronounced at pH 7, 2. Here the interaction coefficient is 3 in the potassium-containing system. A comparison of the $S_{0,5}$ values for F6P in the potassium and ammonium system yields differences of about one order of magnitude at both pH-values. The sigmoidicity of the F6P-dependence curve in the potassium system can by increased elevation of ATP. The $S_{0,5}$-value for F6P will be higher, but the interaction coefficient remains unchanged (Fig. 2).

There are earlier communications (3) on the decrease of F6P affinity by elevation of ATP concentrations and the pH dependence of this effect in ammonium sulfate. The smaller affinity of PFK for F6P indicates a stronger inhibition by ATP in the potassium system. The function of ATP in the PFK-reaction as substrate as well as inhibitor is known for a long time. At pH 7, 2 the PFK is 10 times more sensitive to ATP than at pH 8, 4 (Fig. 3).

Already > 0, 1 mM ATP inhibit the enzyme. It indicates that the enzyme is in an inhibited state under intracellular conditions. The strong inhibition of the enzyme by ATP is therefore assumed to be one of the most important regulation parameters of PFK and glycolysis. The affinity of ATP to the site of inhibition on the enzyme is most dependent on variation of the pH near 7, 0. The conclusion was reached from measurements on F6P dependence in the ammonium sulfate system at constant ATP and varying pH-values. The affinity of PFK for F6P is most changed between pH 6, 8 and 7, 2. $S_{0,5}$-values for F6P are decreased by one order of magnitude in this pH-range. We assumed from these observations changes in the conformational state of the enzyme which are related to the state of ionization of imidazole residues at the site where inhibition of ATP takes place. MANSOUR et al. also reported on such a property of the imidazole groups of the PFK of muscle. Modification of the enzyme by photooxidation (4) or specific acylation of histidine (5) lowered the sensitivity to inhibition by ATP. The cooperativity of the F6P-dependence curve is also decreased. One may conclude that the binding of ATP to the site of regulation at PFK depends on the protonation state of its histidine.

Since the PFK activity is nearly completely inhibited by ATP under intracellular condition at pH 7, 1 positive effectors have to define the actual activity. The influence of inorganic phosphate, ADP, AMP and glucose-1, 6-diphosphate was investigated. Inorganic phosphate diminishes the inhibition by ATP, presumably by competition with ATP at the site for ATP inhibition (6) (Fig. 3).

The influence of AMP was examined at pH 7, 2 at several ATP concentrations (Fig. 4). At pH 7, 2 as well as at pH 8, 4 already 20 μM AMP produces distinct activation. 100 to 200 μM activate the enzyme maximally. The activating effect is strongest when the enzyme is in a inhibited state. The intracellular concentration of AMP is 40 μM (1) and lies, therefore, in the range of regulation.

Low concentrations of G 1, 6 P are able to activate the PFK to a large extent. 1 μM G 1, 6 P increases the activity of PFK at pH 7, 2 by a factor of 2, 2 (Table 1).

Table 1

INFLUENCE OF G 1, 6 P at pH 7, 2

G 1, 6 P μM	PFK-activity μMol F6P
-	1, 0
1	2, 2
6	3, 8
100	5, 0

Incubation medium:
0, 2 mM ATP
1, 25 mM $MgCl_2$
1 mM F6P
1 mM P_i
90 mM KCl
40 mM Trisbuffer
0, 2 mM $NADH_2$
aldolase, triosephosphate-isomerase
α-glycerophosphate dehydrogenase

We could not detect any activating effect of ADP at pH 7, 2.

The influence of temperature on the glycolysis rate offers further insights in metabolic regulations. Examination of the glycolytic metabolite profile by changing the temperature gave evidence for influence of the temperature on the equilibrium of the PFK (7). Above 20^0 a decrease of F 1, 6 Pand a increase of G6P occurs, indicating a cross-over point at the PFK. We assumed a decrease of the cooperativity of the PFK at lower temperatures and an increase of the affinity to F6P. Kinetic studies supported this assumption. The sigmoidicity of the F6P-dependent curve increases by elevation of temperature at pH 8, 4 and pH 7, 2 (Fig. 5).

The maximal value for the interaction coefficient n is 2, 1 at 45^0 at pH 8, 4. HILL' s coefficients and the $S_{0,5}$ values for F6P at pH 7, 2 are summarized in Table 2. Decreasing of temperature from 37^0 to 10^0 changes the $S_{0,5}$-values by one order of magnitude. A decrease of the cooperative properties of the enzymes is also apparent. There is also an effect on the sensitivity towards ATP. The inhibition by ATP is less at lower temperatures.

Table 2

INFLUENCE OF TEMPERATURE ON $S_{0,5}$-VALUES FOR F6P AND INTERACTIONS COEFFICIENTS

pH 7,2			pH 8,4	
temperature ^{o}C	F6P $S_{0,5}$ mM	n	F6P $S_{0,5}$ mM	n
10	0, 14	1, 88	0, 013	1, 19
20	0, 51	2, 60	0, 16	1, 37
37	1, 37	3, 00	1, 29	1, 48
45	-	-	0, 96	2, 15

Incubation medium:
1, 25 mM $MgCl_2$
1 mM P_i
90 mM KCl
40 mM Trisbuffer
0, 2 mM $NADH_2$
aldolase, triosephosphate isomerase,
α-glycerophosphate dehydrogenase
0, 2 mM ATP at pH 7, 2
1, 0 mM ATP at pH 8, 4

One may conclude from these experiments that F6P, G 1, 6 P and the adenine nucleotides ATP and AMP are the most important regulating metabolites of PFK. ATP as well as G6P are also regulators for the hexokinase. So PFK and hexokinase act as a functional control system for the glycolytic rate.

How do the presented properties of the erythrocyte PFK explain the control of reticulocyte glycolysis? The reticulocyte has anaerobic and aerobic glycolysis. It shows a temperature and pH-dependence of the Pasteur effect. It seems that the Pasteur effect is most pronounced under conditions of allosteric control of PFK, i. e. at nearly neutral pH and at temperatures near 37°.

PFK and other glycolytic enzymes show a distinct dependence on maturation. Fig. 6 shows the PFK-activity during an anaemia of rabbits. It may be seen that a cell suspension with 55 % reticulocytes contains three times higher PFK

activity than erythrocytes. Compared to mitochondrial enzymes and PK there is a much lower decrease of PFK during maturation. The stability of the reticulocyte PFK is also clearly visible in the case of density gradient centrifugation and subsequent determination of PFK in the separated steps of maturation (Fig. 7). The first fraction contains 70-90 % reticulocytes, the third one contains about 30 %. Nevertheless the PFK-activity in the third fraction is only 25 % lower as in the youngest cell fraction. In very immature cells there is a relation between RNA-content and PFK-capacity. The RNA-content drops rapidly during maturation in contrast to the PFK capacity which decreases slowly (Fig. 8).
It should be interesting to check whether the PFK of reticulocytes is an isoenzyme with control properties differing from those of erythrocyte PFK. ATP-sensitivity is the same for both PFK's. Citrate which is apparently imporant for the operation of PASTEUR-effect by inhibition of PFK should also be considered. The inhibition depends on the ionic composition of the medium. The inhibition constants are about 8 mM for the erythrocyte enzyme in presence of ammonium sulfate. The inhibition is stronger in the potassium system. Inhibition constants were determined in the range of 0, 2 or 1, 2 mM (Fig. 9). We found that the effects of the negative inhibitors citrate and ATP depends on the age of enzyme preparation. An inhibition constant of 0, 2 mM for citrate was found for the freshly prepared enzyme. Therefore one may assume an age dependence of the allosteric properties of the PFK.
The results obtained up to now by us indicate a resemblance of erythrocyte - and reticulocyte - PFK in their kinetic behaviour. The actual affinity in reticulocytes is controlled by intracellular effector concentrations which differ in erythrocytes and reticulocytes.

REFERENCES

(1) JACOBASCH, G., Dissertation B. Biowiss. Fak. der Humboldt-Universität Berlin 1970

(2) MINAKAMI, S., S. TAKAYASU, U. SUZUKI, H. YOSHIKAWA, Biochem. Biophys. Res. Comm. 17, (1964), 748

(3) KÜHN, B., G. JACOBASCH, S. RAPOPRT, Acta biol. med. germ. 23, (1969), 1

(4) AHLFORS, C. E., T. E. MANSOUR, J. biol. Chem. 244, (1969), 1247

(5) SETLOW, B., T. E. MANSOUR, J. biol. Chem. 245, (1970), 5524

(6) ODEIDE, R. B. DUPULS, M. GUILLOTON, J. ROSENBERG, Bull. Soc. chim. Biol. 51, (1969), 47

(7) HASART, E., G. JACOBASCH, S. RAPOPORT, Acta biol. med. germ. 24, (1970), 725

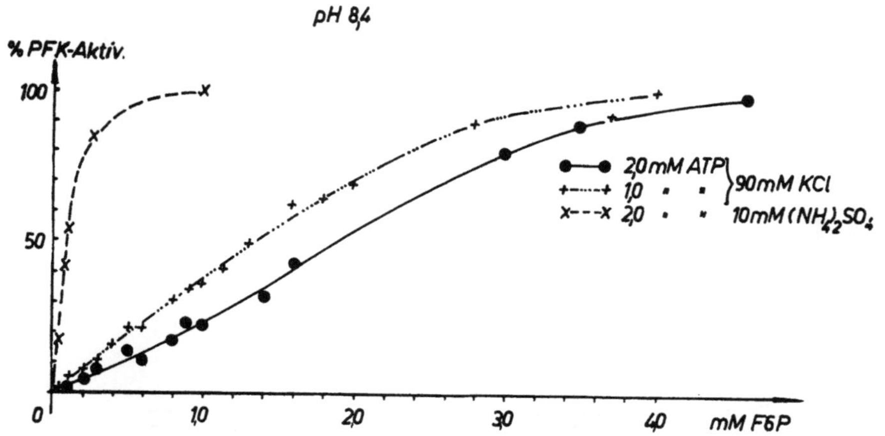

Fig. 1. F6P-dependence of PFK in potassium- and ammonium system.
Incubation medium potassium system: KCl 90 mM, P_i 1 mM, $MgCl_2$ 1, 25 mM
Incubation medium ammonium system $MgCl_2$ 1, 25 mM, $(NH_4)_2SO_4$ 10 mM.

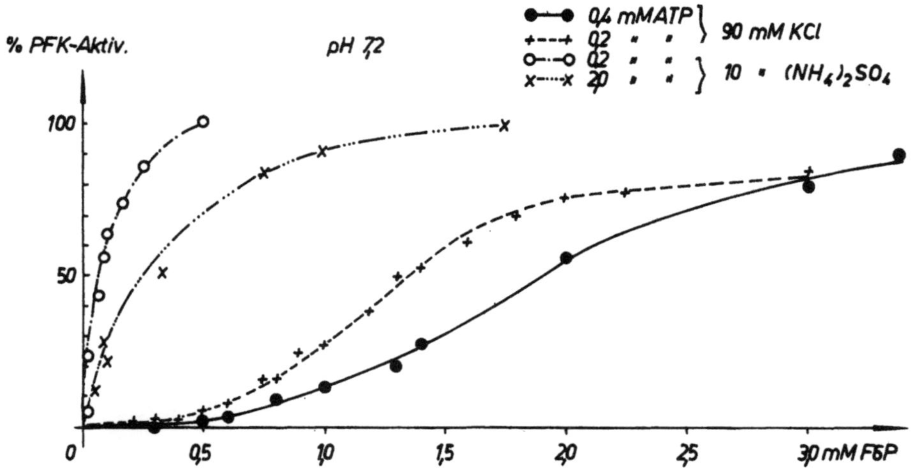

Fig. 2. Influence of ATP on F6P-dependence.

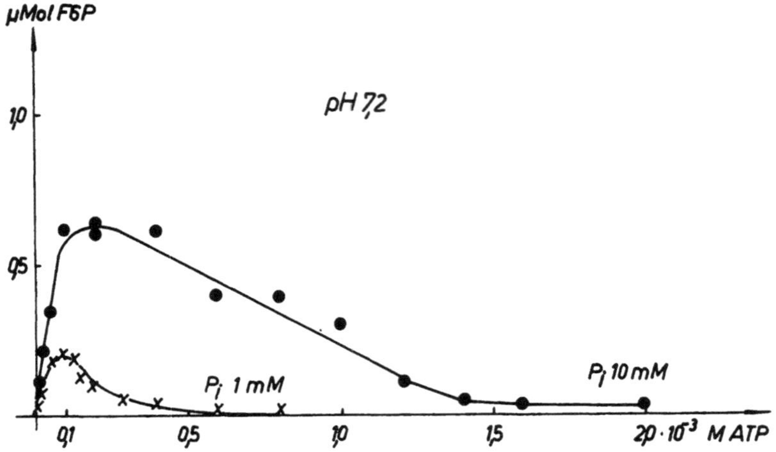

Fig. 3. ATP-dependence at different phosphateconcentrations KCl 90 mM; F6P 1 mM.

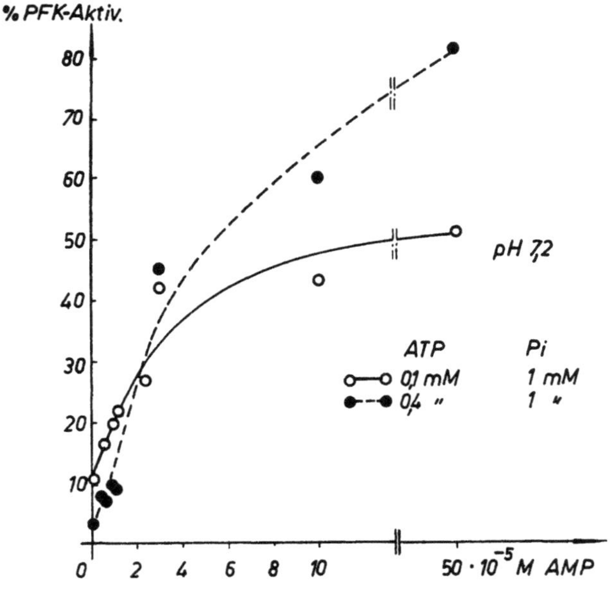

Fig. 4. Influence of 5' AMP on PFK-activity at pH 7, 2 KCl 90 mM; F6P 1 mM.

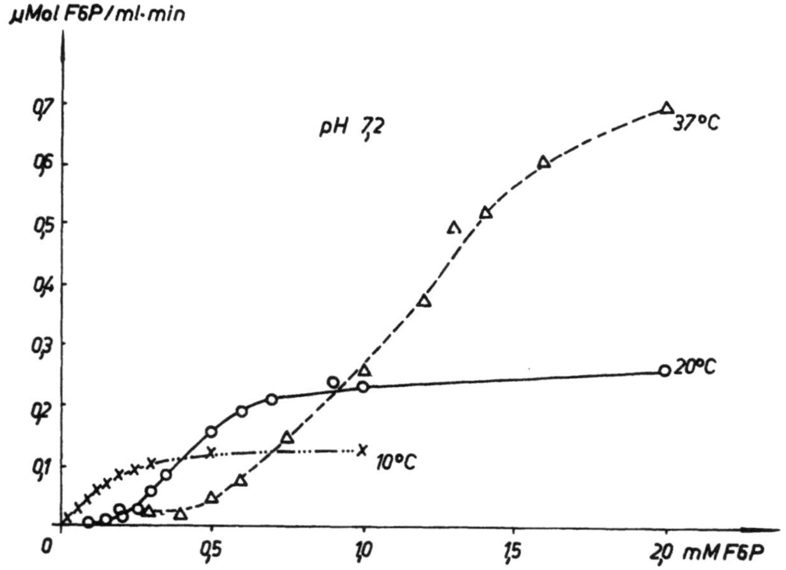

Fig. 5. Influence of temperature on F6P dependence.

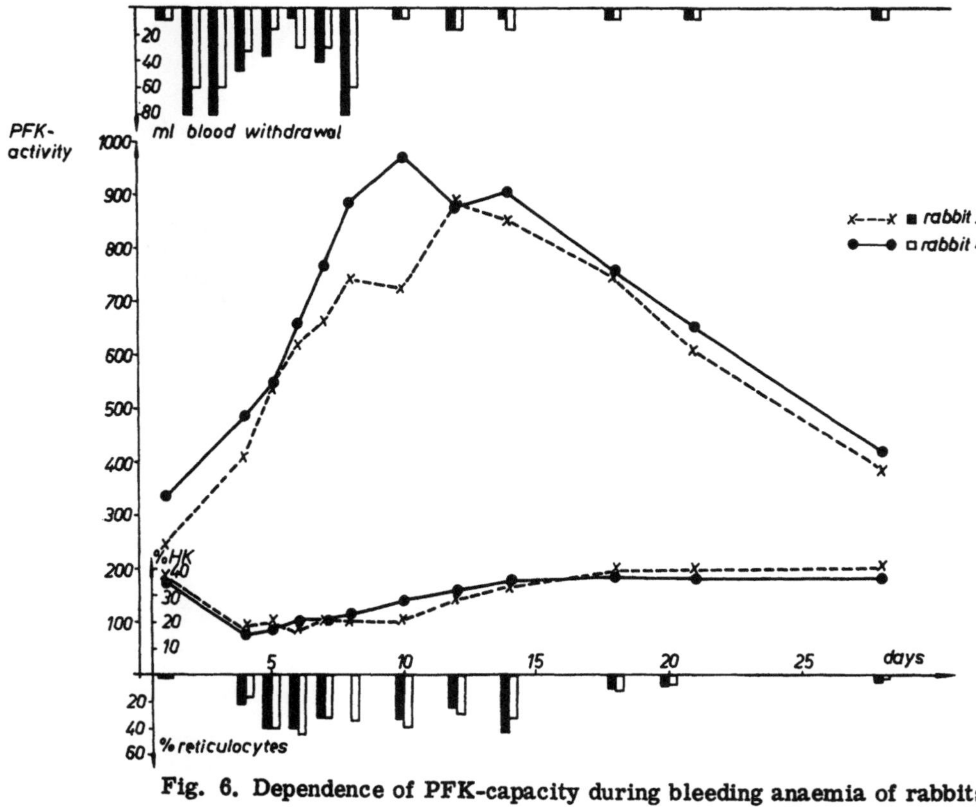

Fig. 6. Dependence of PFK-capacity during bleeding anaemia of rabbits.

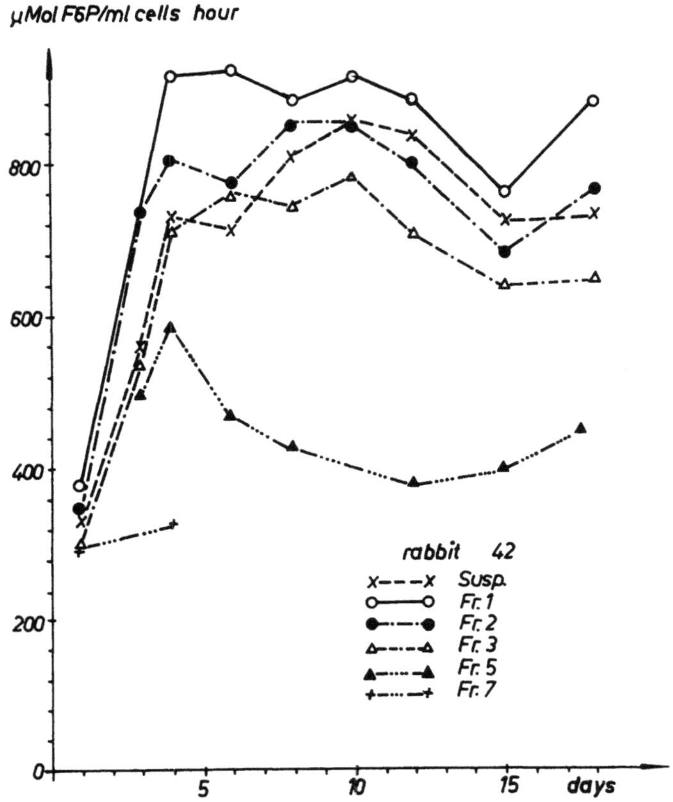

Fig. 7. Dependence of PFK on the degree of maturation. The different fractions were separated by density gradient centrifugation. The content of reticulocytes in the optimum of reticulocytosis in the suspensionmixture is 42 %, in fraction 1 90 %, in fraction 2 83 %, in fraction 3 30 %, in fraction 5 4 %, in fraction 7 6 %.

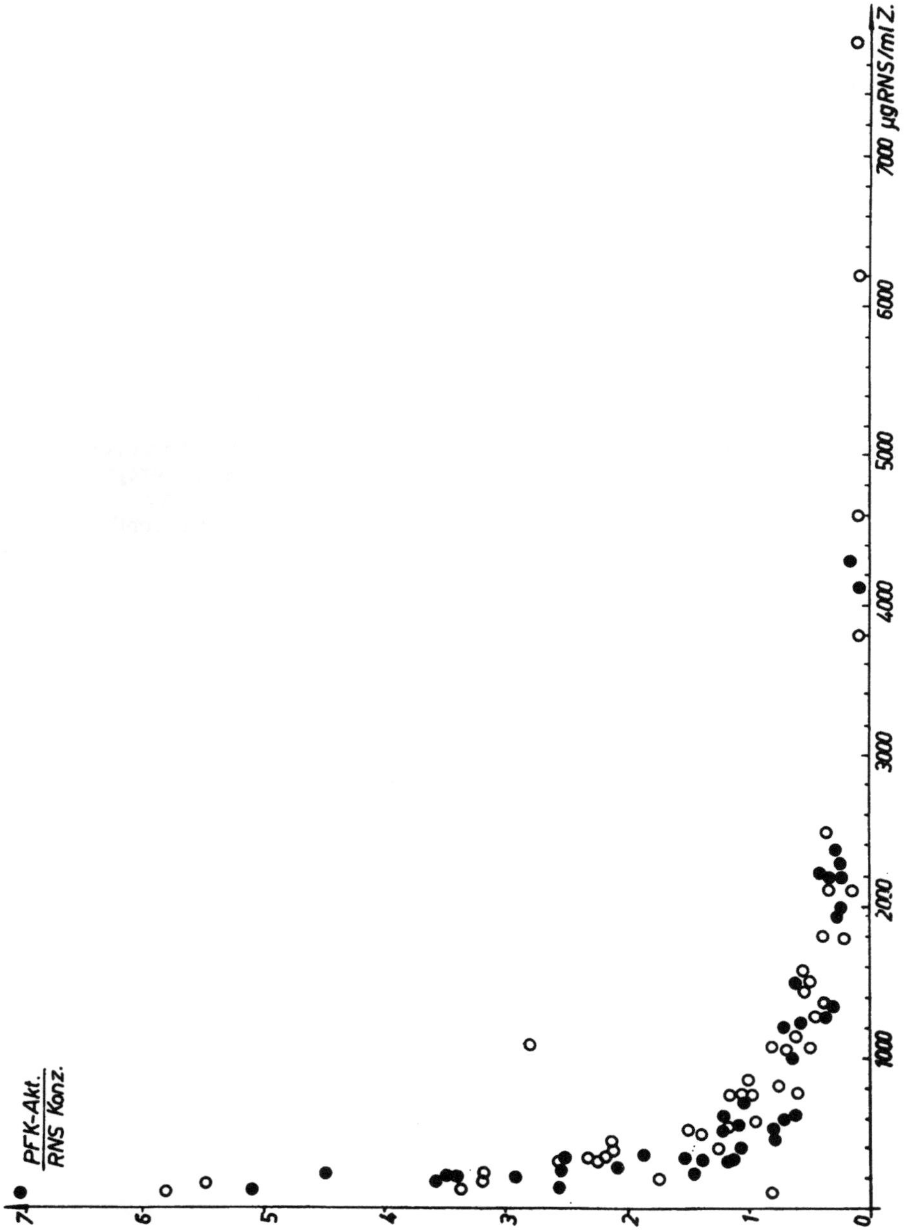

Fig. 8. Dependence of the ratio $\frac{\text{PFK-capacity}}{\text{RNS content}}$ as function of RNS content.

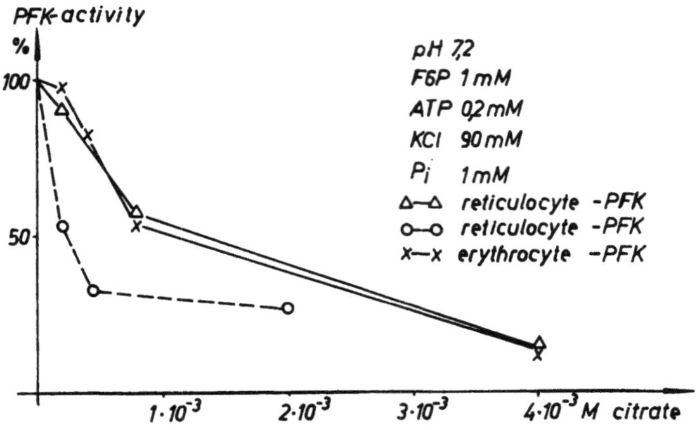

Fig. 9. Influence of citrate at pH 7, 2 on reticulocyte - and erythrocyte - PFK

	age of enzympreparation
o—o reticulocyte PFK	on day
Δ—Δ reticulocyte PFK	one month.

Instituto de Enzimologia, Centro de Investigaciones Biolôgicas,
Conseja Superior de Investigaciones Cientificas, Madrid, Spain

C. GANCEDO

REGULATION OF FRUCTOSE 1, 6-DIPHOSPHATASE IN YEASTS, INACTIVATION BY GLUCOSE

Introduction

The study of metabolic regulation and the discovery of specific mechanisms able to control the activities of an enzyme or a pathway are relatively recent events. In a classical book, some twenty years ago, Warburg wrote: "No matter how complicated a mixture of substances or prosthetic groups may be, every enzymatic reaction will take place as if all the other substances not participating in it were not present. Thus, in life no physiological enzymatic reaction can disturb another"[1] (20). Today we know that in life many physiological enzymatic reactions are influenced by others and indeed this fact is what accounts for the astonishing ability of functional integration found in every cell, especially among microorganisms. Generally, enzymes catalyzing key reactions in metabolic sequences are particularly prone to be sensitive to these influences.

In this report I will focus my attention on the enzyme fructose 1, 6-diphosphatase (FDPase) a key enzyme of gluconeogenesis and try to review some aspects of the regulation of this enzyme in yeasts. In particular I will examine in some detail our last work concerning a new mechanism of regulation which may be of wider extent and eventually prove to be an important factor in the rapid regulation of certain enzymes in microorganisms.

Regulation of fructose 1, 6-diphosphatase

FDPase is the terminal key enzyme of the gluconeogenic pathway in microorganisms. Gluconeogenesis is a metabolic pathway closely overlapping with glycolysis, i. e., sharing with it many enzymatic steps. However, since some glycolytic reactions

---

1 "In einem noch so komplizierten Gemisch von Substanzen und prosthetischen Gruppen wird jede Fermentreaktion so ablaufen, als ob alle in ihr nicht teilnehmenden Stoffe nicht vorhanden wären. So kann im Leben keine physiologische Fermentreaktion die andern stören."

are not physiologically reversible (10) specific enzymes are needed in order to circumvent these barriers when the cell utilizes gluconeogenesis. The relative rates of glycolysis and gluconeogenesis are thus determined by the ratio of glycolysis and gluconeogenesis are thus determined by the ratio of activities of these antagonistic enzymes which will then direct the metabolic flux in one or other direction. The switch over from glycolysis to gluconeogenesis and viceversa faces the cell with multiple regulation problems, and precisely the enzymes which catalyze the one way reactions are likely to be controlled in order to avoid wastage of energy. It is now well established that the reactions involved in the formation of FDP and in its hydrolysis to yield F6P are an important control point in the above mentioned switch over.
The formation of FDP from F6P is catalysed by phosphofructokinase a typical glycolytic enzyme, which has been amply discussed in other sessions of this Symposium. The hydrolysis of FDP to F6P is catalysed by the enzyme FDPase (Fig. 1) an enzyme widely distributed among living organisms.

Repression and allosteric inhibition
FDPase has been found to be adaptive in yeast. High levels of enzyme are found when the yeasts are grown on non-carbohydrate carbon sources (glycerol, ethanol...) while growth on sugars results in barely detectable levels (4, 16, 6); this phenomenon is very general among yeasts as indicated by the results shown in Table 1.
This behaviour could be accounted for by two mechanisms: either the enzyme is induced by some metabolite "below" the level of FDP or it is repressed by some metabolite "above" it. It was found that growth of the yeast[1)] on galactose resulted in significant levels of FDPase (Table 2). This observation rules out the hypothesis of metabolic induction by some metabolite from "below".
Table 2 also presents the values of hexose monophosphates and FDPase in yeast growing on different carbon sources. As can be seen the concentrations of the hexosemonophosphates and of FDPase are antiparallel, i. e., high levels of enzyme are found when the concentrations of hexosemonophosphate are low and

---

1) Unless otherwise stated, the studies on FDPase have been performed with Saccharomyces cerevisiae strain 1714-24A, or commercial baker's yeast "Danubio".

Table 1

SPECIFIC ACTIVITY (MILLIUNITS/mg PROTEIN) OF FDPase IN DIFFERENT YEASTS

Yeast species	Carbon source	
	Ethanol	Glucose
Saccharomyces cerevisiae	40	1
Candida utilis	30	3
Rhodotorula gracilis	63	15
Hansenula anomala	28	5
Debaryomyces hansenii	144	55
Pichia membranaefaciens	52	5
Sporobolomyces pararoseus	80	4

viceversa. This result suggests that the enzyme is repressed by hexose monophosphates. Moreover, in cultures with galactose and 2-deoxyglucose, the levels of FDPase dropped while the yeast accumulated 2-deoxyglucose-6-P (Table 2). This fact can be explained by assuming that 2-deoxyglucose-6-P is acting as an analog of the correpressor, which may therefore be G6P (5).

In addition to this control mechanism, the enzyme is sensitive to allosteric inhibition by AMP (4) an almost general property of FDPases whcih however may not hold true for all yeasts (6).

Inactivation of FDPase

Superimposed to these two "classical" mechanisms of regulation a phenomonon that appears to have also a control function has been observed. This is a rapid decrease in activity when glucose is added to a derepressed culture.

Fig. 2 shows the derepression of FDPase when glucose grown cells are washed and transferred to a minimal medium containing 50 mM acetate. The specific activity raises in a linear fashion and in about four hours it reaches a plateau. This increase in activity is due to actual protein synthesis and not to activation of a precursor as indicated by the action of cycloheximide which completely prevents this increase.

Addition of glucose to a yeast culture where FDPase is fully derepressed produces

Table 2

CONCENTRATIONS OF FDPase, GLUCOSE 6-P PLUS FRUCTOSE 6-P AND 2DG 6-P IN YEAST GROWN ON DIFFERENT CARBON SOURCES

Carbon source	FDPase	Glucose 6-P + fructose 6-P	2 DG 6-P
	milliunits/mg protein	μ moles/g fresh yeast	
Glucose	1	0. 7	-
Fructose	1	-	-
Ethanol	36	0. 05	-
Galactose	9	0. 15	-
Galactose + 2DG	1	0. 14	1. 5

For experimental details see (5).

a drop of the measurable activity to the original low levels in about one hour (Fig. 3). In this time no appreciable growth takes place as determined by measurements of the optical density of the cultures or by dry weight determinations. As the enzyme is stable in resting cells[1], this decrease in activity cannot be accounted for only in terms of repression of synthesis and dilution of the pre-existing FDPase since one hour sufficed to return the enzyme to basal levels. As shown in Fig. 4 dilution only, would need about five generations to reach the same low level.

The situation is highly reminiscent of the inactivation of glutamine synthetase by NH_3 described some years ago by the group of Holzer (for a review see 8). It could be that also FDPase exists in two interconvertible forms. Therefore a variety of attempts were made to reactivate the enzyme in vitro and in vivo. The inactivated enzyme was not reactivated in vitro by controlled heating, dialysis, or changing the pH and/or cation in the assay system. As can be seen in Fig. 5 recovery of activity in vivo is dependent on an energy source and is fully prevented by cycloheximide. These results seem to rule out the existence of interconvertible

1) That the enzyme is stable in resting cells was shown by adding cycloheximide to a derepressed yeast culture. After 2 hours the FDPase activity remained unchanged, thus indicating that the enzyme is not being continuously synthesized and degraded.

forms of the enzyme, since if that were the case reappearance of the enzyme would be independent of protein synthesis. Therefore, this result suggests that the enzyme is irreversibly transformed in a form without (or with very small) catalytic activity. A similar phenomenon has been observed by GÖRTS (7) with the maltose uptake system and by NEUMANN (13) with malate dehydrogenase.
A decrease in the affinity of the enzyme for its substrate as it occurs in the case of the maltose uptake system (7) may be ruled out since assays performed up to 1000 times the Km value did not result in increased activity. A possible mechanism could be the destruction of the enzyme either by proteolysis or by conversion to inactive subunits.
Inactivation of FDPase in situ is not only provoked by glucose: fructose and mannose also produce inactivation while galactose did not affect significantly the enzyme activity. 2-Deoxyglucose did not produce inactivation and did not markedly interfere with glucose. In this respect it is interesting to not that malate dehydrogenase, an enzyme with similar phenomenological behaviour (2) is partially inactivated by 2-deoxyglucose (3, 13).
Since the work of PASTAN and PERLAN (14) it is well known that 3', 5'-cyclic AMP may overcome the catabolite repression caused by glucose in different enzyme systems of E. coli. Very recently Van WIJK and KONIJN (19) have reported that the levels of cAMP in S. carlsbergensis were about eight times lower when growing in glucose than when growing in galactose. We have tested the effect of cAMP on the inactivation of FDPase. The result obtained is shown in Table 3. As can be seen in the conditions used added cAMP did not affect the inactivation of FDPase nor of malate dehydrogenase.
Anaerobiosis, azide, 2,4-dinitrophenol, or iodoacetate were without significant effect on the inactivation process.
Inactivation is not affected by cycloheximide if this antibiotic is added simultaneously with the sugar. However, if the cells are preincubated 30 min with the antibiotic before the addition of the sugar about 70 % inhibition is observed (Fig. 6). That the drug was effectively acting was checked by following in parallel the derepression of acid phosphatase (Fig. 6). It could be thought that the different behaviour of the cultures was due to interference with sugar uptake caused by the exposure to the drug, however, no difference in sugar utilization was found between preincubated and non preincubated cultures as measured by the rate of anaerobic fermentation (Fig. 7).

Table 3

EFFECT OF cAMP ON THE INACTIVATION OF FDPase AND MALATE DEHYDROGENASE (MDH) IN PROTOPLASTS OF S. cerevisiae

Enzyme	No addition	Glucose 1 %	Glucose 1 % + 1 mM cAMP
FDPase (mU/mg P)	70	14	8
MDH (mU/mg P)	500	25	5

Hypothetic mechanism of inactivation

One reasonable way to explain the facts described above may be to assume the existence of a constitutive protein which alters in some way FDPase upon addition of glucose and that turns over quite rapidly (Fig. 8). If cycloheximide is added simultaneously with glucose, the antibiotic would block further synthesis of the protein, but the existing one could still act on FDPase. When the cells are preincubated with the antibiotic before the addition of glucose, the pool of modificatory protein would decrease and inactivation of FDPase could be reduced (Fig. 9). It is conceivable that some metabolite derived from glucose would act allosterically on FDPase changing its configuration in such a way that it becomes more susceptible to degradation (Fig. 8). On the other hand, although less likely, the metabolite could also activate the degradatory enzyme making it able to attack FDPase. Degradation of protein is a significant mechanism for remotion of unneeded enzymes and therefore a means of controlling enzyme levels in higher animals, where cell division is slow or negligible, in contrast with microorganisms where usually this remotion occurs by repression and subsequent dilution in phases of rapid growth. There exist, however, in microorganisms clearly documented instances (9, 11) of selective degradation of enzymes which are superfluous in a concrete metabolic situation. This model, which should be taken only as a working hypothesis, is in some way similar to that proposed by the group of Tomkins for the superinduction of tyrosine amino-transferase by actinomycin in rat hepatoma cells (18).

In vitro experiments on inactivation

FDPase was not inactivated in extracts by addition of glucose. Extracts from glucose inactivated cells did not inactivate FDPase from acetate derepressed cells, neither alone nor in the presence of metabolites likely to act as positive effectors of the inactivation. Among the most obvious reasons that could explain these negative results in vitro are the destruction of the inactivating system during the extraction process or the existence of a requirement not yet found. We are still trying to find conditions allowing the in vitro reproduction of the inactivation.

Distribution of the inactivation of FDPase among yeasts

Specific inactivation of enzymes has appeared to be restricted to limited species of microorganisms. For instance, inactivation of malate dehydrogenase by glucose has been detected only in Saccharomyces (1) and the same has been reported for the elaborate regulatory system of ornithintranscarbamylase (12). A screening among different genera of yeasts has shown that the inactivation of FDPase is widespread among yeasts although it is not general to all yeasts (Table 4). The result could warn against generalization deduced from a screening comprising only a few species.

Final remarks

The immediate question one is prompted to ask is about the real usefulness of such a mechanism of regulation. At first sight the same result of switching off rapidly an enzyme can be achieved (and indeed is achieved generally!) by allosteric inhibition is subject to oscillations depending on changes in the concentrations of the inhibitor, while conversion of the enzyme into an inactive form (which may be recuperable or not) offers the possibility to "freeze" the enzyme in the inactive state. It seems that conversion into an inactive form which may be afterwards reconverted to the active form would be more advantageous for the cell. However, these interconvertible systems often involve additional enzymes and interplay of effectors, while in the case of degradation one enzyme with broad specificity could perform its work on different proteins if the degradability is determined by a particular induced conformation of the protein in question. Summarizing, it appears from the foregoing results that the regulation of FDPase in yeasts in quite complex and that multiple mechanisms exist in order to control its actual activity. It is quite likely that for other key enzymes elaborate regulatory mechanisms, superimposed on the already known control systems will be eventually

Table 4

SPECIFIC ACTIVITY (mU/mg PROTEIN) OF FDPase IN DEREPRESSED AND GLUCOSE TREATED YEAST CULTURES

Yeast species	Derepressed culture	Glucose treated (1 h)
Saccharomyces cerevisiae "Danubio"	38	4
Saccharomyces carlsbergensis	17	1
Saccharomyces oleaginosus	68	1
Saccharomyces norbensis	35	5
Candida krusei	12	1
Candida melibiosi	88	40
Candida pelliculosa	94	13
Candida pulcherrima	45	22
Hansenula anomala	135	20
Pichia membranaefaciens	58	71
Rhodotorula glutinis	38	35
Torulopsis salmanticensis	42	56

found. As SCHIMKE (17) has pointed out "one should keep an entirely open mind as to potential regulatory mechanisms in the system studied".

REFERENCES

(1) DUNTZE, W., D. NEUMANN, H. HOLZER, Eur. J. Biochem. 3, (1968), 326

(2) FERGUSON, J., M. BOLL, H. HOLZER, Eur. J. Biochem., 1, (1967), 21

(3) GANCEDO, C., Unpublished results

(4) GANCEDO, C., M.L. SALAS, A. GINER, A. SOLS, Biochem. Biophys. Res. Commun., 20, (1965), 15

(5) GANCEDO, C., J.M. GANCEDO, A. SOLS, Biochem. Biophys. Res. Commun., 26, (1967), 528

(6) GANCEDO, J.M., C. GANCEDO, Arch. Mikrobiol., 76, (1971), 132

(7) GÖRTS, C.P.M., Biochim. Biophys. Acta, 184, (1969), 299

(8) HOLZER, H., Adv. Enzymol., 32, (1969), 297

(9) JOHN, P.C.L., C.F. THORSTON, P.J. SYRETT, Biochem. J., 119, (1970), 913

(10) KREBS, H.A., H.L. KORNBERG, Energy transformations in living matter. Springer Verlag, Berlin, 1957

(11) LEWIS, C.M., J.R.S. FINCHAM, J. Bacteriol., 103, (1970), 55

(12) MESSENGUY, F., J.M. WIAME, FEBS Letters, 3, (1969), 47

(13) NEUMANN, D., Thesis, Medizinische Fakultät der Universität Freiburg im Breisgau, 1968

(14) PASTAN, I., R. PERLMAN, Science, 169, (1970), 339

(15) PONTREMOLI, S., S. TRANIELLO, B. LUPPIS, A. WOOD, J. biol. Chem., 240, (1965), 3459

(16) ROSEN, S.M., O.M. ROSEN, B.L. HORECKER, Biochem. Biophys. Res. Commun., 20, (1965), 279

(17) SCHIMKE, R.T., Ed. by B.L. HORECKER and E.R. STADTMAN, Academic Press, 1, (1969), 77

(18) THOMPSON, E.B., D.K. GRANNER, G.M. TOMKINS, J. mol. Biol., 54, (1970), 159

(19) VAN WIJK, R., T.M. KONIJN, FEBS Letters, 13, (1971), 184

(20) WARBURG, O., Wasserstoffübertragende Fermente. Berlin, Arbeitsgemeinschaft Medizinischer Verlage, (1948)

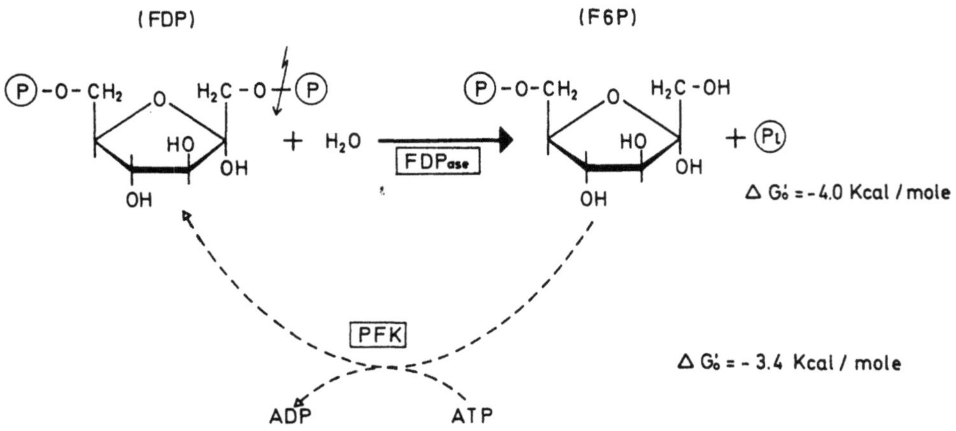

Fig. 1. Reactions catalysed by FDPase and phosphofructokinase. Hydrolysis of the phosphate bond takes place as indicated by the arrow (15).

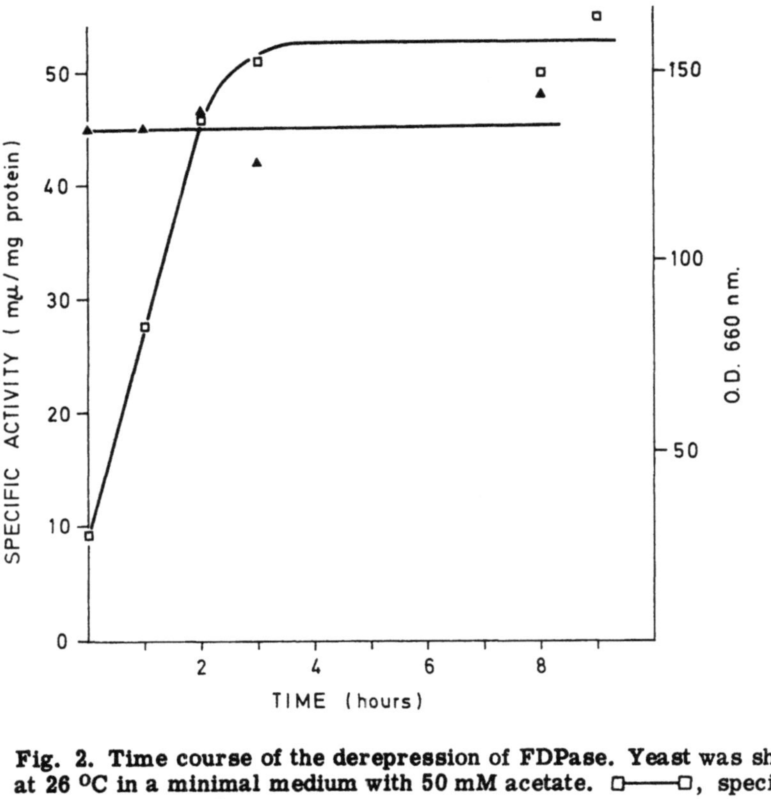

Fig. 2. Time course of the derepression of FDPase. Yeast was shaked at 26 °C in a minimal medium with 50 mM acetate. □——□, specific activity; ▲——▲, optical density.

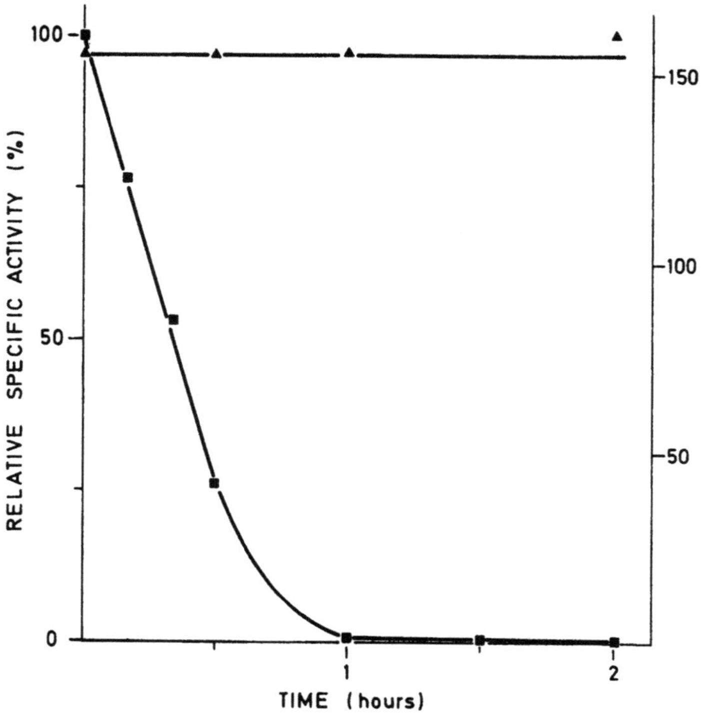

Fig. 3. Effect of glucose additic.. (final concentration 55 mM) on the level of FDPase.

■——■, specific activity;
▲——▲, optical density.

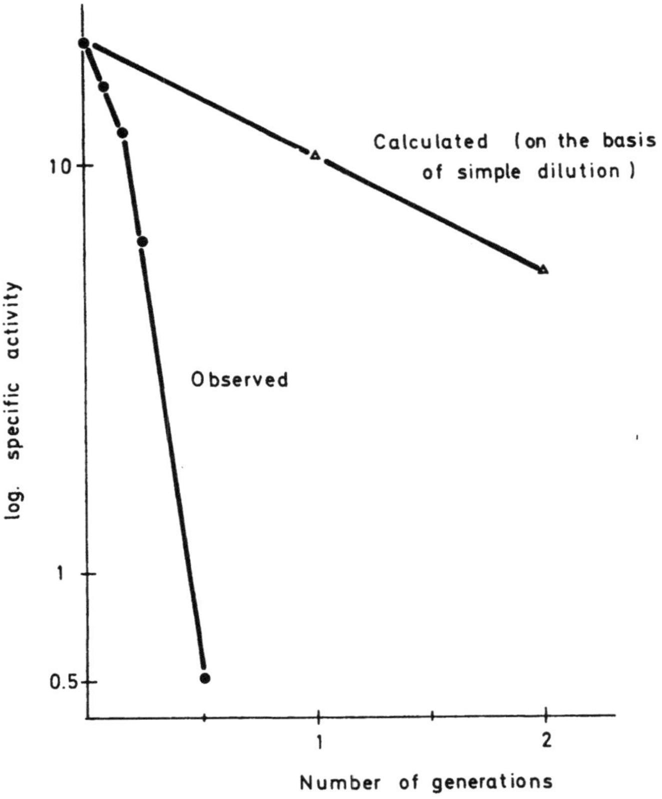

Fig. 4. Theoretical and experimental curves of disappearance of FDPase with time upon addition of glucose.

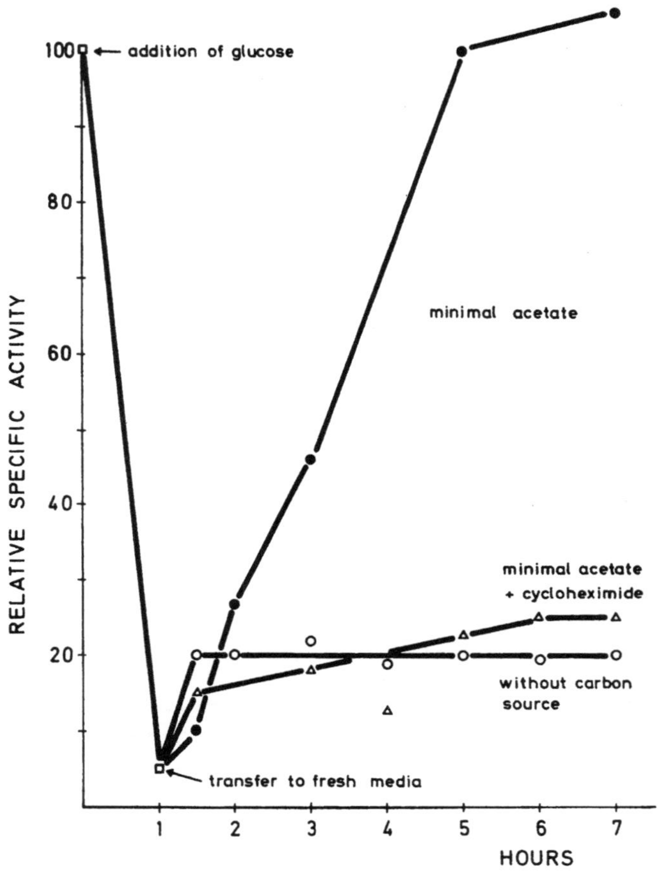

Fig. 5. Reappearance of FDPase activity. After treatment with glucose the yeast was transferred to different media as indicated. (Cycloheximide concentration 1 μg/ml).

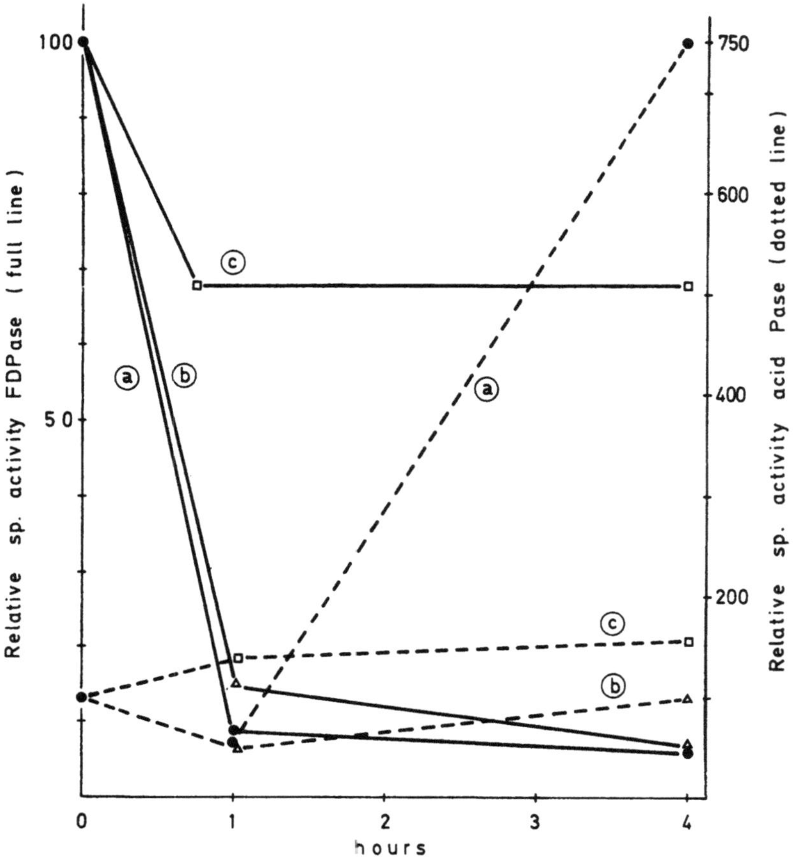

Fig. 6. Effect of cycloheximide on the inaktivation of FDPase and the derepression of the acid phosphatase. The yeast was transferred from an acetate minimal medium to a medium without added phosphate and containing 1 % fructose, a) without further addition, b) with cycloheximide 1 µg/ml added at 0 time, c) treated with cycloheximide (1 µg/ml) 30 min before fructose addition at 0 time. (The use of the medium lacking phosphate allowed to follow the derepression of the acid phosphatase, tested as control of the inhibition of protein synthesis).

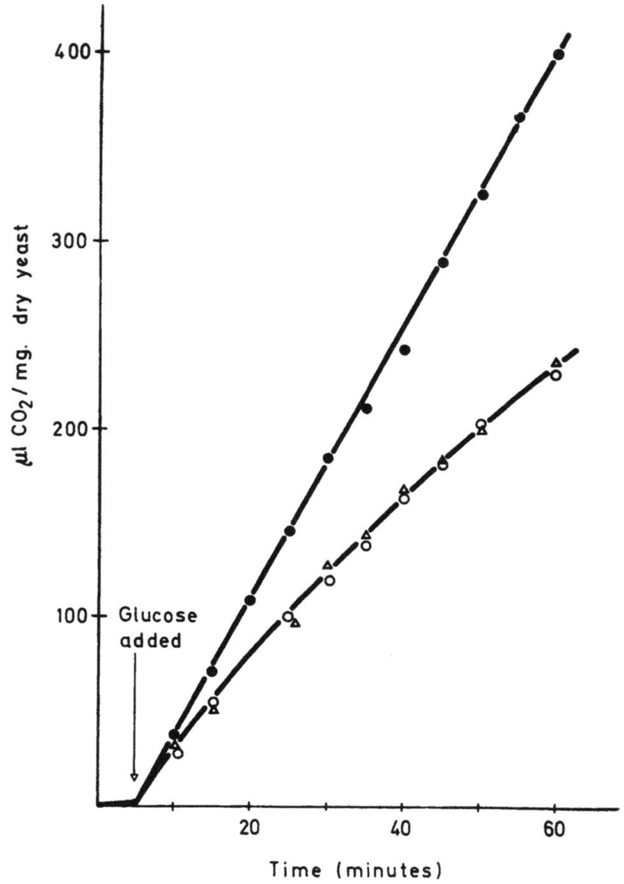

Fig. 7. Glucose fermentation by S. cerevisiae. ●——● , non treated culture, ○——○ + cycloheximide (1 µg/ml), △——△ preincubated 30 min with cycloheximide (1 µg/ml) before addition of the sugar.

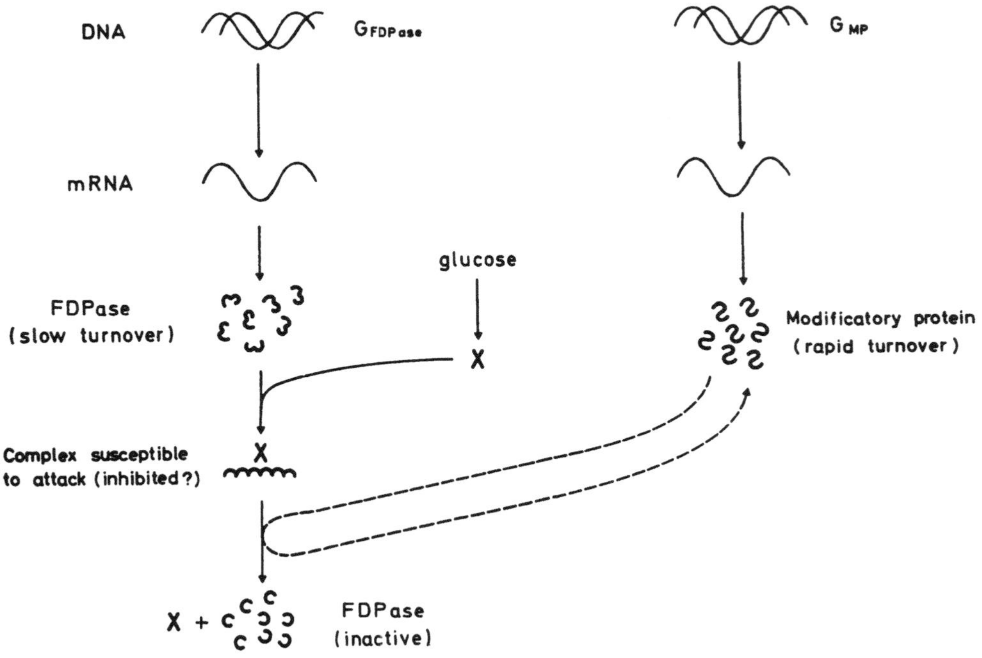

Fig. 8. A working hypothesis on the mechanism of inactivation of FDPase in yeast.

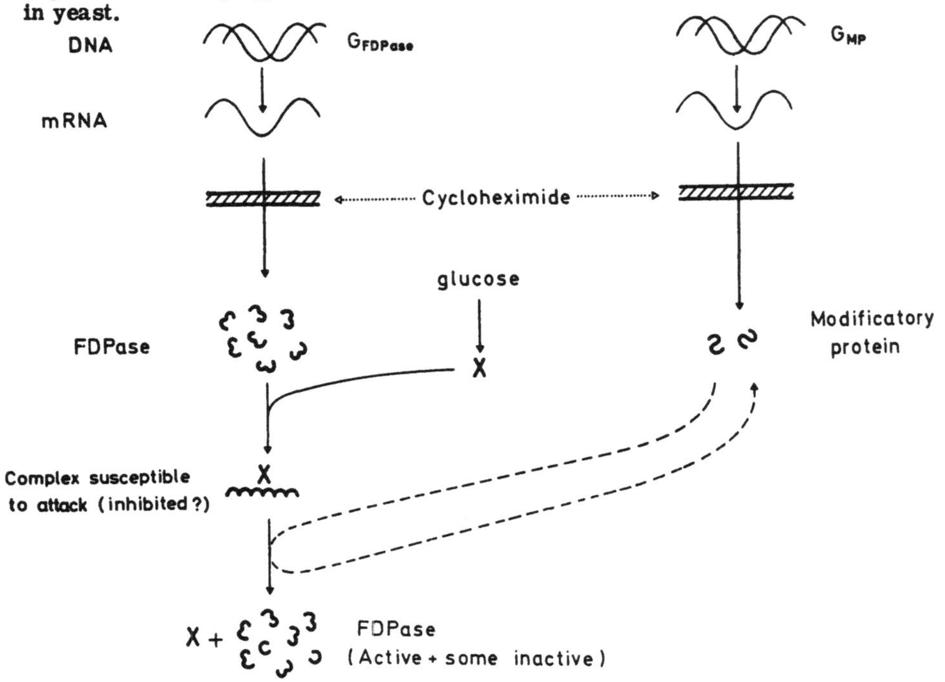

Fig. 9. Effect of cycloheximide on the inactivation of FDPase in yeast.

Max-Planck-Institut für Ernährungsphysiologie, Dortmund

K. BRAND

MECHANISMS OF ENZYMATIC ALDOL CLEAVAGE OF FRUCTOSE 6-PHOSPHATE

Introduction

Perhaps the two most common biological mechanisms for extending and shortening carbon chains are the aldol and ketol condensations. As shown on Fig. 1 the aldolcondensation involves a carbon-bound hydrogen adjacent to a carbonyl group and yields a structure containing the hydroxyl group β to the carbonyl function. In the ketol condensations two carbonyl compounds react and the product contains the hydroxyl group adjacent to the carbonyl function. A number of different types of aldol and ketol condensations are known, and these have now been implicated in fatty acid and amino acid metabolism as well as in carbohydrate metabolism.

In glucose metabolism of higher organisms the most important sites where enzymatic aldocleavage reactions do occur are:

1. Cleavage of fructose 1, 6-diphosphate by aldolase in the glycolytic pathway.
2. Liver aldolase furthermore is able to split also fructose 1-phosphate into dihydroxyacetone-phosphate and glyceraldehyde at a considerable rate.
3. Cleavage of fructose 6-phosphate by transaldolase in the nonoxidative pentose phosphate pathway.

In the present paper the mechanisms involved in the aldolcleavage reaction of fructose 6-phosphate by transaldolase are discussed.

Reaction of transaldolase

Fig. 2 shows the reaction of this enzyme.

Transaldolase, an enzym of the pentose phosphate pathway, catalysis the transfer of "active dihydroxyacetone", derived from fructose 6-phosphate by aldocleavage to the four carbon aldehyde acceptor erythrose 4-phosphate forming a seven carbon sugar, sedoheptulose 7-phosphate and releasing glyceraldehyde 3-phosphate.

Molecular properties of transaldolase

The enzym has been isolated and crystallized from Candida utilis according to the procedure described by PONTREMOLI et al. (1). It has been found, however, that only about one fourth of the enzymatic activity could be obtained in crystalline form, 3/4 of the activity remained in the supernatant and did not crystallize by all means. TSOLAS et al. (2) studied this phenomen more carefully and discovered that the residual enzymatic activity, which did not crystallize, can be attributed to an isoenzyme with slightly different properties and to a hybrid between both isoenzymes. As shown on Fig. 3, three distinct enzymatically active fractions could be obtained by analytical DEAE-Sephadex ion exchange chromatography, which are called type I, II, and III. From hybridization experiments it became clear that type I and III are isoenzymes and type II is a hybrid.
From these studies it cannot be dinstinguished if these isoenzymes are heterog-enzymes or isogenzymes according to the definition proposed by THEORELL (3). These hybridization experiments furthermore indicate that both isoenzymes are composed of two, probably identical subunits ($\alpha\alpha$ and $\beta\beta$), which was confirmed by ultracentriguation studies carried out by SIA (4).
The molecular weight of type III transaldolase was found to be 66.000 and that of the type I enzyme 72.000 (4). For the monomeric units of type III transaldolase a molecular weight of 33.000 has been obtained in the ultracentrifuge at pH 12.4 or with the maleylated enzyme at pH 7.5 (4).
All experimental results reported here were obtained with crystalline type III transaldolase.

Mechanisms effective in the aldocleavage reaction of fructose 6-phosphate by transaldolase

Schiff base mechanism

Irreversible inactivation by reduction with borohydride.
In 1963 already HORECKER (5) could demonstrate that a Schiff base mechanism is involved in the aldolcleavage reaction.
When transaldolase is incubated with fructose 6-phosphate in the absence of the second substrate erythrose 4-phosphate a stable and active transaldolase-substrate complex accumulates, which can be isolated in the active form by gel-filtration or ammonium sulfate precipitation. This complex has been identified by Horecker and coworkers as a Schiff base intermediate, containing dihydroxy-acetone linked to an ε-aminogroup of a lysine residue of the active site of the enzyme (Fig. 4).

As illustrated on Fig. 5 evidence for this Schiff base intermediate has been obtained by borohydride reduction, leading to an inactive secondary amine and by acid hydrolysis of the reduced complex yielding β-glyceryllysine.
In all experiments it was found that the complex was formed only with the triose and all attempts to obtain a Schiff base intermediate with the hexose fructose 6-phosphate failed.

Reversible inactivation by cyanide

More recently we got further information for the ketimine structure of this transaldolase-dihydroxyacetone complex using cyanide as reagent (6). As shown on Fig. 6 cyanide causes a reversible inactivation of the complex to the same extent as borohydrid does in an irreversible way. The inactivation with cyanide occurs only in the presence of substrate fructose 6-phosphate or - as shown on Fig. 6 - with the isolated Schiff base intermediate.
Fig. 7 illustrates the chemical mode of inactivation by this reagent. Cyanide is added to the ketimine forming an inactive aminonitril derivative. The borohydride as well as the cyanide experiments clearly indicate that a Schiff base intermediate is involved in the reation mechanism of transaldolase.

Base catalysis by histidine

On Fig. 8 the overall reaction catalyzed by transaldolase is shown again in more mechanistic detail. The Schiff base intermediate obviously facilitates the cleavage of the carbon-carbon bond between carbon atom 3 and 4 by pulling electrons.
As can be seen from this scheme a proton furthermore has to be removed from the C-4-hydroxylgroup during the aldolcleavage reaction in order to release glyceraldehyde 3-phosphate. Therefore a mechanism involving base (B) catalysis has to be postulated in the transaldolase reaction.

Tritium labelling experiments (7)

To elucidate the mechanism of base catalysis in transaldolase the enzyme was incubated with fructose 6-phosphate in the presence of tritiated water. Of course all the OH-groups of fructose 6-phosphate became labeled with tritium by exchange.
By several ammonium sulfate precipitations in ordinary water, however, this tritium could be washed out. As shown on Fig. 9 in spite of the washing procedure

23 μmoles of tritium remained retained in 40m μmoles of TA-DHA complex, which did not exchange with water.
From the known specific radioactivity of tritium it could be calculated that 0.6 equivalent of tritium retained in the complex. This result is in contrast to that obtained with aldolase where this proton does exchange with water. When the complex was reduced with borohydride and exposed to acid hydrolysis this tritium was found to be attached to β-glyceryllysine. After periodate degradation and dimedone precipitation it could be shown that this tritium was on the C-3 position of the glyceryl-moiety. When the complex was heated to 80 °C this tritium activity was found in the dihydroxyacetone moiety, which was released during the denaturation. In the control experiment, when the complex was formed in ordinary water and then reduced with borohydride in the presence of T_2O, almost no tritium was found to be incorporated into the complex indicating that under the experimental conditions this tritium incorporation into the complex is a specific reaction.
From these tritium labelling experiments it can be deduced that about one equivalent of tritium derived from substrate fructose 6-phosphate remained somewhere in the isolated TA-DHA complex, but no information could be obtained to which group this tritium is attached.

Photooxidation experiments (7)
The most probable amino acid residue acting as a base catalyst and accepting this proton can be expected to be histidine as has been found in a number of enzymes. For this reason photooxidation experiments were carried out to modify the histidine residues of transaldolase.
The results of these experiments providing evidence that histidine in fact is involved in the proton transfer were the following as shown on Fig. 10 - 13.
a. Photooxidation is accompanied by a loss of enzyme activity. As shown on Fig. 10 photoinactivation of transaldolase strongly depends on pH. At acid pH, where the histidine residues are protonated the relative rate of activity loss is very low, while at neutral and alkaline pH the rate is high. The pK value for the residue, which is responsible for the inactivation was found to be 7.2. The solid line shows the inactivation curve of transaldolase, the dashed curve taken from a paper of WESTHEAD (8) shows the oxidation of imidazole. The fact that both curves are congruent suggests that the loss of enzymatic activity is caused by the destruction of histidine residues in transaldolase during photooxidation.

b. Amino acid analyses furthermore revealed that under the conditions employed up to pH 7. 6 only histidine residues were destroyed. At more alkaline pH also tyrosine and tryptophane were oxidized. On Fig. 11 the remaining activity is plotted versus the number of histidines lost by photooxidation. Transaldolase contains only 2 histidine residues and as can be seen from this figure all the activity is lost when one of these histidine residues is destroyed.
The pK value of 7. 2 (Fig. 10) and the amino acid analyses of the photooxidized transaldolase indicate that one histidine residue is critical for the catalytic activity of transaldolase.
c. Evidence that histidine in fact is involved in the proton transfer, accepting the proton of the C-4-hydroxylgroup has been obtained by photooxidation experiments with the isolated transaldolase-dihydroxyacetone complex.
From Fig. 10 it can be seen that transaldolase is more stable to photooxidation at lower pH, where histidine is protonated. If histidine really accepts the proton from the C-4-hydroxylgroup it should be expected to be protonated in the complex also at neutral pH and therefore the complex should be stable to photooxidation also at neutral pH, where free transaldolase is rapidly inactivated. For these experiments transaldolase was incubated with fructose 6-phosphate and the complex formed was rapidly isolated by gelfiltration and exposed to photooxidation.
On Fig. 12 the rate of photoinactivation of native transaldolase is compared to that of a mixture containing 50 % complex and 50 % native enzyme as judged from inactivation with borohydride reduction.
The rate of inactivation by photooxidation is markedly decreased, when 50 % of the activity is present as complex, indicating that the complex is more stable to photooxidation than the free enzyme.
In a series of experiments with different amounts of complex present we got a linear relationship between the amount of complex present and the amount of protection against photooxidation. This is shown on Fig. 13. Here the percentage of complex present is plotted versus the percentage amount of protection against photooxidation. As can be seen the protection is directly proportional to the amount of complex present and if fully complexed enzyme could be made, it should be expected to be stable to photooxidation.
From these results we conclude that the critical histidine in the complex is protonated even at pH 7. 6 and therefore protected against photooxidation as histidine in the native enzyme is only at acid pH.

The fact that one histidine residue is critical for the enzymatic activity and this histidine is protonated in the Schiff base intermediate at neutral pH strongly suggests that base catalysis by histidine does occur in transaldolase.

Conclusion

The experiments reported so far revealed that two mechanisms are effective in the aldol cleavage reaction of fructose 6-phosphate by transaldolase as illustrated in the model proposed on Fig. 14.

1. Schiff base formation between the carbonylgroup of the substrate and an -aminogroup of a lysine residue of the active site of transaldolase.
2. Base catalysis by histidine: the proton of the C-4-hydroxylgroup is taken up by the nucleophilic imidazole nitrogen of the critical histidine.

Both mechanisms facilitate the aldolcleavage of fructose 6-phosphate. Since the carbanion is a stronger base than the imidazole the proton is more attracted by the negative charge of the carbanion on carbon atom 3, interacting, however, with the basic nitrogengroup forming a δ^+-δ^--dipole.
So far all experiments designed to detect other functional groups in transaldolase failed. It is absolutely sure that SH-groups do not play any role for the catalytic activity of transaldolase and tyrosine residues seem to be not essential, too.

Physical and chemical evidences for substrate mediated changes of conformation of transaldolase

Binding studies with 1-anilino-8-naphthalene sulfonate (ANS) (9)

ANS has been reported by a number of investigators (10, 11, 12, 13) to interact with non-polar regions of proteins causing an enhancement of fluorescence intensity and a shift of the emission maximum to shorter wavelengths. In order to obtain more information about the nature of the active site and the mechanism of complex formation binding studies were carried out using 1-anilino-8-naphthalene sulfonate as fluorescent probe. The results of these experiments are summarized on Fig. 15.
When transaldolase was titrated with ANS the fluorescence intensity increased to a saturation level. At 22 °C an apparent equilibrium constant of dissociation (K) of 2.2×10^{-5} M was determined for the transaldolase-ANS complex. The number (n) of ANS bound to transaldolase was calculated according to Klotz et al. (14) to be 2.2. Therefore it can be assumed that each subunit binds one molecule of ANS.

When fructose 6-phosphate was added in about 3-fold excess the fluorescence almost completely disappeared, indicating that the substrate is able either to displace all the dye bound to the enzyme or to induce a conformational change leading to a decrease of hydrophobicity around the ANS binding site and therefore to a quenching of fluorescence.

The fact, however, that a 3-fold excess of fructose 6-phosphate, which has a tentimes lower affinity to the enzyme ($K_M = 3x10^{-4}$ M) than ANS is able to abolish the fluorescence intensity completely cannot be explained by simple displacement. It is more likely that the quenching of fluorescence is due to a substrate mediated conformational change affecting the hydrophobic nature of the ANS binding site.

When the second substrate erythrose 4-phosphate was added to the same mixture, the fluorescence intensity returned to the original level. This is expected from the reaction since erythrose 4-phosphate accepts dihydroxyacetone from the complex and therefore free transaldolase is regenerated, which is able now to bind ANS again.

The native enzyme obviously has two hydrophobic regions to bind ANS, while the transaldolase-dihydroxyacetone complex is not able to interact with ANS in a hydrophobic way. From these data it appears that there are different conformational states of transaldolase in the absence and presence of substrate fructose 6-phosphate.

<u>Measurement of the tryptophane fluorescence emission spectrum of transaldolase (15)</u>

Further evidence for a substrate induced conformational change has been obtained from measurements of the fluorescence emission spectrum of transaldolase in the absence and presence of substrates. The results of these measurements are shown on Fig. 16.

On the left the fluorescence emission spectrum of native transaldolase is shown. The fluorescence maximum is at a wavelength of 330 nm, the fluorescence intensity is 1. 05 volt. On the right the upper curve shows the fluorescence emission spectrum of transaldolase in the presence of fructose 6-phosphate. The fluorescence maximum is shifted to 326 nm and the fluorescence intensity increased to 1. 1 volt. When the second substrate erythrose 4-phosphate was added to the same mixture, the original spectrum of free transaldolase (330 nm; 1. 05 volt) was obtained again. This can be expected, since the complex becomes discharged by erythrose 4-phosphate and free transaldolase is released. We interpret these differences in the tryptophane fluorescence emission spectrum also in terms of conformational changes of transaldolase induced by fructose 6-phosphate, which affect the tryptophane region of the enzyme. It is known that transaldolase contains two tryptophane residues (4), but until now we have no information about the position of tryptophane in the peptide chain.

Studies with tetranitromethane

Reaction with tetranitromethane in the absence and presence of substrates

In collaboration with Dr. Kobashi from the University of Toyama, Japan, chemical studies on transaldolase were carried out using tetranitromethane (16). This reagent was introduced to enzyme chemistry by SOKOLOVSKY et al. (17) and in the past 2 years widely used by a number of enzymologists to modify tyrosine, cysteine, and tryptophane residues in enzymes and to nitrate carbanion intermediates (18, 19, 20, 21, 22, 23).

Since an intermediate carbanion formation is assumed in the reaction of transaldolase we used tetranitromethane to study this mechanism. RIORDAN (18) has shown in case of aldolase that the kinetic of the nitroformate production, which can be measured at 350 nm in the presence of substrate fructose 1, 6-diphosphate was consistent with the kinetic of the catalytic reaction. Therefore he concluded that in case of aldolase the carbanion intermediate becomes nitrated and concomitantly nitroform is produced.

We confirmed this result with aldolase but observed a completely different type reaction with tetranitromethane in case of transaldolase.

As can be seen from Fig. 17 almost no nitroformate is produced when tetranitromethane was added to transaldolase in the absence of substrate fructose 6-phosphate. In the presence of substrate fructose 6-phosphate and transaldolase, however, a very rapid formation of nitroformate was observed. The fast reaction was completed in less than one minute, followed by a slow reaction, in which nitroformate is produced at a rate comparable to that observed without substrate. From the right ordinate it can be seen that around four equivalents of nitroformate were formed during the fast reaction, which was accompanied with a complete loss of enzymatic activity.

On the Table 1 the molar ratios of nitroformate to transaldolase obtained in 4 experiments are shown. As can be seen around 4 moles of nitroformate are formed per mole of inactivated transaldolase. At more alkaline pH this ratio is increased up to 30. From this data one should expect that four critical amino acid residues have been modified in transaldolase by tetranitromethane causing a complete loss of the enzymatic activity.

Table 1

NUMBER OF MOLES OF NITROFORMATE PRODUCED PER MOLE OF TRANSALDOLASE

Exp. No.	Reaction Conditions						Inactivation %	Molar ratio of $C(NO_2)_3^+$	
	TA	F6P	TNM	ml	°C	min		to TA	to Inact. TA
	M	M	M						
I	$9.28x10^{-6}$	$9.2x10^{-4}$	$5.0x10^{-4}$	1.0	15	1.0	86.5	3.37	3.90
II	$9.28x10^{-6}$	$1.8x10^{-4}$	$5.0x10^{-4}$	1.0	12	1.0	83.5	3.37	4.04
III	$9.28x10^{-6}$	$1.8x10^{-4}$	$5.0x10^{-4}$	1.0	15	1.0	80.3	3.25	4.04
IV	$1.72x10^{-5}$	$3.0x10^{-4}$	$8.3x10^{-4}$	3.0	10	2.0	84.3	3.84	4.53

All the reactions were carried out in 0.05 M phosphate buffer, pH 6.0. The order of addition is described in the legend to Fig. 17.

In order to examine, which amino acid residues have been modified by tetranitromethane, the inactivated enzyme and the corresponding native enzyme were hydrolyzed under the same conditions and analyzed with an amino acid analyzer. With respect to the amino acids histidine, serine, proline, methionine, tyrosine, and phenylalanine no difference betwen the modified enzyme and the control was found.

The same was true for tryptophane, which was determined spectrophotometrically at a wavelength of 280 nm in an alkaline solution of native and modified transaldolase.

The number of other amino acids such as lysine, aspartic acid, threonine, glutamic acid, glycine, alanine, valine, isoleucine, and leucine were determined to be nearly equal to those of the control. Though the coincidence of the data in these cases is within the error of the instrument, it does not necessarily mean that these amino acid residues were not modified by tetranitromethane.

Since it is expected that less than 4 moles of certain amino acid residues have been modified, causing a full inactivation of the enzyme, such a small difference cannot be detected on the chromatogram of the amino acid analysis when the enzyme contains more than forty moles of this amino acid per mole of enzyme. From our experiments arginine seems to be the most probable candidate for a modification by tetranitromethane since it was found that compared to the control ground four residues were lost in the inactivated transaldolase. Native transaldolase contains 18 arginine residues (4).

Using N, N' -diphenylbenzidine, which is known to be extremely sensitive for the detection of nitro groups we were not able to find any nitro group in the solution of the inactivated enzyme as well as in the hydrolysate.

Protecting effect of erythrose 4-phosphate against the reaction with tetranitromethane (16)

Fig. 18 illustrates that the presence of the second substrate erythrose 4-phosphate prevented the formation of nitroformate when a mixture of transaldolase and fructose 6-phosphate was treated with tetranitromethane. This finding is in good agreement with all the results obtained from physical studies about substrate mediated conformational changes of transaldolase. As has been pointed out above, erythrose 4-phosphate discharges the transaldolase-dihydroxyacetone complex and releases free enzyme and from Fig. 17 it can be seen that tetranitromethane reacts only with the Schiff base intermediate not with free transaldolase. This result clearly indicates that fructose 6-phosphate causes a conformational change of transaldolase by means of which certain groups become exposed to the reagent.

Effect of tetranitromethane on the transaldolase-dihydroxyacetone complex (16)

The effect of tetranitromethane on the transaldolase-dihydroxyacetone complex is shown on Fig. 19.

In this experiment transaldolase was incubated with ^{14}C-labeled fructose 6-phosphate and then treated with tetranitromethane under the usual conditions. The mixture was then passed through a Sephadex G-50 column to separate the enzyme from excess substrate. Without treatment with tetranitromethane radioactivity can be found in the protein fraction. This radioactivity can be attributed to ^{14}C-labeled dihydroxyacetone, which is complexed with transaldolase in the Schiff base form. The molar ratio of dihydroxyacetone to complex has been found

in many experiments to be 1 (24). When the Schiff base, however, was treated with tetranitromethane the complex became decomposed as judged from the fact that no radioactivity could be detected in the protein fraction.
All attempts to reactivate the modified enzyme failed. From these results it is concluded that due to a change of conformation induced by fructose 6-phosphate, catalytically active groups of transaldolase became accessible to tetranitromethane and modified. This modification is accompanied by an irreversible inactivation and a decomposition of the transaldolase-dihydroxyacetone complex.

Possibilities of nitration of the transaldolase-dihydroxyacetone complex (16)

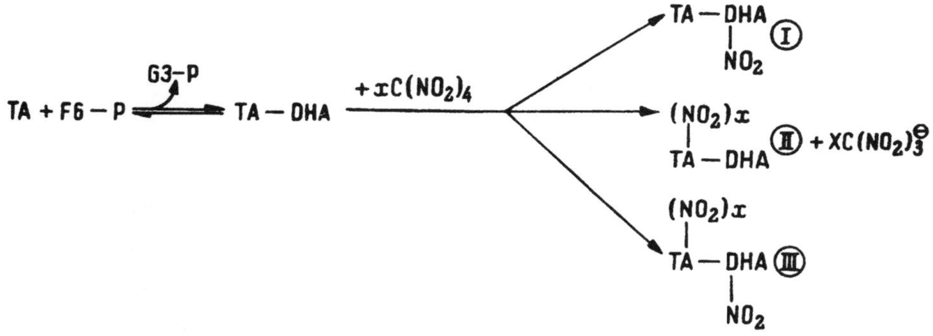

As can be seen from the scheme there are several possibilities (I, II, III) for the nitration of the Schiff base by tetranitromethane. The first possibility, where only the carbanion becomes nitrated, has been excluded by the finding in all experiments that more than one mole of nitroformate was produced per mole of inactivated enzyme and by the finding that the complex became decomposed after nitration without getting reactivated.
The second possibility is a nitration or oxidation only of the protein moiety and the third possibility provides that tetranitromethane attacks both, the carbanion and the enzyme protein.
From the present results it is not possible to decide whether the carbanion is attacked by tetranitromethane or not.
The results obtained, however, clearly indicate that the enzyme protein is modified probably being nitrated or oxidized by tetranitromethane only in the presence of substrate fructose 6-phosphate. This modification leads to an irreversible loss of the enzymatic activity and to the decomposition of the complex.

Another possible reaction with tetranitromethane is a nitration of the Schiff base leading to a cleavage of the C = N-bond:

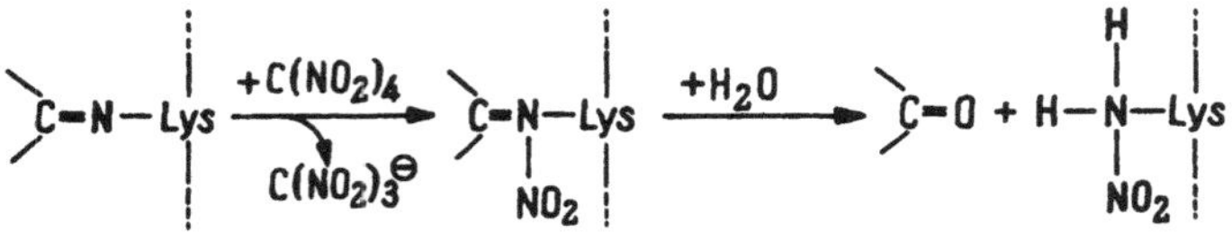

The nitramine derivative can be expected to decompose spontaneously yielding molecular nitrogen and oxidized lysine similar to a VAN SLYKE reaction:

The oxidation of the lysine residue at the active site could cause the irreversible loss of enzymatic activity. Furthermore this reaction would explain why the complex became decomposed by treatment with tetranitromethane. This reaction, however, accounts only for a ratio of one mole of nitroformate produced per one mole of inactivated Schiff base not for a ratio of 4:1 as has been obtained in the experiments.

Conclusion

The physical and chemical studies reported here clearly indicate that fructose 6-phosphate causes a change of conformation of transaldolase, which affects the active site of the enzyme, the non polar region of the ANS binding site and the region around the tryptophane residue.
Since, however, the amino acid sequence of transaldolase has not been established, we have no information about the complete nature of the active site and the extent of change of conformation caused by substrate fructose 6-phosphate.

Summary

The chemical and physical studies on transaldolase, reported here, provided informations about the molecular properties of the enzyme and the mechanism of action. Transaldolase type III has a molecular weight of 66. 000 and is composed of 2 subunits. Two mechanisms are effective in the aldocleavage reaction of

fructose 6-phosphate: formation of a Schiff base intermediate between the carbonyl-group of the substrate and the ε-aminogroup of a lysine residue of the active site of the enzyme and base catalysis by histidine. The fact that one proton protects both histidine residues in the complex from photooxidation suggests that the two are closely linked in the protein, possibly through hydrogen bonding as shown on the model proposed on Fig. 20. In this case the second histidine would be expected to have an abnormal ionization constant.
The formation of the Schiff base is accompanied by a conformational change induced by substrate fructose 6-phosphate.
Transaldolase has been shown to consist of two subunits (4) but to contain a single active site (5, 24) and obviously only one of the two histidine residues is critical for the catalytic activity (7). The proposed model on Fig. 20 attempts to explain the contradictious results by the concept of both subunits contributing to one active site in transaldolase. The relation of this subunit structure to the active site is under further investigation.

REFERENCES

(1) PONTREMOLI, S., E.D. PRANDINE, A. BONSIGNORE, B.L. HORECKER, Proc. Natl. Acad. Sci., U.S., 47, (1961), 1942

(2) TSOLAS, O., B.L. HORECKER, Arch. Biochem. Biophys., 136, (1970), 287

(3) THEORELL, H., 8th International Congress of Biochemistry, Switzerland, 1970, Abstract p. 75

(4) TSOLAS, O., B.L. HORECKER, with an Appendix y C.L. SIA, Arch. Biochem. Biophys., 136, (1970), 303

(5) HORECKER, B.L., P.T. ROWLEY, E. GRAZI, T. CHENG, O. TCHOLA, Biochem., Z. 338, (1963), 36

(6) BRAND, K., B.L. HORECKER, Arch. Biochem. Biophys., 123, (1968), 312

(7) BRAND, K., O. TSOLAS, B.L. HORECKER, Arch. Biochem. Biophys., 130, (1969), 521

(8) WESTHEAD, E.W., Biochemistry, 4, (1965), 2139

(9) BRAND, K., FEBS Letters, 7, (1970), 235

(10) WEBER, G., L.B. YOUNG, J. Biol. Chem., 239, (1963), 1415

(11) STRYER, L., J. Mol. Biol., 13, (1965), 482

(12) McCLURE, G.M. EDELMAN, Biochemistry, 5, (1966), 1908

(13) THOMPSON, W., K.L. YIELDING, Arch. Biochem. Biophys., 126, (1968), 399

(14) KLOTZ, I.M., J.M. WALKER, R.B. BIVAN, J. Am. Chem. Soc., 68, (1946), 1486

(15) BRAND, K., unpublished results

(16) KOBASHI, K., K. BRAND, submitted for publication

(17) SOKOLOVSKY, M., J.F. RIORDAN, B.L. VALLEE, Biochemistry, 5, (1966), 3582

(18) RIORDAN, J.F., P. CHRISTEN, Biochemistry, 8, (1969), 2381

(19) SHLYAPNIKOV, S.T., M.Y. KARPEISKY, Europ. J. Biochem., 11, (1969), 424

(20) FEMFERT, U., G. PFLEIDERER, FEBS Letters, 4, (1969), 262

(21) KASSAB, R., A. FALTOUM, L.A. PRADEL, Eur. J. Biochem., 12, (1970), 264

(22) SOKOLOVSKY, M., D. HARELL, J.F. RIORDAN, Biochemistry, 8, (1969), 470

(23) SOKOLOVSKY, M., J.F. RIORDAN, FEBS Letters, 7, (1970), 167

(24) BRAND, K., B.L. HORECKER, Z. Anal. Chem., 243, (1968), 640

(25) BRAND, K., B.L. HORECKER, in press

(26) TSOLAS, O., B.L. HORECKER, Arch. Biochem. Biophys., 136, (1970), 303

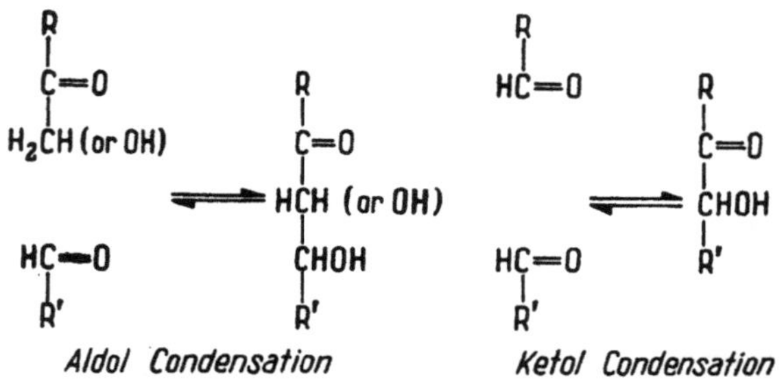

Fig. 1. General reactions of aldol and ketol condensations.

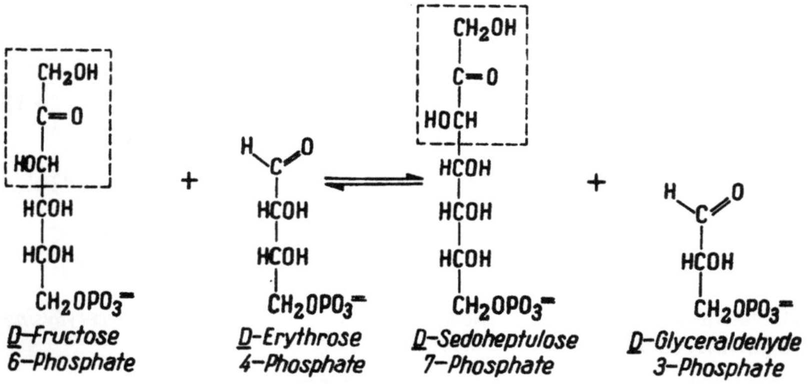

Fig. 2. The reaction catalyzed by transaldolase.

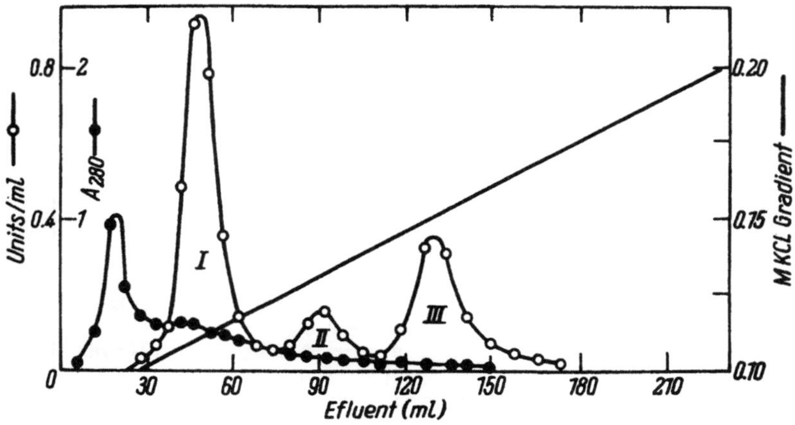

Fig. 3. Separation of transaldolase isoenzymes from an extract of dried Candida utilis by analytical DEAE-Sephadex chromatography. For experimental conditions see 2.

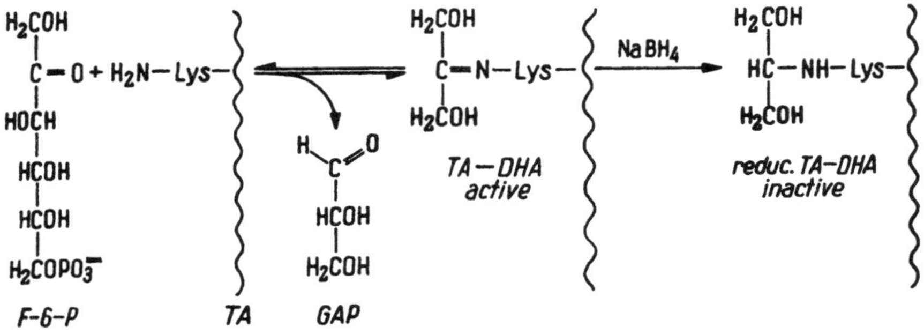

Fig. 4. Formation of the transaldolase dihydroxyacetone complex (TA-DHA).

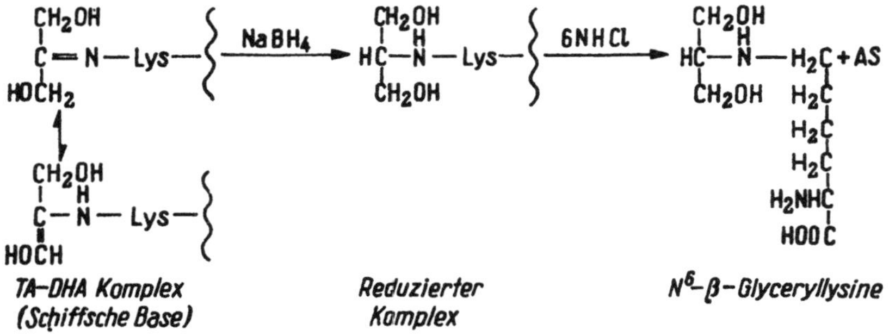

Fig. 5. Identification of the Schiff base intermediate.

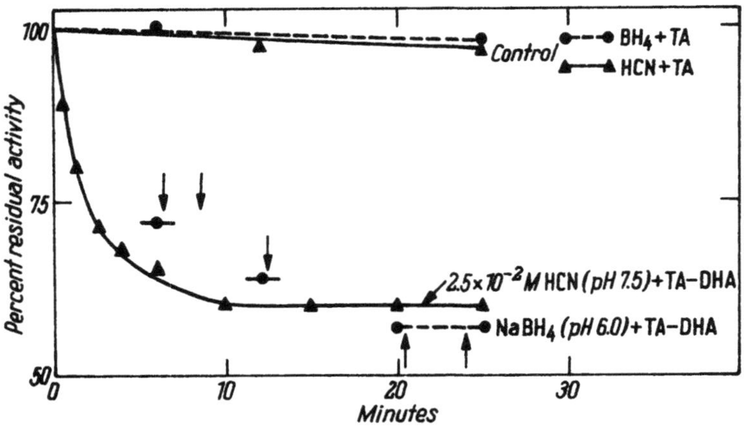

Fig. 6. Inhibition of the isolated TA-DHA complex by cyanide. For experimental conditions see 6.

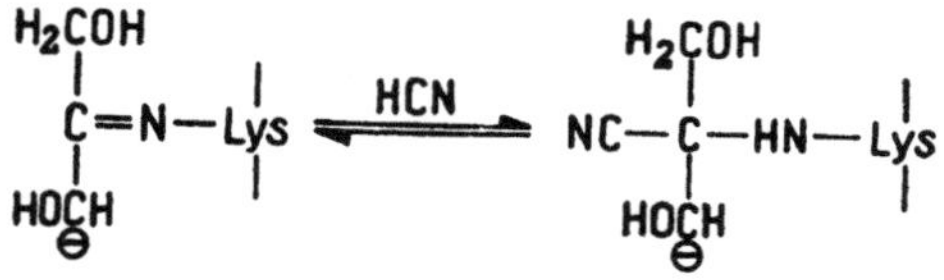

Fig. 7. Possible reaction of the transaldolase-dihydroxyacetone complex with cyanide.

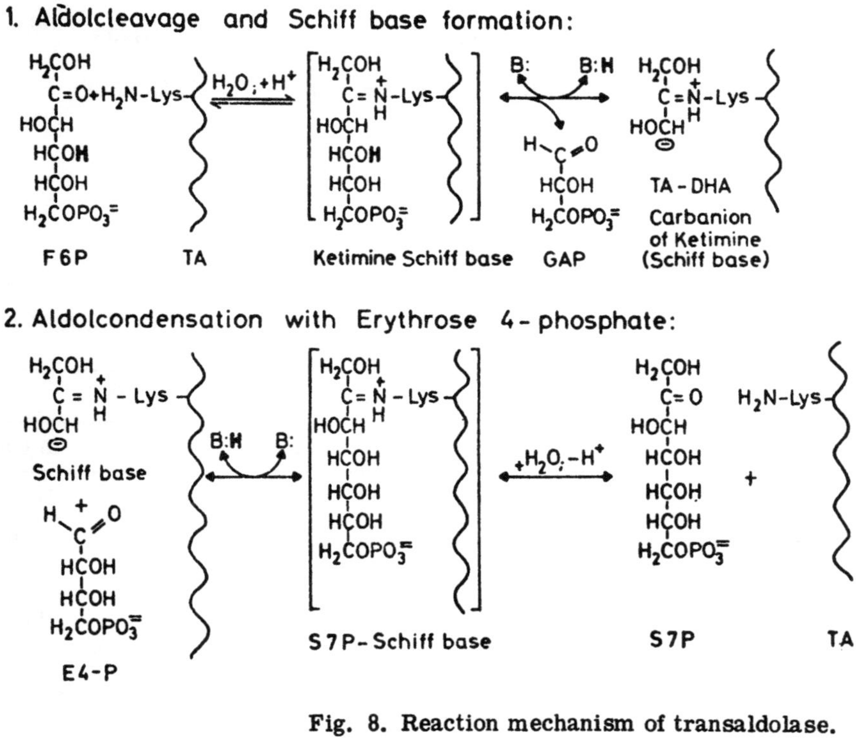

Fig. 8. Reaction mechanism of transaldolase.

EXPERIMENTAL

TA + F-6-P TA-DHA(T) + Ga3P

EXPERIMENTAL

$$TA + F\text{-}6\text{-}P \underset{}{\overset{T_2O}{\rightleftharpoons}} TA\text{-}DHA(T) + Ga3P$$

$$TA\text{-}DHA(T) + NaBH_4 \longrightarrow \beta\text{-}GLYCERYL\text{-}TA(T)$$

CONTROL

$$TA + F\text{-}6\text{-}P \rightleftharpoons TA\text{-}DHA + Ga3P$$

$$TA\text{-}DHA + NaBH_4 \xrightarrow{T_2O} \beta\text{-}GLYCERYL\text{-}TA$$

SAMPLE ANALYZED	EXPERIMENTAL		CONTROL	
	cpm	m μmoles	cpm	m μmoles
TOTAL TA-DHA		40		78
TITRIUM INCORPORATED	15, 500	23	1060	1. 6
β-GLYCERYLLYSINE	12, 500	19	-	-
PERIODATE-DIMEDON	7, 200	11	7	0. 01

Fig. 9. Incorporation of tritium from tritiated substrate into the TA-DHA complex. For experimental conditions see 7.

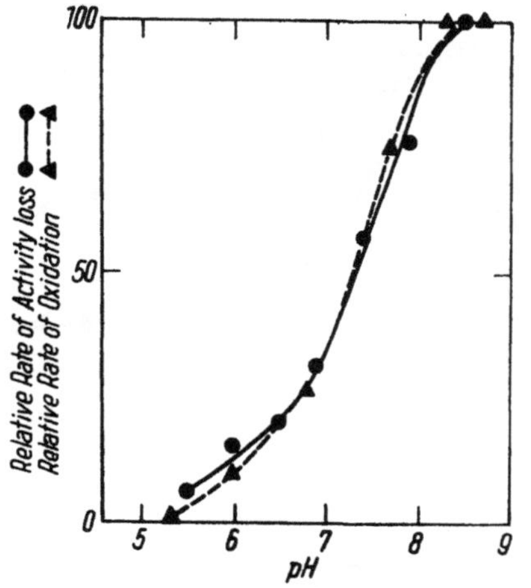

Fig. 10. The effect of pH on the rate of photooxidation of transaldolase (7) and imidazole (8). For the experimental procedure and conditions see references 7 and 8.

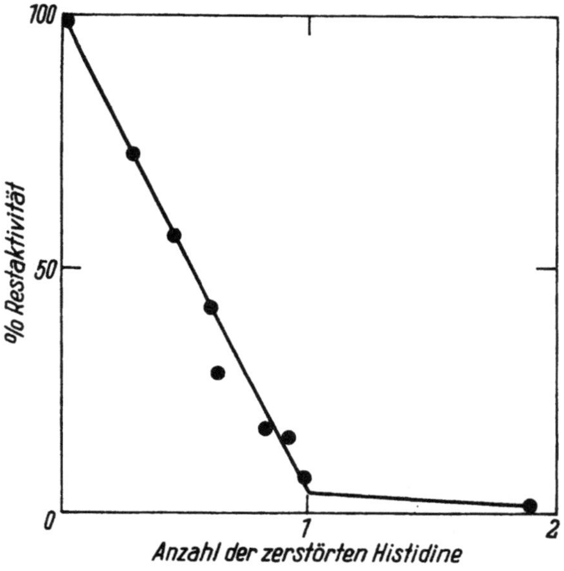

Fig. 11. Correspondence between loss of histidine residues in transaldolase and the degree of photoinactivation. For experimental conditions see 7.

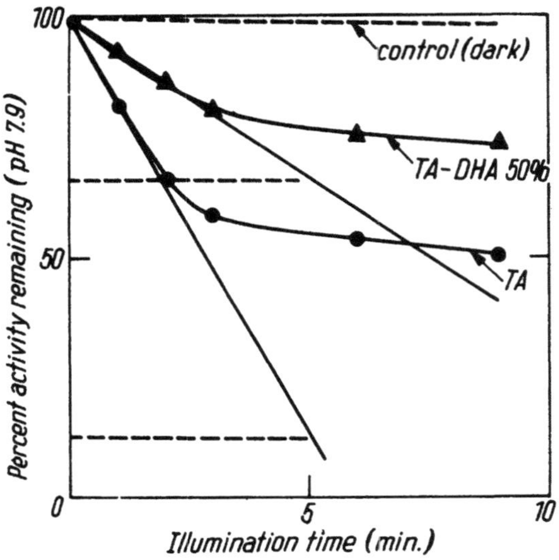

Fig. 12. Comparison of the rate of photoinactivation of native and a mixture containing 50 % transaldolase-dihydroxyacetone complex and 50 % free enzyme. For experimental conditions see 7.

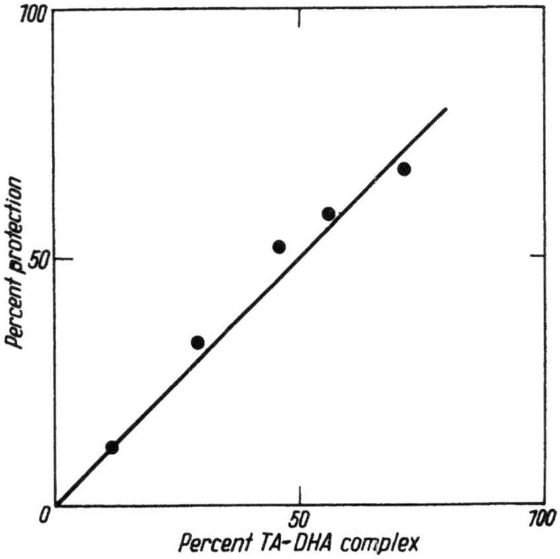

Fig. 13. Linear relationship between the amount of TA-DHA complex present and the extent of protection against photooxidation. For experimental conditions see 7.

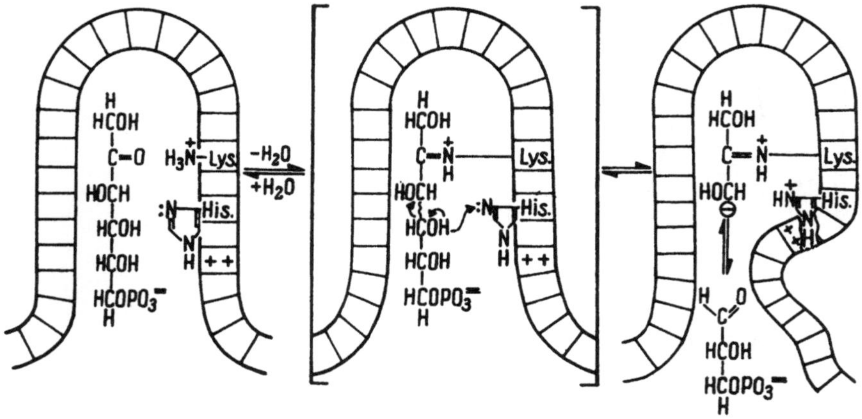

Fig. 14. Scheme of complex formation and aldolcleavage of fructose 6-phosphate by transaldolase.

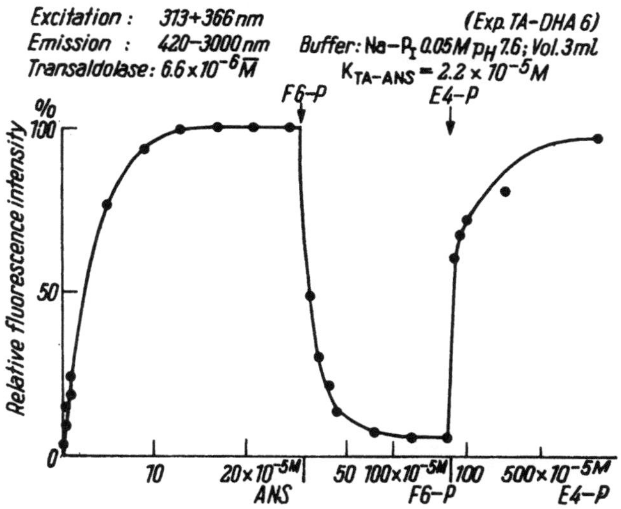

Fig. 15. Measurement of fluorescence intensity of ANS bound to transaldolase in the absence and presence of substrates. Experimental conditions are described in 9.

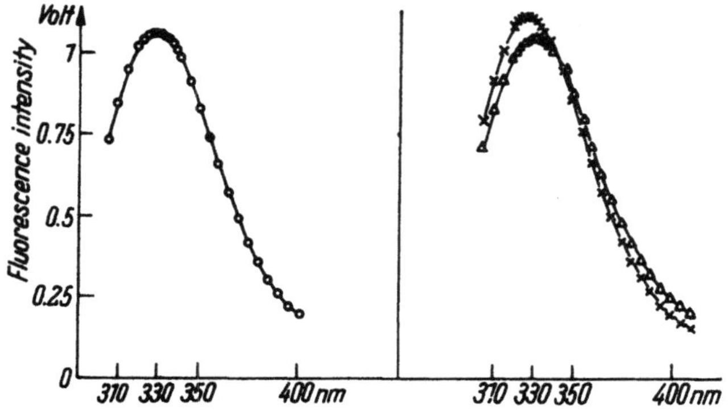

Conditions: Excitation 290 nm; Transaldolase: 3.3 mg Protein in 1 ml 0.05 M Na-phosphate buffer pH 7.4 = 4.1×10^{-5} M; Temp. 20°C

Left: o—o Transaldolase 4.1×10^{-5} M; E_{max} = 330 nm; Fluorescence intensity 1.05 Volt

Right: ×—× Transaldolase 4.1×10^{-5} M + Fructose 6-phosphate 6.7×10^{-4} M; E_{max} = 326 nm; Fluorescence intensity 1.1 Volt

△—△ Transaldolase 4.1×10^{-5} n + Fructose 6 phosphate 6.7×10^{-4} n + Erythrose 4-phosphate 5.6×10^{-4} M; E_{max} = 330 nm; Fluorescence intensity 1.05 Volt

Fig. 16. Tryptophane fluorescence emission spectrum of transaldolase in the absence (left curve) and presence of substrates (right curve).

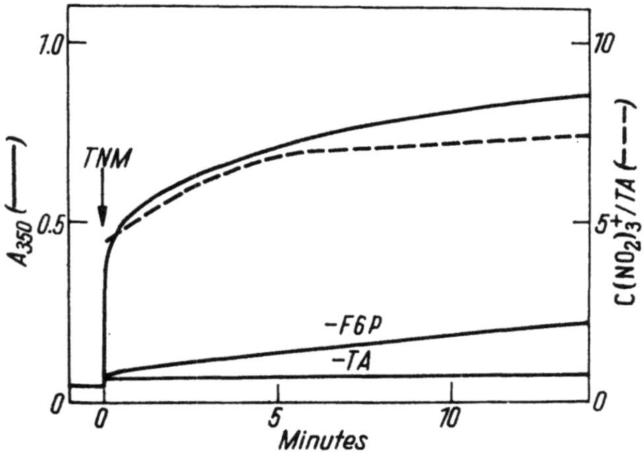

Fig. 17. Production of nitroformate in the presence of fructose 6-phosphate. Tetranitromethane (TNM) (10 µl of a 5×10^{-2} M alcoholic solution) was added to 0.99 ml of 0.05 M phosphate buffer, pH 7.7, in a 1 cm semimicro-cuvette, containing either 5.94×10^{-6} M yeast transaldolase (TA) (Q = 59) and 9.2×10^{-4} M fructose 6-phosphate (F6-P), 5.94×10^{-6} M enzyme and no substrate, or 9.2×10^{-4} M substrate and no enzyme, respectively. The reaction mixtures were preincubated for 5 min at 25 °C prior to the addition of tetranitromethane. The solid lines show the time course of absorbance at 350 nm. The broken line shows the molar ratios of nitroformate to enzyme after substraction of the values obtained in the absence of substrate.

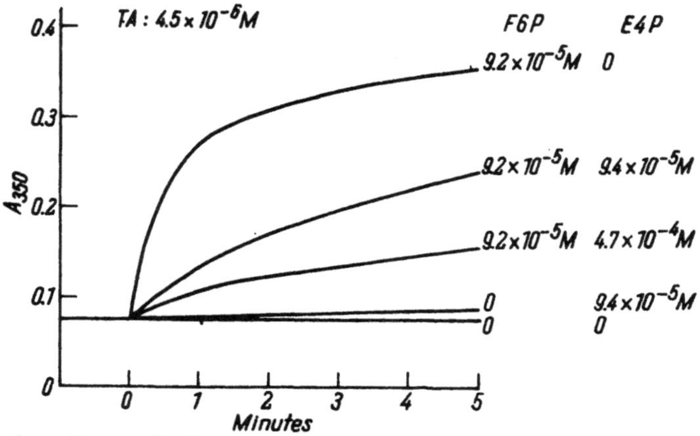

Fig. 18. Effect of erythrose 4-phosphate on the nitroformate production. The reaction mixtures (1.0 ml) contained 4.5 x 10^{-6} M transaldolase (TA) (Q = 61.4), fructose 6-phosphate (F6-P) and erythrose 4-phosphate (E4-P) in concentrations as shown in the figure, and 0.05 M phosphate buffer, pH 6.0. After preincubation for 5 minutes at 25 °C, 5 x 10^{-4} M tetranitromethane was added at the time 0.

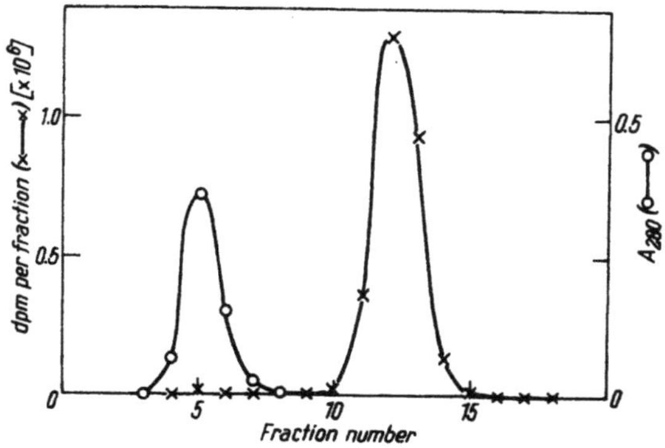

Fig. 19. Effect of modification of transaldolase with tetranitromethane on the Schiff base intermediate.
The reaction mixture contained 1.72 x 10^{-5} transaldolase (Q = 65.2), 3.0 x 10^{-4} M ^{14}C-fructose 6-phosphate (uniformly labeled, total d.p.m. = 2.99 x 10^{6}) in 0.05 M phosphate buffer, pH 6.0. After preincubation for 5 minutes at 10 °C, 8.34 x 10^{-4} M tetranitromethane was added to the reaction mixture in a final volume of 3.0 ml. The number of moles of nitroformate produced within one minute was 3.84 per mole of enzyme. The reaction mixture immediately was passed through Sephadex G-50 column (1.65 x 47 cm, bed volume 100 ml) which was equilibrated with 0.01 M phosphate buffer, pH 6.0, and eluted with the same buffer. Fractions of 10 ml volume were collected except of the protein fractions (fraction number 4, 5, 6 and 7) where the fraction volume was 5 ml. The fractions were analyzed for protein, transaldolase activity, and radioactivity. The main protein fraction (no. 5) contained 0.376 mg of protein/ml (= 5.70 nmoles), 23.2 units/ml (Q = 9.93, = 15.2 % of the original Q value) and 1.930 d.p.m./ml (= 0.56 nmole of fructose 6-phosphate).

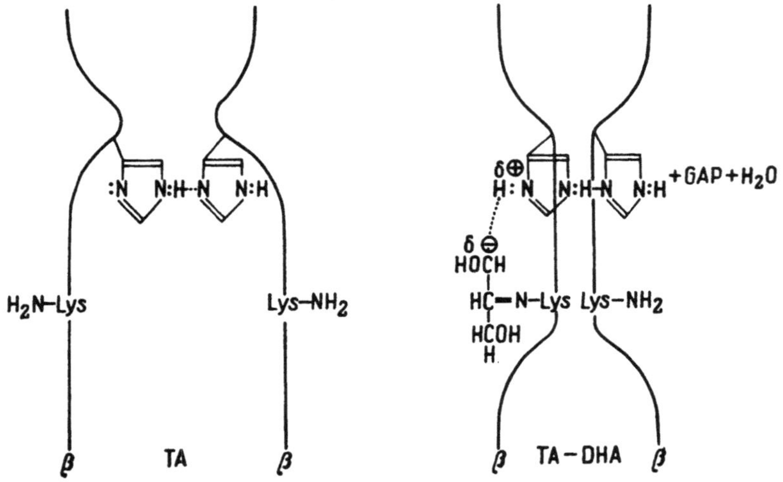

Fig. 20. Model of the subunit structure of transaldolase and the mechanisms of TA-DHA complex formation.

Enzymology Department, Institute of Biochemistry, Hung. Acad. Sci. Budapest

J. OVADI

INTERACTION AND DISSIMILARITY OF SUBUNITS OF GLYCERALDEHYDE-3-PHOSPHATE DEHYDROGENASE

As dr Keleti has already said, in the case of oligomeric enzymes regulation can be considered as an interaction between subunits as well as interaction within a single subunit.

We have examined these two different types of regulation in the case of D-glyceraldehyde-3-phosphate dehydrogenase prepared from pig muscle. As shown in several laboratories this tetrameric enzyme can be dissociated into dimers and monomers. The subunits of the enzyme are chemically identical but some data indicate that the subunits associate assymetrically.

Earlier data have shown that great excess of ATP causes the dissociation of rabbit muscle glyceraldehyde-3-phosphate dehydrogenase. We have observed the same phenomenon with the enzyme isolated from pig muscle. Moreover we found that after gel-filtration 4 moles of ATP per mole of protein were bound to the enzyme. Therefore we examined the effect of 4 moles of ATP per mole of protein on the rate of inactivation of the NAD-free enzyme in 0, 1 M tris-HCl buffer pH 8. 5, at $0^{o}C$.

Fig. 1 curve b shows the inactivation of the ATP-treated enzyme. The activity of the untreated enzyme only slightly decreased during incubation. The ATP-treated enzyme was much faster though it was still a rather slow process.

We found that 2 moles of ATP per mole of protein can be tightly bound to the enzyme within a few minutes, whereas the further 2 only after prolonged incubation. In spite of striking difference in the binding we found the same time course of inactivation whether 2 or 4 mole equivalents of ATP were added. This means that out of the 4 ATP' s the 2 unbound or loosely bound ATP' s do not influence the effect of the 2 firmly bound ones.

The same time course of inhibition caused by 2 or 4 mole equivalents and tight binding of only 2 mole equivalents of ATP to the tetrameric enzyme indicate that the subunits are functionally dissimilar and interact with each other within

the tetramic molecule. As curve b shows, the inactivation of enzyme-ATP complex does not follow simple first order kinetics, which indicates that the inactivation is a rather complex process.
As I already mentioned not more than 4 equivalents of ATP can be bound to the enzyme. This suggests that dissociation into subunits is effected only by the 4 bound ATP' s and the excess is, in fact, not necessary for the process.
Dissociation was tested under conditions similar to that of the inactivation experiments. 4 mole equivalents of ATP were added to the protein and incubated for 5 or 20 hours. The sample was gel-filtered on a calibrated Sephadex G-100 column to separate the fractions of different molecular weight (Fig. 2).
As the broken-line curve in Fig. 2 shows, one peak appeared in the elution profile at the position corresponding to the molecular weight of the dimeric form of the enzyme. It follows that during inactivation with ATP irreversible dissociation into dimers takes place. The complex inactivation curve presented in Fig. 1 might be explained if we assume that the inactivation of dimer-ATP complex is faster than that of the tetramer-ATP complex. The superimposition of these processes and of the tetramer-dimer transformation is reflected in the complex kinetics of inactivation. We have shown by difference spectrophotometry that only slight steric changes occur during this process, i. e. dissociation into subunits is not accompanied by the unfolding of the molecule.
I should like to emphasize that in this case dissociation into dimers was accompanied by inactivation.
However, under certain conditions the enzyme could be dissociated into active dimers. We found that mercaptoethanol influenced the effect of ATP both on the inactivation and dissociation of the enzyme. When the enzyme was incubated with 4 mole equivalents of ATP in the presence of mercaptoethanol, the activity remained practically unchanged for 2 hours, which indicates that SH-groups play a role in the secondary process of inactivation. After a 2 hour incubation the sample was gel-filtered on a Sephadex G-100 column. The column was not equilibrated with mercaptoethanol.
In this case two protein peaks appeared in the gel-filtration profile (Fig. 2, solid line), the position of which corresponded to that of the tetrameric and dimeric enzyme. Apparently in this case dissociation was an irreversible process. The isolated lower molecular weight fraction was enzymically active. Only 50 % activity could be recovered probably because during gel-filtration the enzyme was separated from mercaptoethanol.

If gel-filtration was performed on a column equilibrated with 0.001 M mercapto-ethanol, only one peak appeared in the elution profile between the positions of the tetrameric and dimeric forms. This finding suggests that in the presence of mercaptoethanol there is a reversible equilibrium between the tetrameric and dimeric forms.

On the basis of the above results it is reasonable to assume that ATP can be bound very specifically and tightly to the enzyme. We have shown that ATP is not bound to the binding site of NAD in the active center. In the following I am going to present experiments which suggest that NAD is bound to a regulatory site. According to earlier data excess of NAD and inorganic phosphate inhibit the enzymic activity. It has been assumed that the enzyme has a regulatory NAD-binding site, the liganding of which by excess of NAD descreases enzymic activity. Since ATP can be regarded as an analogue of NAD and also contains phosphate groups, we examined the effect of ATP on the inhibition by excess NAD and phosphate (Fig. 3).

Fig. 3 shows that immediately after the addition of 2 or 4 equivalents of ATP, when inhibition by the nucleotide cannot yet be demonstrated, there is a partial protection against inhibition by excess NAD. As the inactivation caused by 4 moles of ATP proceeds the protecting effect increases and it becomes complete after about 1 hour of incubation. It has to be mentioned that the protecting effect of 2 moles of ATP remains partial even after several hours of incubation.

This suggests that for the complete suspension of NAD-inhibition ATP must be bound to each subunit of the enzyme and that ATP is bound to a regulatory NAD-binding site.

Examining the effect of high concentrations of inorganic phosphate on the enzymic activity we obtained the following results.

Fig. 4 shows the effect of 4 mole equivalents of ATP on the inhibition by excess of phosphate. Immediately after the addition of ATP there was no protection against inhibition, only after 2 hours of incubation, when also inactivation was demonstrated.

We also studied the effect of ADP, AMP and adenin on the inhibition caused by excess substrate in order to examine the role of phosphate moiety in the binding of nucleotide. We found that ADP and AMP, similarly to ATP, suspended the inhibition caused by both NAD and phosphate excess. However, adenin did not protect instantaneously against NAD-inhibition, only after about 3 hours of incu-bation, and inhibition by phosphate was not influenced by adenin even after prolonged incubation.

Since the protective effect of ATP, ADP and AMP against inhibition by excess NAD manifests, itself instantaneously after the binding of nucleotide, whereas that of adenin does not, it seems probable that an ionized phosphate group plays an important role in the orientation and binding of the nucleotide to the protein. As adenin can also bind to the enzyme, even if much more slowly than the nucleotides, and suspend NAD-inhibition, it seems probable that adenyl moiety of the nucleotides interact with the regulatory NAD-binding site. We assumed that each of the three adenyl nucleotides is bound at two points in the regulatory center of the enzyme, though the adenyl phosphate groups.
Summarising, we conclude that both intra- and intersubunit interactions may exist in the tetrameric enzyme and the regulation of enzymic activity might be realized through these interactions.

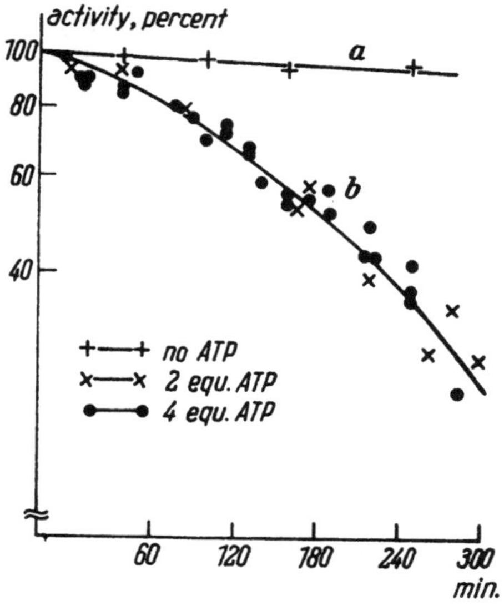

Fig. 1. Effect of ATP on the activity of glyceraldehyde-3-phosphate dehydrogenase
Enzyme freed of NAD (1.4 mg/ml) was incubated with 2 or 4 moles of ATP per mole of protein in 0.1 M tris-HCl buffer, pH 8.5, in the absence of 2-mercaptoethanol. (+) control, (o) and (x) incubation with 4 and 2 moles of ATP per mole of protein, respectively.

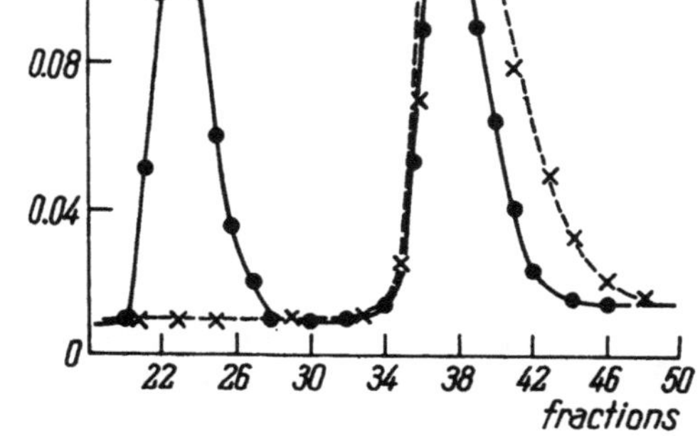

Fig. 2. Dissociation of ATP-treated glyceraldehyde-3-phosphate dehydrogenase
Enzyme, freed of NAD (1.4 mg/ml) was incubated with 4 moles of ATP per mole of protein in 0.1 M tris-HCl buffer, pH 8.5, in the presence and absence of mercaptoethanol. Samples of 1.5 ml were withdrawn and gel-filtered on a Sephadex G-100 column (130 x 1 cm) equilibrated with 0.1 M tris-HCl buffer, pH 8.5.

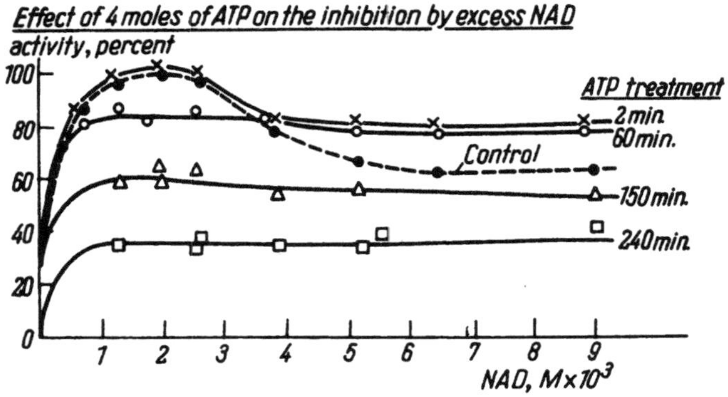

Fig. 3. Effect of ATP on the inhibition by excess NAD
Enzyme freed of NAD (1.4 mg/ml) was incubated with 4 moles of ATP per mole of protein in 0.1 M tris-HCl buffer, pH 8.5, at 0 °C. The enzymic activity was measured with 1 μg/ml enzyme in the following mixture: 0.1 M glycinbuffer pH 8.5, 1.10^{-2} M disodium hydrogenphosphate, $2.5.10^{-3}$ M GAP, and different concentrations of NAD.
(●) untreated enzyme; (x), (o), (△) and (□) enzyme incubated with ATP for 2, 60, 150 and 240 minutes, respectively.

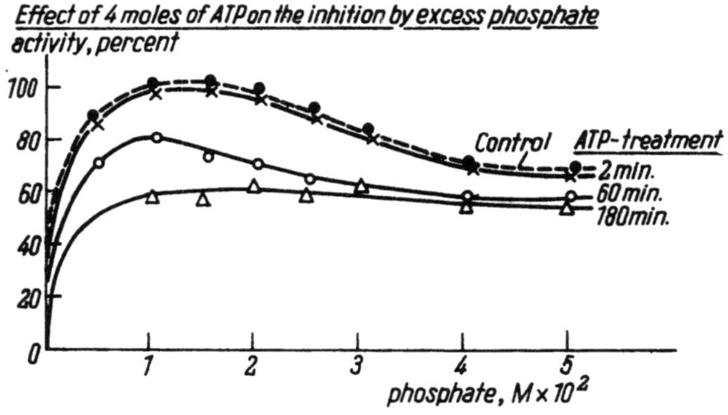

Fig. 4. Effect of ATP on the inhibition by excess phosphate
The same as in Fig. 3 but instead of NAD-concentration that of phosphate was varied. NAD concentration: 2.10^{-3} M.
(●) untreated enzyme, (x), (o) and (△) enzyme incubated with ATP for 2, 60 and 180 minutes, respectively.

Institute of Physiological and Biological Chemistry, Humboldt-University, Berlin

D. MARETZKI

KINETIC PROPERTIES OF GLYCERALDEHYDE-PHOSPHATE DEHYDROGENASE (GAPD) FROM ERYTHROCYTES

In connection with investigations of the effect of inorganic phosphate on the glycolysis of red blood cells we examined the physico-chemical and kinetic properties of a purified glyceraldehydephosphate dehydrogenase from human erythrocytes. The enzyme was isolated and crystallized according to a procedure given in Table 1.

The GAPD was separated from hemoglobin with DEAE-cellulose by a batch-operation and further purified with CM-Sephadex C-50, also batchwise. 1300 fold purification with a yield of about 40-50 % was achieved.

Under optimal conditions the molecular activity of the purified erythrocyte GAPD so obtained was 20000 according to a molecular weight of 144000. This value is equal to the activity of the enzyme from rabbit muscle.

From the ratio of absorption at 280 and 260 nm the NAD^+ content was calculated (2).

In the course of the preparation NAD^+ is partially lost from the enzyme. Dialysis for 24 hours yielded a ratio of absorption of 1, 75 - 2, 0 which corresponds to the value obtained for the apoenzyme. The erythrocyte-GAPD may also be crystallized without NAD^+ and has probably a lower affinity to the coenzyme than the muscle enzyme. Addition of NAD^+ is necessary to stabilize the enzyme. The enzyme recrystallized in 50 mM Trisbuffer containing a 20fold excess of NAD^+ showed from spectral data that 4 mole NAD^+ per mole enzyme were bound. Polyacrylamid-gel-electrophoresis with MATSON-buffer at pH 9, 2 yielded one activity band with nitro-tetrazolium-blue coincident with the protein band. We have looked for other GAPD activities by activity-gel-electrophoresis at different steps of the preparation and have found always a single band. Utilizing a slit-gel-electrophoresis technique we found a higher mobility toward the anode for the erythrocyte enzyme than for the muscle enzyme.

Isoelectric focusing with an ampholine between pH 3-10 reveals a single peak

Table 1

TYPICAL ISOLATION PROCEDURE OF HUMAN ERYTHROCYTES GLYCERALDEHYDEPHOSPHATE DEHYDROGENASE. FOLLOWING FORWARD TEST WAS USED TO DETERMINE ENZYME ACTIVITY

Procedure	Volume ml	Total activity U	Total protein mg	Spec. activity $U \cdot mg^{-1}$	Recovery %	$\frac{A_{280}}{A_{260}}$
Hemolysate (760 ml cells)	4600	6900	180 000	0, 038	100	-
DEAE-Cellulose-batch	2760	9000	1 460	6, 1	130	1, 1
CM-Sephadex C-50-batch	203	6500	114	57	94	1, 3
Dialysis						1, 75-2, 0
NAD^+-addition						0, 95
Crystallization	19	4400	90	49	64	1, 0
2. Recrystallization	8	3200	64	50	46	1, 04

Following forward test was used to determine enzyme activity
1, 76 ml Tris (50 mM) + EDTA (5 mM) + Na_2HAsO_4 (3 mM), pH 8, 6, onl ml NAD^+ (0, 5 mM), 0, 1 ml FDP (10 mM) and 0, 02 ml aldolase. Protein concentrations were estimated from hemoglobin content in the hemolysate and at 280 nm in the other steps assuming the extinction coefficient $E^{0,1\,\%}_{280\,nm}$ was 1, 06 (1).

of enzyme activity with an isoelectric point at about pH 8, 3.
Gelfiltration on Sephadex G-150 column resulted in a single peak containing the enzyme activity. Therefore it is evident that the enzyme is highly purified.
We have estimated from the molecular activity and the enzyme activity in hemolysates that the erythrocyte-GAPD amounts to about 0, 1 % of the total soluble protein of the cells. This corresponds to an intracellular concentration of about 1 μM.
In muscle and yeast the enzyme concentrations are 250fold and in liver 3 fold higher.
In the forward reaction, oxidation of D-glyceraldehydephosphate, the pH-optimum was at pH 8, 5 to 8, 9 in triethanolamine buffer. In glycyl-glycine and tris-buffer it was close to pH 8, 2 to 8, 8.

These results are in agreement with data for the muscle enzyme (3). The MICHAELIS and inhibitor constants for the forward reaction were determined graphically according to LINEWEAVER-BURK. We used a procedure analogous to the given by FURFINE and VELICK (4) for the muscle enzyme.
The activity was either assayed at 366 nm spectrophotometrically or fluorimetrically with an Eppendorf-fluorometer. Initial velocities were obtained by extrapolation.
While varying the concentration of one substrate in the three-substrate reaction the concentration of the second was held optimal at a series of different but constant high concentrations of the third.
Table 2 shows a comparison between kinetic parameters of GAPD of various sources.
It may be seen that all constants for the erythrocyte enzyme unter these conditions are higher than those for the muscle enzyme. Distinct differences are found for the K_m of GAP at pH 7, 2 and for the K_m of arsenate. In the arsenate dependence there is an increase of the K_m values by lowering the pH for the erythrocyte and liver enzymes. This is in contrast to the muscle enzyme.
The inhibitor constants for 1, 3 DPG and NADH were determined by direct measurement of the NADH fluorescence.
These constants are higher for the erythrocyte enzyme than for the muscle enzyme at pH 7, 2 to 7, 4.
We found a 40-fold lower inhibition by NADH.
For the muscle enzyme the inhibition by NADH is more efficient than by 1, 3 DPG which is just the opposite to the erythrocyte enzyme.
For inorganic phosphate the K_m values are similar for the liver and the erythrocyte enzyme respectively.
Maximal activity with inorganic phosphate was achieved at pH 7, 2 with concentrations of 20 - 25 mM and at pH 8, 5 with concentrations of 10-15 mM.
Fig. 1 shows a reciprocal plot of the initial velocities against phosphate concentration. Inorganic phosphate produces a secondary enzyme activation.
At pH 7, 2 this activation occurs above a concentration about 1 mM phosphate, at pH 8, 5 above 0, 5 mM.
With arsenate as the third substrate we found an activation at pH 7, 2 and pH 8, 5 above concentrations of 100 and 50 μM respectively.
The influence of phosphate on the K_m values of glyceraldehydephosphate was studied in the activated and non-activated ranges. Fig. 2 shows a primary plot at pH 7, 2 and 8, 5 at optimal NAD^+-concentrations.

Table 2

MICHAELIS - AND INHIBITOR CONSTANTS (μM) FOR RABBIT MUSCLE (4), HUMAN ERYTHROCYTES AND BOVINE LIVER (7) GLYCERALDEHYDEPHOSPHATE DEHYDROGENASE

		K_m			K_i		
Source	pH	NAD^+	D-GAP	P_i	arsenate	NADH	1, 3 DPG
muscle	7, 4	90	2, 5	-	28	0, 3	1
	8, 6	13	90	290	69	2	2 - 4
erythrocyte	7, 2	130	130	3400	500	12, 5	2 - 3
	8, 5	96	500	970	200	21	13
liver	8, 0	190	240	7150	1700	96	-
	8, 9	31	385	3400	1540	1, 7	-

In the non-activated range the K_m values are independent from the phosphate concentration. At pH 7, 2 the K_m value for GAP is 33 μM, at pH 8, 5 this K_m value is 37 μM.
With arsenate the K_m values for GAP are half a high as for phosphate. These findings confirm previous results on the muscle enzyme indicating stronger binding of arsenate to the enzyme (5). From the kinetic data reported here, the capacity of GAPD in hemolysates and the intracellular concentrations of 2 μM GAP, 0, 37 μM 1, 3 DPG (6) and 0, 33 mM inorganic phosphate we estimated the actual activity of GAPD within the erythrocytes which agrees fairly well with the glycolytic rate at pH 7, 2
According to investigations by the groups of HEINZ (7) and of PAPADOUPOULOS (8) on liver-GAPD as well as from our results on the enzyme of erythrocytes the existence of several variants of this enzyme in various animal tissues may be inferred.

REFERENCES

(1) FAHIEN, A. L., J. biol. Chem. 241, (1966), 4115

(2) FOX, J. B. Jr., W. B. DANDLIKER, J. biol. Chem. 221, (1956), 1005

(3) CORI, G. T., M. W. SLEIN, C. F. CORI, J. biol. Chem. 173, (1948), 605

(4) FURFINE, C. S., S. F. VELICK, J. biol. Chem. 240, (1965), 844

(5) KELETI, T., In: T. DEVENY, P. ELÖDI, T. KELETI, G. SZABOLCSI, "Strukturelle Grundlagen der biologischen Funktion der Proteine", Budapest, Akademiai Kiado 1969

(6) SCHEWE, CH. person. comm.

(7) HEINZ, F., K. D. KULBE, Hoppe-Seyler's Z. physiol. Chem. 351, (1970), 249

(8) PAPADOPOULOS, C. S., S. F. VELICK, Feder. Proc. 26, (1967), 557

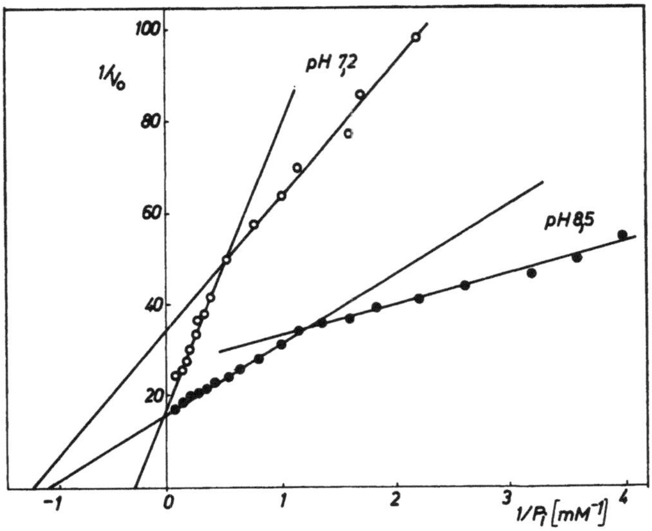

Fig. 1. Activation of GAPD by phosphate at pH 7, 2 and 8, 5
Conditions: pH 7, 2; 0, 06 M imidazole
0, 7 mM GAP, 0, 98 mM NAD^+
8 x 10^{-9} M GAPD
pH 8, 5; 0, 06 mM triethanolamine
0, 7 mM GAP, 0, 98 mM NAD^+
2 x 10^{-9} M GAPD

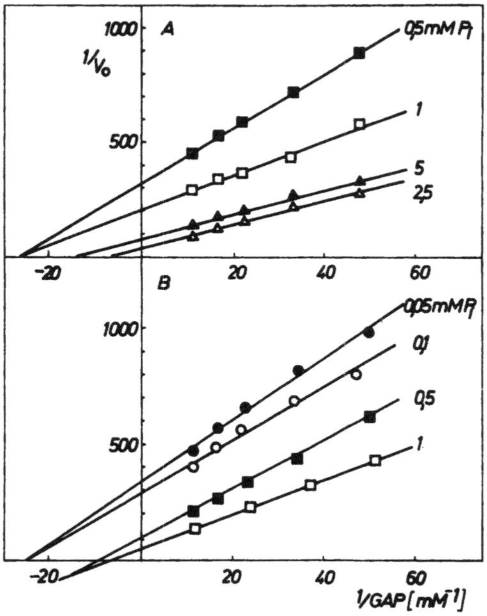

Fig. 2. Influence of phosphate on the kinetics of GAP-oxidation
A - pH 7, 2; 0, 1 M triethanolamine
0, 49 mM NAD^+, 2, 2 x 10^{-9} M GAPD
B - pH 8, 9; 0, 1 M triethanolamine
0, 49 mM NAD^+, 1, 1 x 10^{-9} M GAPD

Max-Planck-Institut für Ernährungsphysiologie, Dortmund

B. HESS, K. -J. JOHANNES, C. KUTZBACH, H. -P. BISCHOFBERGER, C. -J. BARWELL, P. RÖSCHLAU

STRUCTURE AND FUNCTION OF PYRUVATE KINASE

Introduction

Pyruvate kinase (E. C. 2. 7. 1. 4.) catalyzes the formation of pyruvate and ATP from phosphoenolpyruvate and ADP, in an irreversible key-reaction of glycolysis, with an equilibrium constant of $K_{eq} = 2 \times 10^3$ (30°, pH 8, 0) (1). The reaction needs the presence of K^+ and Mg^{++}. The enzyme is ubiquitously distributed as a component of the glycolytic enzyme sequence of all living organisms down to simple cellular species like Clostridium kluyvery (2). Its activity is controlled by the concentration of its own substrate phosphoenolpyruvate as well as its product ATP, furthermore by
1. FDP in E. coli (3), yeast (4, 5, 6), rat liver (7, 8), pig liver (9), rat epididymal adipose tissue (10), erythrocytes (11), 2. Enzyme induction (pyruvate kinase type I) in E. coli (12), 3. Hormonal and nutritional control (13-16).
Since the enzyme is involved in the control of pyruvate and ATP production, two of the most common intermediates in energy metabolism, and since it is important in the switch from glycolysis to gluconeogenesis, the molecular nature of its control mechanism is of highest interest.
The following lecture summarizes a number of recent data on the structure, function and control of the enzyme, as observed primarily in our laboratory on pyruvate kinase of S. carlsbergensis and pig liver. The data are compared to pyruvate kinase of other sources.

Structure

Hydrodynamic data on pyruvate kinase of S. carlsbergensis and pig liver, obtained as highly purified homogeneous preparations in our laboratory, are summarized in Table 1. In both preparations the subunit structure indicates the presence of four subunits per native enzyme. The data on the molecular weight of native pyruvate kinase and its subunits are compared with other species, as reported in the literature, in Table 2, which indicates that a sub-structure of four units is found in the cases tested so far. Furthermore, it should be noticed, that the molecular weight of the native enzyme of S. cerevisiae is clearly lower than that of S. carlsbergensis. It is of interest to see that pyruvate kinase of muscle, which cannot be activated by FDP, has a significantly larger molecular weight than the pyruvate kinase of other species which can be activated by this ligand. It is not known, whether this difference has any mechanistic significance (17).

In the aminoacid composition of S. cerevisiae (30), S. carlsbergensis (31) or pig liver (19), there are no appreciable differences. The composition of the latter two species is summarized in Table 3. The properties in absorption and fluorescence spectroscopy of the latter two enzymes are similar (31-33).

Table 1

HYDRODYNAMIC PROPERTIES OF PYRUVATE KINASE OF S. CARLSBERGENSIS AND PIG LIVER

Parameter	Medium	YPK	LPK	Method
$s^{o}_{20,w}$	Dilute Buffer	8.69	8.10	Ultracentrifuge
$D^{o}_{20,w}$	"	4.15	3.93	"
$M_{S/D}$	"	191 000	192 300	Svedberg-Equation
M_w	"	190 400	202 400	Equilibrium Centrifugation
M	"	200 000	265 000	Gelfiltration on G-200
$s_{20,w}$	6M G-HCl	2.83	1.59	Ultracentrifuge
M_w	6M G-HCl	51 000	49 900	Equilibrium Centrifugation
M_w	succinyl-Enzym	47 500	-	"
n		4	4	

Table 2

STRUCTURE OF PYRUVATE KINASE

Source	FDP activable	molecular weight native	dissociated	n	Ref.
Rabbit	-	237000	57200	4	(18)
Rat muscle	-	250000	-	?	(14)
Rat liver	+	208000	-	?	(14)
Pig liver	+	202000	50000	4	(19)
Saccharomyces cerevisiae	+	162 - 168000	42 - 45000	4	(20)
Saccharomyces cerevisiae	+	161000	20000	8 (?)	(21)
Saccharomyces Carlsbergensis	+	190000	45 - 51000	4	(17)
Human erythrocytes	+	205000 $\pm$ 5000	-	?	(22)
Human erythrocytes	+	195000 $\pm$ 6000	-	?	(23)
Mucor rouxii	+	182000 $\pm$ 9000	-	?	(24)

Function

The mechanism of the reaction catalyzed by pyruvate kinase is obscure, there is evidence that ADP and ATP reacts in form of their magnesium-complexes (26). Furthermore, free magnesium ions are essential in the reaction. A relaxation study of rabbit muscle pyruvate kinase points to the existence of multiple conformation equilibria (27). On the basis of a study of the Mn-enzyme, Mildvan and Cohen suggest a multiple subsite binding structure of the active center involving a site-site interaction among Mn and phosphoenolpyruvate binding sites effecting the enzyme affinity for phosphoenolpyruvate under control of fructose-diphosphate in the case of the yeast enzyme (for summary see (28)). The authors indicate the participation of the histidine residue the binding of the metal (Mn)-ion.

In our laboratory, a function of histidine in the phosphoenolpyruvate binding process was recently demonstrated by diazotization of approximately 1 histidine residue/subunit, which is completely protected by phosphoenolpyruvate added prior to the diazotization reaction (29). From this experiment, we conclude that phosphoenolpyruvate directly interacts with a histidine residue or is at least in its immediate neighbourhood.

Table 3

AMINOACID COMPOSITION OF PYRUVATE KINASE OF S. CARLSBERGENSIS AND PIG LIVER

	Yeast-PK		Pig liver PK	
AS	%	Mol/50. 000	%	Mol/50. 000
Asp	13. 1	56	7. 4	32
Thr	8. 5	41	4. 6	23
Ser	4. 6	26	4. 8	27
Glu	10. 0	38	14. 2	55
Pro	4. 0	20	3. 6	19
Gly	3. 7	32	4. 8	42
Ala	5. 8	40	7. 4	52
Cys H	0. 9	5	1. 2	6
Val	8. 7	43	8. 3	42
Met	2. 1	8	2. 4	9
Ile	6. 5	28	8. 0	35
Leu	7. 4	32	8. 8	39
Tyr	4. 5	14	2. 4	7
Phe	4. 5	15	4. 1	14
Lys	7. 2	28	4. 7	18
His	2. 3	8	2. 5	9
Arg	6. 5	20	9. 6	31
Try	0. 9	3 (2)	1. 1	3
$-NH_2$	-	-	1. 2	-
total SH	-	5	-	7. 2 aus DTNB
p I	6. 08		6. 15 ± 0. 1	
$E^{0.1\%}_{280}$	0. 76		0. 68	

Table 4

KINETISCHE PARAMETER VON LEBER-PYRUVATKINASEN

Metabolite	Bedingungen	Schweineleber [a]		Rattenleber 1) Mäuseleber 2) [b]	
		$K_{0.5}$ [mM]	n_H	$K_{0.5}$ [mM]	n_H
PEP	-FDP	0.5	2.5	0.84[1] 1.0[2]	>1[1] 1.5[2]
PEP	+ 20 μM FDP	0.15	0.95-1.05	0.08[1] 0.3[2]	~1[1] 1.0[2]
ADP	-FDP, 0.2 mM PEP	0.1[x]	1.1	-	-
ADP	+FDP, 0.2 mM PEP	0.20	1.0	0.15[1] 0.21[2]	1[2]
FDP	0.2 mM PEP	0.02×10^{-3}	1.2	0.9×10^{-3} [1]	1.15[1]
ATP	0.2 mM PEP	0.18	1.3	-	-
ATP	1.0 mM PEP	0.67	1.5	-	-
Mg^{++}	-FDP, 0.2 mM PEP	1.0	-	5	-
K^{+}	-FDP, 0.2 mM PEP	2.5	-	-	-

Bedingungen: a 0.2 mM PEP, 0.5 mM ADP, 25 mM K^{+}, 5 mM Mg^{++}, pH 7.0, 37°C gekoppelter Test mit LDH

b 1) siehe (13)
2) siehe (25)

x Hemmung > 0.5 mM

The function of SH-groups will be discussed in a later presentation of the symposium (34). The location of the few tryptophane residues within the enzyme molecule is of great interest because of their fluorescent properties which can be used as a built-in probe, sensing the environment of the tryptophane residues and indirectly the conformational state of the intact molecule. In a chemical study of the reactivity of the tryptophane residue we have found that tryptophane of the native enzyme does not react with N-brom succinimide. The tryptophane molecules are attacted only if the reaction is carried out in 8 M urea, indicating that an unfolding of the native enzyme is necessary for interaction of the reagent and tryptophane, and furthermore, that the tryptophane residues are deeply buried in the intact molecule. This result is in accord with the observations of fluorescence emission changes in rabbit muscle pyruvate kinase (35) and yeast pyruvate kinase (33) in response to various conditions, some of them will be discussed later on (Fig. 7).

Control

Following the discovery of FDP-activation of yeast pyruvate kinase in 1965 (4, 5), the complex and allosteric control properties of pyruvate kinase of a great number of species became known, leading to a greater understanding of the control mechanism of glycolysis. Whereas pyruvate kinase of rabbit muscle follows classical MICHEALIS-MENTEN-kinetics, the enzyme from a number of yeast, Fungi and the organs of higher organisms interacts with different controlling ligands. In dependence of the ATP and FDP concentration, pH and ionic strength, the dissociation constants of phosphoenolpyruvate of various enzymic species varies between the mmolar- and almost nmolar-range with HILL-coefficients between 2. 9 and 1, reflecting the delicate molecular control mechanism with which the enzymes operate in various species.

In the following discussion I would like to focus the attention on pyruvate kinase of liver and the yeast Saccharomyces carlsbergensis and discuss their respective kinetic parameters.

A summary of $K_{0.5}$ and n_H of the pig liver enzyme as obtained in our laboratory (9, 19) and of rat and mice liver (13, 25) is given in Table 4, demonstrating the activity of the enzymes from three different sources on variation of the ligands. The affinity of the enzyme of pig and rat liver for FDP is very high. In kinetic experiments as well as in binding studies we have found a variable K_D for FDP depending in the state of the SH-groups, the ionic strength and the molarity of phosphoenolpyruvate (36). In binding experiments using the gelfiltration technique a maximum binding of 4 moles FDP per mole pyruvate kinase was found. A K_D of 1 mole FDP in the presence of 1 mmolar phosphoenolpyruvate and a K_D of 20 moles FDP in the absence of phosphoenolpyruvate were measured. On variation of the ionic strength K_D' s for FDP in the range of 20 nM were detected. The high affinity of liver pyruvate kinase for fructosediphosphate is a clue to an understanding of the apparent macro-heterogeneity of liver pyruvate kinase as observed in various laboratories (36). Indeed, a differentiation of two types of liver pyruvate kinase could be achieved by isoelectric fractionation using radioactive FDP as well, as the activity of the enzyme as markers. When radioactive FDP and phosphoenolpyruvate are added to a sample of purified liver pyruvate kinase, an isoelectric pattern with two peaks was found as shown in Fig. 1. A larger activity fraction with an isoeletric point of pH 6.1 contained no radioactive FDP, whereas in the other activity peak, with an isoeletric point of pH 5.33, the radioactivity of ^{14}C-FDP was clearly accumulated. The excess of radioactivity was found in the anode solution.

In another experiment FDP or FDP + phosphoenolpyruvate were added to the sample prior to the electrofocussing procedure and also to the whole gradient and again only the more acid peak was detected carrying the radioactivity, as shown in Fig. 2, even if the electrofocussing was prolonged to 36 or 48 h. The presence of phosphoenolpyruvate did not seem to influence or to be necessary for the formation of the enzyme form with a low isoelectric point. From these experiments we conclude that the "5.33 - 5.35"-species is probably a FDP-pyruvate kinase complex, whereas the "6.0 - 6.25"-species is free pyruvate kinase. In parenthesis we like to point out that also a microheterogeneity within each peak could be demonstrated (36).

Our conclusions on the nature of at least two types of liver pyruvate kinase are well supported by variations of the kinetic properties of the enzyme in raw extracts of liver, which has been described by several authors and was also observed in our laboratory (36). Here, we even would like to suggest that the same phenonmena might explain macro-heterogeneities of pyruvate kinase from kidney and erythrocytes, as described recently in other laboratories (37, 38).

The experiments clearly identify two forms of liver pyruvate kinase as defined by electrofocussing experiments and FDP-binding. The designation of these two forms as isoenzymes no longer seems to be justified. A third form of an enzyme with an isoelectric point of pH 5.5 still needs to be characterized. However, since the transformation of the two forms is a slow process, which can be trapped even during isoeletric focussing, a definition of a third form might also be possible in the near future.

A more rigorous analysis of the kinetic mechanism of pyruvate kinase has recently been achieved in case of the yeast enzyme. A computer analysis of the experimental steady-state kinetics clearly showed, that the relationship between homotropic and heterotropic interactions observed with the enzyme fulfils the criteria of the model of Monod, Wyman and Changeux (39) as a two state equilibrium system. The parameters of the model were obtained with the procedure developed by Blangy, Buc and Monod in case of phosphofructokinase of

E. coli (40). The logic of this procedure is given in Fig. 3 demonstrating the connection between raw data and the finally yielded constants of the model: L_o, K_R, K_T and n.

The substrate saturation curves obtained with varying modifier concentrations of FDP and ATP are used to compute the constants V_{max}, $K_{0.5}$ (for half-saturation) and the HILL-coefficient n_H with an iteration procedure described earlier (41). An optimized plot of the Hill-coefficients against the logarithm of the half-saturation concentration yields the dissociation constants for the substrate and the non-exclusive binding coefficient c. This coefficient and the maximum HILL-coefficient lead to the number of protomers n. With the help of the non-exclusive binding coefficient c and the value of n L' can be calculated for the maximum HILL-coefficient, which leads to a value of the allosteric constant L_o. Remember that the L' are a function of the activator resp. inhibitor concentration, which gives the dissociation constants for the activator and inhibitor. Also, the non-exclusive binding coefficient and the number protomers deliver, with the help of the half-saturation concentrations, L' -values. Finally, again, a value for the allosteric constant L_o is obtained. Unfortunately, a detailed discussion of this technique is out of the scope of this paper (42).

The theoretical and experiment saturations of yeast pyruvate kinase are plotted in Fig. 4 in a computer printout. The closed points are experimental, the stars are computed points, both clearly indicate the very good fit obtained between theory and experiment. The half-saturation concentration ranges over almost two orders of magnitude. The limiting curvatures on the left and right side reflect resp. the R-state and the T-state of the enzyme. The steepness of the curves is greatest in the middle of the field, where the Hill-coefficient approaches the value of 2.45 decreasing to the left and the right side.

The plot of the Hill-coefficient n_H against the logarithm of the half-saturation concentration $K_{0.5}$ is given in a computer printout of Fig. 5. Since the HILL-coefficient passes through a maximum, the non-exclusiveness of the substrate binding is clearly indicated. This maximum is obtained with a substrate saturation curve in the absence of modifiers. Increasing activator concentration shifts the points to the left side finally reaching a HILL-coefficient of 1. The dissociation constant for the substrate phosphoenolpyruvate of the R-state of the enzyme is given by the intercept on the abscissa. A shift to the right is obtained with increasing inhibitor concentration, again reaching a HILL-coefficient of 1 if the enzyme is completely in its T-state and displaying the dissociation constant

of phosphoenolpyruvate for the T-state. The quotient of both dissociation constants gives the non-exclusive binding coefficient. Between these two extreme conditions cooperativity with a maximum HILL-coefficient of 2.45 is observed.
There are two curves in this plot: the star curve is the optimized curve fitted on the basis of the experimental points, which are closed. The +-curve, a little bit higher, is the theoretically expected curve for n = 3 and a deviation factor of 0.07 from the theoretically expected curvature is observed. However, I would like to mention that the three protomers as computed by this technique are not unequivocal and will be discussed later in this lecture.
Fig. 6 demonstrates the linearisation of the L' -values for the activator fructose-diphosphate as well as for the inhibitor ATP. Here again optimization was obtained by a least square fitting procedure and the dissociation constants for both modifiers are obtained from the abscissas. It should be noticed, that also this function gives the best fit with n = 3.
There are other techniques which can be used to determine the various model parameters. I mention, the Q-function (43), the linearized abbreviated Y-function of HORN and BÖRNIG (44) as well as the linearized Y-function of Wong and his collaborators (45). These techniques are supplementary, however of less rigidity and efficiency since either the accuracy is not high enough or not generally applicable e.g. if in the first and second case c does not equal zero.
The constants of the model are summarized in Table 5. These values and the corresponding L' -values were used for the computation of the theoretical saturation functions which are presented in Fig. 4.
The dissociation constant for FDP of the R-state is supported by two supplementary techniques. Since the fluorescence spectrum of the enzyme is sensitive to FDP addition, indicating a conformational change in the environment of the tryptophane residues, a titration of the fluorescence emission by FDP was carried out.
Fig. 7 demonstrates the fluorescence spectrum in the presence and in the absence of FDP. A clear shift of the maximum of the emission spectrum as well as a quenching is observed on addition of FDP. The fluorescence emission change was used for a titration of the FDP affinity and a curve was obtained as given in Fig. 8, plotted and fitted by the computer. In agreement with the steady-state data a half maximal constant for FDP of 0.23 mmolar was found indicating the R-state of the enzyme (33). Furthermore, FDP binding studies were carried out with the Colowick technique and a dissociation constant in the range of 0.2 mmolar FDP was obtained using protein concentrations between 8 and 17 mg/ml.

Table 5

Ligand	K_R	K_T
PEP	1.9×10^{-4}M	4.8×10^{-2} M
FDP	2.0×10^{-4}M	$> 2 \times 10^{-2}$ M
ATP	$> 1 \times 10^{-1}$M	9.3×10^{-3} M
n = 3	L_0 = 3900	$c = 4 \times 10^{-3}$

On the basis of the molecular weight of 190000 a Scatchard plot calculates approximately 3.6 binding sites for FDP per native enzyme which in extrapolation points to four binding sites (42).

Here, we encounter a serious discrepancy between the structural and binding data and the result of the steady-state analysis. The first technique clearly indicates that we are dealing with an enzyme composed of four subunits and probably of four active binding sites for FDP. The latter technique yields only three. Indeed, we are dealing with the same problem which is observed in the case of hemoglobin, in which a $n_{H,max}$ of approximately 2.6 and 4 protomers are reported (46).

In the logic of our analytical procedure of fig. 3 the HILL-coefficient as a function of the model parameter is a basic and quantitative relationship, which is unique only in the case of ideal two-state systems. Since a deviation of the number of protomers (n) is observed, the presence of hybride states is suggested. In order to cover the occurance of hybride states such as R_iT_{n-i}, which are not recognized in the original model of MONOD, WYMAN and CHANGEUX (39), the equation, which relates the well-known HILL-coefficient with the model parameter $\alpha_{1/2}$, n and c was modified by addition of a weighting for a number of protomers (n) by a quotient σ^2/σ^2_{max} (42):

$$n_H = 1 + \left(n\frac{\sigma^2}{\sigma^2_{max}} - 1\right)\frac{(\alpha_{1/2}-1)(1-c\alpha_{1/2})}{(\alpha_{1/2}+1)(1+c\alpha_{1/2})}$$

here, σ^2 is related to the frequency of occurance of the conformations R_iT_{n-i}, and $\alpha_{1/2} = K_{0.5}/K_{R(S)}$. Thus, the equation relates the experimental Hill-coefficient and $K_{0.5}$ with the model parameter n, $K_{R(S)}$ and $K_{T(S)}$. Only if $\sigma^2/\sigma^2_{max} = 1$ is all or non behaviour observed.

As shown in fig. 5a deviation factor of 0. 07 from the ideal curvature was found in the case of n = 3, i. e. with three protomers per enzyme. If the same fit is carried out with four protomers per mole enzyme, a factor of 0. 38 pointing to the form of R_iT_{n-i} were computed from the experimental analysis of the steady-state kinetic in agreement with the chemical data. Since the trimeric model does not agree with the chemical properties of the enzyme it has no priority.

Thus we postulate an additional hybride state participating within the equilibrium, as demonstrated in the following reaction scheme:

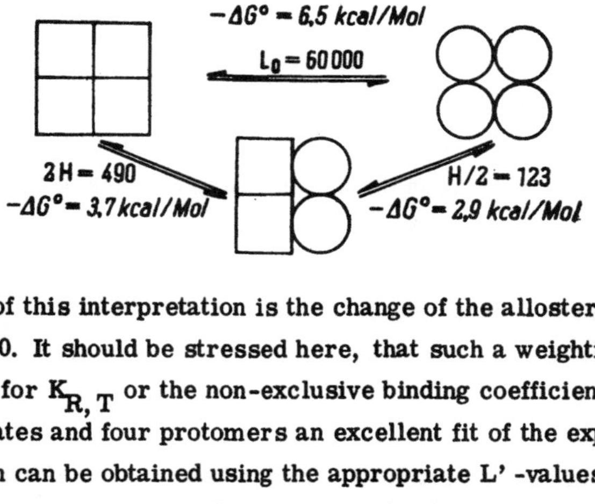

The consequence of this interpretation is the change of the allosteric constant from 3900 to 60000. It should be stressed here, that such a weighting does not change the values for $K_{R,T}$ or the non-exclusive binding coefficient. Allowing for the hybride states and four protomers an excellent fit of the experimental saturation function can be obtained using the appropriate L' -values as given in Fig. 9. Thus, macroscopically no difference can be detected in case of three protomers compared to the case of four protomers including hybride states. Thus, interpretation of steady-state kinetics by classical means do not allow unique conclusions about molecular mechanisms involved, if no complementary information on the molecules are available.

The thermodynamic paramters derived for the models are summarized in Table 6 for the case of n = 4 (L_o (4)) as well as n = 3 (L_o (3)). It is obvious from the ΔG^o-values that the transition from the T-state to the R-state is an extremely large step, which is expected to be carried through in a series of small transitions such as indicat ed by the assumption of a hybride state, and even more intermediary states might be discovered in the future.

Table 6

pH = 7.0	T = 25°C
L_o-Werte	$-\Delta G^o$ kcal/Mol
$L_o^{(4)}$ = 60000	6.5
$L_o^{(3)}$ = 3900	4.9
2 H = 490	3.7
H/2 = 123	2.9

Within such a time-scale inhomogeneities of the transition can scarcely be detected. However, it is hoped that more sophisticated techniques will differentiate the conformation change in more detail (33).
A discussion of the structural consequences of this model is out of the scope of this presentation. However, I would like to mention a kinetic experiment, which demonstrates the time-scale of the transient of the enzyme shifting from the T- to the R-state. As demonstrated above the various states of pyruvate kinase of yeast can directly be recognized by recording the tryptophane fluorescence of the native enzyme in the presence and absence of ligands. Recent experiments with a stoppedflow technique recording fluorescence changes on rapid mixing of native pyruvate kinase with FDP revealed, that the activation time of the fluorescence change is completed whithin the mixing time of the instrument of about 1.5 msec.

Physiology

An evaluation of the properties of pyruvate kinase under physiological conditions is important for our understanding of the operation of the enzyme within the intact glycolytic system. We therefore analyzed overall kinetic parameters of pyruvate kinase over a large concentration range. In such a study it was found that a clear proportionality of pyruvate kinase of rabbit muscle in the concentration range between 10^{-9} - 10^{-6} M is found. There is no significant concentration dependency over a wide concentration range up to physiological levels of the activity of the enzyme (47). Furthermore, the transient time of pyruvate kinase related to the Michaelis-constant for ADP and V_{max} of the enzyme was found concentration-independent in the range between 10^{-8} - 10^{-6} M (47). Finally a recent analysis of the yeast enzyme again demonstrated a concentration independency of the specific activity of the enzyme in the range between

2.5×10^{-10} M up to 5×10^{-6} M pyruvate kinase of yeast (Fig. 10 (48)). These results demonstrate that there is a good reason to assume that analytical studies on enzymes in diluted state can be extrapolated to the highly concentrated state of the enzyme under physiological conditions and no dissociation - association equilibria need to be considered with this enzyme.

Finally I would like to point to the physiological significance of a regulation of pyruvate kinase by fructose-1, 6-diphosphate in S. cerevisiae. Barwell and coworkers recently have analyzed the conditions under which a crossover behaviour of phosphoenolpyruvate and pyruvic acid together with the inverse relationship in companion with a changing concentration of fructose-1, 6-diphosphate can directly be related to the activity of pyruvate kinase (49).

REFERENCES

(1) McQUATE, J.T., M.F. UTTER, J. Biol. Chem. 234, (1959), 2151

(2) DECKER, K., C. BARTH, H. METZ, Biochem. Z. 345, (1966), 472

(3) MAEBA, P., B.D. SANWAL, J. Biol. Chem. 243, (1968), 448

(4) HESS, B., In: B. CHANCE, R.W. ESTABROOK, J.R. WILLIAMSON, Control of Energy Metabolism, Academic Press New York, (1965)

(5) HESS, B., R. HAECKEL, K. BRAND, Biochem. Biophys. Res. Communs. 24, (1966), 824

(6) HAECKEL, R., B. HESS, W. LAUTERBORN, K.H. WÜSTER, Hoppe-Seyler's Z. Physiol. Chem. 349, (1968), 699

(7) TAYLOR, C.B., E. BAILEY, Biochem. J. 102, (1967), 32C

(8) TANAKA, T., F. SUE, H. MORIMURA, Biochem. Biophys. Res. Commun. 29, (1967), 444

(9) KUTZBACH, C., B. HESS, Hoppe-Seyler's Z. Physiol. Chem. 351, (1970), 272

(10) POGSON, C.I., Biochem. Biophys. Res. Communs. 30, (1968), 297

(11) MUNRO, G.F., D.R. MILLER, Biochemica et Biophysica Acta 206, (1970), 87

(12) MALCOVATI, M., H.L. KORNBERG, Biochemica et Biophysica Acta 178, (1969), 420

(13) TANAKA, Y. HARANO, F. SUE, H. MORIMURA, J. Biochem. Tokyo, 62, (1967), 71

(14) TANAKA, T., Y. HARANO, H. MORIMURA, R. MORI, Biochem. Biophys. Res. Commun., 21, (1965), 55

(15) KREBS, H.A., L.V. EGGLESTON, Biochem. J., 94, (1965), 3C

(16) WEBER, G., Regulation of Pyruvate Kinase in Advances in Enzyme Regulation 7, (1969), 15

(17) BISCHOFBERGER, H., B. HESS, P. RÖSCHLAU, Hoppe-Seyler's Z. Physiol. Chem. 352, (1971), 1139

(18) STEINMETZ, M.A., W.C. DEAL, Biochemistry 5, (1966), 1399

(19) HESS, B., C. KUTZBACH, H. BISCHOFBERGER, in preparation

(20) KUCZENSKI, R.T., C.H. SUELTER, Biochemistry 9, (1970), 2043

(21) ASHTON, K., A.R. PEACOCKE, personnel communication

(22) STAAL, G.E.J., J.F. KOSTER, H. KAMP, L. van MILLIGEN-BOERSMA, C. VEEGER, Biochem. Biophys. Acta 227, (1971), 86

(23) BLUME, K.G., R.W. HOFFBAUER, D. BUSCH, H. ARNOLD, G.W. LÖHR, Biochem. Biophys. Acta 227, (1971), 364

(24) TERENZI, H.F., E. ROSELINO, S. PASSERSON, Europ. J. Biochem. 18, (1971), 342

(25) CARMINATTI, H., L. JIMENEZ DE ASUA, E. RECONDO, S. PASSERU, E. ROZENGURT, J. Biol. Chem. 243, (1968), 3051

(26) WIEKER, H.J., K.J. JOHANNES, B. HESS, Abstract FEBS-Congress of Madrid, (1969), 138

(27) HAMMES, G.G., J. SIMPLICIO, Biochemica et Biophysica Acta 212, (1970), 428

(28) MILDVAN, A.S., M. COHN, Advances in Enzymology 33, (1970), 1

(29) HESS, B., P. RÖSCHLAU, L. BORNMANN, unpublished experiments

(30) HUNSLEY, J.R., C.H. SUELTER, J. Biol. Chem. 244, (1969), 4815

(31) BISCHOFSBERGER, H., B. HESS, P. RÖSCHLAU, H.-J. WIEKER, H. ZIMMERMANN-TELSCHOW, Hoppe-Seyler's Z. Physiol. Chem. 351, (1970), 401

(32) SUELTER, C.H., personnel communication

(33) HESS, B., R. MÜLLER, unpublished experiments

(34) WIEKER, H.-J., B. HESS, This symposium

(35) SUELTER, C.H., Biochemistry 6, (1967), 418

(36) HESS, B., C. KUTZBACH, Hoppe-Seyler's Z. Physiol. Chem. 352, (1971), 453

(37) JIMENEZ DE ASUA, L., E. ROZENGURT, H. CARMINATTI, FEBS-Letters 14, (1971), 22

(38) IBSEN, K.H., K.W. SCHILLER, T.A. HAAS, J. Biol. Chem. 246, (1971), 1233

(39) MONOD, J., J. WYMAN, J.P. CHANGEUX, J. Mol. Biol. 12, (1965), 88

(40) BLANGY, D., H. BUC, J. MONOD, J. Mol. Biol. 31, (1968), 13

(41) WIEKER, H.-J., K.-J. JOHANNES, B. HESS, FEBS-Letters 8, (1970), 178

(42) HESS, B., K.J. JOHANNES, H. BUC, in preparation

(43) BUC, H., Biochem. Biophys. Res. Commun., 28, (1967), 59

(44) BORN, A., H. BÖRNIG, FEBS-Letters 3, (1970), 73

(45) ENDRENYI, L., M.S. CHAN, J. TZE-SEI WONG, personnel communications

(46) PERUTZ, M.F., Nature 228, (1970), 726

(47) HESS, B., B. WURSTER, FEBS-Letters 9, (1970), 73

(48) BARWELL, C., B. HESS, unpublished experiments

(49) BARWELL, C., B. WOODWARD, R.V. BRUNT, Europ. J. Biochem. 18, (1971), 59

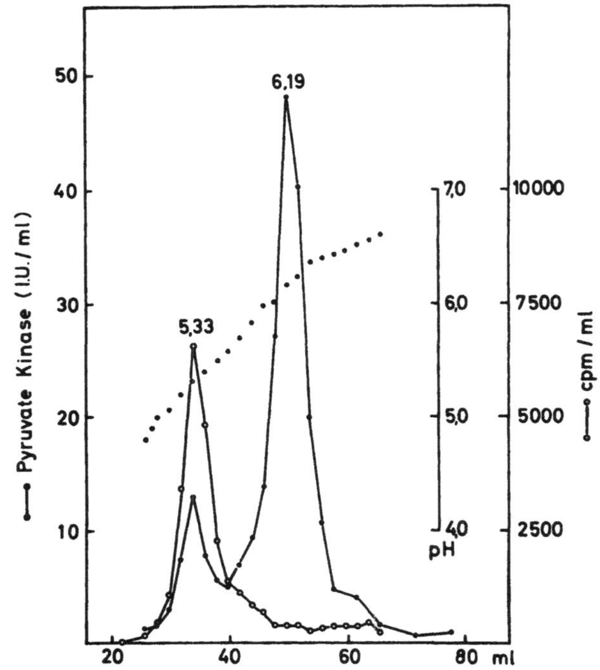

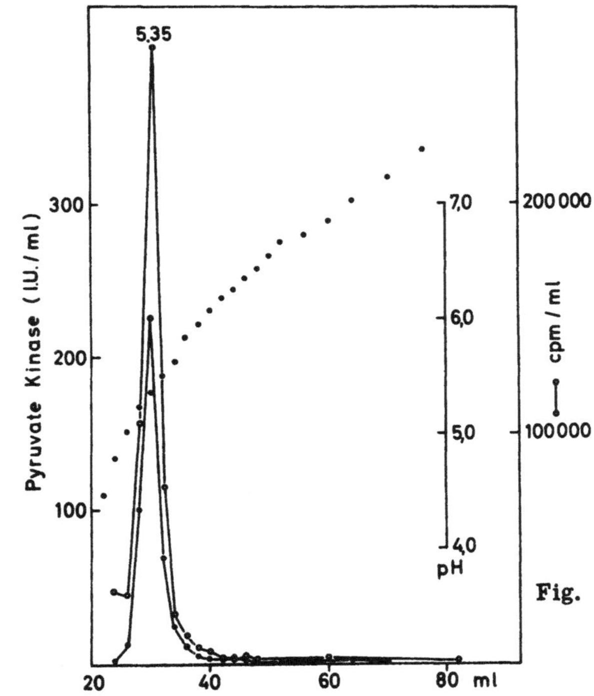

Fig. 1 and 2. For explanation see text

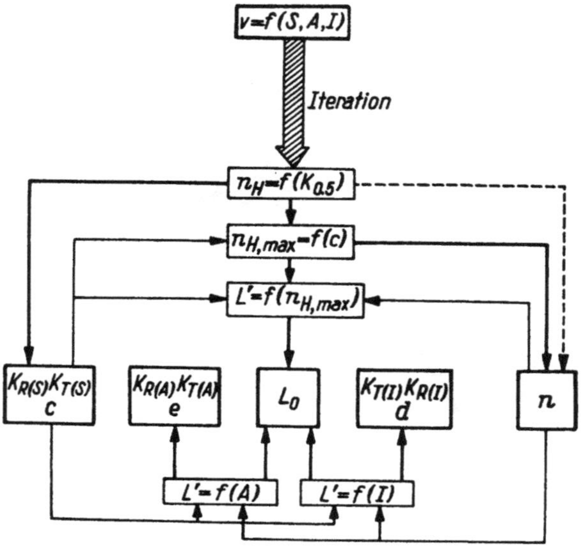

Fig. 3. Logic of kinetic analysis.

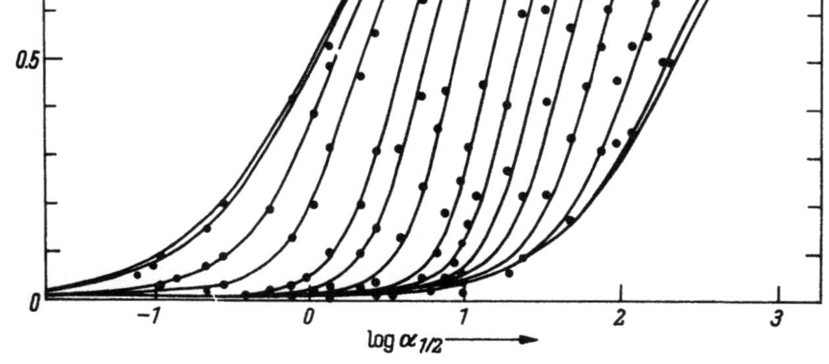

Fig. 4. Y-function.

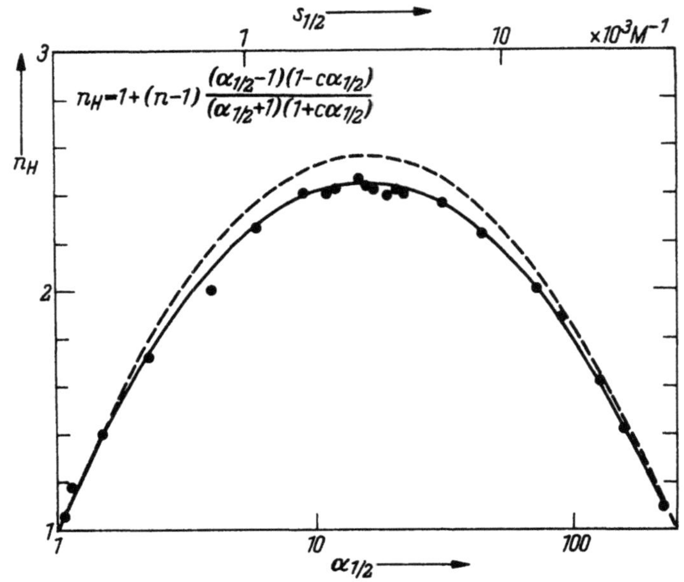

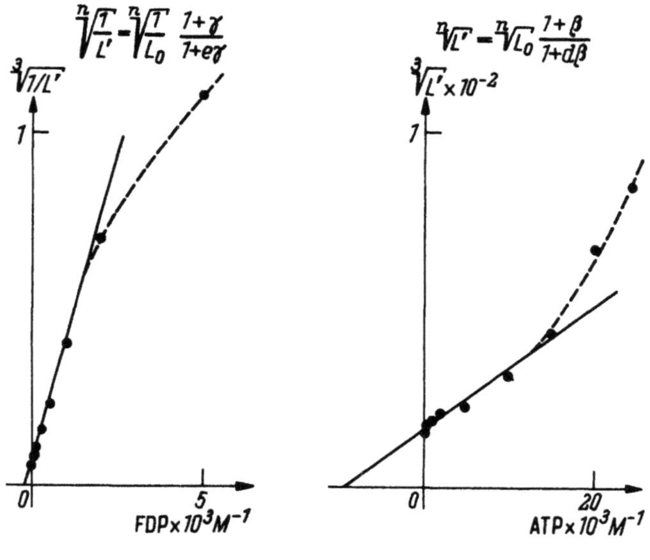

Fig. 5 and 6. For explanation see text.

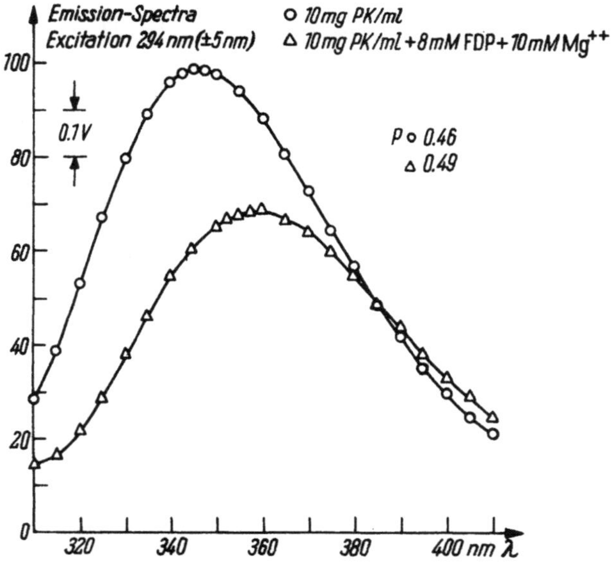

Fig. 7. Fluorescence emission spectra of pyruvat kinase of S. carlsbergensis. p = polarisaison. The incident and scattered light is vertically polarized by means of a filter.

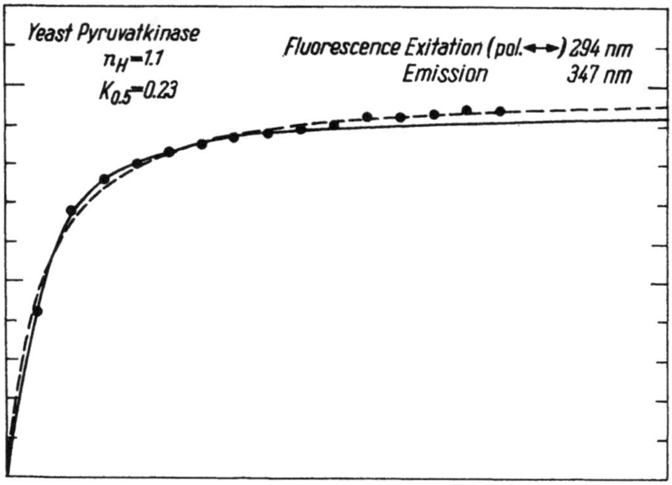

Fig. 8. For explanation see text. Ordinate: relative units, abscissa: FDP concentration in nmolar.

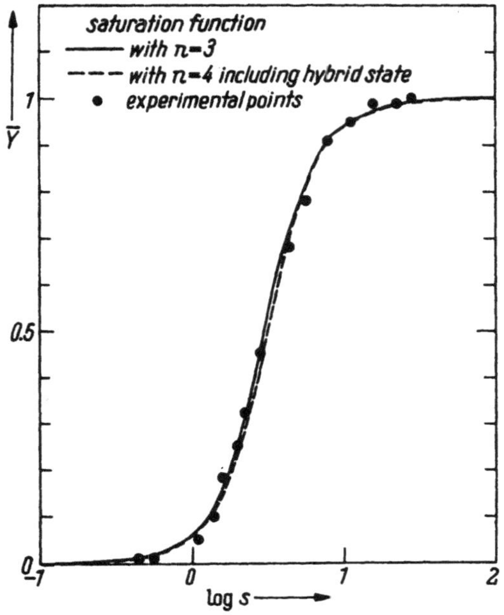

Fig. 9. Computer fit(—and---)of a saturation curve of pyruvate kinase of S. carlsbergensis. For explanation see text.

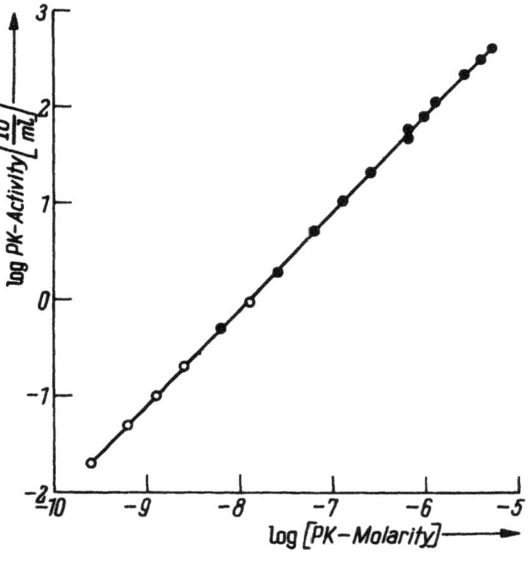

Fig. 10. Relation between activity and molarity of pyruvate kinase of S. carlsbergensis. The o and • -values are obtained with an Eppendorf-photometer and stoppend-flow technique respectively.

Max-Planck-Institut für Ernährungsphysiologie, Dortmund

H. J. WIEKER, B. HESS

pH - FUNCTION AND ALLOSTERIC CONTROL OF YEAST PYRUVATE KINASE

In another presentation (see B. HESS et al.) the effects of some metabolites on the allosteric properties of yeast pyruvate kinase were demonstrated, especially the activation by fructose - 1, 6 -diphosphate (FDP). Here the effects of another ligand which is an essential factor in enzymic reactions will be discussed, the H^+-ion.

The aim of our studies on the pH-dependence of yeast pyruvate kinase reaction was to differentiate the effects of H^+-ions into their effects on the catalytic process, on the binding of substrate and especially on the allosteric properties of this enzyme. Such a differentiation will be possible by determining the pH-functions of the kinetic parameters "maximal velocity" (V_{max}), reflecting catalysis, "half saturation constant" ($K_{0.5}$), correlated to substrate binding, and "interaction coefficient" or "Hill coefficient" (n_H) reflecting cooperativity. In order to obtain these parameters and functions, reaction velocities were measured as a function of the concentration of phosphoenolpyruvate (PEP) between pH 5 and 9.5. The concentrations of the second substrate, ADP, and the essential Mg^{2+}- and K^+-ions were kept constant. In addition, the pH-dependence of the activated enzyme was measured in the presence of FDP. (The corresponding parameters are symbolyzed by the index "(FDP)", e.g. $K_{0.5(FDP)}$.) Thus, a series of twenty substrate saturation curves (v vs. s) were obtained, each consisting of twenty-five v-values. These curves were sigmoidal in the absence and hyperbolic in the presence of FDP.

Curves of these types can be described with a minimum of parameters and without any preinterpretation by models for allosteric enzymes according to

$$v = \frac{V_{max}}{1 + \left(\frac{K_{0.5}}{s}\right)^{n_H}}$$

Using a special computer program (for details see WIEKER et al.; FEBS-Letters 8 (1970), 178) the kinetic parameters n_H, $K_{0.5}$ and V_{max} were determined for each saturation curve with a mean standard deviation of $\pm$ 5 %. They are plotted as n_H (Fig. 1), $pK_{0.5}$ (Fig. 2) and log V_{max} (Fig. 3) as functions of pH. These functions have now to be analyzed in terms of acid dissociation constants (pK_a-values). The commonly used graphical method of Dixon was not applicable to our data, because it leads to very erroneous results if the pK_a-values are not sufficiently seperated from each other. Therefore, on the basis of the Dixon-scheme, we developed a computer program, by which the pK_a-values could be determined by a trial and error-procedure with a variation of $\pm$ 0.05 pH-units. (For a detailed description see H.-J. WIEKER and B. HESS: Biochemistry 10 (1971), 1243).

The most striking difference in the pH-dependencies of the nonactivated and the activated enzyme is found for the "interaction coefficient" (Fig. 1): while $n_{H(FDP)}$ is pH independent with a mean value of $n_{H(FDP)} = 1.05$ n_H increase with pH and reaches above pH 7 a limiting, pH-independent value of $n_H = 2.95$. The curve was obtained by computing a somewhat modified pH-function of a monobasic acid with a $pK_a = 5.35$.

As n_H and $n_{H(FDP)}$, the "half-saturation constants" (Fig. 2) also merge to the same value at low pH. Below pH 7, where n_H is pH-dependent, the curves of $K_{0.5}$ and $K_{0.5(FDP)}$ are divergent, but they run parallel above pH 7, where n_H becomes pH-independent. The curves are computed according to the Dixon-scheme with the pK_a-values given in the figure. The straight lines with zero or $\pm$ one unit slope are constructed with these pK_a-values, and it becomes obvious that - in this case - the graphical method of Dixon would lead to very erroneous results and is not applicable. These results show a striking analogy to the activation by FDP: with increasing concentration of H^+-ions n_H decreases and $K_{0.5}$ and $K_{0.5(FDP)}$ merge to the same value. As a merely qualitative result we can conclude that H^+-ions diminish or even abolish the cooperative behaviour of yeast pyruvate kinase.

In contrast to these figures the pH-dependencies of V_{max} (Fig. 3) are not significantly altered by the allosteric activator FDP. Both curves are not computed and no pK_a-values are determined, for the asymmetries necessitate more complex pH-functions for their computation, but the results would not be interpretable at the moment.

Nevertheless, these curves lead to the following informations on the catalytic process: The alterations of the allosteric properties caused by FDP are not accompanied by alterations at the catalytic site of the enzyme. At least two ionizing groups are involved in the catalytic process, one in its deprotonated form, the other in its protonated form; their pK_a-values can be estimated to be in the region of 5.5 and 8, resp.

In case of an allosteric enzyme, $K_{0.5}$ (Fig. 2) cannot be simply interpreted with respect to the binding process, because it also reflects cooperativity. However, from $n_{H(FDP)} = 1$ it follows, that there is no cooperativity in the presence of FDP, and $K_{0.5\ (FDP)}$ only reflects the binding properties of the activated enzyme. Therefore, we can apply the method of Dixon: according to the mathematical treatment, two pK_a-values, namely $pK_{a1} = 5.45$ and $pK_{a2} = 9.0$, correspond to ionizing groups of the free enzyme, and two pK_a-values, namely $pK_{a3} = 4,8$ and $pK_{a4} = 8.35$, correspond to the enzyme substrate complex. (It can be excluded that pK_{a1} and pK_{a2} refer to dissociation constants of the free substrate PEP, for they are found to be 3.4 - 3.5 (carboxyle) and 6.35 - 6.40 (phosphate).)

Thus, two groups are involved in the binding of PEP. Bouth groups can also ionize in the enzyme substrate complex, and their acidity is increased, for their pK_a-values are decreased from 5.45 to 4.8 and from 9.0 to 8.35.

Furthermore, as n_H becomes pH-independent above pH 7, the pH-dependence of $K_{0.5}$ is not influenced by cooperativity at higher pH. The existence of a group with $pK_{a2} = 9.0$, shifted to $pK_{a4} = 8.4$, indicates that the ionization properties of this group are not influenced by the allosteric activator FDP.

As demonstrated in another communication (see B. HESS), FDP shifts the equilibrium between R- and T-state to the more activ R-state and this shift is accompanied by a decrease of n_H to the limiting value of $n_{H(FDP)} = 1$. Thus, n_H can be regarded as a quantity reflecting such allosteric transitions. Therefore we conclude from the pH-dependence of n_H (Fig. 1) that H^+-ions also shift this equilibrium to the state of higher affinity and that the T-state is converted into the R-state by protonation of a group with a $pK_a = 5.35$:

$$R \xrightleftharpoons{pK_a\ =\ 5.35} T + H^+$$

This conversion decreases the cooperativity as well as that caused by FDP, i.e. H^+-ions must be regarded as allosteric activators of yeast pyruvate kinase. Furthermore, as $n_{H(FDP)}$ is pH-independent and $K_{0.5}$ and $K_{0.5(FDP)}$ merge to the same value at low pH, we can conclude that the enzyme is converted both by FDP and by H^+-ions into states which are identical with respect to the binding of PEP. One may argue that it is not allowed to interpret n_H in such a molecular way. However, the same conclusions are obtained if the analysis of the pH-dependence of n_H is based on the intrinsic parameters determining the "interaction coefficient" in terms of the model of Monod, Wyman and Changeux.

If these conclusions are valid, there must exist a strong correlation between the effects of FDP and H^+-ions. In Fig. 4 negative logarithms of FDP concentrations (p(FDP)) which cause a certain n_H-value are plotted versus those pH-values which cause the same n_H. The straight line obtained in fact indicates a 1:1-correlation between the allosteric effects of FDP and H^+-ions.

After those groups which are essential for enzyme activity are characterized by their dissociation constants, we are now trying to identify the corresponding aminoacid residues by chemical modifications. At the moment we can present only some preliminary results of our studies on the function of thiol groups. Yeast pyruvate kinase contains five cysteine residues per subunit, all being in the reduced thiol form. In the absence of denaturing agents only three of them can be substituted by Ellman' s reagent, and this substitution leads to a total loss of activity.

However, one SH-group exerts an appreciable higher reactivity towards ELLMAN' s reagent, iodoacetate or iodoacetamide than the other two. Thus, by reaction with iodoacetamide a substituted enzyme (CAM-PK) was prepared which contains 0. 8 to 1. 0 carboxamidomethylated thiolgroups per subunit. Although it is not exactly proved now, it can be assumed that this highly reactive SH-group is substituted above all.

Fig. 5 presents the saturation curves of the native and the carboxamidomethylated enzyme obtained in the absence and presence of FDP. As the greatest alteration of the kinetic parameters is only about 15 %, we can colclude that the highly reactive SH-group is not involveld in enzyme activity.

With ELLMAN' s reagent more than one SH-group react. If one plots the fractional inhibition (i) versus the number of substituted SH-groups per subunit (Fig. 6), it can be seen that there is no loss of activity if less than 0. 8 to 1. 0 SH-groups have reacted and that inactivation is complete if 2. 5 SH-groups are substituted. These results demonstrate that one SH-group is situated in or near the active site of yeast pyruvate kinase, and we conclude that the pK_a-value of 8. 4 to 9. 0 corresponds to this SH-group.

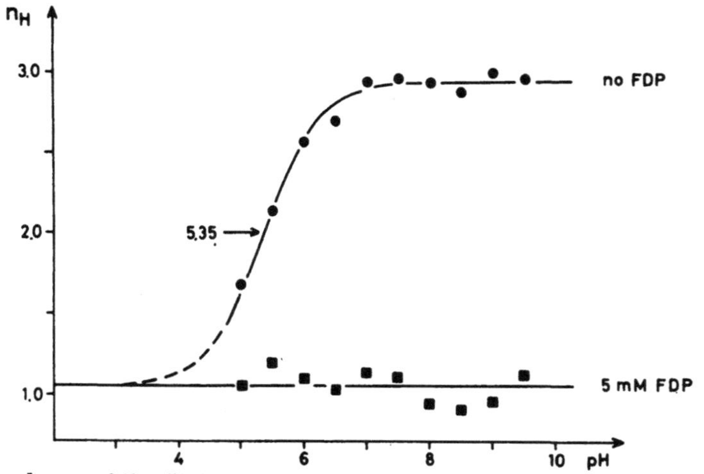

Fig. 1. pH-dependence of the "interaction coefficient" n_H in the absence and presence of fructose -1, 6-diphosphate (FDP).

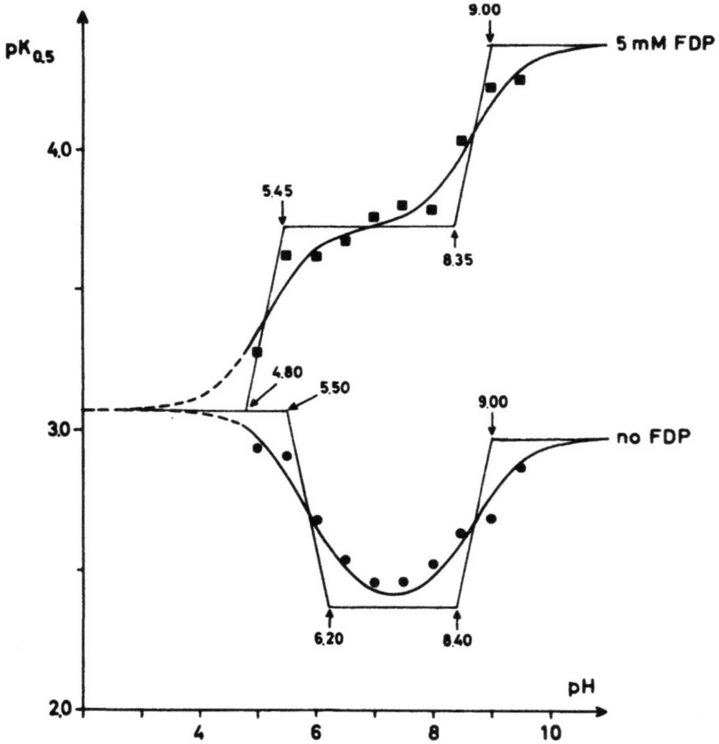

Fig. 2. pH-dependence of the "half saturation constant" $K_{0.5}$ in the absence and presence of fructose -1, 6-diphosphate (FDP).

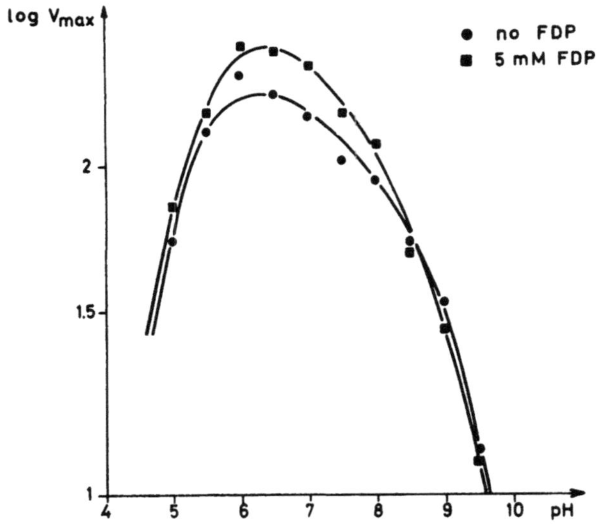

Fig. 3. pH-dependence of the "maximal velocity" V_{max} in the absence and presence of fructose -1, 6- diphosphate (FDP).

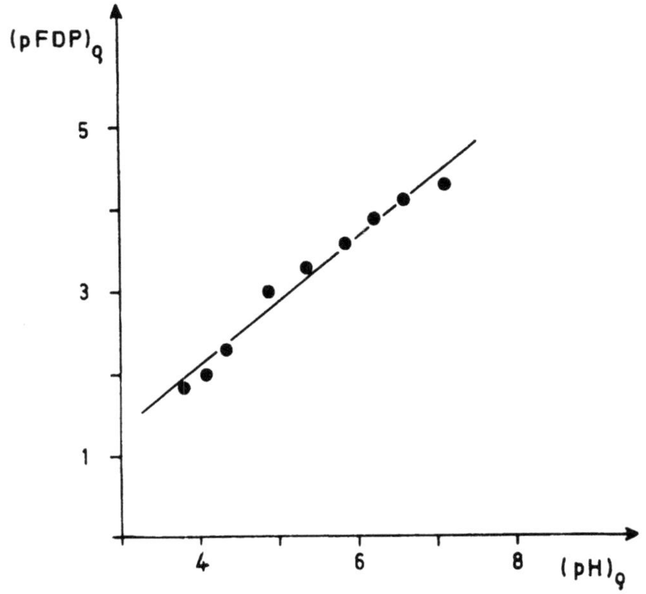

Fig. 4. Correlation between fructose -1, 6- diphosphate (FDP) and pH. For details see text.

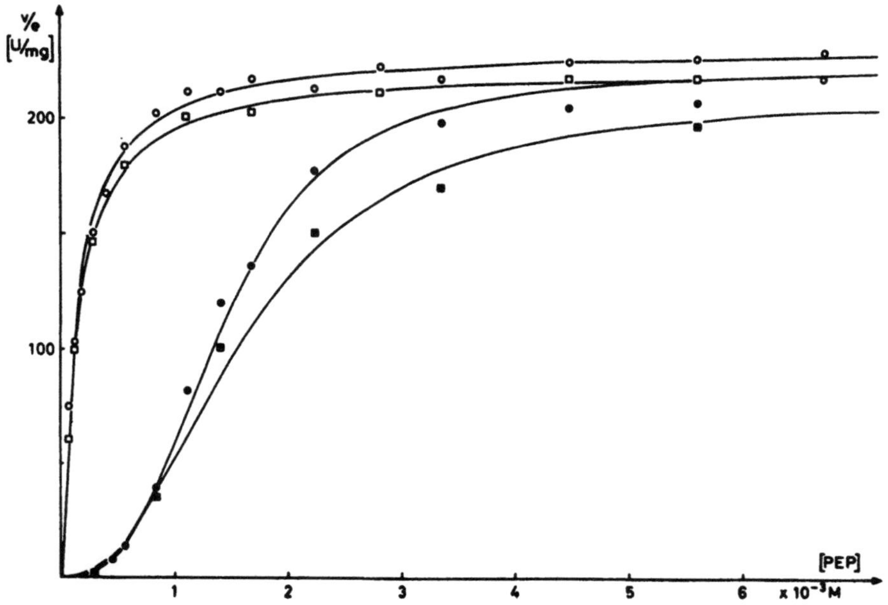

Fig. 5. Substrate saturation curves of native (●, o) and carboxamidomethylated (■ , □) yeast pyruvate kinase in the absence (o, ■) and presence (o, □) of fructose -1, 6- diphosphate ([FDP] = 5 mM).

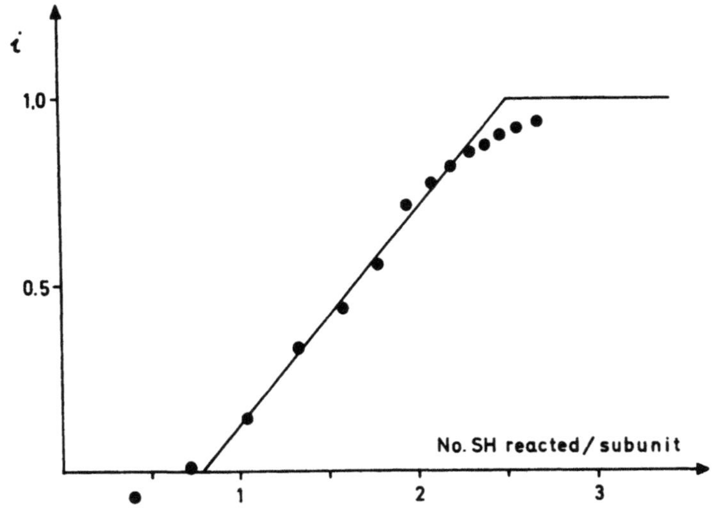

Fig. 6. Fractional inhibition (i) as a function of the number of SH-groups per subunit substituted by 5, 5' -dithiobis (2-nitrobenzoate) (Ellman's reagent).

Department of Clinical Biochemistry, Hospital of Medical Faculty,
University J. E. Purkyne, Brno

V. HULE

THE ALTERATIONS OF ISOENZYMOGRAM OF LACTATE DEHYDROGENASE PRODUCED IN RABBITS BY SYNTHETIC ESTROGEN

The changes following estrogen administration on serum total protein concentrations and serum albumin are reported although the reports are contradictory. MUSA et al. (8) from your observations on 11 healthy female volunteers concluded that the action of ethinyl estradiol alone or in combination with norethynodrel, a synthetic progestional agent, caused changes in the concentration of six of the serum proteins. Serum albumin and total protein were significantly depressed. Two weeks cessation of therapy was insufficient time to allow the protein changes induced by estrogen to return to normal. I could be supposed that also the highly specified proteins of enzymes could change on molecular level. Previously (3) we have shown the influence of single injection of 15 mg diethylstilbestrol dipropionate (Agostilben) on lactate dehydrogenase isoenzymogram of rabbit thrombocytes. In present study we have investigate the influence of repeated injections of Agostilben on the activity of isoenzymes of lactate dehydrogenase (LDH) (L-lactate: nicotinamide-adenine dinucleotide oxidoreductase, E. C. 1. 1. 1. 27) in serum of rabbits.

Materials and methods

20 young adult female rabbits of mixed breeding and weighing 3-4 kg, 10 control test rabbits and 40 normal rabbits were used. Each rabbit was caged individually for at least 1 month at 18°C. 20 rabbits received 15 mg of Agostilben in 1 ml oil daily for 6 days, 10 control test rabbits 1 ml of pure oil, 40 normal rabbits were only 4times venepunctured.

5 ml of blood was taken with a silconed syringe into a siliconed vesell. The tube was stood for about half an hour until the coagulation was completed. Then the tube was centrifuged at 15000 r. p. m. for 20 min. in a high speed centrifuge (Janetzki). The serum was aspirated and immediatly examined.

Electrophoretic separation was carried out applying 800 V to six slides for 55 min. using barbital buffer (I 0.05; pH 8.6) at 10°C. The material was applied in quantities of 0.025 ml. Isoenzymes separated were detected by staining the agarose (Biol. franc. Ind., Gennevilliers, Seine) with nitroblue tetrazolium salt and phenazine methosulphate according to the previously published method (4). The total LDH activity was determined using the reduction of pyruvate to lactate with concomitant oxidation of a molecule of reduced NAD. The reagents were supplied as Boehringer sets, all absorbance measurements were made at 340 nm in a Spectromom 251 spectrophotometer. For the quantitation of the isoenzymes activity the slides were scanned in a Vitatron densitometer. The total protein was determined by a biuret method (7). Paper (Whatmann 1) electrophoresis in barbital buffer at pH 8.6. The samples were electrophoresed at 250 V, 30 mA for 4 hrs. Protein was stained with bromphenol blue, and the strips were analyzed in ERI recording densitometer.

Results

Total protein and paper electrophoresis. The effects of estrogen upon total protein and paper electrophoresis are shown in Tab. 1.

Table 1

PAPER ELEKTROPHORESIS OF SERUM PROTEINS

	Total protein	Albumin	α_1-Globulin	α_2-Globulin	β-Globulin	γ-Globulin
Before estrogen	6,5 ± 0,3	69,8 ± 2,6	5,3 ± 0,5	6,1 ± 1,0	8,2 ± 0,7	10,8 ± 2,1
2 weeks	6,3 ± 0,6	64,9 ± 3,7	5,5 ± 1,1	6,2 ± 1,1	9,6 ± 0,5	13,7 ± 1,5
4 weeks	7,0 ± 0,8	68,1 ± 1,1	5,8 ± 0,6	5,9 ± 0,6	7,0 ± 1,7	10,0 ± 2,7

A depression of the total protein and albumin occurred but was not statistically significant. Changes of beta globulins and gamma globulins were statistically significant in gamma globulins - $0.02 > p > 0.01$.

LDH isoenzymograms

The effects of estrogen upon activity of LDH isoenzymes show Tab. 2. Mean value for LDH isoenzymes expressed as percentage of total activity and standard errors in examined group of 20 rabbits before the injections of Agostilben are compared with values after 2 and 4 weeks. Highly significant elevation of activity

Table 2

ISOENZYMOGRAM OF LACTATE DEHYDROGENASE (LDH) AFTER SYNTHETIC ESTROGEN

	Before estrogen	After estrogen	
		2 weeks	4 weeks
LDH - 1	67, 5 ± 7, 4	52, 5 ± 10, 5	58, 8 ± 14, 5
LDH - 2	25, 5 ± 6, 6	24, 1 ± 6, 6	30, 3 ± 10, 4
LDH - 3	7, 1 ± 3, 0	23, 3 ± 15, 2	9, 6 ± 3, 3
LDH - 4	±	+	+

of LDH - 3 occurred, $0.005 > p > 0.001$. Still 4 weeks did not allow the changes of the LDH isoenzymes induced by estrogen to return to normal. The Fig. 1 shows the slides with LDH isoenzymes, at the top isoenzymogram before estrogen injections, midle after 2 weeks, at the bottom isoenzymogram after 4 weeks. The isoenzymes are called 1 to 5, starting from the anode, according to the conventional nomenclature by WIELAND and PFLEIDERER (9). From left to right LDH-1, LDH-2, LDH-3, LDH-4. LDH-4 activities were rather low. In rabbits we found practically only 4 LDH isoenzymes.

Discussion

MUSA et al. (8) found a significant depression of the total protein in 11 women receiving 20 days ethinyl estradiol, almost entirely as a consequence of a lowering of the serum albumin by the same amount. Slight changes in the four globulin fraction were not statistically significant and did not contribute to changes in the total protein concentration. Ceruloplasmin, corticosteroid binding globulin, thyroxine binding globulin and G-c components were consistently elevated by estrogen treatment. Haptoglobins were decreased by estrogen in all cases.
In rabbits we have also found a lowering of the total protein and albumin but the

changes were not statistically significant. On contrary to MUSA et al. (8) we observed changes also in globulins. Beta and gamma globulins were elevated, the last mentioned with statistical significance. In beta globulins is present the LDH-3 isoenzyme which showed highly significant elevation of activity by estrogen treatment. It is well established that steroid hormones may affect enzymes. GOODFRIEND et al. (2) found that estradiol preferentially increased LDH-5 (M) subunits in uterus of rat and rabbit and they forther showed that this effect was abolished by Actinomycin D which suggested that formation of new enzyme molecules was involved. Estrogen induced changes of activity of LDH in serum of man described COHEN et al. (1) but we could not verify these results (5, 6).

Summary

In serum of female rabbits a shift of activity of isoenzymes of lactate dehydrogenase from anodic isoenzymes towards slow moving isoenzymes by treatment by diethylstilbestrol dipropionate 15 mg daily for 6 days was observed. Statistically significant changes occurred in control test rabbits after injections of pure oil and in normal rabbits repeatedly examined. Hence the increased M subunits may be a hormone-induced one.

A depression of the total protein and albumin occured. The elevation of gamma globulins was statistically significant.

REFERENCES

(1) COHEN, L., J. BLOCK, J. DJORDJEVICH, Proc. Soc. Exp. Biol. Med. 126, (1967), 55

(2) GOODFRIEND, T. L., N.O. KAPLAN, J. Biol. Chem. 239, (1964), 130

(3) HULE, V., Symposium on Thrombosis, Plzen, (1970)

(4) HULE, V., FEBS Symposium, 18, (1970), 329

(5) HULE, V., Clin. Chim. Acta 31, (1971), 298

(6) HULE, V., Acta bioch. Bohemoslovaka 1, in press

(7) LOWRY, O.H., N.J. ROSENBROUGH, A.L. FARR, R.J. RANDALL, J. biol. Chem. 193, (1951), 265

(8) MUSA, B.U., R.P. DOE, U.S. SEAL, J. Clin. Endocr. Metab. 27, (1967), 1463

(9) WIELAND, T., G. PFLEIDERER, Ann. N.Y. Acad. Sci. 94, (1961), 898

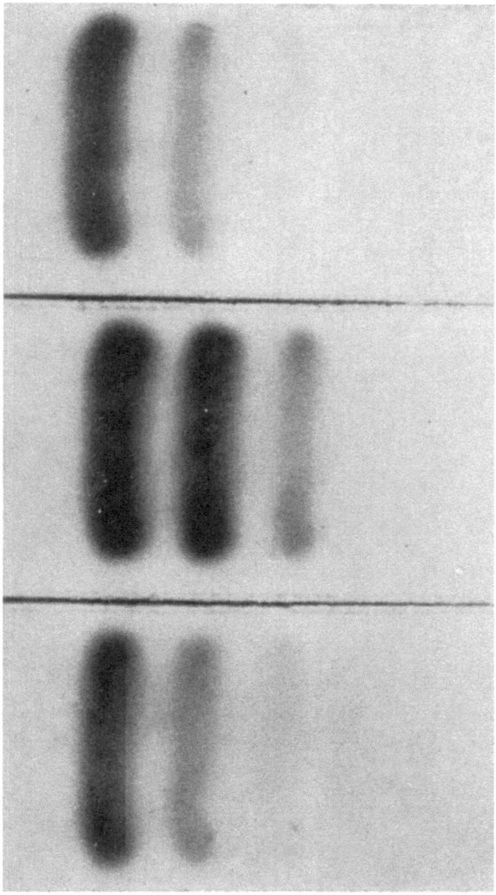

Fig. 1. Isoenzymograms after estrogen injections At the top: LDH isoenzymogram before estrogen injections, midle: after 2 weeks, at the bottom: after 4 weeks.

4. ENZYMATIC SYNTHESIS AND SPLITTING OF RIBONUCLEIC ACID

Institut für Molekulare Genetik der Universität Heidelberg, Heidelberg

EKKEHARD K. F. BAUTZ, JOHN J. DUNN, FRIEDLINDE A. BAUTZ

POSITIVE CONTROL OF RNA SYNTHESIS

The adaptability of micro-organisms to changes in environmental conditions offer a unique system of analysis for gene activation and repression. The control of gene activity was generally believed to occur according to the model of JACOB and MONOD (1) which is based on the regulation of the lac operon in E. coli, in which the product of a regulatory gene binds to an operator site and thereby prevents the DNA dependent RNA polymerase from initiating transcription at a nearby transcription signal, the promoter site. This type of control has been called negative control as it operates through the inhibition of a process which the RNA polymerase is always ready to perform. During the last few years, a second type of control has been suggested, functioning through the activation of the transcribing enzyme by making it recognize a transcription signal, thereby positively controlling the transcription of an operon through altering the affinity of the enzyme to the transcription signal (promoter) (2). The discovery of a protein subunit of RNA polymerase, which is involved in the initiation process of transcription, has shown that RNA polymerase can itself be activated by a protein factor (3). In this paper we discuss some studies on the role of the σ-factor in the initiation process as well as transcriptional controls as they occur after infection of bacteria with DNA phages, which in one case have been shown to involve the production of an entirely new RNA polymerase (4).

RNA synthesis is a complicated process entailing a set of a least three different and complex steps as outlined in table 1. The complete enzyme consists of 5 protein subunits which are all required in the initiation step. Once initiation has taken place chain elongation is continued by the core enzyme with a concomitant loss of the σ-factor. There are also terminating signals which release the finished RNA chain.

An additional protein factor (ρ) has been implicated in this step (5). The initiation reaction can be subdivided again into at least four different steps involving:

Table 1

DNA-DEPENDENT RNA SYNTHESIS

$$(n + 1)\ pppX \xrightarrow[\text{enzyme}]{\text{DNA, } Mg^{++}} pppPu\ (pX)_n + nPPi$$

	Step	Protein involved
1)	Initiation	$\alpha_2\sigma\beta\beta'$ = complete enzyme
2)	Chain elongation	$\alpha_2\beta\ \beta'$ = core enzyme
3)	Chain termination	core enzyme + ρ -factor

1) binding of polymerase to the promoter site
2) partial melting of the promoter site involving a conformational change of the enzyme
3) addition of the first and second nucleoside triphosphates
4) formation of the first phosphodiester bond

Using the drug rifampicin, an inhibitor of RNA polymerase, HARTMANN and colleagues have shown that during incubation of polymerase with template DNA a binary complex forms which is then resistant to the inhibitor (6). We have analyzed how many of such resistant complexes are formed per DNA molecule and we have concluded from our data that every promoter site can allow one polymerase molecule to become rifampicin resistant. Moreover, it was possible to show that rifampicin resistant complexes could only form in the presence of the initiation factor σ. If core enzyme was incubated with template DNA and if σ was then added together with rifampicin, no resistant complexes were found. This suggests, that the σ-factor is either involved in the recognition of the promoter site by the polymerase or that it effects a conformational change of the enzyme at this site (7).

Using the formation of these rifampicin resistant complexes we have tried to estimate for several DNA templates how many RNA chains are produced by measuring the incorporation of either γ-labeled P^{32}-ATP or γ-P^{32}-GTP and have calculated how many molecules of 5' ends of RNA were produced per DNA genome. The data are summarized in table 2. They show that DNA templates such as λ or T7 have very few promoter sites which are recognized by the host RNA polymerase, whereas a larger number of promoters, most of which are promoting the incorporation of ATP as the first nucleotide, are seen in T4 and T5 DNA. These in vitro data agree with the findings obtained by several laboratories through in vivo studies which suggest that λ DNA has 2 - 3 promoters, T7 1 - 2, T4 20 - 25, and the even larger number of promoters found on T5 DNA is in agreement with the results of DNA and RNA hybridization

Table 2

RIFAMPICIN-RESISTANT INCORPORATION OF γ-LABELED NUCLEOSIDE TRIPHOSPHATES INTO RNA CHAINS+)

Labeled PuTP	^{32}pppA				^{32}pppG			
DNA (14μg) Number of genomes μμmole	λ 0.42	T4 0.11	T5 0.16	T7 0.56	λ 0.42	T4 0.11	T5 0.16	T7 0.56
γ-^{32}P incorporated μμmole	0.68	2.0	4.2	0.46	0.33	0.30	0.39	0.40
Number of promoter sites per genome	1.6	18	26	0.8	0.8	2.7	2.4	0.7
Corrected for λ to next integral number	2	22	32	1	1	3	3	1

+) The reaction conditions were as in ref. 7

experiments suggesting that, in contrast to the three other phages, transcription of two groups of T5 genes is under the control of the E. coli σ-factor (8). Infection of bacteria by all these DNA phages, however, results in the transcription of only a fraction of their genes, namely of those genes that are expressed immediately after infection. Transcription of the rest of the viral genome appears to be under the control of a phage specific protein which could either be a new σ-factor, redirecting the core enzyme to recognize another set of promoters, or, as discussed by Zillig (this volume), a modification protein could be produced which would alter the host polymerase to change specificity. A third alternative found recently by CHAMBERLAIN et al. (4) for phage T7 is that the virus produces a new RNA polymerase which specifically recognizes a DNA sequence which is unique to that phage. This new RNA polymerase appears to be much less complicated in structure than the E. coli enzyme, and, therefore, should become an extremely useful tool to study the enzymatic mechanism of RNA synthesis. Such studies should be further facilitated through the availability of T7 mutants producing an impaired RNA polymerase. It should therefore become possible to find variant classes of RNA polymerase with altered structural and functional properties. In this respect, the enzyme of phage T3 appears to be a natural variant of the T7 enzyme, since both enzymes after purification are almost entirely undistinguishable from each other and yet they appear to recognize slightly different promoter sites as each enzyme prefers its own DNA template (table 3) (9).

DNA templates of unrelated phages, such as phage T5 or T4, appear not to be transcribed at all by either T3 or T7 polymerases. This lack of activity is not due to a lack of binding of the enzymes to DNA, but rather due to an inability of the enzymes to initiate transcription. This can be shown by the fact that excess T4 DNA inhibits the template activity of T7 and T3 DNA, and also that T3 polymerase can be purified by affinity chromatography on a T4-DNA cellulose column.

The RNA produced by T7 polymerase is identical with RNA produced in vivo late after phage infection, suggesting that the viral enzyme transcribes those genes which are not read by the host RNA polymerase (9, 10).

Several lines of evidence have suggested that the number of late promoter sites recognized by the phage enzymes is rather small. It also appears that the 5' ends are unique as evidenced by the fact that the only γ-P^{32}-labeled nucleoside triphosphate incorporated into RNA is GTP. We have used the reverse reaction of RNA synthesis, namely the exchange of inorganic pyrophosphate with nucleoside triphosphates as an assay to monitor the synthesis of the first dinucleotide in the RNA chain (11). As only two out of the four nucleoside triphosphates are maximally needed for the formation of a dinucleotide we have determined which combination of 2 substrates gives maximal PPi exchange. The pertinent data are summarized in table 4.

Whereas combination of G+U and A+U give optimal exchange with E. coli enzyme on T3 DNA, T3 enzyme shows optimal activity on T3 DNA with any mixture of

Table 3

TEMPLATE SPECIFICITY OF PHAGE RNA POLYMERASES[1)]

Template DNA	Source of Enzyme mμmoles ^{3}H-UMP incorporated / 0.25 ml T3	 T7
T3	3.50	1.18
T7	0.60	2.0
ΦX 174	0.44	0.45
Calf thymus	0.18	0.35
T5	0.03	0.05
T4	<0.01	<0.01
poly d-AT	<0.01	<0.01

1) T3 and T7 polymerases were assayed for activity under standard assay conditions with the various templates listed. All DNA templates were at a concentration of 60 μg/ml and were highly efficient in supporting the transcription of E. coli RNA polymerase. T5 DNA and ΦX174 DNA were gifts of Drs. H. BUJARD and H. HOFFMANN-BERLING respectively.

nucleoside triphosphates containing G, suggesting that the first two nucleotides incorporated are G-G. On T7 DNA template, no discernible exchange is seen with GTP alone; only a combination of G + A gives a reasonably high activity. This difference may provide the clue to the strongly reduced synthetic activity of T3 enzyme on T7 DNA, suggesting that the enzyme cannot initiate with the same dinucleotide as on T3 DNA. T7 enzyme does not appear to discriminate between T3 and T7 DNA in the PPi exchange assay, whereas a slight preference for T7 DNA is seen in the synthetic reaction (table 3).

The comparison of the template specificities of T3 and T7 RNA polymerases has shown that they differ somewhat in their ability to recognize a certain DNA sequence. It is hoped that further work on the protein structure of the two enzymes will provide crucial information on how promoter sites are recognized by the enzymes. This problem includes the most basic question of how genes are regulated, namely how does a protein recognize a given nucleic acid sequence?

Table 4

EXCHANGE OF ^{32}P-PYROPHOSPHATE WITH NUCLEOSIDE TRIPHOSPHATES BY DIFFERENT RNA POLYMERASES ON T3 and T7 DNA TEMPLATES

Enzyme	E. coli	T3		T7	
DNA	T3	T3	T7	T3	T7
Nucleoside triphosphate added					
GG	0.03	0.59	0.03	0.24	0.24
GA	1.1	0.45	0.31	0.22	0.18
GC	1.3	0.61	0.06	0.27	0.20
GU	2.8	0.55	0.07	0.36	0.30
AU	3.2	0.01	0.01		
AA				0.05	0.05

Experimental details as in ref. 12

REFERENCES

(1) JACOB, F., J. MONOD, J. Mol. Biol. 3, (1961), 318

(2) SHEPPARD, D. E., E. ENGLESBERG, J. Mol. Biol. 25, (1967), 443

(3) BURGESS, R. R., A. A. TRAVERS, J. J. DUNN, E. K. F. BAUTZ, Nature 221, (1965), 43

(4) CHAMBERLAIN, M., J. McGRATH, L. WASKELL, Nature 228, (1970), 227

(5) ROBERTS, J. W., Nature 224, (1969), 1168

(6) LILL, H., U. LILL, A. SIPPEL, G. HARTMANN, Proc. First Lepetit Colloquium on RNA-Polymerase and Transcription, (North-Holland, Amsterdam 1970)

(7) BAUTZ, E. K. F., F. A. BAUTZ, Nature 226, (1970), 1219

(8) BAUTZ, E. K. F., J. J. DUNN, F. A. BAUTZ, D. A. SCHMIDT, A. J. MAZAITIS, Proc. First Lepetit Colloqium on RNA-Polymerase and Transcription, (North-Holland, Amsterdam 1970)

(9) DUNN, J. J., E. K. F. BAUTZ, F. A. BAUTZ, Nature New Biology, 230, (1971), 94

(10) SUMMERS, W.C., R.B. SIEGEL, Nature 228, (1970), 1160

(11) KRAKOW, J.S., E. FRONK, J.Biol.Chem. 244, (1969), 5988

(12) DUNN, J.J., E.K.F. BAUTZ, Biochem.Biophys.Res.Commun., 36, (1969), 925

Max-Planck-Institut für Biochemie, München

W. ZILLIG, A. HEIL, P. PALM, D. RABUSSAY and R. MAILHAMMER

STRUCTURE, FUNCTION AND REGULATION OF DNA DEPENDENT RNA POLYMERASE FROM E. COLI

DNA dependent RNA polymerase from E. coli is a complex enzyme consisting of four different subunits (β', β, α_2) σ with a complex function consisting of primary binding of the enzyme to the template, a number of steps leading to the initiation of RNA synthesis, the synthesis or gene elongation phase and finally termination. The role of subunits in the transcription process may be investigated either by biochemical analysis of single native subunits or better by the construction of mixed reconstituates from one subunit out of a mutant enzyme with a characteristically changed function together with the other subunits from normal enzyme. The changed property will be found in those of all possible reconstituates which contain the subunit involved in the corresponding function.

A method was developed for the preparation of highly pure single subunits, for their renaturation and for their reconstitution into active enzyme (1). The results obtained by both procedures may be summarized as follows:

1. β' is involved in the primary binding of the enzyme to the template.
2. α is absolutely required for activity. It appears that one of the two subunits suffices.
3. β is involved in the binding of both α and initiation factor σ and contains the binding sites for rifampicin and therefore is involved in the initiation step requiring σ and also in the initiation step specifically inhibited by rifampicin. Evidence for this conclusion is also furnished by mixed reconstitution experiments using subunits out off rifampicin resistant and normal enzymes. Experiments of the same type show that the function in gene elongation specifically blocked by streptolydigine also involves subunit β.
4. σ is required for the formation of the heparine- and rifampicin-resistant initiation complex of the enzyme with the promotor. It is suggested that σ is not the element recognizing the promotor but performs an essential task in a limited strand separation in the promotor under the enzyme before synthesis can begin (2).

The changes of the transcription system of E. coli after infection with bacteriaphage T_4 have been analysed as a model for the regulation of transcription (3, 4, 5, 6, 7, 8). A first fast event termed alteration not requiring protein synthesis and induced even by phage ghosts appears to be a potential of the normal cell and leads to a phosphorylation of part of the σ subunit and to the incorporation of one adenosine and two phosphate residues into one of the 42 tryptic peptides of one of the two α subunits. As a result host transcription is abolished whereas T_4 early transcription remains possible.

A second change requiring synthesis and proceeding with the velocity of early gene expression finally leads to a chemical modification of all subunits including α in which the peptide carrying the A-and the two P-residues is further modified. Modification leads to a strong reduction of the affinity of the core enzyme towards factor σ which therefore is lost on purification but has been shown to be present in the crude extract. The physiological meaning of modification is still unknown. Both alteration and modification provide evidence for the conclusion that the specifity of transcription is not only regulated by factors but also by chemical changes of the enzyme itself.

The possible relation of these results to the finding of an in vitro adenylylation of E. coli polymerase (9), of an AMP residue in T_4-modified α (10) and of a change of enzyme activity by a phosphorylation of factor σ using mammalian protein phospho kinase (11) is discussed.

REFERENCES

(1) HEIL, A., W. ZILLIG, FEBS-Letters, 11, (1970), 165

(2) ZILLIG, W., K. ZECHEL, D. RABUSSAY, M. SCHACHNER, V. S. SETHI, P. PALM, A. HEIL, W. SEIFERT, Cold Spring Harbor Symposia Quant. Biol., XXXV (1970), 47

(3) WALTER, G., W. SEIFERT, W. ZILLIG, Biochem. Biophys. Res. Commun. 30, (1968), 240

(4) SEIFERT, W., P. QASBA, G. WALTER, P. PALM, M. SCHACHNER, W. ZILLIG, Europ. J. Biochem. 9, (1969), 319

(5) SCHACHNER, M., W. SEIFERT, Hoppe-Seyler's Z. Physiol. Chem. 352, (1971), 734

(6) SCHACHNER, M., W. SEIFERT, W. ZILLIG, Europ. J. Biochem. submitted

(7) SCHACHNER, M., W. ZILLIG, Europ. J. Biochem., submitted

(8) SEIFERT, W., D. RABUSSAY, W. ZILLIG, FEBS-Letters in press

(9) CHELALA, C.A., L. HIRSCHBEIN, H.N. TORRES, Proc. Natl. Acad. Sci. U.S., 68, (1971), 152

(10) GOFF, C.G., K. WEBER, Cold Spring Harbor Symp. Qant. Biol. 35, (1970), 101

(11) ORLANDO, J.M., S.L.C. WOO, E.M. REIMANN, E.W. DAVIE, Biochemistry 9, (1970), 4807

Institute of Organic Chemistry and Biochemistry, Czechoslovak Academy of Sciences, Prague

K. SEBESTA, K. HORSKA

DNA-DEPENDENT RNA POLYMERASE: PROPERTIES OF THE RIBONUCLEOSIDE TRIPHOSPHATES BINDING SITE AS REVEALED BY INHIBITION STUDIES WITH SUBSTRATE ANALOGUES

Synthetic and naturally occurring analogues of normal substrates are a useful tool for the study of the mechanism of action of DNA-dependent RNA polymerase. Results obtained with some analogues in the study of the binding site for the ribonucleoside triphosphate to be added to the growing chain permitted us to propose a scheme of this binding site which is assumed to be composed of a number of recognition sites providing for the recognition of the correct substrate (Fig. 1). The main purpose of this study was an estimation of the relative specificity of these recognition sites.

Analogues used

Hydroxymethanephosphonate analogue of ATP (1) (Fig. 2) possesses the same structure as ATP but for one hydroxylic group of the terminal phosphate replaced by the hydroxymethyl group. The negative charge of this compound is thus lowered whereas the hydrophilic character of the substituent at γ-phosphorus is preserved.

L-ATP is an enantiomer of ATP. The structure of L-ATP and its implications are discussed in detail in the paper by HORSKA, HOLY and SEBESTA, p. 543

Exotoxin produced by Bacillus thuringiensis (2, 3) (Fig. 3); this compound is composed of an adenosine moiety and a phosphorylated dicarboxylic polyhydroxy-acid linked together by a glucopyranose system.

Requirements of the single recognition sites for the structure of substrates

For the determination of the specificity of the recognition site for the acidic moiety of the substrate molecule the phosphonate analogue of ATP and exotoxin of B. thuringiensis were used. It was found that the first analogue is without effect on RNA polymerase catalyzed reaction, whereas the exotoxin is a competitive inhibitor of this enzyme. Consequently, from this result it is evident that a free γ -phosphate group of ATP is indispensable for the binding. The specificity of this recognition site is comparatively low as the triphosphate group can be replaced by a negatively charged monophosphate group if located at a suitable distance from the adenosine moiety. Moreover, the distance of the phosphate group need not be exact: this follows from the fact that the distance of this group in exotoxin is probably greater than that in ATP (Fig. 4). On the other hand, even ADP in which the polyphosphate residue is shorter than that in ATP, can act as a weak inhibitor of the enzyme (4).

Regarding the recognition site for the sugar moiety it is known that only the 2' -hydroxylic group is essential for the binding. This follows from the ability of 3' -deoxyribonucleoside triphosphate which is a polymerase substrate and can be attached to the end of the growing chain (5). An indication of a strong specificity of the sugar-recognition site was further furnished in experiments with L-ATP (see paper HORSKA, HOLY and SEBESTA, p. 543). In this case the displacement of 2' -hydroxylic group of the ribose moiety merely by the width of the ribose ring has a profound effect: the substance reacts irreversibly with the enzyme-template complex. The base recognition site is responsible for exact copying of the template. Therefore, it should recognize the base moieties of ribonucleoside triphosphates according to the rules of base-pairing. It is not quite clear whether this recognition is brought about by direct base-pairing of the substrate base to the base of the template (direct checking) or by binding to a recognition site on the enzyme (indirect checking) (6) (Fig. 5). In the latter case the base of the template would have to change this site in a way which allows the complementary base as the only one to be recognized. The results obtained with the exotoxin derivatives modified in the base moiety pertain to this question.

The adenosine moiety of exotoxin was either deaminated by nitrous acid to inosine ("inosine-exotoxin") or oxidized by perphtalic acid to adenosine-N_1-oxide ("exotoxin-N-oxide"). The inhibitory effect of inosine-exotoxin upon RNA polymerase was analogous to the effect of the parent substance. The inhibitory effect of exotoxin (containing an adenosine moiety) was counteracted by an increased concentration of ATP, whereas the inhibition caused by inosin-exotoxin could be counteracted only by an increased concentration of GTP (Fig. 6) (7). This is apparently due to the fact that both guanosine and inosine can pair with cytidine. The inhibitory effect of exotoxin-N-oxide was lower than that of exotoxin by 75 % (7). This may be in connection with the base pairing again; the oxidation takes place at the N_1-position which is involved in hydrogen bonding. These results demonstrate the close connection between base-pairing properties and the extent of inhibition and are thus compatible with the direct checking hypothesis. In the indirect checking hypothesis a superficial duplication of the base selection mechanism by means of a protein recognition site would have to be assumed. The problem of base recognition is not yet clear though it is of major importance.

REFERENCES

(1) HOLY, A., to be published

(2) SEBESTA, K., K. HORSKA, J. VANKOVA, Collection Czechoslov. Chem. Commun. 34, (1969), 891

(3) FARKAS, J., K. SEBESTA, K. HORSKA, Z. SAMEK, J. DOLEJS, F. Sorm, Collection Czechoslov. Chem. Commun. 34, (1969), 1118

(4) STERNBACH, H., V. NEUHOFF, W.B. SCHILL, Abstracts of Commun., 6th FEBS Meeting, Madrid (1969), 224

(5) SENTENAC, A., A. RUET, P. FROMAGEOT, Eur. J. Biochem. 5, (1968), 385

(6) CRAMER, F., A. LEZIUS, H. MATZURA, Bull. Soc. Chim. Biol. 50, (1968), 255

(7) SEBESTA, K., K. HORSKA, Biochim. Biophys. Acta 209, (1970), 357

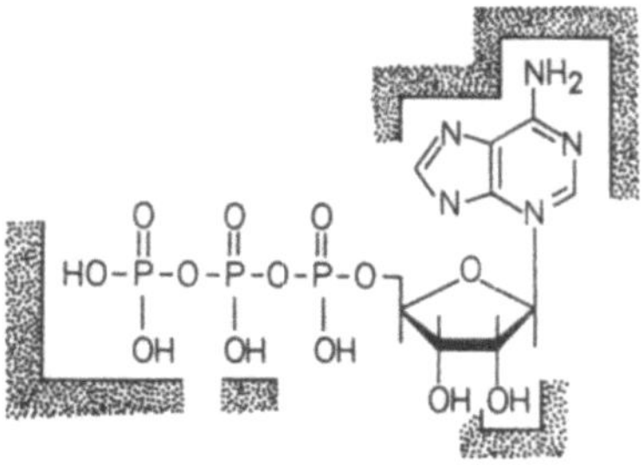

Fig. 1. Scheme of ribonucleoside triphosphate binding site.

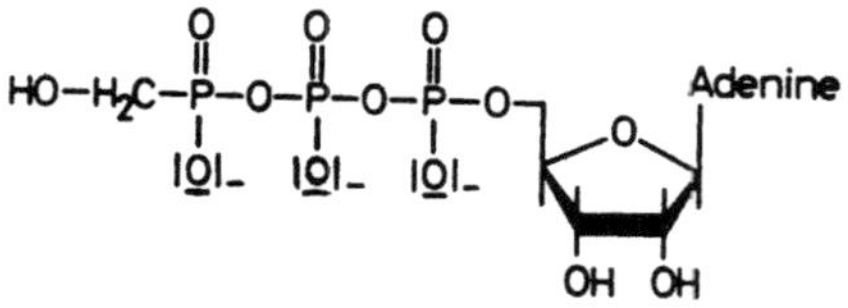

Fig. 2. Structure of hydroxymethanephosphonate analogue of ATP.

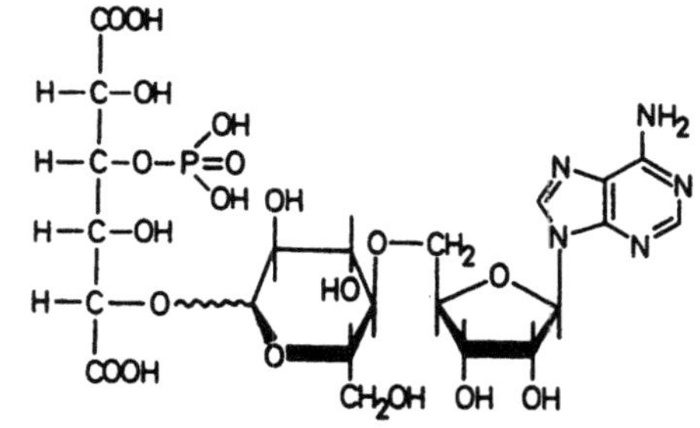

Fig. 3. Structure of B. thuringiensis exotoxin.

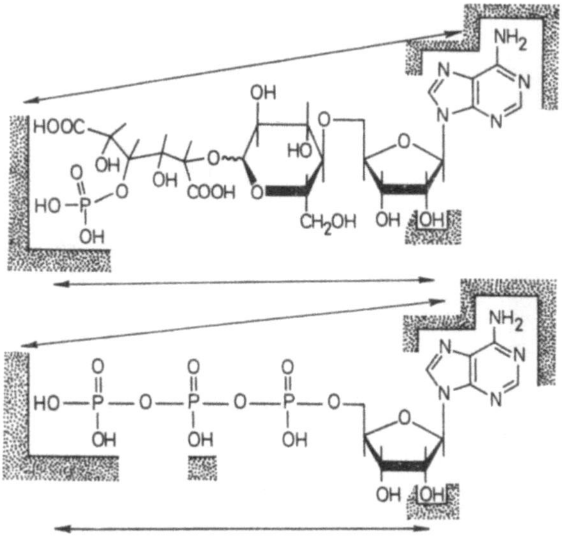

Fig. 4. Proposal of replacement of ATP by exotoxin in ribonucleoside triphosphate binding site.

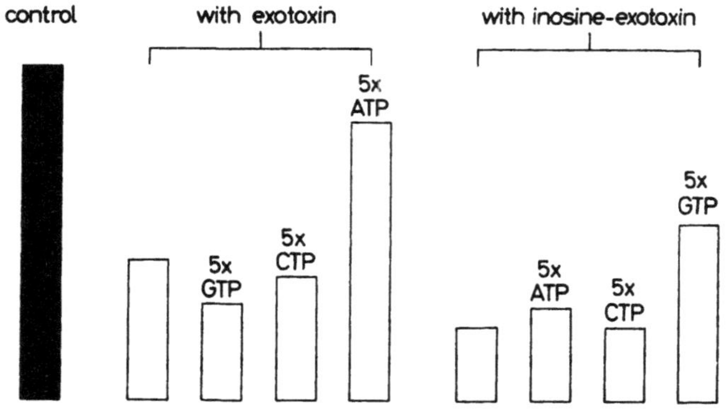

Fig. 5. Scheme of base recognition in the binding site.

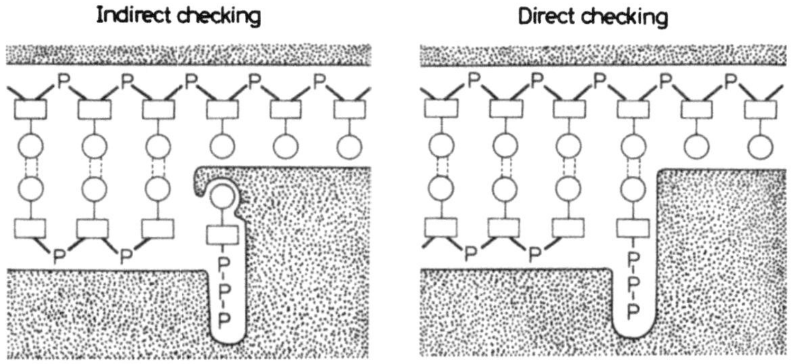

Fig. 6. Inhibition of RNA polymerase activity by exotoxin and inosine-exotoxin and its reversion by increased concentrations of ribonucleoside triphosphates[6]. Columns: enzyme activity.

Institute of Organic Chemistry and Biochemistry, Czechoslovak Academy of Sciences, Prague

K. HORSKA, A. HOLY, K. SEBESTA

DNA-DEPENDENT RNA POLYMERASE INHIBITION BY L-ATP

DNA-dependent RNA polymerase is an enzyme which is being studied at present intensively. Many laboratories are interested in the subunit structure of this enzyme and in the mechanism of its action.

The use of modified substrates represents a classical tool for the study of the enzyme-substrate interaction. In this respect the ribonucleoside triphosphates which are substrates of the RNA polymerase offer broad possibilites. The substrate molecule can be modified either in the base or sugar moiety, or even in the triphosphoric acid residue. Representatives of all these groups have been already investigated as substrates or inhibitors of this enzyme. In this connection we have investigated the behaviour of L-ATP, an enantiomer of ATP which contains L-ribose instead of D-ribose (1) (Fig. 1). As follows from our experiments, the interactions with the heterocyclic base, the triphosphate residue and the sugar residue are necessary for the binding of the substrate (see paper of SEBESTA and HORSKA, p. 537). As to the structure of the sugar moiety it has been shown (2) that the 3' -hydroxyl group is probably not essential for binding (cordycepine triphosphate can be attached by the enzyme to the end of the growing chain). On the other hand, the presence of the 2' -hydroxyl group in natural substrates seems to be important for the complex formation. Therefore, the experiments with L-ATP could afford more information as to the character of the sugar moiety to be recognized by the enzyme.

First experiments performed with L-ATP revealed an inhibitory effect of L-ATP towards DNA-dependent RNA polymerase. The extent of the inhibition depends on the concentration of L-ATP in the reaction mixture (Fig. 2). A 50 % inhibition is produced by a 10fold excess of L-ATP with respect to natural substrate. This inhibition by L-ATP cannot be reverted by the increased concentration of any of the ribonucleoside triphosphates. Apparently there is no competition between the two enantiomers.

In the RNA polymerase catalyzed reaction, ATP has a dual function participating both in initiation and in chain growth. Therefore we tried to find out to what extent L-ATP inhibits in either case. The set of experiments devised for this purpose is shown in Fig. 3. Two series of experiments were performed with the use of L-ATP and labelled triphosphate. In the first series, L-ATP and labelled substrate were present in the reaction mixture from the beginning, whereas in the other series both compounds were added to the reaction mixture after a delay of 3 or 6 minutes, respectively. A set of experiments without L-ATP was performed as a control. It was thus possible to compare the effect of L-ATP on initiation and polymerisation (L-ATP present from the beginning) with its effect mainly on polymerisation (L-ATP added at later time intervals). It follows from Fig. 3 that the inhibition is considerably more pronounced in the presence of the inhibitor from the beginning of the reaction than at later stages (3 or 6 minutes after starting the reaction). This result indicates that in the presence of L-ATP the initiation step is affected more considerably than chain elongation. Most interesting feature of our results is the irreversible character of the inhibition caused by L-ATP. A possible explanation for this fact is a strong binding of the inhibitor to the enzyme-template complex. Therefore, we examined the behaviour of L-ADP, L-AMP and L-adenosine in the same system. L-ADP and L-AMP were found to have the same inhibitory effect as L-ATP, while L-adenosine was without any influence (Fig. 4). One phosphate group in the molecule is therefore sufficient for binding to the enzyme - in contrast to the behaviour of the corresponding derivatives in D-ribo-series. Only D-ADP is known to cause a weak inhibition of a competitive type, whereas D-AMP is without any inhibitory effect. The results obtained with L-derivatives are thus in agreement with the presumption of an irreversible binding.

REFERENCES

(1) HOLY, A., F. SORM, Collection Czechoslov. Chem. Commun. in the press

(2) SENTENAC, A. RUET, P. FROMAGEOT, Eur. J. Biochem. 5, (1968), 385

(3) SEBESTA, K., K. HORSKA, Biochim. Biophys. Acta 209, (1970), 357

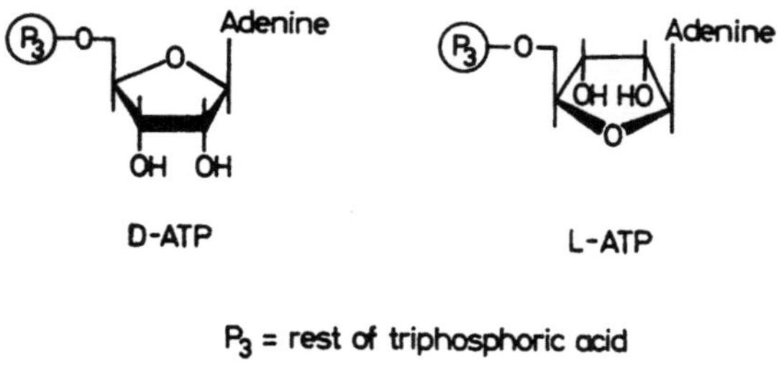

Fig. 1. Structure of ATP enantiomers.

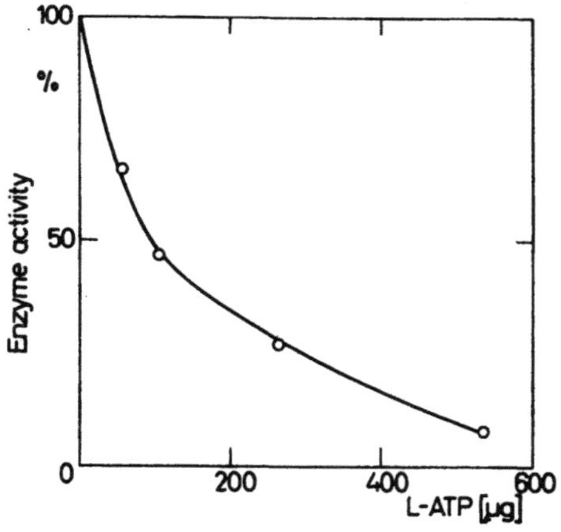

Fig. 2. Inhibition of RNA polymerase activity by L-ATP Composition of the reaction mixture (3).

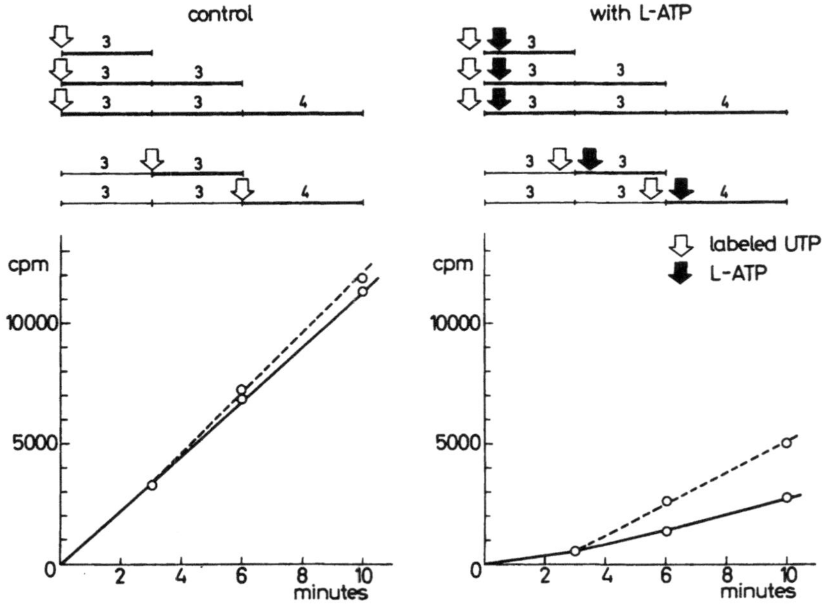

Fig. 3. The effect of L-ATP on initiation and polymerisation Composition of the reaction mixture (3). Addition of labelled UTP (50 000 c.p.m.) and L-ATP (266 μg) is indicated by arrows. Full line: substance(s) added present from the beginning of the reaction. Broken line: substance(s) added after 3 or 6 minutes, respectively; plotted values obtained by summation of values for relevant time intervals.

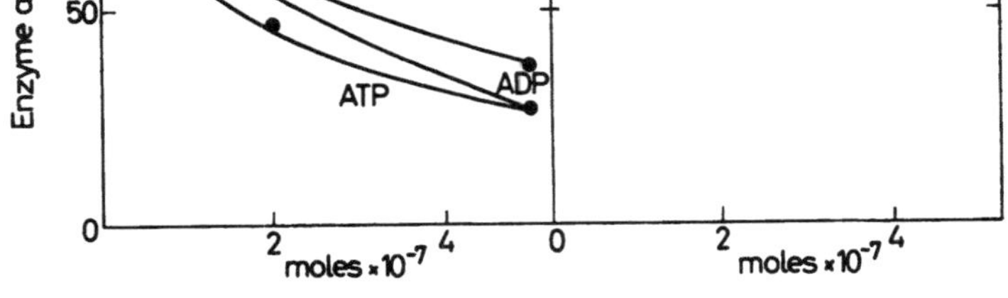

Fig. 4. Inhibition of the RNA polymerase activity by L- and D-adenosine derivatives.

Slovak Academy of Sciences, Biological Institute, Department of Biochemistry of Microorganisms, Bratislava and Institute of Organic Chemistry and Biochemistry, Czechoslovak Academy of Sciences, Prague

J. SIMUTH and A. HOLY

THE STUDY OF SUBSTRATE SPECIFICITY OF POLYNUCLEOTIDE PHOSPHORYLASE (Streptomyces aureofaciens)

Within this work, a polynucleotide phosphorylase isolated from Streptomyces aureofaciens (chlorotetracycline producing microorganism) has been used. It was found previously (1) that the maximum enzyme activity is achieved at the end of the logarithmic phase of cultivation. The isolated enzyme is capable of polymerizing the natural diphosphates UDP, CDP, ADP and IDP, catalyzing the phosphate exchange reaction in the same 5' -diphosphates and catalyzing a phosphorolysis reaction of polyribonucleotides. (2) In the polymerization of CDP, $V_{max} = 1.67 \times 10^{-8}$ mol/min and $K_m = 2.7 \times 10^{-4}$ M (as estimated by the incorporation of ^{3}H -CMP into the polymer). With all the reactions mentioned, the optimal pH value is 8.9. A 200-fold enrichment of the enzyme has been achieved with the use of the usual purification procedures. The final specific activity was 185 μmol [^{3}H]-CMP incorporated into the polymer in 1 hour for 1 mg protein.

To obtain additional data on the substrate specificity of the enzyme mentioned, two types of synthetic analogues of natural ribonucleoside 5' -diphosphates have been investigated with respect to their behaviour to the enzyme: enantiomeric L-ribonucleoside 5' -diphosphates derived from L-ribose (L-ADP, L-UDP) and 6-methyluridine 5' -diphosphate (6-MeUDP). It has been pointed out (3) that on investigation of the behaviour of the enzyme towards enantiomers of natural substrates, it can be decided whether the enzyme-substrate interaction includes an interaction with the nucleoside moiety as a whole (or its heterocyclic base), or whether the asymmetric part of molecule does not enter into it. Moreover, the study of enantiomeric polynucleotide chains. 6-MeUDP was chosen as a model of a nucleotide analogue in which the nucleoside conformation is changed due to the substitution of the base. (3, 4)

Both polymerization reactions and phosphate exchange reactions catalyzed by the enzyme were used for the investigation of the both substrate types mentioned above.

The presence of the L-derivative inhibits the polymerization of a natural substrate (Fig. 1). A similar effect was observed with 6-MeUDP as an inhibitor (Fig. 2). Analogously, the phosphate exchange reaction of natural substrates is inhibited in the presence of both L-derivative (Fig. 3a) and 6-MeUDP (Fig. 3b).

The active role of the L-nucleotide derivative in the reaction with PNPase can be demonstrated by the phosphate exchange in L-UDP which takes place in the presence of ADP (Fig. 4). A similar exchange reaction occurs in the molecule of L-UDP in the presence of the dinucleoside phosphate ApC (Fig. 5). In this connection, it is interesting to emphasize that no primer dependance was observed with natural substrates. (5) There has been no significant phosphate exchange observed with L-ADP alone. On the other hand, after preincubation with the enzyme before addition of radioactive phosphate and the other components of the reaction mixture, L-ADP undergoes a significant phosphate exchange reaction (Table 1). An additional amount of the enzyme enhances this reaction.

Table 1

P_i - EXCHANGE IN L-ADP. THE EFFECT OF PREINCUBATION

L-ADP + PNPase	40' 37°C Preincubation	+ Incub. mixt.	2.2 % exchange
		+ Incub. mixt. + 0.04 units PNPase	4.8 % exchange
control (ADP)			32.6 % exchange

Incubation mixture (0.1 ml): 0.1 M TRIS-HCL pH 7.4, $5 \cdot 10^{-3}$M $MgCl_2$, $5 \cdot 10^{-4}$M EDTA, $5 \cdot 10^{-3}$M K_2HPO_4-^{32}P (as. act. $1 - 5 \cdot 10^6$ ipm/μmol) $2 \cdot 10^{-3}$M L-ADP, enzyme 0.12 e.u.[1]

1 e.u. catalyses a polymerisation of 1 μmol ^3H -CDP/min. under the above conditions.

Under the usual conditions, there is no polymerization of L-UDP or L-ADP. Therefore, the incorporation of 8-[^{3}H]-L-ADP into the polymer in the presence of ADP has been followed in connection with the enhancing effect of a natural substrate in an exchange reaction (see above). In the presence of a 5-fold excess ADP, 8 % of the L-derivative was incorporated into the polymeric fraction (Fig. 6). The polymerisation of L-ADP was also achieved in the presence of CpA as a primer. The yield of the polymerization reaction is dependent on the pH and metal ion as shown in Table 2. The presence of Mn^{++} ions strongly inhibits the polymerization reaction. In contrast to the optimal conditions for natural substrates, the optimal pH value for L-derivatives is 7.4.
The preincubation with enzyme (see above) also improves the polymerization of L-derivatives. The over-all effect, however, is lower if compared with the reaction primed by CpA.

Table 2

POLYMERIZATION OF 8-[^{3}H]-L-ADP IN THE PRESENCE OF CpA

pH	Metal ion	Polymerization, %
8.9	Mg^{++}	7.0
8.9	Mn^{++}	1.7
7.4	Mg^{++}	14.7
7.4	Mn^{++}	1.1

Incubation mixture (0.05 ml): 0.1 M TRIS-HCl, 0.5 mM EDTA, 2.5 mM [^{3}H]-L-ADP (spec. act. 2 . 10^5 imp/µmol), 2.5 mM CpA; enzyme: 0.05 e.u., 5 mM Me^{++}; 24 hours, 37°C.

Separation technique: P. C. 2-propanol-NH_4OH-H_2O = 55:10:35, Whatman No. 1.

It is evident, that L-5'-diphosphates are both inhibitors and substrates of polynucleotide phosphorylase from S. aureofaciens. Therefore, the interaction with the heterocyclic base of the substrate does not seem to be the necessary conditions for the affinity of the enzyme towards the substrate. Nevertheless, its presence does enhance this affinity. The effect of natural primers, or the preincubation with L-derivative (which could afford short L-oligonucleotides) is in agreement with the results of PNPase investigations as published recently by Grunberg-Manago. (6)

REFERENCES

(1) SIMUTH, J., J. ZELINKA, Biológia (Bratislava) 26, (1971), 239

(2) SIMUTH, J., J. ZELINKA, unpublished results.

(3) HOLY, A., "The Use of Modified Nucleotide Derivatives for the Investigation of Some Nucleolytic Enzymes": This book.

(4) KAPULER, A.M., C. MONNY, A.M. MICHELSON, Biochim. Biophys. Acta 217, (1970), 18

(5) KAUFMANN, G., U.Z. LITTAUER, FEBS Letters 4, (1969), 79

(6) THANG, M.N., R.A. HARVEY, M. GRUNBERG-MANAGO, J. Mol. Biol. 53, (1970), 261

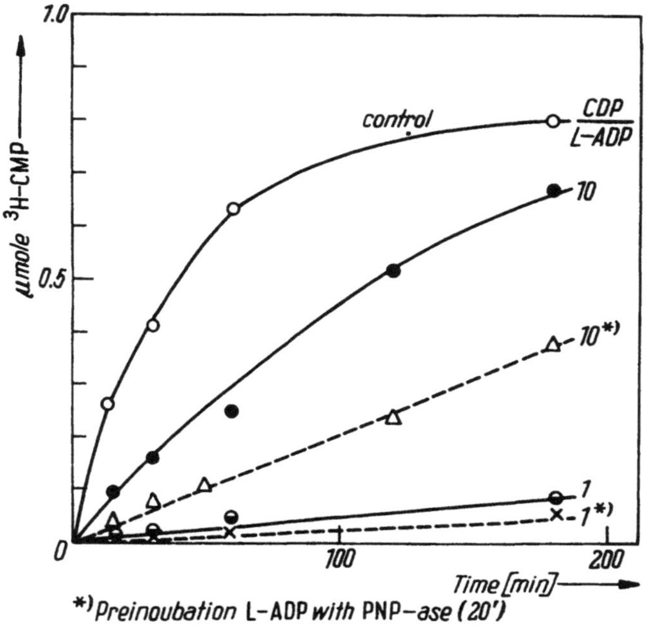

Fig. 1. Effect of L-ADP on the polymerization CDP.
Standard assay for enzymatic polymerization:
The incubation mixture contained 0.1 M Tris HCl buffer (pH 8.9); 5 mM $MgCl_2$; (^3H) CDP (specific activity 1 mC/µmole) 5.10^5 counts/min per ml; (^{14}C)-L-ADP (specific activity 0.5 mC/µmole) 5.10^5 counts/min per ml; 0.5 mM EDTA and 1,2 enzyme unit polynucleotide phosphorylase per ml. The substrate concentration is specified in the corresponding figures. The assay was carried out as reported by SIMUTH, J., SCHEIT, K.H. and GOTSCHALK, E.M., Biochim. Biophys. Acta 204 (1970) 371.

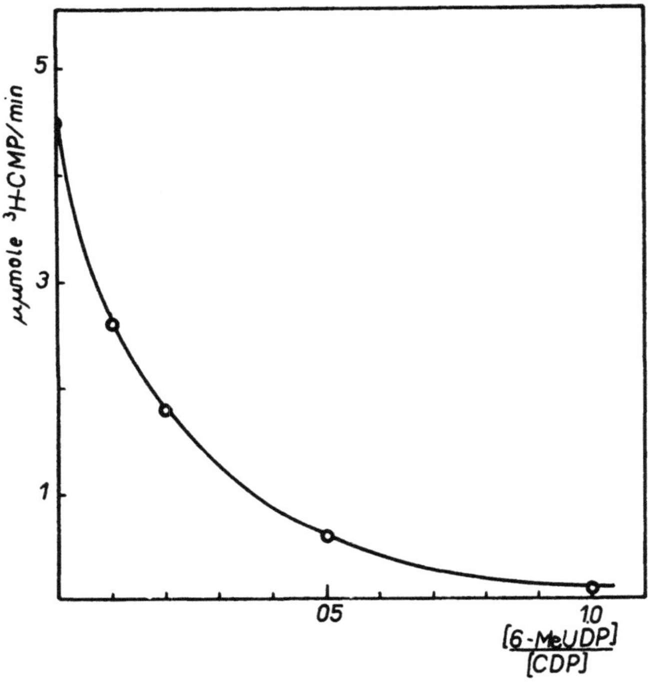

Fig. 2. Effect of 6- MeUDP on the polymerization CDP. Experimental details of the assay procedure are given under Fig. 1.

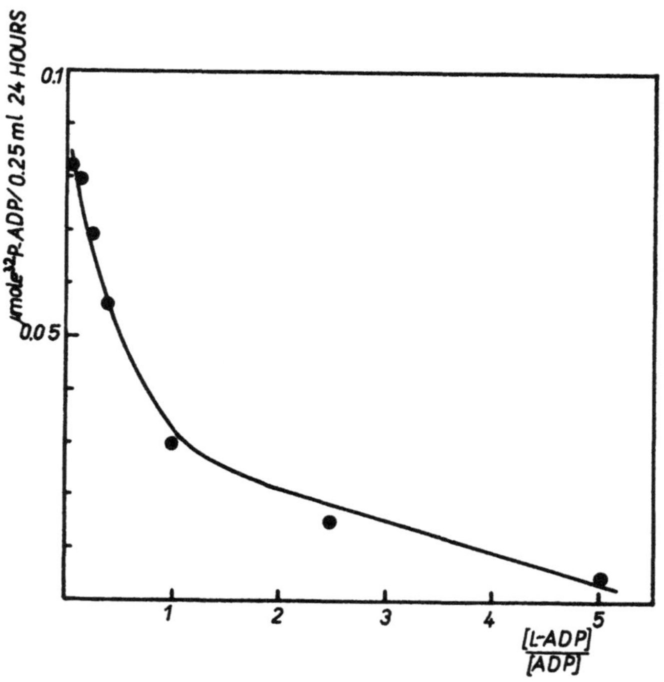

Fig. 3a. Effect of L-ADP upon the P_i exchange in ADP.
Standard assay for P_i exchange reaction:
The incubation mixture contained 0.1M Tris HCl buffer (pH 8.9); 5 mM $MgCl_2$; 0.5 mM EDTA; 5 mM (^{32}P) K_2HPO_4 (approx. 5.105 counts/min) and 2.5 enzyme units polynucleotide phosphorylase in total reaction volume of 0.25 ml. The substrate concentration is specified in the corresponding figures. The assay was carried out as reported by HSIEH, W. I. and BUCHANAN, J. M., Biochemistry 58 (1967) 2468.

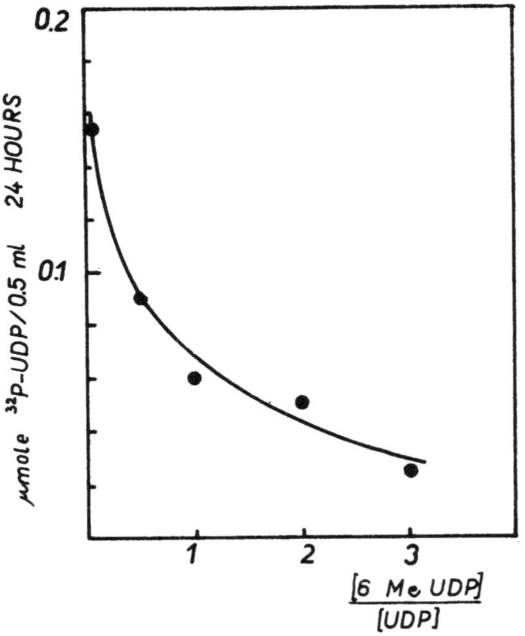

Fig. 3b. Effect of 6-MeUDP upon the P_i exchange in UDP. Experimental details of the assay procedure are given under Fig. 3a.

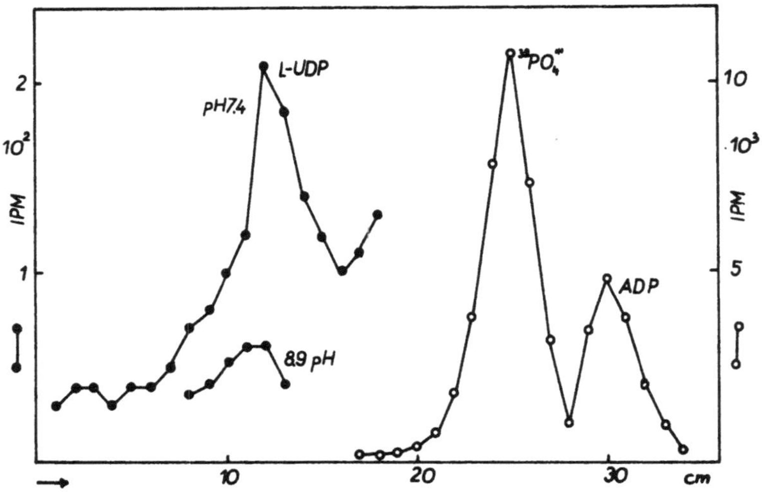

Fig. 4. L-UDP P_i exchange in the presence ADP. (ADP): (L-UDP) = 5 : 1. Experimental details of the assay procedure are given under Fig. 3a. Separation technique: P. C. : isobutyric acid - 1 M NH_4OH - 0.1 M EDTA (50 : 30 : 0.8), WHATMANN No 3 MM.

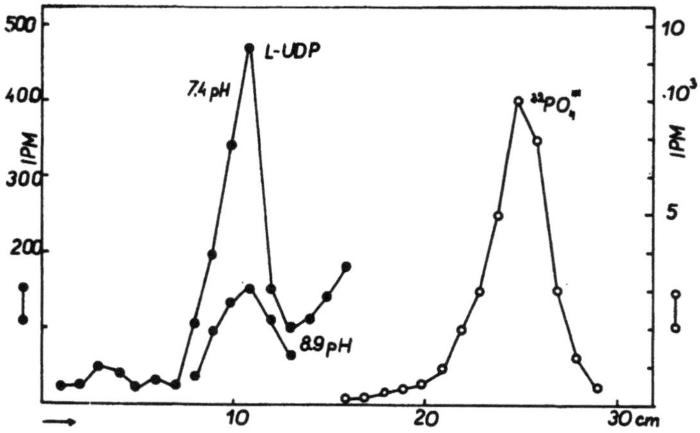

Fig. 5. L-UDP P_i exchange in the presence C_pA. (C_pA) : (L-UDP) = 1 : 1. Experimental details of the assay procedure are given under Fig. 3a and

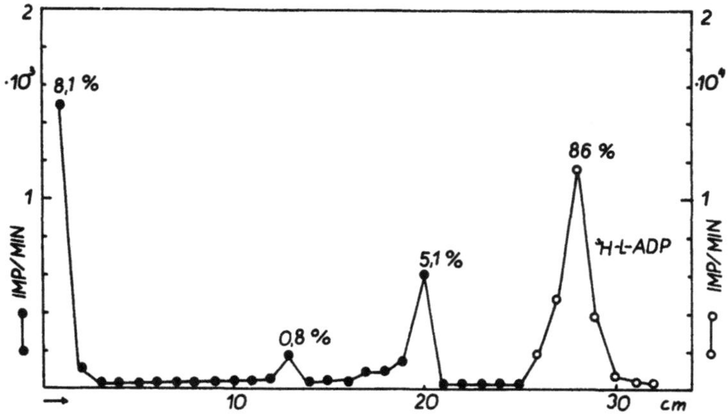

Fig. 6. Copolymerization of L-ADP with ADP. Experimental details of the assay procedure are given under Fig. 8.

5. SPECIAL TOPICS

Institute of Organic Chemistry and Biochemistry, Czechoslovak Academy of Science, Prague

A. HOLY

THE USE OF MODIFIED NUCLEOTIDE DERIVATIVES FOR THE STUDY OF SOME NUCLEOLYTIC ENZYMES

During the investigation of nucleolytic enzymes,[1] the main effort has been concentrated on the degradation studies of natural nucleotide derivatives. These results make possible a directed application of these enzymes to the degradation of nucleic acid derivatives and, thereby, to the solution of primary and secondary structure problems. A more detailed study of the specificity of nucleolytic enzymes could afford important information on the character of forces taking part in an enzyme-substrate interaction and, consequently, stimulate a synthesis of compounds which have a specific inhibitory activity or which are resistant to the action of the corresponding enzymes. Following this line of investigation, we have been mainly concerned with the study of the affinity of modified nucleotide derivatives towards nucleolytic enzymes. A nucleotide derivative is composed of three parts: a heterocyclic base, a sugar moiety (ribofuranose or 2-deoxyribofuranose ring) and a phosphoric acid ester residue. Therefore, there are three poential groups for modification, one or more of which can be applied to the substrate molecule.

The affinity of nucleolytic enzymes towards the enantiomers of natural substrates

The enantiomers of natural nucleotide derivatives are now synthetically accessible (1-4) and posses chemical and physico-chemical properties identical with those of the natural compounds (with the exception of optical rotation), being mirror-images of the latter (Formulae 1, 2).[2] On the basis of their interaction

1) Throughout this review, the term nucleolytic enzymes refers to enzymes which split phosphodiester or phosphomonoester linkages of nucleotide derivatives.

2) In these and following formulae, the phosphoric ester residue is not involved. Its assignment follows from the discussion in text, with the nomenclature as given in formula (1). The symbol B stands for a heterocyclic base residue to sugar moiety by means of N-glycosidic linkage.

with L-nucleotide derivatives, the nucleolytic enzymes can be separated into two groups (Table 1). In an enzyme-substrate complex (ES-complex), the mutual orientation of two three-dimensional bodies is exactly defined by at least three points of their mutual interaction. The two classes of enzymes mentioned differ in their affinity towards L-enantiomers due to the different involvement of the nucleoside moiety (or mostly, the heterocyclic base), in this interaction. Supposing

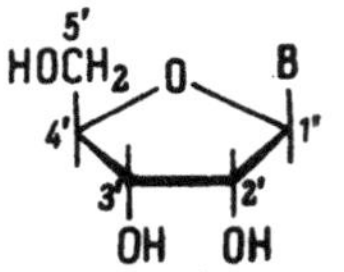

(1)
β-D-Ribonucleoside derivative

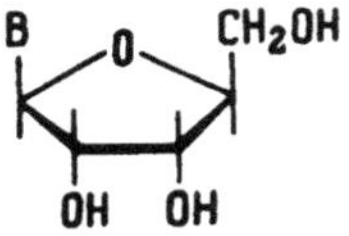

(2)
β-L-Ribonucleoside derivative

that at least one (mostly two) centre is responsible for the phosphate group interaction (or binding), the lack of activity of the enzyme towards an L-enantiomer is due to the necessary interaction with the nucleoside part of the substrate molecule. As a rule, the interaction with the base itself would be necessary. This follows from the case of basespecific enzymes, but may be demonstrated even with some non-specific enzymes. On the other hand, in enzymes which split both D- and L-enantiomeric nucleotide derivatives, the ES-complex does not include that part of the molecule with the chirality changed.

Anomers of natural nucleotides

In close analogy to the conclusions drawn from the preceding paragraph, the affinity of enzymes towards anomeric α-nucleotide derivatives (3) must be different in those cases where the interaction with the heterocyclic base forms a necessary condition for the enzyme activity. Moreover, the mutual orientation of the parts of the nucleotide molecule is drastically changed. Some data on the affinity of α-nucleotide derivatives are summarized in Table 2 (4).

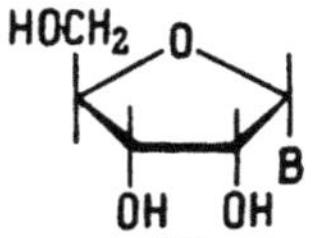

(3)
α-D-Ribonucleoside derivative

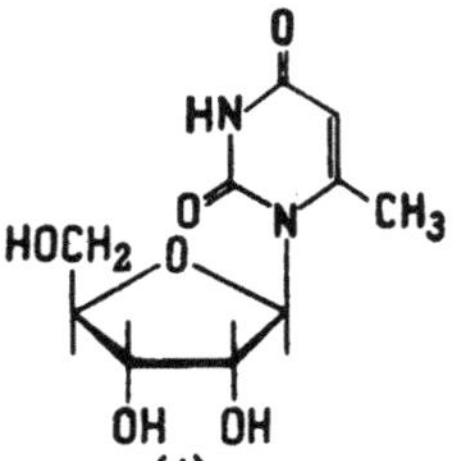

(4)
6-Methyluridine

Table 1

SPECIFICITY OF NUCLEOLYTIC ENZYMES TOWARDS ENANTIOMERIC DERIVATIVES

Enzyme	Splitting
Pancreas ribonuclease	-
Ribonuclease T1	-
Ribonuclease T2	-
Ribonuclease U2	-
Phosphodiesterase (calf spleen)	-
Phosphodiesterase (snake venom)	-
Cyclic phosphodiesterase (beef heart)	-
Cyclic phosphodiesterase (rat kidney)	+
Phosphomonoesterase alkaline (E. coli)	+
Phosphomonoesterase alkaline (calf intestine)	+
Phosphomonoesterase acid (wheat germ)	+
5' -Nucleotidase (snake venom)	-
Polynucleotide phosphorylase	+
RNA-polymerase (DNA-dependent)	-
Nucleotidase (rat liver)	-
Other enzymes	
Phosphotransferase (carrot)	+
Phosphotransferase (prostatic)	+
Adenosine deaminase (calf intestinal)	-
Adenosine deaminase (takadiastase)	+
Adenosine deaminase (snake venom)	-
Adenylate deaminase (bacterial)	-
Pyrophosphorylase (bacterial)	+
Leucyl-TRNA-synthetase	-

Table 2

AFFINITY OF NUCLEOLYTIC ENZYMES TOWARDS α-NUCLEOTIDE DERIVATIVES

Enzymes	Activity
Ribonuclease (pancreas)	-
Ribonuclease T1	-
Ribonuclease T2	-
Phosphodiesterase (snake venom)	-
5' -Nucleotidase (snake venom)	-
Phosphomonoesterase (bacterial)	+
Phosphomonoesterase (calf intestine)	+
Phosphomonoesterase (wheat germ)	+
Polynucleotide phosphorylase	-

The modification of the heterocyclic base

The effects of this modification can be separated into various groups.

A steric effect of substitution can influence the interaction of the heterocyclic base with the enzyme molecule where necessary. A typical representative of this effect is a substitution of the uracil ring in the N^3-position (or pyrimidone-(2) in C_3-position) which results in an inhibition of pancreatic ribonuclease activity (5). The substituent changes the conformation of the nucleoside moiety and, thereupon, the ES-complex formation or a catalytic step cannot take place. This may be demonstrated by the affinity of enzymes towards 6-methyluridine derivatives (4) (Table 3) (6, 7) in which the conformation of the nucleoside residue is syn. Similarly, the C_8-substitution of guanosine causes a change of nucleoside conformation and a lowering of the ribonuclease T1 activity towards modified nucleotides (8). The introduction of a substituent changes the distribution of electron density in the base system. The character of the grouping required for the binding of for the catalytic step can be substantially changed. This effect is demonstrated by the behaviour of some enzymes towards nucleotides of 2-thiouracil (5, 6) 6-thiopurine (6, 9) and 3-(β-D-ribofuranosyl)uracil (7) (Table 4). The change in the π-electron distribution is also caused by the introduction of other heteroatoms into the heterocyclic base system, or by its omission (aza-, or deazaanalogues of

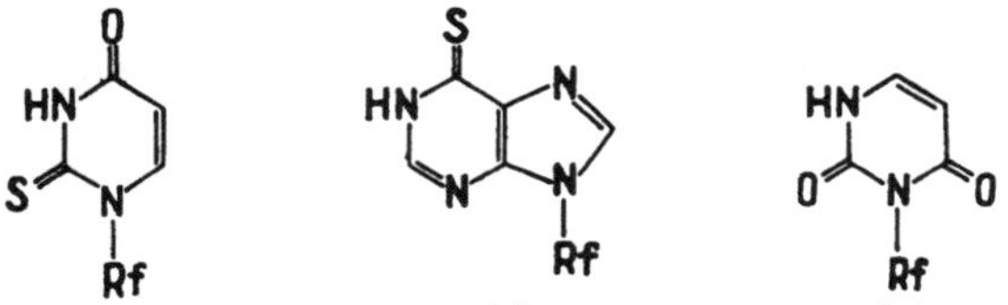

(5) (6) (7)

Rf = β–D–Ribofuranosyl residue

Table 3

AFFINITY OF SOME NUCLEOLYTIC ENZYMES TOWARDS 6-METHYLURIDINE DERIVATIVES

Enzyme	Activity
Ribonuclease (pancreas)	-
Ribonuclease T2	-
Phosphodiesterase (snake venom)	-
5' -Nucleotidase (snake venom)	(+)
Polynucleotide phosphorylase	+
Phosphomonoesterase (bacterial)	+
Phosphomonoesterase (calf intestine)	+

nucleotides). As a result of such changes, the pK-values of dissociable groups are usually shifted, and the interaction of the base with the enzyme molecule can be affected. Such situations are to be expected in those cases where the modification results in an acidic shift of the pK-value (6-azauridine derivatives). The introduction of an electronegatively charged substituent (carboxylic group) into the base results in a change of the mutual orientation of the heterocyclic base and the active site responsible for its binding or interaction. Such deviation is due to the electrostatic interaction of the substituted base with a region of positive charge concentration inside or near to the active centre. As a result of this effect, the affinity of the enzyme to the substituted derivatives is strongly affected in those cases where the interaction with the base is a necessary condition (Table 1). This effect is shown with 5- or 6-carboxyuridine nucleotide derivatives

Table 4

THE AFFINITY OF SOME ENZYMES TOWARDS NUCLEOTIDES OF 2-THIOURIDINE (5), 6-THIOPURINE RIBOSIDE (6) AND 3-(β-D-RIBOFURANOSYL)URACIL (7)

Enzyme	Affinity towards the compounds type (5)	(6)	(7)
Pancreatic ribonuclease	-		-
Ribonuclease T1		-	
Ribonuclease T2	-	-	-
Phosphodiesterase (snake venom)			-
5'-Nucleotidase (snake venom)	+	+	(+)
Phosphomonoesterase (bacterial)	+	+	+
Phosphomonoesterase (calf spleen)	+	+	+

(8a, b) and supported by the comparison with the ester of the carboxy derivative (8c), the character of which is neutral again (Table 5) (10). It is to be pointed out that the substitution of a uracil ring by a carboxy group in position 5- or 6 does not affect the conformation of the nucleoside and, therefore, the resulting effect cannot be due to conformational changes. A nonspecific binding of the charged substrate molecule to any part of the enzyme with a concentration of basic amino acids does not explain the situation; there are enzymes which are evidently not affected by the presence of the acidic substituent.

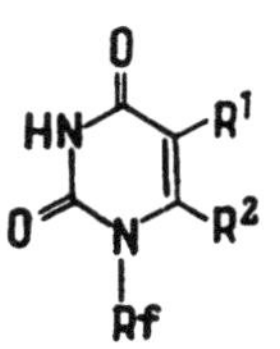

(8)

a R^1 = COOH, R^2 = H
b R^1 = H, R^2 = COOH
c R^1 = $COOC_2H_5$, R^2 = H

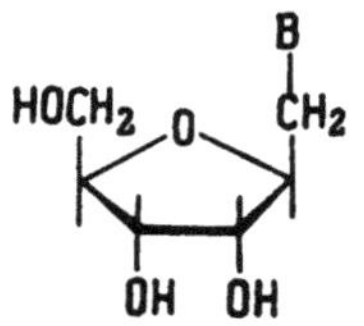

(9)

„1'-homonucleoside"

Table 5

THE AFFINITY OF SOME NUCLEOLYTIC ENZYMES TOWARDS 5- AND 6-CARBOXYURIDINE DERIVATIVES (8) AND L-NUCLEOTIDES (2)

Enzyme	Compound type			
	(8a)	(8b)	(8c)	(2)
Pancreatic ribonuclease	(+)	(+)	+	-
Ribonuclease T2	(+)	(+)	+	-
Phosphodiesterase (snake venom)	-	-	+	-
5' -Nucleotidase (snake venom)	-	-	+	-
Phosphomonoesterase (bacterial)	+	+	+	+
Phosphomonoesterase (calf intestine)	+	+	+	+

By the introduction of a methylene group between the sugar ring system and a heterocyclic base in the 1' -homonucleoside derivatives (9), (11) the freedom of rotation of the base is substantially increased. The different affinity of enzymes towards such nucleotide derivatives could be due to the differences in the area (or space) which is responsible for the interaction with the base in one particular enzyme. In other cases (pancreatic ribonuclease) the lack of activity can be ascribed to the impossibility of the catalytic process due to geometrical factors. The data on various enzymes are summarized in Table 6.

Table 6

THE SPLITTING OF 1' -HOMONUCLEOTIDE DERIVATIVES (9) BY SOME NUCLEOLYTIC ENZYMES

Enzyme	Affinity
Pancreatic ribonuclease	-
Ribonuclease T1	-
Ribonuclease T2	+
5' -Nucleotidase (snake venom)	-
Phosphomonoesterase (bacterial)	+
Phosphomonoesterase (calf spleen)	+

The alteration of the sugar moiety

The changes in the sugar moieties of nucleotide derivatives include a variety of transformations on the aldopentafuranose ring system. The mutual orientation of hydroxylic functions or their respective position to the heterocyclic base can be altered, or some of the hydroxyl groups may be ommited. The possible transformation of β-D-ribofuranosides is shown in formulae (10) - (13). With the phosphate group bound to the hydroxylic function at the altered position, the change is usually too drastic not to affect the enzyme activity. A more specific change can be produced if the conformational change takes place on carbon atoms adjancent to that bearing a phosphate ester function. As a typical example, the properties of α-L-lyxofuranosyl derivatives (14) can be discussed. The nucleotides ıcludes the transformation of the ribofuranose ring to a ribopyranose system.

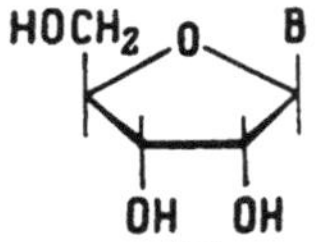

(10)
β-D-Ribofuranoside

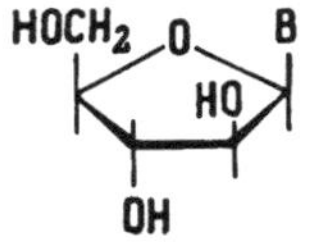

(11)
β-D-Arabinofuranoside

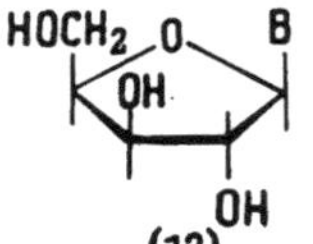

(12)
β-D-Xylofuranoside

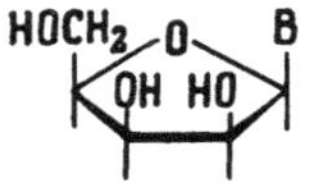

(13)
β-D-Lyxofuranoside

with a phosphate ester bound to the 2' or 3' position differ from those in a β-D-ribofuranose series (10) only in the configuration at the $C_{4'}$-atom. Evidently, ribonucleases A, T1 and T2 (12, 13) are not affected by this change, whereas with ribonuclease U2 an accelerating effect has been observed (14). On the other hand, the enantiomeric derivatives (15) are resistant to those enzymes which do not split the L-ribonucleotide derivatives (Table 1).

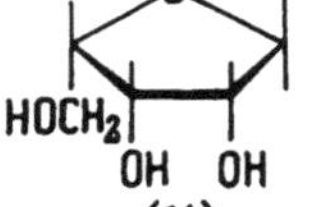

(14)
α-L-Lyxofuranoside

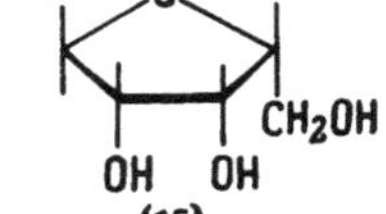

(15)
α-D-Lyxofuranoside

In other cases, the proper orientation of particular hydroxylic functions towards the phosphate group and towards the active site seem to be a necessary condition for the ES-complex formation. Such a situation is demonstrated by snake venom 5'-nucleotidase which specifically hydrolyses the phosphomonoester linkage of 5'-nucleotides in the ribo- and 2-deoxyribo-series. This enzyme requires the presence of a hydroxylic group in the 3'-position with the configuration ribo- (trans to the phosphate group); the 5'-nucleotides with the configuration xylo- (12) and lyxo- (13), (14) are resistant to the enzyme action. As it will be shown later, the interaction with the hydroxylic group represents one of three necessary points of substrate fixation to the enzyme molecule (the heterocyclic base and phosphorus atom being the remaining two). Another modification of the sugar ring includes the transformation of the ribofuranose ring to a ribopyranose system. In this process, the mutual orientation of the hydroxylic function changes; all three of the hydroxylic functions are now cis-orientated and posses the same character as a secondary hydroxylic group. This brings about a greater lability

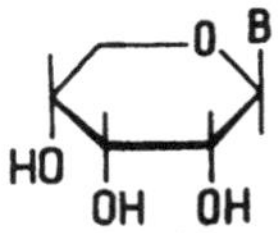

(16)

β-D-Ribopyranoside

of the phosphoric acid ester bonds linked to these hydroxyl groups to non-enzymatic hydrolysis. In the β-D-ribopyranosides (16), the mutual orientation of the heterocyclic base and the hydroxylic groups is also changed, due to the different conformation of ribofuranose and ribopyranose rings. Consequently, with enzymes in which both species play a role at the ES-complex formation, the derivatives of type (16) cannot interact properly. As a matter of fact, the 2', 3'-cyclic phosphates of compounds (16) are resistant towards ribonucleases A and T2 (15). This field has not been sufficiently investigated yet, owing to experimental difficulties in gaining the compounds needed.

The great variety of other modifications of the sugar system include substitution, introduction of other assymetric centres, and other variations. Not much work has been published on these topics in the behaviour of the respective nucleotide derivatives. Some special problems of this type (as the studies on UDP-sugar analogues) are out of scope of this review. Nevertheless, new interesting aspects may be well uncovered by the study of such compounds.

The modification of the phosphoric acid ester residue in nucleotides
The investigation of the topic has been comparatively neglected, although, on its application, some interesting conclusions of enzyme requirements can be drawn. The substantial property of a phosphoric acid ester residue is its electronegative charge (or charges). Therefore, its modification is limited to such changes which save at least one of negative charges. The first possibility involved is a transformation to nucleoside phosphonate derivatives (17), (18) containing a carbon-phosphorus linkage attached directly to the sugar moiety of the nucleoside. Such linkages are quite resistant to enzyme action (16). These derivatives, however, are interesting as inhibitors, being bound to enzymes in a manner similar to the natural nucleotides (17). The phosphodiester analogues containing one phosphonate bond are comparatively resistant to the action of snake venom phosphodiesterase, (18) evidently due to the shortening of the linkage (the corresponding homophosphonates are split by this enzyme (19). The resistance of such compounds against ribonucleases is evident.
The modification leading to derivatives with a phosphorus acid bound to the nucleoside by an ester linkage leaves only one electronegative charge; therefore, such compounds can be treated as analogues of mononucleotides. The exclusion is represented by the thio analogues studied exhaustively by Eckstein at el. (20, 22).

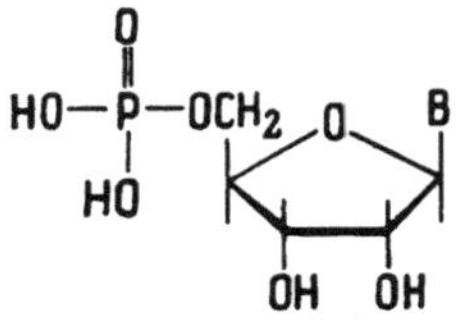

(17)

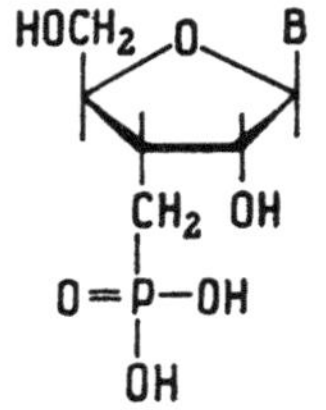

(18)

On diminishing the negative charge, the activity of some nonspecific phosphomonoesterases drops. Therefore, these enzymes seem to require a double binding of the phosphate group. An exceptional situation was observed with snake venom 5' -nucleotidase which splits nucleoside 5' -phosphites (19b) very well (23). Consequently, this enzyme requires one electro-negative charge only. With the same enzyme, no activity has been found with alkane phosphonates (24), 5' -nucleotide alkyl esters or oligonucleotides. Therefore, the reason might be due either to the steric or the hydrophobic character of the substituent at the phosphorus atom (neither of which are important with the phosphite derivate (19b)) (Table 7). It has been found that the hydroxymethane phosphonate (19c) is a good substrate for this enzyme (25), whereas the hydroxyethane phosphonate (19c) or 2-hydroxyethyl phosphate (19d) are resistant to the enzyme action (26). Therefore, the presence of a hydrophilic group in the nearest neighbourhood of

Table 7

THE EFFECT OF MODIFICATION OF URIDINE 5'-PHOSPHATE ANALOGUES (19, B = URACIL) TOWARDS SNAKE VENOM 5'-NUCLEOTIDASE

R	Affinity	k (rel. to UMP)
OH	+	1.00
H	+	0.49
$HOCH_2$	+	0.39
$HOCH_2CH_2$	-	-
CH_3	(+)	>0.01
$HOCH_2CH_2O$	-	-
CH_3O	-	-

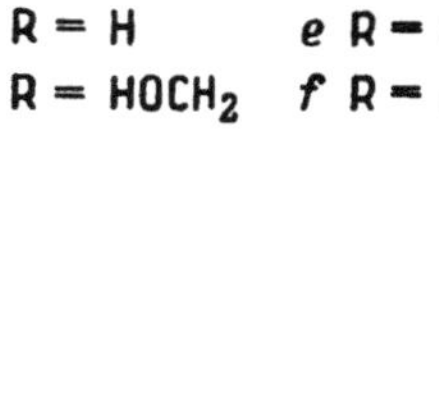

(19)

a R = OH *d* R = $HOCH_2CH_2$
b R = H *e* R = CH_3
c R = $HOCH_2$ *f* R = $HOCH_2CH_2O$

the phosphorus atom is necessary to accomplish the binding. The requirement of the hydrophilic nature of the region surrounding the active site in this enzyme also follows from its specificity for derivatives with 3' -hydroxylic group in a ribo configuration.

In this review, only some typical effects of nucleotide modification have been summarized, without mentioning the great variaty of special cases. On the basis of the understanding of substrate requirements of the enzymes of nucleic acid metabolism, a synthesis of specific inhibitors can be expected in near future. Thus, the theoretical value of our present temptations can turn in a broad applicability in molecular biology.

REFERENCES

(1) HOLY, A., F. SORM, Collection Czechoslov. Chem. Commun. 34, (1969), 3383

(2) HOLY, A., F. SORM, Collection Czechoslov. Chem. Commun., in press

(3) HOLY, A., Tetrahedron Letters (1971), 189

(4) HOLY, A., unpublished results

(5) GASSEN, H.G., H. WITZEL, European J. Biochem. 1, (1967), 36

(6) HOLY, A., R.W. BALD, Collection Czechoslov. Chem. Commun., in press

(7) HOLY, A., R.W. BALD, F. SORM, Collection Czechoslov. Chem. Commun. in press

(8) PONGS, O., Dissertation. Marburg 1969

(9) HOLY, A., Collection Czechoslov. Chem. Commun. 33, (1968), 2245

(10) HOLY, A., Collection Czechoslov. Chem. Commun., in press

(11) HOLY, A., Collection Czechoslov. Chem. Commun. 35, (1970), 81

(12) HOLY, A., F. SORM, Biochim. Biophys. Acta 161, (1968), 264

(13) HOLY, A., F. SORM, Collection Czechoslov. Chem. Commun. 34, (1969), 3523

(14) HOLY, A., F. EGAMI, in preparation

(15) HOLY, A., Collection Czechoslov. Chem. Commun. 34, (1969), 3510

(16) HOLY, A., Tetrahedron Letters (1967), 881

(17) YENGOYAN, L., D.H. RAMMLER, Biochem. 5, (1966), 3629

(18) JONES, G.H., J.G. MOFFATT, J. Am. Chem. Soc. 90, (1968), 5337

(19) JONES, G.H., H.P. ALBRECHT, N.P. DAMODARAN, J.G. MOFFATT, J. Am. Chem. Soc. 92, (1970), 5510

(20) ECKSTEIN, F., Tetrahedron Letters (1967), 1157

(21) ECKSTEIN, F., J. Am. Chem. Soc. 92, (1970), 4718

(22) ECKSTEIN, F., H. GINDL, European J. Biochem. 13, (1970), 558

(23) KUCEROVA, Z., J. SKODA, A. HOLY, F. SORM, Collection Czechoslovak Chem. Commun. 33, (1968), 4350

(24) HOLY, A., Collection Czechoslov. Chem. Commun. 32, (1967), 3713

(25) HOLY, A., N. D. HONG, Collection Czechoslov. Chem. Commun. 36, (1967), 3713

(26) HOLY, A., N. D. HONG, in preparation

Biochemisches Institut der Universität Münster

H. WITZEL

CATALYTIC PRINCIPLES USED IN THE REACTIONS OF RNase A, T_1 AND T_2 WITH SUBSTRATES

Studies on enzyme mechanisms should demonstrate how enzymes lower the activation energy for a definite reaction. We studied this problem on a relatively simple reaction - the hydrolysis of the 3' -5' -phosphodiester bond of ribonucleic acids, which is catalysed by different ribonucleases. Our results permit us now to compare the prinziples used by the pancreatic RNase A, by the RNase T_1 and the RNase T_2, both isolated from aspergillus orycae (1).

The overall reaction from the 3' -diester (I) to the 3' -monoester (III) passes a 2', 3' -cyclic diester (II), an intermediate which can be isolated.

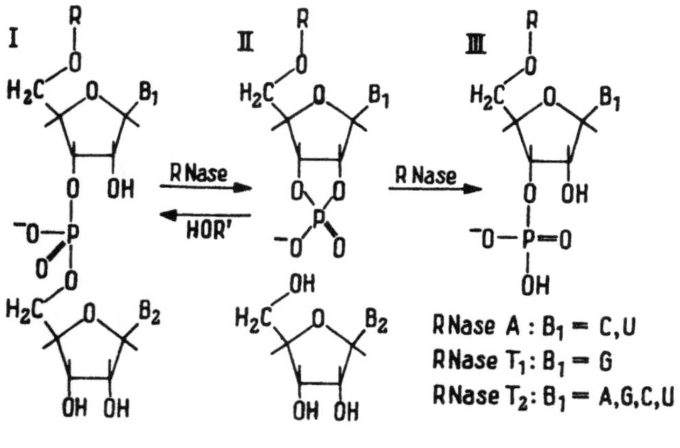

This indicates that all three enzymes catalyse in a first step a transesterification in which a 2' 0-P bond is formed while the 5' 0-P bond is split. In a second step the just formed 2' 0-P bond is split again, while a water molecule enters the phosphate group. The result is a 3' -monoester.

Instead of a water molecule an alcohol molecule can react with II yielding a new 3' -diester. In presence of high concentrations of nucleosides the original 3'. 5' -diesters can be obtained.

It should be justified to assume the same mechanism and the same type of activation for HOH and HOR, when both react as nucleophiles. Therefore we can conclude that the second step is exactly the reverse of the first step and that analogous transititition states are passed in both steps.

The three enzymes, however, differ in their specificity. RNase A splits only, when B_1 is a pyrimidine base, the T_1 enzyme requires a guanine base and the T_2 enzyme splits all internucleotide bonds. This indicates that the catalytic process is complex and contains elements in the activation energy ΔE which depend on a specific function of the base, or better said, the base has functions which contribute to the lowering of the overall ΔE.

Without any catalysis at pH 7 the process should follow an S_N2 type mechanism for both steps. The "in line" variant (IVa, b) is expected to need a higher activation energy than the "adjacent" variant (Va, b).

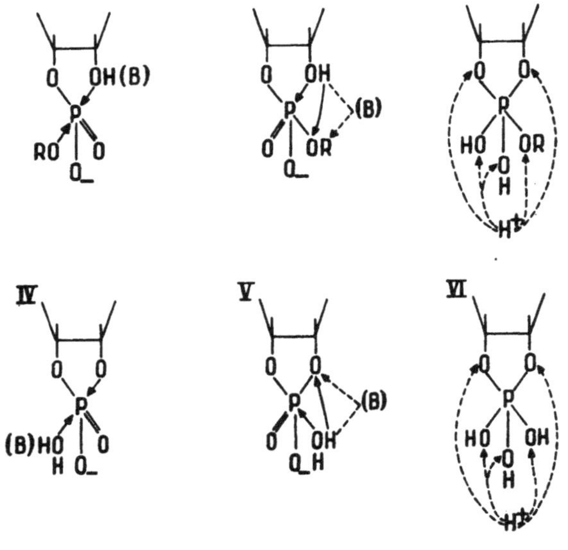

IV V VI

Base catalysis at the 2' -OH group or at the water molecule should follow the same mechanisms (IV or V). Here again the "adjacent" mechanism should be preferred, since the base taking up the proton can immediately transfer the proton to the leaving group. Thus complete separation of charges and conversion of configuration are not required. The problem is not yet decided (2). The "in line" mechanism, on the other hand, requires two different bases for the two reaction, steps while the "adjacent" variant requires only one base.

Proton catalysis leads to an addition mechanism with the formalition of a proton stabilized intermediate state with pentacovalent phosphorus (VIa, b). The reaction follows exactly second order with the proton concentration indicating that only the simultaneous attack of two protons leads to the formation of the pentacovalent intermediate.

The breakdown of the intermediate needs an additional catalysis by protons which might act at the 2' -o (back reaction) or at the 3' -0 (formation of a 2' -5' -diester) or at the 5' -0 (formation of the 2', 3' -cyclic product).

Here also an "in line" or an "adjacent" mechanism can be discussed, the latter connected with the problem of pseudorotation which is a relatively fast process and therefore does not influence the rate (3).

We see that proton catalysis acts in two ways. Protonation at the free oxygens leads to an increase of electrophilicity at the phosphorus and facilitates the attack of the nucleophil, while protonation of the ester oxygens leads to a polarisation of the bond and lowers the activation energy to cleave it. This additional proton catalysis therefore is responsible for a selective cleavage among equivalent bonds.

The enzymes have to use the same priciples. In addition, they can also lower the overall activation energy by the entropy term which can be completely neglected in the nonenzymatic catalysis. Theoretically, there is no reason, why they should not use base and proton catalysis simultaneously, together with a further decrease of the activation energy by an increase in the probability factor. therefore, we expect that the catalytic process in our enzyme reactions might use completely different contributions and combinations to lower the activation energy. This is roughly indicated by the different base specificity.

We discuss first our results on RNase A:

It could be demonstrated that there exists no evidence that the nucleophilic site, i. e. the 2' -OH group or the water molecule can be activated by the enzyme (4). The only argument to support such a concept might be derived from the bell shaped pH-dependence curve of the kinetic parameters K_m and k_2 (5). But other concepts are in agreement with bell shaped curves as well (6).

On the other hand, we found that all substrates need the C-2 oxygen of the pyrimidine base which we had shown some years ago to accelerate the rate in the nonenzymatic hydrolysis, too (7).

Even the simple 2-oxo-pyridone nucleotide (VII) or pyridazine nucleotide (VIII) diesters are substrates, demonstrating that the position 3, 4, 5 and 6 of the pyrimidine base does not play a specific role in the catalytic process. Many other compounds confirm this result.

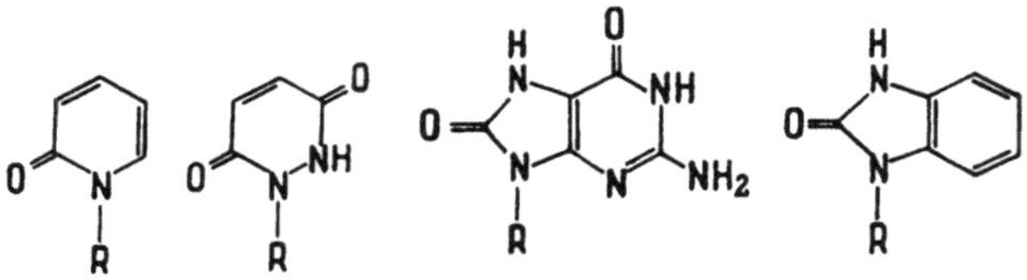

VII VIII IX X

Our first conclusion that the specificity of the enzyme for the pyrimidine base does not depend on a specific binding but on a specific base catalysis by the C-2 oxygen on the 2'-OH group has been strongly supported by our further findings that purine bases with an oxo group in position 8 (e. g. IX) are also substrates and even the benzimidazolone nucleotide diester (X) is slowly split. As a consequence of these results the conclusion was justified: If the enzyme does not interact at the nucleophilic site and also does not interact specifically with the base, it must activate the electrophilic site. That means: Two protonating groups of the enzyme have to attack simultaneously the free phosphate oxygens and an intermediate with a pentacovalent phosphorus must be formed during the reaction. This intermediate breaks down by a further proton catalysis at the oxygen of the leaving group. Supported by other results, particularly binding studies with inhibitors, we could now postulate a mechanism according to the scheme XI - XVI, in which the enzyme attacks the phosphate group and the base the 2'-OH group. This leads in a k_{+1} step to the enzyme stabilized intermediate XIIa. - The breakdown of the intermediate should be catalysed by the same base now acting in form of the conjugate acid (XII) in two alternative steps k_{-1} and k_{+2}.

The hydrolysis follows after an exchange of alcohol with water. XIV - XVI is exactly the reverse of XI - XIII. The intermediate XV in equilibrium with XVa is identical with XII and XIIa, when R is replaced by H.

Both reaction steps require a typical Michaelis-Menten kinetics in which XIIa and XVa are the ES-complexes.

We could demonstrate further that the two constants k_{-1} and k_{+2} are coupled and change by the same factor when something is altered on the base. Thus we have the chance to get informations about alterations of k_{+1} from the

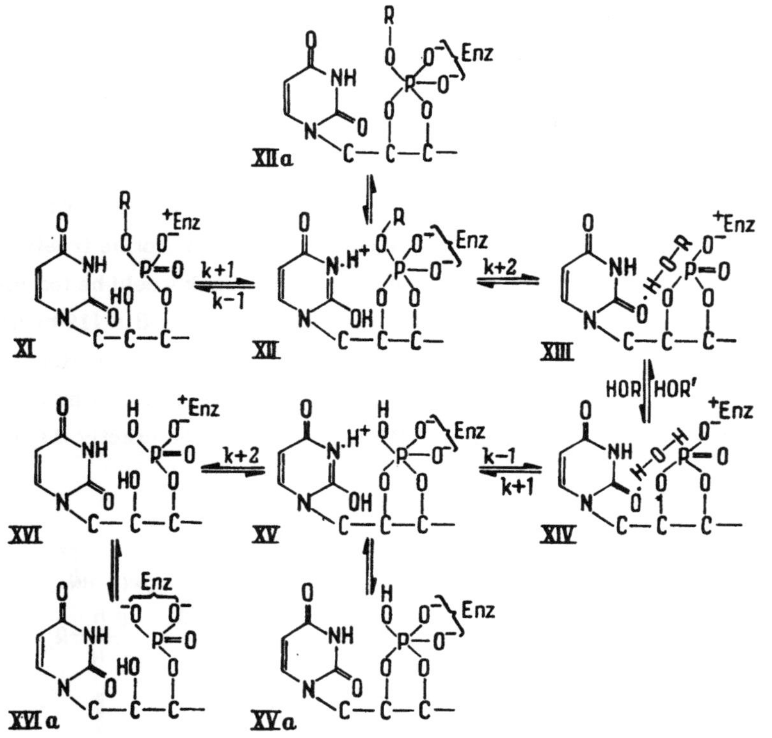

measured K_m- and k_{+2}-values (8) according to

$$K_m = \frac{k_{-1} + k_{+2}}{k_{+1}}$$

How is the activation energy lowered for these steps and what are the arguments and evidences?

Four factors (A, B, C and D) play a role for the k_{+1} step (4, 8, 9):

A contains the collisions of the phosphate group with the two protonating groups of the enzyme leading to an increase of electrophilicity at the phosphorus in the time of interaction.

B contains the probability of contact between the 2' oxygen and the phosphorus atom, when it is activated.

C contains the probability for the base catalysis when conditions A and B are given.

D is related to an additional binding of the substrate increasing the number of collisions in factor A.

We consider first factor A:
The bell chaped pH-dependence of the reaction is determined by two imidazoles, one of them acting in the base form, the other one in the protonated form (5). If one accepts our mechanism, these two groups cannot be used for the protonation. We therefore have to postulate one pH-independent group - Lys 41 - and one imidazolium ion - His 119 - for the binding and activation process (6). We have to postulate furthermore that these two groups cannot be freely rotating on the enzyme surface, otherwise the simultaneous attack would be too improbable. Therefore we claimed the second imidazole - His 12 - to fix His 119 by an H-bridge in a definite position to Lys 41. It corresponds to a reduction of the reaction from one of third to one of second order. This triprotonated diimidazole system XVIII - in equilibrium with XVII and XIX - requires exactly the found pH-dependence:

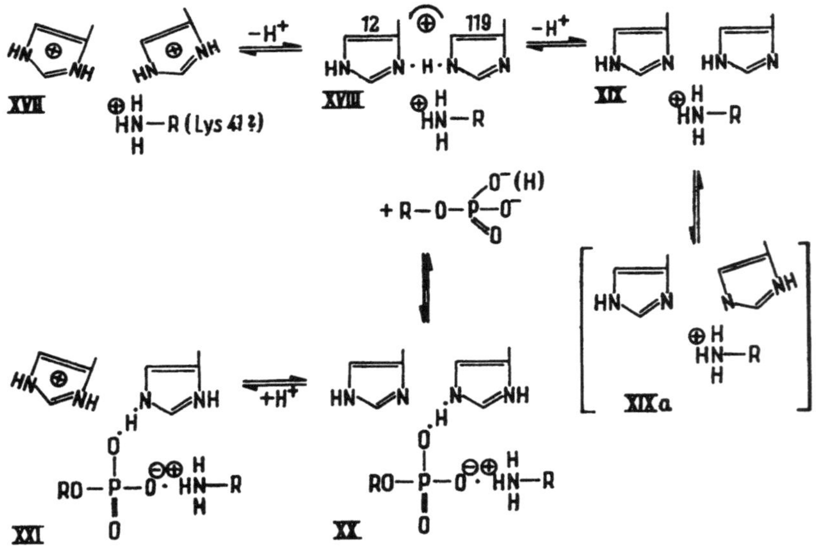

On the acid side it is tetraprotonated and therefore the two protonating groups are not fixed, on the alkaline side XIX cannot protonate the phosphate group. The interaction of a dianionic phosphate group with this system leads to the formation of a stable complex in which His 119 and Lys 41 bind, while His 12 is released (XX ⟶ XXI) (10). Both His get higher pK-values explaining the shift of the pH-dependence of the breakdown reaction ($k_{-1} + k_{+2}$) to the alkaline side, (we see later that the breakdown is initiated by a deprotonation of His 12 or 119).

In this process the energy to lower the entropy of the enzyme-substrate-complex is produced by the liberation of His 12 which was fixed before in another low-entropy complex. This is a classical case for an entropy compensation mechanism. Strong evidences besides the chemical and kinetic arguments for the existence of thé triprotonated diimidazole system XVIII were obtained by our studies of the pH-dependence of the proton uptake and release after binding of inhibitors to the enzyme (10) and by our data on the pH-dependence of the proton release after addition of silver ions (11). The final proof could be given in a direct way by our NMR-studies with Dr. Rüterjans. The titration curves for His 12 and 119 are complex and show in the first derivative a common minimum (12).

The next finding is very important for the interpretation of the binding process: Dianionic monoesters like 3' deoxy-Cp are strong inhibitors, while the corresponding substrate analogous diesters like 3' -5' -deoxy-CpC, which are monoanionic, are not. No exception from this rule has been found in a long series of experiments. (Data on a 3' -thymidylic acid benzyl ester cannot be confirmed 13, 14) This indicates that the collisions of the monoanionic substrates with the active site do not lead to the formation of a stable complex. The basicity of the diester phosphate group (XXII) is obviously too low compared with the monoester phosphate group (XXIII), which due to its pK around 6 offers electrostatic and H-bond elements for the formation and the stability of the complex.

It follows from these observations that the kinetically relevant enzyme-substrate complex is formed after the catalysis step, when the dianionic pentacovalent intermediate (XXIV) has been built up.

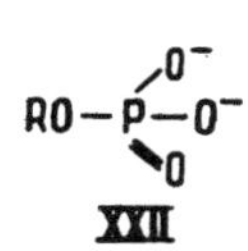

XXII

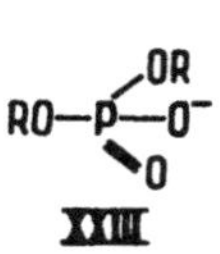

XXIII

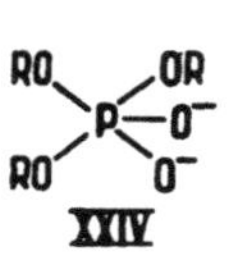

XXIV

This intermediate should behave in basicity and H-bonding abilities similar to the dianionic monoester phosphate group.

The increase of the electrophilicity at the phosphorus by the collisions with the enzyme superimposes the original electrophilicity given by the substituents. Thus the 3' -naphthyl ester (15) is 10^2 times faster in the k_{+1} step, compared with the methyl ester, and the 2', 3' cyclic phosphonate is 10^4 slower in k_{+1} compared with the corresponding cyclic phosphate (16).

The stability of the complex, however, does not only depend on the interaction of the phosphate group with the enzyme. The contribution of the base to the stability and its influence on the k_{+1} step will be discussed later.

The factor B is a probability factor. In the Cp methyl ester (Cpm - XXVa) the phosphate group is freely rotating, in CpA (XXVb) the phosphate group is fixed by a strong base stacking effect.

In this conformation the 2' oxygen is pressed on the phosphorus.

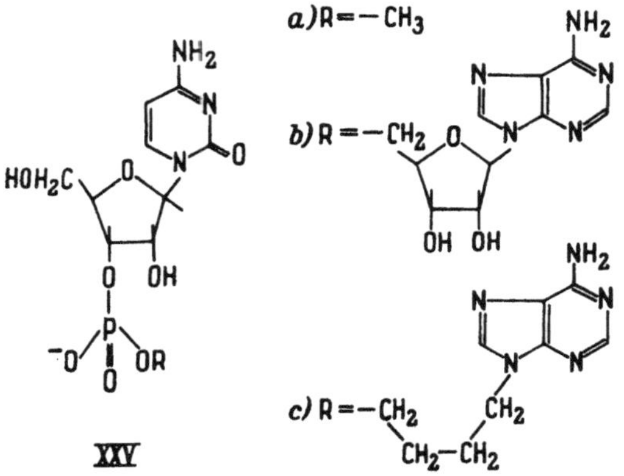

XXV

We found a 1000 fold higher rate for CpA compared with Cpm and between these extremes a series of about 25 compounds with increased rates, the increase always paralleling the base stacking effect (9).

For analogy reasons we compared CpA (XIXb) and Cp butyladenine (XIXc), which differ by a factor of 50. In the temperature dependence it can be seen that the difference lies only in the entropy term of the activation energy (9).

	k_{+2}	$\Delta G^{\ddagger}$	$\Delta H^{\ddagger}$	$\Delta S^{\ddagger}$
CpA	2370	12, 8	3, 5	- 31, 4
Cp butyl A	50	15, 1	3, 5	- 39, 0
	sec^{-1}	Kcal/M	Kcal/M	e. u.

Since the contribution by factor A and C should be constant in both substrates, we conclude that in our series this fixation effect can lower the activation energy by further 4 kcal in addition to that of the factors A and C.

The factor C is related to the base catalysis and includes two elements:

The first is the probability that the base is in the correct position to take up the proton of the 2' -OH group. This means that it has to offer a free orbital of its C-2 oxygen to the proton. If the base were freely rotating around the glycosidic bond, the probability would be very low.

From CD and NMR measurements in several laboratories (17, 18) we are sure that even free in solution the nucleotides and diesters have the base preferentially in the anti position, i. e. the C-2 oxygen turned to the 2' -OH. Bound to the enzyme we find similar Cotton effects. (In fact we find small differences just supporting our concept). This indicates that the conformation of the nucleotides has not altered.

In further studies together with Dr. Bald and Dr. Holy we could demonstrate that the base is fixed at the enzyme by an interaction with the aromatic ring of Phe 120. UV-difference spectroscopy and comparison with binding constants

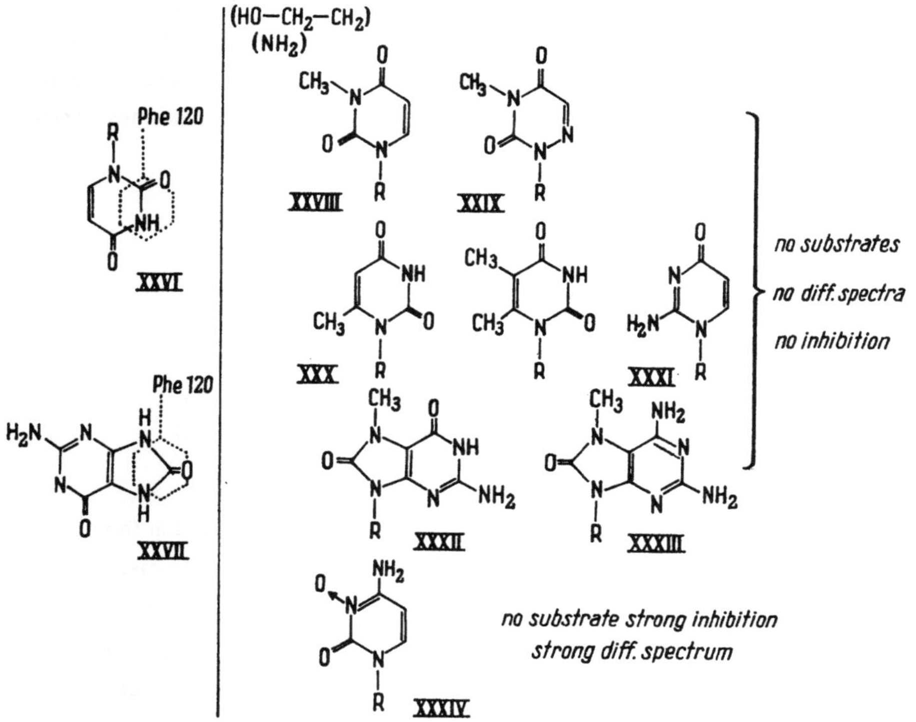

indicate that this interaction depends on induced dipol forces between the two π-systems (19, 21).
The overlapping is only partial (XXVI, XXVII) and involves mainly position 2 and 3 of the pyrimidine base (19). A methyl or amino group in one of these positions of the pyrimidine ring or at position 7 or 8 of the purine ring prevents the induced dipol forces to operate. These compounds are not substrates since the base is obviously too mobile. The corresponding 2' or 3' mononucleotides are only very weak inhibitors by the same reason. An oxygen in the position 3, seen on Cp-N-oxid however increases the interaction (XXVIII - XXXIV).
From inhibition constants which are true binding constants without a catalytic step, we can estimate that the additional fixation of the base lowers the activation energy by more than 5 kcals. When nucleosides are addes to the enzyme, such a binding does not occur. The effect is found only with dianionic phosphate monoesters which bind primarily by the phosphate group. The relatively high value of 5-6 kcal should therefore be based on a cooperativity effect.
On the basis of these results, we can now explain easily why the β-L-ribotides, the mirror images of β-D-substrates, or the α-D- and α-L-ribotides are not substrates nor - in form of the mononucleotides - inhibitors nor do they show UV-difference spectra (20).

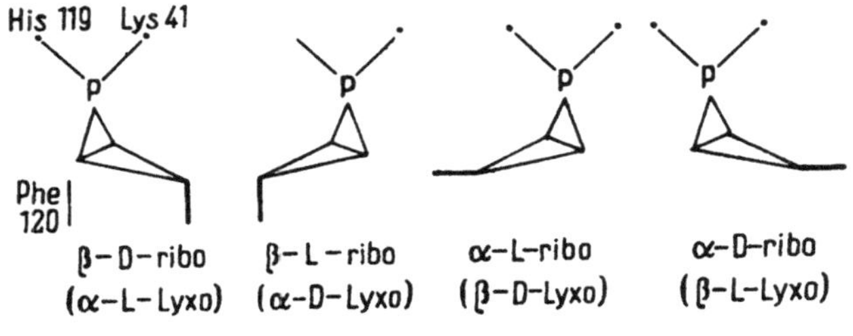

The second element in factor C is the basicity of the base, on which the activation energy depends to take over the proton in the k_{+1} step. In order to clear this part we need substrates in which factor A and B and the base interaction remained constant, the basicity however is different. This effect can be seen when we compare 6 Aza-Up with Up, 6 Aza-Up is 50 times slower in the k_{+1} step.
On the other hand, in compounds with comparable basicities we found that the rate is governed by the polarisability (4). In such cases, k_{+1} (the base catalysis) and k_{+2} (the proton catalysis) increase by the same factor, K_m thus remaining

constant. This indicates that the rates for both steps depend on the rearrangement of the electronic system in the base. Increased polarisability lowers the activation energy to take up the proton and to release the proton (XXXV, XXXVI) by the same factor. Lower basicity can be compensated by higher polarisability as seen in Up and 5-Nitro-Up, which have the same rate.

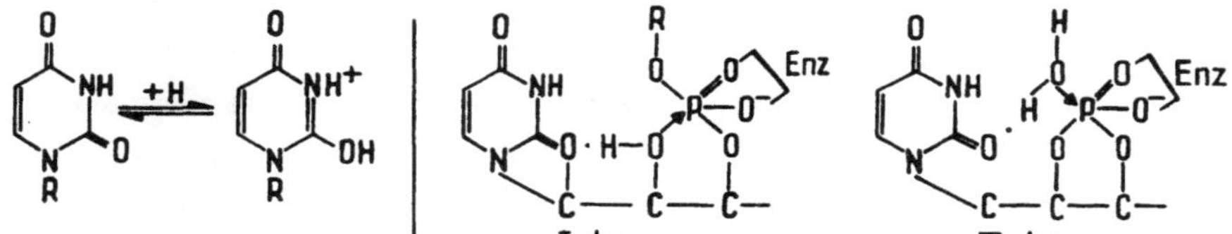

I step — II step

XXXV — XXXVI — XXXVII — XXXVIII

What about k_{+1} in the second reaction step, the hydrolysis of the 2', 3' cyclic diesters? There is no principle difference. The only alteration is caused by the fact that in the first reaction step the phosphate group is mobile (XXXVII) and the nucleophile is fixed, while in the second reaction step the phosphate group is fixed and the nucleophile, the water molecule, is mobile (XXXVIII). Therefore factor B does not work in the hydrolysis reaction.
The activation energy for the breakdown of the intermediates XIIa and XVa is reflected in the k_{+2} value, which parallels the k_{-1} value too. From the pH-dependence we know that the breakdown of the enzyme stabilized intermediate is initiated by a deprotonation of the complex. This occurs at the two imidazolium ions (see XXI) indicated by the finding that the k_{+2} value drops by more than a factor of 10 in the pH range between 7 and 8. Our NMR-data support this concept (12). This is an equivalent to the factor A in the k_{+1} step.
The final breakdown, however, is proton catalysed by the pyrimidine base, now in form of the conjugate acid (XII and XV). It corresponds to factor C. - Here we get another interesting inside look at the nature of the successive base-proton catalysis. Increased basicity should result in <u>inc</u>reased rates for k_{+1} and <u>dec</u>reased rates for k_{+2}, i. e. the proton catalysis. Therefore K_m should be smaller. Instead, we found k_{+2} for the more basic cytidine 3 times higher compared with the weaker basic uridine diester, which should be the stronger conjugate acid and therefore faster.
We found in temperature dependence studies (9) that the stronger conjugate acid of uridine has a 5 times faster rate in the enthalpy term, but is 15 times

slower in the entropy term. It results the 3 times faster rate for the weaker conjugate acid. This indicates that indeed the higher acidity causes a higher rate for the proton transfer, but it also causes a lower probability for the base to be in the form of the conjugate acid.

We realize here: for the base catalysis with a subsequent proton catalysis by the conjugate acid there exist some limitations. We need a relatively weak base. A stronger base would give a too weak conjugate acid, - on the other hand, a weaker base is not so often in the protonated form for the proton catalysis. However, the low basicity we have to take in account can be compensated by a strong fixation of the base in a position to transfer the proton directly. This is a classical case that entropy compensates for enthalpy in the activation energy.

The next and the most important problem is the question: Can we replace one factor by another? If we take the reaction of 2', 3' -cyclic Cp as a standard in the order of 1 per second, then this value is reached when factor A and C operate together. Factor B can be added or not. There is a continuous increase in the series of diesters, e.g. Cp-methyl, Cp-benzyl, CpU, CpC, CpA or Cp-deoxy A up to 1000 sec^{-1} (9).

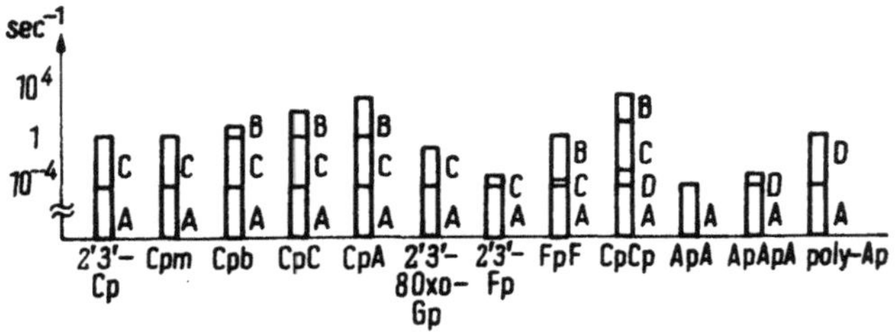

Base catalysis can be less efficient when for steric reasons the probability factor in it is lowered. Thus the cyclic phosphates of 8-oxopurine nucleotides are much slower, since the oxygen at position 8 is more distant from the 2' -OH group. If we replace the C=O group in position 8 by a -N= group as in formicine, we get a rate in the range of 10^{-3} sec^{-1}. (22) If however we add the base stacking effect of factor B as in the dinucleosidephosphate FpF, we get a rate in the order of the Cp-methyl ester, but still 1000 times smaller than CpA. This explains the results of WARD and others (23).

The base catalysis in factor C, therefore, can be very weak, when its contribution is compensated by factor B. The π -interaction between the base and Phe 120, however, cannot be compensated. The base catalysis can be compensated also by factor D which is related to factor A. If we have a second phosphate group on the dinucleoside phosphate, i. e. a dinucleotide, and we compare it with the dinucleoside phosphate (e. g. CpC and CpCp) we find a further increase of the rate by a factor of 15 (15). It is located in K_m, i. e. only in the k_{+1} step. This effect of an additional binding retaining the diester phosphate group longer in the active center can be better studied in the Ap-series. 2', 3'-cyclic Ap and ApA without base catalysis are not substrates and do not inhibit. ApApA with an additional phosphate group is split very slowly, ApApApA a little faster. They

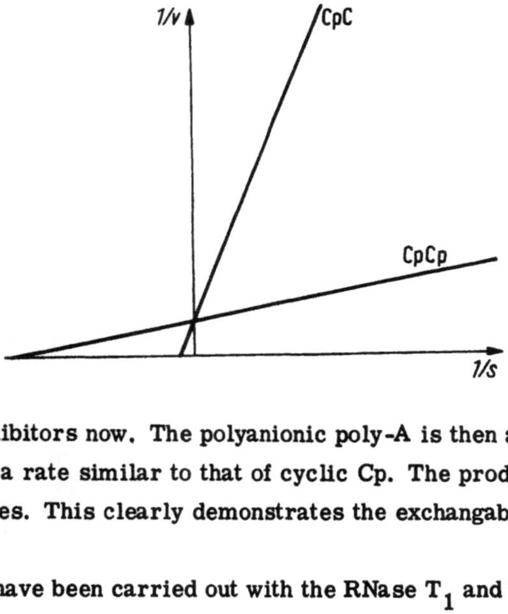

become weak inhibitors now. The polyanionic poly-A is then a strong inhibitor and is split with a rate similar to that of cyclic Cp. The products are a series of oligonucleotides. This clearly demonstrates the exchangability of the different principles.

Similar studies have been carried out with the RNase T_1 and T_2. The T_1-enzyme splits only diesters of purinenucleotides with a carbamide structur in position 1, 6 (1, 24). This position is too distant to influence directly the bond exchange at the phosphate group (XXXVI). This indicates that the T_1 enzyme has a binding site and a separate catalytic site in its active center. Kinetic results (25) and NMR measurements (26) on the free enzyme and the enzyme inhibitor complexes indicate that both sites consist of an imidazolium-carboxylate system. They also indicate a positively charged group (prabably Arg 77) which seems to be involved in the fixation of the phosphate group. Since the C-2 proton signal of the imidazole at the binding site does not alter its position markedly comparing the free enzyme with the complex, we conclude that binding occurs by two H-bonds. Here no

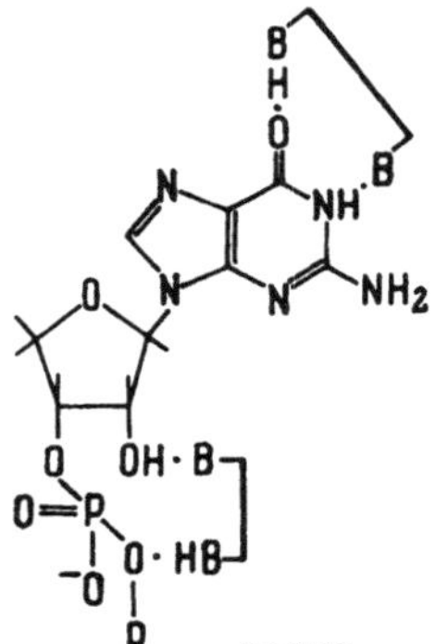

XXXIX

charge separation is necessary, the activation energy is low, since entropy compensation assists the reaction. The consequences of our kinetic results are summarized in the following scheme XXXVII - XXXXI: They indicate at the catalytic site that in the first step the carboxylate anion performs a base catalysis at the 2' -OH group and the imidazolium ion a proton catalysis at the leaving ester oxygen. The rate drops when the imidazolium is deprotonated. The optimum is at pH 5.

In the second step imidazole is used in the base form to activate the water molecule. Therefore the rate increases with the deprotonation of that imidazole and the pH-optimum is found at pH 7. The drop at higher pH is caused by the deprotonation of the histidine at the binding site.

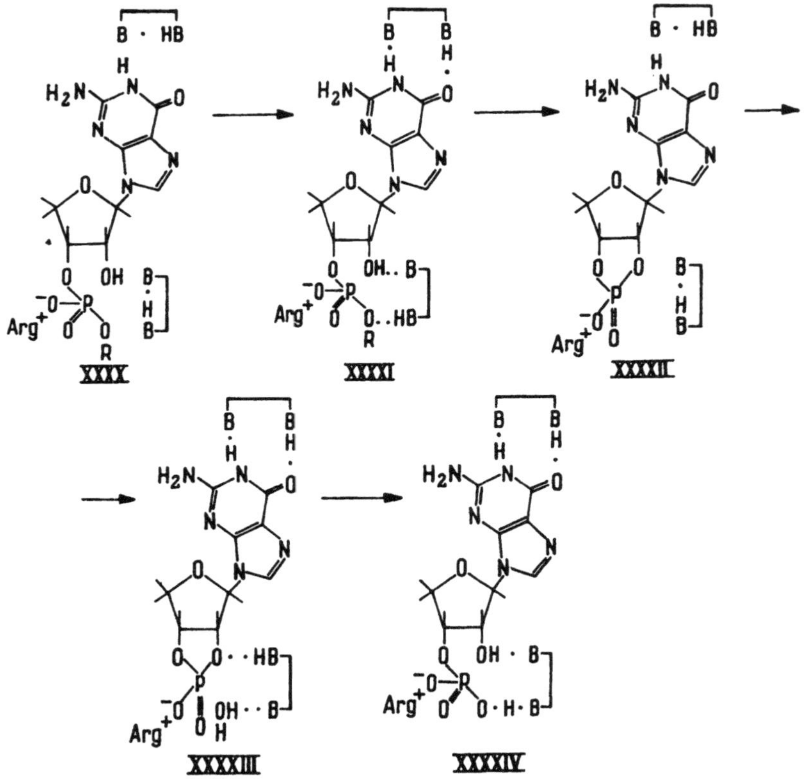

The K_m values for the 3' diesters with a free 2' OH group are 50 times lower than that of the corresponding 2', 3' cyclic diesters. Furthermore K_m is a function of k_{+2}. This indicates that the catalytic site contributes to the binding strength as well. K_m is therefore a complex constant with elements from both sites. It cannot be treated as in the RNase A reaction.

The 2' and 3' mononucleotides are strong inhibitors. The pH-dependence of K_i however, differs. The 3' nucleotide follows the K_m curve of the 3' diesters. Here the binding should be composed by the interaction of the binding site, by the positively charged Arg 77, and by the carboxylate anion, while the imidazolium binds now to the other negatively charged phosphate oxygen, however looses its strength, when the phosphate is protonated. Thus a pH-optimum in the range of 6, 5 results.

2' mononucleotides should also interact with the binding site and Arg 77. The carboxylate group however cannot interact with the 2' oxygen which now carries the phosphate group. But it can interact with the monoprotonated phosphate group and the imidazolium ion with the phosphate oxygen. Therefore the pH dependence curve is shifted to the acidic side.

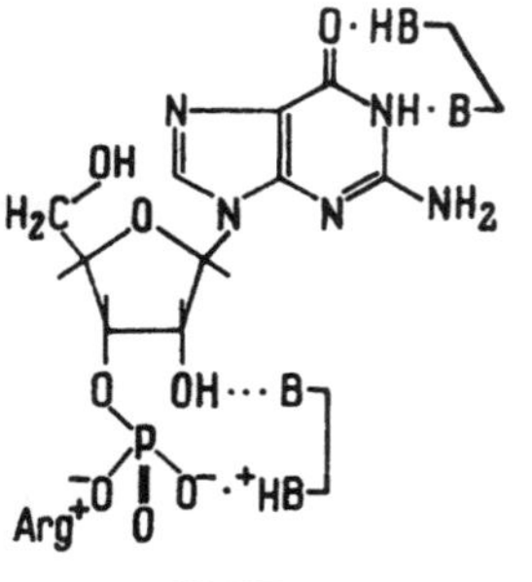

XXXXII

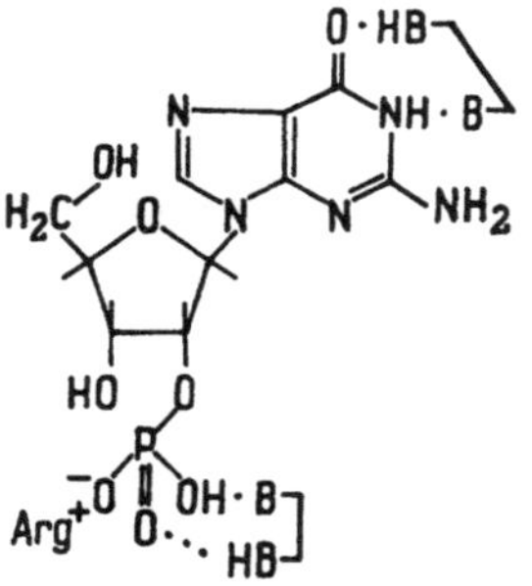

XXXXIII

Our NMR results demonstrate clearly this concept of binding.

The T_2 enzyme does not show a base specificity, however the rate depends strongly an the nature of the base. The results with more than 50 substrates indicate that the base is bound to a hydrophobic area without any specific group requirements. Even the 3' diester of a benzyl-o-glycoside is a substrate, while the corresponding methyl glycoside does not react (27).

The optimum for both steps is at pH 5. Cytidine diesters have an optimum at pH 6, due to the protonation of the base with a pK of 4. 2, while 5-nitro-uridine

diesters have the optimum at pH 3, due to the deprotonation of the base with a pK of 5,4. Thus both types of charges drop the rate considerably.
From the pH-optimum at pH 5 in both steps we conclude that two carboxylate groups - held together by an H-bond - act at the catalytic site, in the first step in form of B_1 and HB_2 and in the second step as B_1H and B_2 (XXXXIV - XXXXVII).

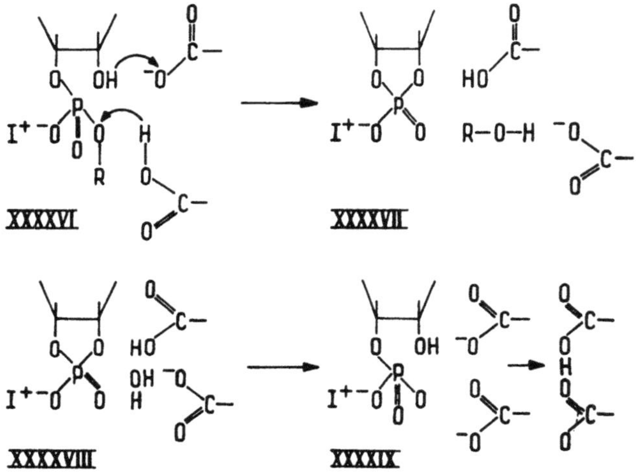

XXXXVI XXXXVII XXXXVIII XXXXIX

K_m is a function of k_{+2}. This again indicates that the catalytic site contributes to the binding strength. We see again that K_m is higher that means the binding is weaker in the second step when a water molecule is the nucleophile.
The third fixation of the substrate occurs obviously by the interaction of the phosphate group with an imidazolium ion, indicated by the increase of K_m at pH 7, when k_{+2} drops down.

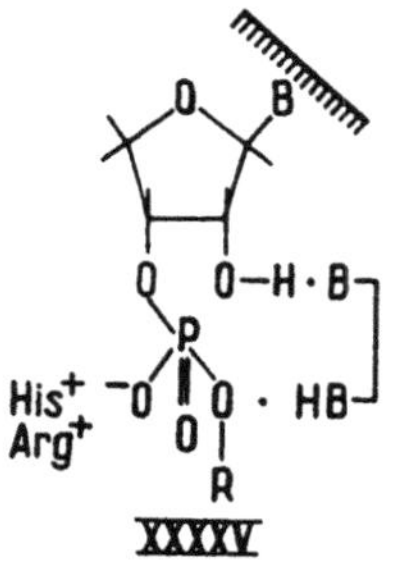

XXXXV

Both enzymes follow obviously the same type of catalysis in form of an S_N2 type mechanism in the "adjacent" variant. The additional binding, at the base and at the phosphate group, plays here a major role. We see that hydrophobic interactions in T_2 and specific H-bonds of the base in T_1 can replace each other.
At the catalytic site we find in both cases a B·HB system in which B_1 is a carboxylate group, B_2 however might be another carboxylate or an imidazol. They determine the different pH-dependence.

Comparing these two enzymes with the RNase A, we see that the base binding site is related to factor D. Instead of substrate binding performs the base in RNase A now a base and subsequent proton catalysis for the bond exchange according to an addition mechanism.

Although the three enzymes catalyse the same bond exchange, it is clear from our present data, that they use a series of principles in different combinations to lower the activation energy. They can be interchanged. This facilitates the thinking of the construction of enzyme analogous catalysts considerably.

REFERENCES

(1) EGAMI, F., K. NAKAMURA, in Microbial Ribonucleases, Springer Verlag Berlin, Heidelberg, New York 1969

(2) RICHARDS, F.M., H.W. WYCKOFF, in The Enzymes, Ed. by BOYER, Academic Press, in press

(3) USHER, D.A., Proc. Natl. Acad. Sci., 62, (1969), 661

(4) GASSEN, H.G., H. WITZEL, Europ. J. Biochem., 1, (1967), 36

(5) HERRIES, D.G., A.P. MATHIAS, B.R. RABIN, Biochem. J., 85, (1962), 127

(6) WITZEL, H., Progress in Nucleic Acid Research, 2, (1963), 221

(7) WITZEL, H., Ann. Chem., 635, (1960), 191

(8) WIEKER, H.J., H. WITZEL, Europ. J. Biochem., 1, (1967), 251

(9) FOLLMANN, H., H.J. WIEKER, H. WITZEL, Europ. J. Biochem., 1, (1967), 243

(10) HUMMEL, J.P., H. WITZEL, J. Biol. Chem., 241, (1966), 1023

(11) BARNARD, E.A., H. WITZEL, unpublished, see ref. 12

(12) RÜTERJANS, H. ,H. WITZEL, Europ. J. Biochem., 9, (1969), 118

(13) DEAVIN, A., A.P. MATHIAS, B.R. RABIN, Nature, 211, (1966), 252

(14) DEAVIN, A., R.C. FISHER, C.M. KEMP, A.P. MATHIAS, B.R. RABIN, Europ. J. Biochem., 7, (1968), 21

(15) WITZEL, H., unpublished results

(16) HARRIS, M.R., D.A. USHER, H.P. ALBRECHT, G.H. JONES, J.G. MOFFATT, Proc. Natl. Acad Sci., 63, (1969), 246

(17) MILES. D.W., M.J. ROBINS, R.K. ROBINS, M.W. WINKLEY, H. EYRING, J. Am. Chem. Soc., 91, (1969), 824

(18) TS'O, P.O.P., N.S. KONDO, M.P. SCHWEIZER, D.P. HOLLIS, Biochemistry, 8, (1969), 997

(19) BALD, W., Thesis, Marburg 1969

(20) WITZEL, H., A. HOLY, Abstract 103, 6. FEBS meeting, Madrid 1969

(21) WITZEL, H., W. BALD, A. HOLY, Abstract A96, 7. Symposium on the Chemistry of Natural Products, Riga 1970

(22) IKEHARA, M., K. MURAO, S. NISHIMURA, Biochim. Biophys. Acta, 182, (1969), 276

(23) WARD, D.C., W. FULLER, E. REICH, Proc. Natl. Acad. Sci., 62, (1969), 581

(24) WHITFELD, P.R., H. WITZEL, Biochim. Biophys. Acta, 72, (1963), 338

(25) PONGS, O., Thesis, Marburg 1968

(26) RÜTERJANS, H. WITZEL, O. PONGS, Biochem. Biophys. Res. Comm., 37, (1969), 247

(27) KAISER, P., Diplomarbeit, Marburg 1971

Slovak Academy of Sciences, Biological Institute, Department of Biochemistry of Microorganisms, Bratislava

E. ZELINKOVA, M. BACOVA, J. ZELINKA

THE SPECIFIC RIBONUCLEASE FROM STREPTOMYCES AUREOFACIENS

A ribonuclease from the cultural medium of Streptomyces aureofaciens, a chlortetracycline producing strain, was isolated and purificated. The purified enzyme was chromatographically and electrophoretically homogeneous and it was found to be free of deoxyribonuclease, non-specific phosphodiesterase and monoesterase activities.

It was shown by the analysis of the products resulting from the action of Str. aureofaciens RNase on yeast transfer RNA that this enzyme, like the well studied ribonuclease T_1, is an endonuclease splitting the ester bond between the guanosine 3' -phosphate and the -OH group at 5' -position of adjoining nucleotide under the formation of guanosine-2', 3' - cyclic phosphate as an intermediate. This intermediate is then hydrolyzed by the enzyme to guanosine 3' -phosphate. Therefore Str. aureofaciens ribonuclease can be regarded as a guanylic acid-specific endonuclease.

The properties of Str. aureofaciens RNase are very similar to those of RNase T_1 (1). This RNase is also an acidic protein, but it contains lower amounts of aspartic acid, tyrosine and serine. The content of basic amino acids lysine and arginine, is higher than in the RNase T_1 (2, 3). The isoelectric point lies around pH 4, 3, while the RNase T_1 has its isoelectric point around pH 2, 9 (3).

The specificity of Str. aureofaciens RNase was demonstrated by the cleavage experiments of synthetic homopolynucleotides. Using 10 units of ribonuclease, poly A, poly C and poly U did not appear to be attacked even after 20 hour incubation the test samples remaining at the origin of the chromatograms. Poly I was hydrolyzed at a higher rate than poly G. Using the ribonuclease in excess (approximately 3000 units) both poly A and poly U were hydrolyzed besides poly G and poly I. IRIE (4) reported also the splitting of poly A, poly C and poly U by ribonuclease T_1. A similar phenomenon was observed with pyrimidine specific pancreatic RNase which hydrolyzed also poly (5, 6) and poly I (6) if used in excess.

Using dinucleosidemonophosphates as substrates, it was found, that only GpC, GpU, GpA and GpG were cleaved. The splitting of these substrates was measured spectrophotometrically by the change of optical density at appropriate wavelength in an automatic recording spectrophotometer with the constant temperature bath (37°). Time course of the splitting of some dinucleosidemonophosphates is shown in Fig. 1.

The relative rates of splitting of these substrates in comparison with those of RNase T_1 (Table 1) are not quite in agreement with results of WHITEFELD and WITZEL (7) for the RNase T_1.

Table 1

RELATIVE RATES OF SPLITTING OF DIFFERENT DINUCLEOSIDEMONO-PHOSPHATES BY RIBONUCLEASE FROM STR. AUREOFACIENS AND BY RIBONUCLEASE T_1

Substrate	Str. aureofaciens RNase	ribonuclease T_1
	$\frac{V_o\ GpX}{V_o\ GpU}$	$\frac{V_o\ GpX}{V_o\ GpU}$
GpC	2, 68	3, 2
GpG	2, 40	1, 8
GpA	1, 33	2, 2
GpU	1	1

(Experimental conditions are described in Fig. 1.)

The difference between splitting rates of various dinucleosidemonophosphates found with Str. aureofaciens RNase, RNase T_1 (7) and RNase from Actinomyces aureoverticillatus (8), indicate that the reaction rate is influenced by the nucleoside bound to the 3' -guanylic acid. WITZEL and BARNARD (9) observed the influence of second nucleoside on the reaction rate also with pancreatic RNase.

Kinetic parameters, K_m and V_{max} of Str. aureofaciens RNase with different

Table 2

KINETIC CONSTANTS OF RNase FROM STR. AUREOFACIENS USING DIFFERENT SUBSTRATES AT pH 7, 0

Substrate	K_m (x 10^3M)	V_{max} (mole/min/10000 U)	$\frac{V_{max} \text{ of GpX}}{V_{max} \text{ of GpU}}$
Commercial yeast RNA	2, 09 1)	1, 65. 10^{-6} 1)	
yeast high polymerized RNA	2, 44 1)	1, 33. 10^{-6} 1)	
GpC	1, 428	2, 86. 10^{-7}	4, 84
GpG	1, 176	2, 00. 10^{-7}	3, 39
GpA	0, 909	1, 25. 10^{-7}	2, 12
GpU	0, 526	0, 59. 10^{-7}	1, 0
G-cyclic-p (pH 7)	11, 5	8, 13. 10^{-9}	
G-cyclic-p (pH 5, 7)	3, 6	3, 7. 10^{-9}	

1) in M of average nucleotide

(The reaction mixture contained in 1, 5 ml dinucleosidemonophosphates in the final concentrations 0, 5 - 2, 0. 10^{-4}M in 0, 05 M phosphate buffer (pH 7, 0) and enzyme solution (25 μl). G-cyclic -p (pH 7) - titration method: the reaction mixture contained 0, 1M NaCl, 1 mM phosphate buffer (pH 7), 5 -15. 10^{-3}M substrate and 25 μl enzyme in 1, 6 ml. Titrant: 0, 02N NaOH. Incubation temperature: 37°C.
G-cyclic-p (pH 5, 7) - electrophoretic assay: the reaction mixture (0, 4 ml) contained 0, 1M phosphate buffer (pH 5, 7), 1, 8 - 28. 10^{-4}M substrate and RNase (25 μl). Paper electrophoresis (filterpaper WHATMAN No 3): Electrophoresis was run at 18 V/cm for 1, 5 hour in 1/150 M phosphate buffer pH 7, 4).

substrates are summarized in Table 2. For better comparison, all measurements were carried out at pH 7, 0 i. e. at the pH optimum for the splitting of the internucleotide bond (1). The K_m values as well as the V_{max} values are decreasing in the following order GpC > GpG > GpA > GpU. The same effect was observed by IRIE with RNase T_1 but only for the V_{max} values (10). The K_m values are higher than those of RNase T_1 . The magnitude of V_{max} values in comparison with RNase T_1 is lower.

The kinetic parameters of hydrolysis of guanosine 2', 3' -cyclic phosphate were determined at pH 7 by pH stat method and at pH optimum 5, 7 an electrophoretic assay was used. The finding that the lower K_m values was obtained at pH 5, 7 rather than at pH 7 and that the V_{max} value was higher at neutral pH is in accordance with the results obtained by IRIE (11) for RNase T_1, while its pH optimum for 2', 3' cyclic GMP hydrolysis was found to be pH 7, 0.

LITERATURE

(1) ZELINKOVA, E., M. BACOVA, J. ZELINKA, Biochim. Biophys. Acta, 235, (1971), 343

(2) BACOVA, M., E. ZELINKOVA, J. ZELINKA, Biochim. Biophys. Acta, 235, (1971), 335

(3) TAKAHASHI, T., J. Biochem. 51, (1962), 95

(4) IRIE, M., J. Biochem. 58, (1965), 599

(5) IMURA, N., M. IRIE, T. UKITA, J. Biochem. 58, (1965), 264

(6) BEERS, R. F., J. Biol. Chem. 235, (1960), 2393

(7) WHITFELD, P.R., H. WITZEL, Biochim. Biophys. Acta 72, (1963), 338

(8) ABROSIMOVA - AMELJANCHIK, N.M., R.I. TATARSKAYA, T.V. VENKSTERN, V.D. AKSELROD, A.A. BAEV, Biochimiya 30, (1965), 1269

(9) WITZEL, H., E.A. BARNARD, Biochem. Biophys. Res. Commun. 7, (1962), 295

(10) IRIE, M. , J. Biochem. 63, (1968), 649

(11) IRIE, M., J. Biochem. 61, (1967), 550

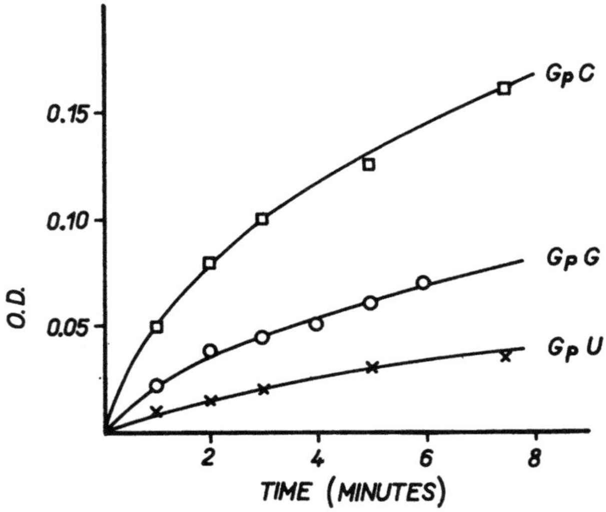

Fig. 1. Time course of the splitting of some dinucleosidemonophosphate by Str. aureofaciens RNase.
(Reaction mixture (1, 5 ml) contained 0, 05 M phosphate buffer pH 7, 0 0, 2 mM substrate and 13000 units of RNase).

Institute of Biochemistry and Physiology of Microorganisms;
Academy of Sciences of the USSR, Puschino on the Oka, Moscow Region, USSR

S. I. BEZBORODOVA

SOME PROPERTIES OF EXTRACELLULAR RNases OF FUNGI

EGAMI and NAKAMURA in the monograph "Microbial Ribonucleases" (1969) concluded that molecular weight of base specific RNases is 11000-13000 and that of nonspecific ones 30000-40000 with the exception of RNase of Bac. subtilis Marburg. Our investigations on extracellular RNases of Fungi revealed that fungi were able to synthesize and secrete nonspecific RNases, different in molecular weight and optimum pH, into the culture medium.
RNases with molecular weight of about 11000 have optimum pH in weak alkali (Asp. clavatus, Asp. pallidus, P. chrysogenum 152A, F. semitectum) and RNases with molecular weight of 30000-40000 have optimum pH 4 to 6.
RNases were isolated from the filtrate of the culture fluid and purified to the chromatography homogeneous state. The steps of the purification were as follows:

1. Ammonium sulfate precipitation or fractionation with acetone.
2. DEAE- and CM-cellulose column chromatography.
3. Gel-chromatography on the Sephadexes G-100 and G-75. RNases were purified about 500-2500 times with the yield from 2 to 60 %.

Homogeneity was studied with disc electrophoresis in polyacrylamide gel under different pH (Fig. 1) and ultracentrifugation. Figures 2a and b show UV spectra of some RNases. Some RNases were obtained in the amount to 10 mg and more with the result that their amino acid composition was determined. The presence of histidine was ascertained in all the RNases under investigation.
There is some analogy in general amino acid composition of "alkali" RNases and that of RNase T_1 Asp. oryzae. For this reason we consider the investigations on the relation between the structure and the function of this group of RNases to be very interesting for the establishing of the cause of base specificity.
Some properties of RNases studied are given in Table 1 and 2. Aotivity of enzymes depending on pH is expressed in the bellshape curve (Fig. 3). It may by proposed

Table 1

PROPERTIES OF RNASES (SUBSTRATE - RNA)

Culture	pH opt.	Temp. opt.	Activity remaining (%) 100°	Mol. w. x 10^{-3}
Asp. clavatus, RNase C_2	ph - 7, 8 t-a - 8, 8	45	50 (pH 7, 0)	10-12
Asp. pallidus	ph - 7, 8 t-a - 8, 8	45	50 (pH 7-9)	10-12
Pen. brevicompactum	5, 0	50	44 (pH 2, 8)	30-40
ac. RNase	4, 5	55	32 (pH 6-7)	30-40
Pen. chrysogenum 152A alc. RNase	ph - 7, 2 t-a - 7, 8	25	60 (pH 7-8)	10-12
Pen. claviforme ac. RNase	ph - 5, 3 t-a - 6, 5	50	60 (pH 7-8)	30-40
Fus. semitectum	ph - 7, 6 t-a - 8, 0	35	66 (pH 8, 0)	10-12

Table 2

PROPERTIES OF RNASES (SUBSTRATE - X > p)

Culture	pH opt.	Temp. opt.	Specificity (pH=7, 0)	"Molecular" activity min^{-1}
Asp. pallidus	4, 75	34	A> C>U>G	4, 2 (U > p)
Pen. brevicompactum	4, 64	18-20	A> C>U>G	1330 (C > p)
Pen. chrysogenum 152A ac. RNase	6, 50	-	C>A>U>G	900 (C > p)
Pen. chrysogenum 152A alc. RNase	5 - 6	44-48	A>C>U>G	51, 6 (C > p)
Pen. claviforme ac. RNase	4, 3-4, 7	27	U>A>C>G	1160 (U > p)
Fus semitectum	7, 7-8, 7	-	G>C; A=U=O	3, 56 (C > p)

that general-acid - general base catalysis is of common occurence in splitting RNA and X > p by different RNases.

The pH and temperature optima of the transferase and hydrolytic steps of the reaction do not always coincide. Notwithstanding the absence of pronounced specificity to chemical nature of bases in RNA. RNase C_2 Asp. clavatus has preference to guanylic and adenylic acid on the early stages of degradation. Some preference to purines was noted in RNase of F. semitectum (Table 1). Hydrolysis X>p is catalyzed with RNases more selectively (Table 2).

Many RNases hydrolyze A>p with the highest rate G>p is hydrolyzed very slowly.

"Molecular" activities of RNases calculated on the base of initial rates are quite different. RNase of P. brevicompactum has the highest activity.

"Alkaline" RNases show weak hydrolase activity and possess marked endonucleolytic type of degradation of RNA (Fig. 4), poly A and poly AU.

"Acidic" RNases degrade RNA to mononucleotides very quickly. Endonucleolytic type of degradation of RNA was marked only under relation enzyme to substrate 1:2000 and more in early periods of splitting.

Nucleotides and nucleosides are competitive inhibitors of RNase P. brevicompactum. The inhibitory effect of the nucleotide isomers having the same base was found to be decreased in the following order: 2> 3> 5 (Fig. 5). M. IRIE in 1964 reported that "the highest inhibitory effects of 2' -phosphates in RNase A and T_1 strongly suggest that the presence of similar steric specificity in the binding of nucleotides with these enzymes. It may be interesting to know whether or not such a steric specificity exists in all the known RNases, since it is reasonable to assume that such a steric specificity is related to the mechanism in hydrolytic cleavage of cyclic phosphate at 2' -side in the ribonucleases".

Nowadays similar phenomenon has been reported for RNase A (pyrimidinspecific), RNase T_1 (guanylicacidspecific) as well as RNases T_2, M and P. brevicompactum (the last three enzymes are non-specific RNases). Hence, we have more reasons in order that such a steric specificity can be assumed to be common for all RNases.

The inhibitory effect of the nucleotides having the phosphate group at 2' (3') position of ribose increased in the order U< G< C <A. The similar dependence was noted by IRIE (1969) for RNase M, Asp. saitoi. Nucleotides with the phosphate group at 5' position have the following dependence in their inhibitory effect: G> C> I> A> U (on C>p as a substrate).

Cytidine-5' -di- and 5' -triphosphates inhibited RNase P. brevicompactum stronger than cytidine-5' -monophosphate did (Fig. 6).

Among four common ribonucleosides, the inhibitory effect was in the following order: A> G> C> U (Fig. 7), but that of the RNase M was A> C> G > U.

The present report is a brief account of investigation carried out by the author in collaboration with I. V. ILYINA, V. G. MOROSOVA, V. I. KRUPIANKO, V. M. GRISTCHENKO, L. I. BORODAEVA, R. B. MARKAUSKAJTE, N. J. MARKELOVA, Z. N. BAGDASARJAN.

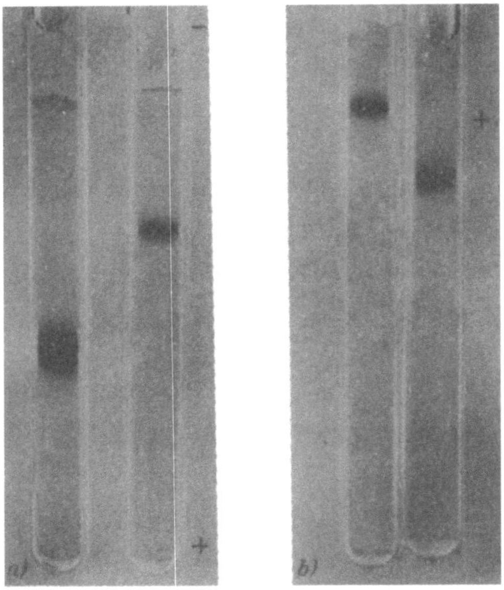

Fig. 1. Disc electrophoresis of RNases.
a) at pH 8, 3 1 - RNase P. claviforme
2 - RNase P. brevicompactum.
b) at pH 3, 0 1 - RNase P. claviforme
2 - RNase P. brevicompactum.

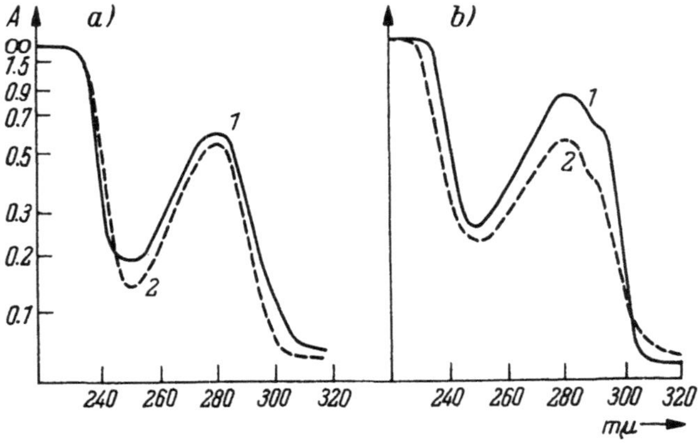

Fig. 2. UV spectra of RNases.
a) alcaline RNases 1. C_2 Asp. clavatus
2. P. chrysogenum 152 A
b) acidic RNases 1. P. claviforme
2. P. brevicompactum.

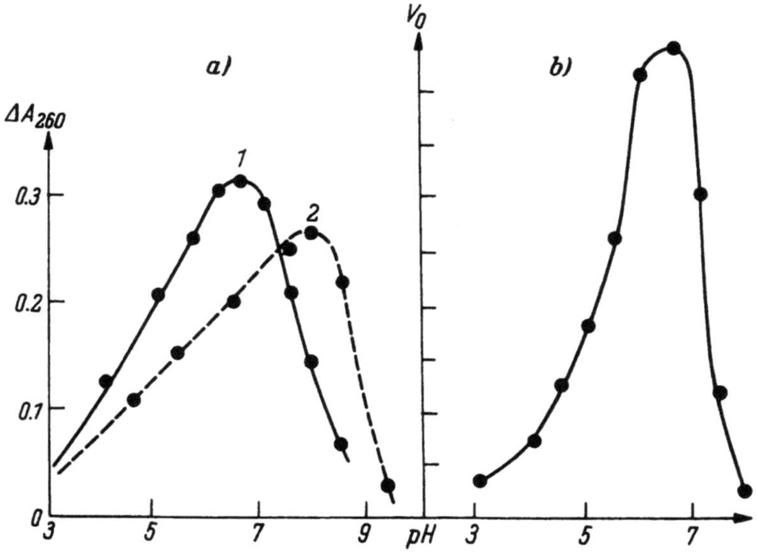

Fig. 3. Effect of pH on the activitv of RNases.
a) Substate - RNA 1. acidic RNase P. claviforme
2. alcaline RNase P. chrysogenum 152 A
b) Substrate - U>p, acidic RNase P. chrysogenum 152 A.

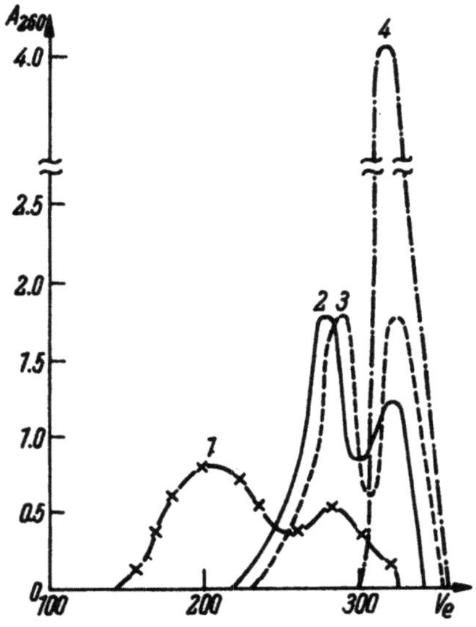

Fig. 4. Gel-chromatography on the Sephadex G-100 of products of degradation RNA by RNase F. semitectum.
1 - RNA
2 - 30-min. hydrolysate
3 - 24-hr. hydrolysate
4 - AMP

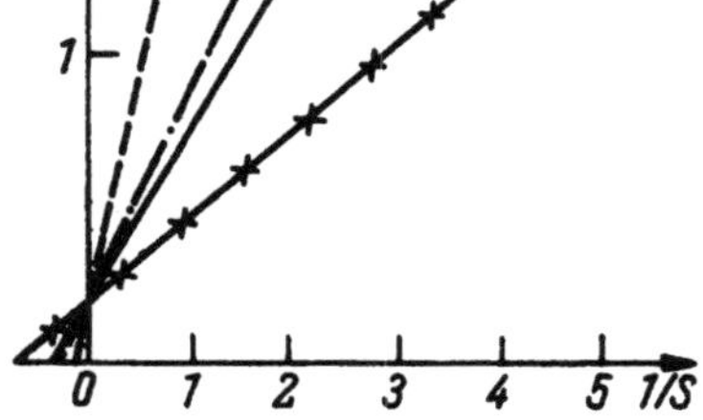

Fig. 5. Competitive inhibition of RNase P. brevicompactum by cytidine nucleotides.

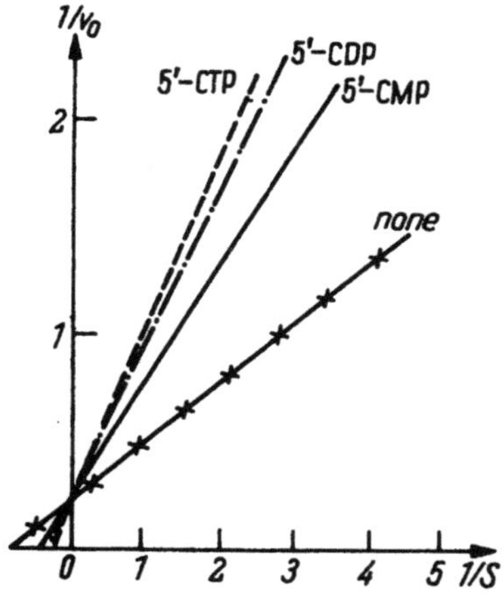

Fig. 6. The inhibition of RNase P. brevicompactum by CMP, CDP and CTP.

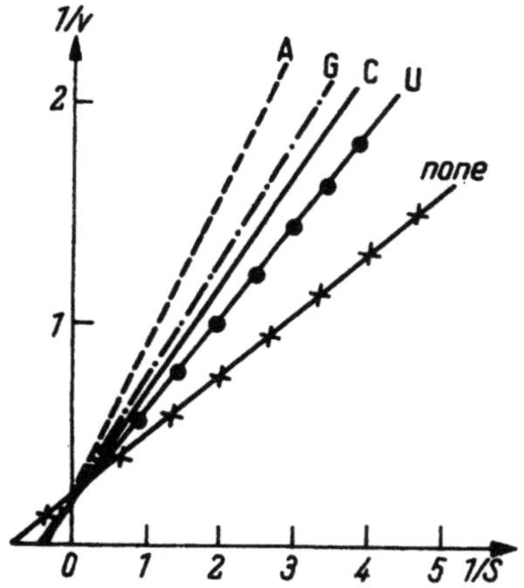

Fig. 7. The inhibition of RNase P. brevicompactum by various nucleosides.

5. SPECIAL TOPIES

Biochemisches Institut der Universität Freiburg im Breisgau
and Gesellschaft für Strahlen- und Umweltforschung München

R. WOHLHUETER, H. HOLZER

THERMODYNAMICS OF THE ADENYLYLATION OF GLUTAMINE SYNTHETASE

In a previous paper Prof. Holzer has discussed the regulatory significance of the adenylylation of glutamine synthetase in E. coli. Attempts to define the regulatory event as fully as possible have unearthed a number of interesting features of the enzymology and chemistry of adenylyl transfer. In this paper we want to emphasize some aspects of the reaction novel to biological chemistry, in particular the generation of a new kind of "high-energy phosphate bond". At the same time we will point out some of the problems we have encountered, how we have coped with them, and how such problems are of general relevance to the study of nucleotidyl transfer reactions.

Nucleotidyl group transfer is well established in the repertoire of enzymatic reactions. In Table 1 we have listed some example and have classified them into

Table 1

Reaction Type	Examples	Equilibrium	Reference
ATP + R-O-P $\rightleftharpoons$ A-P-P-R + PP_i	ATP: nicotinamide mononucleotide adenylyltransferase	K_{eq} = 0.45 (yeast, 3mM Mg)	(1)
	ATP: pantheteine-4'-phosphate adenylyl-transferase	reversible	(2)
NTP + glycosyl-P $\rightleftharpoons$ NDP-glycoside + PP_i	UTP: α-D-glucose-1-phosphate uridylyl transferase	K_{eq} = 0.3	(3) (4)
NTP + 3'-OH-NA $\rightleftharpoons$ NA_{n+1} + PP_i	DNA Polymerase	reversible	(5)

three categories, based on the type of acceptor. The first type occurs in the synthesis of the coenzyme dinucleotides, in which the acceptor is phosphate esterified with a primary alcohol. Reactions of this type have proven to be readily reversible. A second sort of nucleotidyl transfer is that found in the synthesis of the various nucleotide-diphosphate sugars, where the acceptor is a glycosyl-1-phosphate. These reactions, too, involve little free-energy change.

A very important category is the nucleic acid polymerases. In contrast to our first two categories, the acceptor here is the bare 3' -hydroxyl of the (deoxy)ribosyl group. Although this reaction too has been shown to be reversible in the presence of high pyrophosphate concentrations (5), one must recognize that the free-energy of the products includes some 7.5 kcal due to base-pairing of the newly added nucleotide with its template, and that this hydrogen-bond energy is presumably available to drive the reverse reaction. The transfer of the adenylyl group from ATP to glutamine synthetase fits none of these prototypes and thus represents a new category of biological nucleotidyl group transfer:

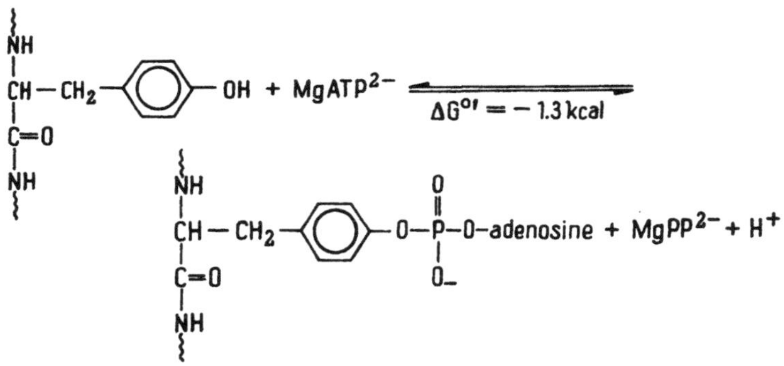

Here the acceptor group is the phenolic hydroxyl of a tyrosyl residue in the glutamine synthetase peptide chain (6); the products are a protein-bound phosphodiester and inorganic pyrophosphate (7); the reaction is reversible (8). Furthermore, in the present paper we will present evidence a) that the magnesium complexes of ATP and pyrophosphate are the actual reactants, b) that the glutamine synthetase subunit reacts independently of the state of adenylylation of its neighbors, and c) that the reaction is only slightly exothermic.

The first problem that confronts one who wishes to determine the thermodynamic parameters of such a reaction is the multiplicity of ionic forms in which ATP and pyrophosphate exist in neutral solutions containing magnesium. Pyrophosphate, in addition to its various protonation levels, forms single and double complexes with magnesium. ATP, of course, also exists as MgATP complex, and when both ATP and pyrophosphate are present in solution simultaneously they compete for available magnesium ions. Dissociation constants for the various individual metal-ligand pairs are available from the literature (9), and the combination of these equations with the conservation equations for total magnesium, pyrophosphate, and ATP permits calculation of the composition of a mixture of all the various ionic forms in equilibrium with one another. The solution of such a system of equations is arduous, so we have enlisted a computer's aid to make the calculations (10).

The results obtained by such calculations are illustrated in Fig. 1. The upper part shows how the distribution of the major forms of ATP change as a function of total magnesium concentration in a solution containing 1 mM total ATP at pH 7.4. At low magnesium concentrations the changes are very steep, but above 5 mM magnesium the change becomes slight, since ATP is present nearly completely as MgATP. The more complicated picture with pyrophosphate is shown in part B, for pH 7.4 and 1 mM total pyrophosphate. The relative proportions of $MgPP^{2-}$ and Mg_2PP^{o} vary over a wide range of magnesium concentrations, and it may be appreciated, for instance, that, if the actual product of a reaction is a specific ionic form of pyrophosphate, an equilibrium constant calculated using the total concentration of pyrophosphate, will be a) an overestimate, and b) dependent upon the magnesium concentration. How these considerations work in practice we will show below, but first we should describe how we measure equilibrium of the reaction.

Glutamine synthetase is composed of 12 identical subunits, each of which may be adenylylated. Thus, on a population average at least. there may be from 0 to 12 moles of AMP incorporated per mole of enzyme. Reaction mixtures containing pure glutamine synthetase, pure adenylyltransferase and various ratios of pyrophosphate to ATP are allowed to attain equilibrium. The ratio of adenylylated to non-adenylylated enzyme subunits are computed from linear extrapolation between the two extremes of fully adenylylated and non-adenylylated enzyme. Pyrophosphate and ATP are used in sufficiently high concentration (compared to glutamine synthetase) so that their concentrations at equilibrium do not differ appreciably from their initial concentrations. An example of the sort of results obtained is given in Table 2. Although there is some disagreement over the matter of linearity (11), this and many similar sets of data give us reasonable confidence in the aptness of our assumption of the linear relationship between activity and adenylylation, since the constants thus calculated are indeed constant. They suggest one further important conclusion, namely that each subunit of glutamine synthetase behaves as a thermodynamically independent entity. We see that with each concentration term at unit exponent there is equivalence among the constants found at various pyrophosphate/ATP ratios. If cooperative effects existed in the adenylylation of the subunits, one would expect that the expression for the equilibrium constant would involve exponents greater than one, in which case the essential property of equivalence would be lost. Let's return now to the question of the distribution of pyrophosphate ions and its influence upon equilibrium calculations. We will consider two possible expressions for an apparent equilibrium, one based on the total concentrations of pyrophosphate and ATP (K_T), the other on the concentrations of the magnesium complexes (K_{ML}):

Table 2

Reaction	$(PP)_{total}$	$(ATP)_{total}$	Adenylylation Ratio	K_{eq} 1)
	(mM)	(mM)		
a	2.0	0.15	0.883	11.8
b	2.0	0.30	1.81	12.1
c	1.0	0.15	2.21	14.8
d	1.0	0.30	4.15	13.8
e	1.0	0.45	5.35	11.9
average ± standard error			12.9 ± 0.6	

1) $$K_{eq} = \frac{(\text{subunit-AMP})\ (PP)_{total}}{(\text{subunit})\ (ATP)_{total}}$$

The reaction mixtures contained 0.1 M imidazole/HCl buffer, pH 7.0, 20 mM L-glutamine, 10 mM $MgSO_4$, 130 µg glutamine synthetase, 35 µg adenylyltransferase, and ATP and pyrophosphate as noted in a total volume of 200 µl. Activity corresponding to completely adenylylated enzyme is that found when (PP_i) = O and (ATP) = 1 mM. Activity was measured in a similar reaction where pyrophosphate and ATP were replaced by 10 μg phosphodiesterase. Incubation was at 30°; 10 μl samples were removed periodically for glutamine synthetase assay (12).

$$K_T = \frac{(\text{subunit-AMP})\ (PP)_{total}}{(\text{subunit})\ (ATP)_{total}}$$

$$K_{ML} = \frac{(\text{subunit-AMP})\ (MgPP^{2-})}{(\text{subunit})\ (MgATP^{2-})}$$

For each reaction mixture we calculate the concentration of $MgPP^{2-}$ and $MgATP^{2-}$ from the combined dissociation equations, as outlined above, and use these calculated concentrations in turn to calculate K_{ML}.

Fig. 2 shows the average equilibrium constants measured at pH 7.4 and four different magnesium concentrations. K_T (the constant calculated with total pyrophosphate and ATP) shows a strong dependence upon the magnesium concentration. When the same data, however, are used to calculate the concentrations of $MgPP^{2-}$, $MgATP^{2-}$, and the constant K_{ML}, the latter is seen to be largely independent of magnesium concentration, particularly in the range from 5 to 20 mM. This we interpret to mean that K_{ML} is the more realistic expression for the equilibrium, and that the ionic forms used in its calculation more accurately correspond to the state of the reactants which actually participate in the reaction, i. e. that the reaction should be written with the magnesium complexes, as we had done above.

Taking the value of K_{ML} (8.73 at pH 7.4, 30°, (Mg) = 10 mM), we calculate a standard free energy change of - 1.3 kcal $mole^{-1}$ for the adenylylation reaction. The free energy of hydrolysis of the pyrophosphate bond of ATP under comparable conditions (9) is - 11.4 kcal; the free energy of the hydrolysis of the tyrosyl-AMP bond in glutamine synthesis is thus - 10 kcal $mole^{-1}$.

This adenylyl group transfer, so important for the regulation of glutamine synthesis, thus constitutes a new category of biological nucleotidyl group transfer reaction and results in the production of a new type of "high-energy phosphate bond".

Table 3

Reaction	Observed Standard Free Energy, $\Delta G^{o\prime}$
	(kcal $mole^{-1}$)
$ATP + H_2O \rightleftharpoons AMP + PP$	-11.4 (from ref. 9)
$AMP + GS_{subunit} \rightleftharpoons GS_{subunit}\text{-}AMP + H_2O$	+10.1
$ATP + GS_{subunit} \rightleftharpoons GS_{subunit}\text{-}AMP + PP$	-1.3 (measured)

REFERENCES

(1) KORNBERG, A., J. Biol. Chem. 182, (1950), 779

(2) IMSANDE, J., P. HANDLER, The Enzymes (P. D. BOYER, H. LARDY and K. MYRBÄCK, eds.), Academic Press, New York, 5 (1961), 281

(3) GINSBURG, V., Adv. in Enzymology 26, (1964), 35

(4) TURNER, D. H., J. F. TURNER, Biochem. J. 69, (1958), 448

(5) RICHARDSON, C. C., C. L. SCHILDKRAUT, H. V. APOSHIAN, A. KORNBERG, J. Biol. Chem. 239, (1964), 222

(6) HEINRIKSON, R. L., H. S. KINGDON; J. Biol. Chem. 245, (1970), 138

(7) HEINRICH, C. P., F. A. BATTIG, M. MANTEL, H. HOLZER, Arch. Mikrobiol. 73, (1970), 104

(8) MANTEL, M., H. HOLZER, Proc. Nat. Acad. Sci. U. S. A. 65, (1970), 660

(9) ALBERTY, R. A., J. Biol. Chem. 244, (1969), 3290

(10) WOHLHUETER, R. M., manuscript in preparation

(11) GINSBURG, A., J. YEH, S. B. HENNIG, M. D. DENTON, Biochemistry 9, (1970), 633

(12) KOHLHAW, G., W. DRÄGERT, H. HOLZER; Biochem. Z. 341, (1965), 224

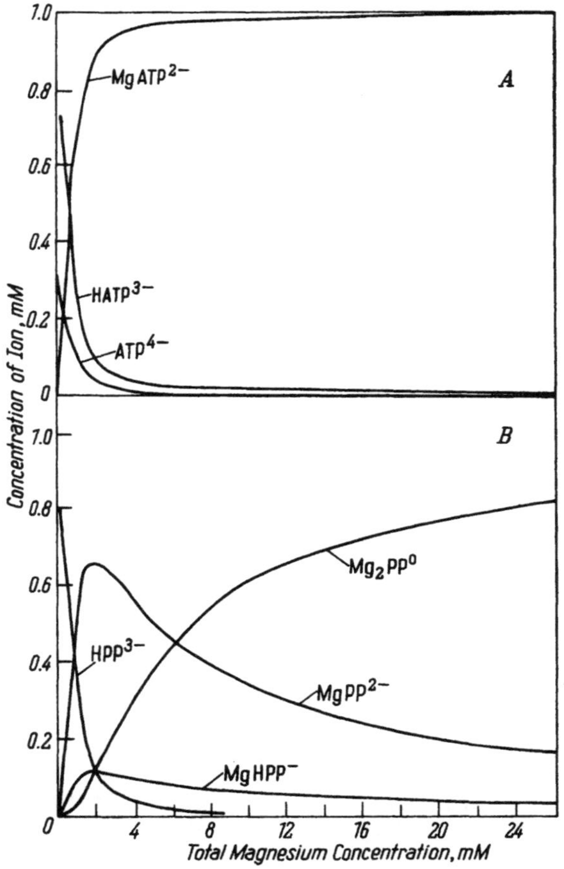

Fig. 1. A. Concentrations of some ionic forms of ATP in solution at pH 7.4 and $(ATP)_{total}$ = 0.001 M as a function of total magnesium concentration. B. Same for pyrophosphate, pH 7.4 and $(PP)_{total}$ = 0.001 M. The curves are theoretical, computed as described in ref. (10).

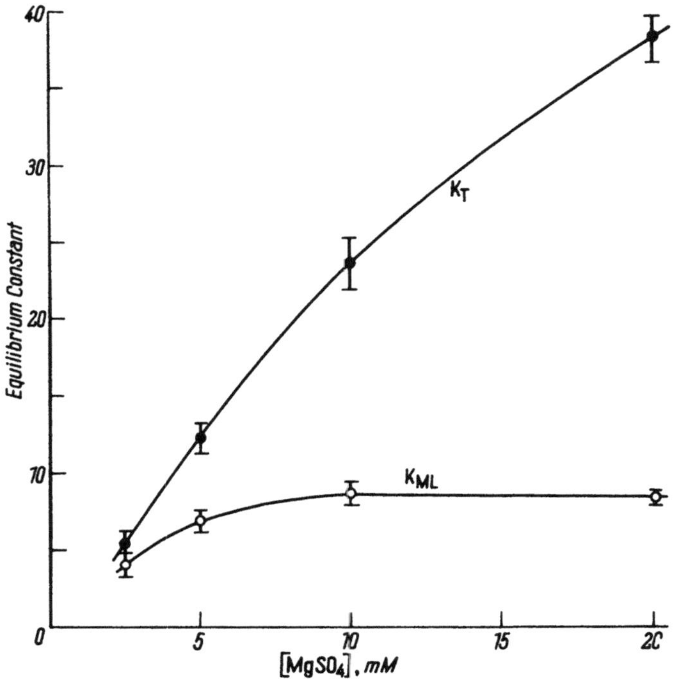

Fig. 2. Variation of equilibrium with concentration of magnesium (pH 7.4). Filled circles indicate values of K_T, open values of K_{ML}, as defined in the text. Each point is the average of five determinations, each with different $(PP)_{total}/(ATP)_{total}$ ratio, similar to examples given in table II.

Department of Biochemistry, State University Leningrad

S. N. LYSLOVA

TEMPERATURE DEPENDENCE OF CREATINE KINASE REACTION RATE

Introduction

Creatine kinase (ATP: creatine-phosphotransferase, E. C. 2. 4. 3. 2.) may be placed among the proteins that exist as different catalytically active conformers (1, 2). In this connection it is of interest to investigate the temperature influence on catalytical properties of enzymes. Some authors report on variations in thermal stability of creatine kinase from homologous tissues of various species and heterologous tissues of one species of animals (3, 4).
The purpose of this investigation was to study how the temperature effected the velocity of the creatine kinase reaction of grown up and developing animals. The enzyme was taken from tissues of frogs (Rana ridibunda, Rana temporaria), tortoise (Test. horsf.), lizard (Agama sang.), white mouse and tadpoles of Rana temporaria.

Methods

Creatine kinase activity was determined by the formation of creatine in the conditions optimal for the action of the enzyme (5, 6).
The enzyme activity of tissue extracts and that of purified enzymes was compared in the samples similar in composition but incubated at various temperatures. The purified enzyme was obtained after separating the tissue extracts by starch gel electrophoresis. The same techniques were used for determining creatine kinase isoenzyme variation of tissues.

Results and discussion

The investigation has revealed that the temperature dependence of creatine kinase reaction rate has essential peculiarities connected with temperature conditions in which the species inhabit. Fig. 1 represents the temperature dependence of the reaction rate obtained for the enzymes of tissues from various animals. The curves show that creatine kinase activity from skeletal muscles increases with the rise of temperature: for white mouse and Rana ridibunda - up to 37^{0}, for Rana temporaria - up to 30^{0}, for tortoise and lizard - up to 37^{0}-43^{0}. The general character of these curves for the skeletal muscles of tortoise and lizard, for the animals that are more acustomed to warm surroundings than all the other studied species, differs from the curves obtained for frogs and mice. It is interesting to note that the temperature optimum of creatine kinase reaction in Rana temporaria brain and cardiac muscle (in opposite to skeletal muscle) lies not at 30^{0} but at 37^{0}. The thermal stability of creatine kinase from various tissues of Rana temporaria is also different. Thus it is much lower for the enzyme from skeletal muscle than for that from brain and cardiac muscle. When heated for 30 min. at 35^{0} the skeletal muscle enzyme is 90 % inactivated, whereas the brain and cardiac muscle enzyme is inactivated only by 40 %. The question then arises as to whether such differences in teperature dependence of creatine kinase reaction rate may exist in tissues of developing frog. To clear it out creatine kinase activity and isoenzyme variation were studied at the Rana temporaria tadpole metamorphosis. We investigated the head, body and tail of tadpoles of different age, and beginning with the 40th day of their growth skeletal muscles and internal organs. In all the parts of tadpoles at early stages of their development when animals are almost immovable creatine kinase is distinctly detected, though the enzyme is present only in trace amounts. As the tadpoles grow and their mobility increases the creatine kinase activity rises too at a rate different for various parts. But there are two falls of creatine kinase activity. One fall occurs on the 12th-15th day of development, the other - on the 50th day (in the head and tail). The reason for the first fall of creatine kinase activity remains obscure. A possible explanation for the second fall may lie in the fact that it coincides with the last stages of metamorphosis when the animal organism undergoes complex morphological and biochemical change. In the course of development there is a change not only in creatine kinase activity, but also in how the enzyme reacts to temperature changes. Up to the 13th-15th day of development the creatine kinase activity is higher at 37^{0} than at 20^{0}. In the tail this rise is not great, but in the head and body the creatine kinase activity increases by 2 - 2.5 times. After the 13th-15th day of development the creatine kinase activity reacts to temperature changes vica versa: in all the parts of tadpoles the creatine kinase activity is higher at 20^{0} than at 37^{0}. (Fig. 2). The observed phenomenon might be attributed to the change of isoenzyme variation at tadpole metamorphosis. But the study of creatine kinase isoenzymes from head and tail on the 8th, 13th, 15th and 30th day of development has shown that during this period of tadpole metamorphosis the isoenzyme variation remains constant, although it is not uniform in all the analysed parts of tadpoles. In the tail there have been found two fractions: one is the enzyme of muscle type (MM, III-isoenzyme), it predominates in amounts; the other, which moves towards the cathode, is in trace amounts. Only one fraction - the enzyme of brain type (BB, I-isoenzyme) - has been detected in the head of tadpoles (Fig. 3). From what has been said above it is possible to conclude that: (a) there is a certain correlation between the functional state of tadpoles and creatine kinase activity in their different parts. The same parallelism was observed with homoiothermic animals (6, 7). (b) The fact that there is no change in isoenzyme variation at the investigated stages of tadpole development makes it possible to assume that frog tissue creatine kinase is able to undergo a peculiar conformation change of the molecule ("metamorphosis") connected with the conditions in which the enzyme exists in the cell. As a result of this process the creatine kinase becomes more active at 20° than at 37°. It is known that in solutions the enzyme molecules are also able to undergo structural change with the change of temperature (8, 9). BEHRISCH (10) reports that when temperature alters there is a change in fructose diphosphatase affinity to ions-activators and substrates with the poikilothermic animals. A number of authors provide other examples of alterations taking place in the structure of the protein with the aging of the animals. Thus FORNAINI et al. (11) assume that there appear some structural modifications of glucose-6-phosphatdehydrogenase when human and rabbit erythrocytes are aging in vivo. The authors come to this conclusion on the basis of the observed alteration of the enzyme properties with the age (decrease of activity, increase of thermolability and some others) while electrophoretic mobility and pH-optimum remain constant.

Summary

Creatine kinases from tissues of various species of animals have different optimal temperatures of action and thermal stability. At metamorphosis of Rana temporaria tadpoles creatine kinase is affected by temperature changes, whereas isoenzyme variation remains constant.

REFERENCES

(1) EPPENBERGER, M. E., H. M. EPPENBERGER, N. O. KAPLAN, Nature, 214, (1967), 239

(2) MARKERT, C. L., G. S. WHITT, Experientia, 24, (1968), 977

(3) DAWSON, D. M., J. Biol. Chem., 242, (1967), 204

(4) DMITRENKOW, M. P., Ukr. Biochem. J., 40, (1968), 44

(5) ENNOR, A. H., H. ROSENBERG, Biochem. J., 57, (1954), 203

(6) LYSLOVA, S. N., A. K. DONDUA et al., J. Evol. Biochem. Physiol., 4, (1968), 3

(7) EPPENBERGER, H. M., M. E. EPPENBERGER, R. RICHTERICH, H. AEBI, Develop. Biol., 10, (1964), 1

(8) LEHRER, G. M., R. BARKER, Biochemistry, 9, (1970), 1553

(9) MASSEY, V., J. Biol. Chem., 241, (1966), 234

(10) BEHRISCH, H. W., Biochem. J., 121, (1971), 399

(11) FORNAINI, G., G. HEONCINE, P. SEGNI, G. A. CALABRIA, M. DASHA, Europ. J. Biochem., 7, (1969), 214

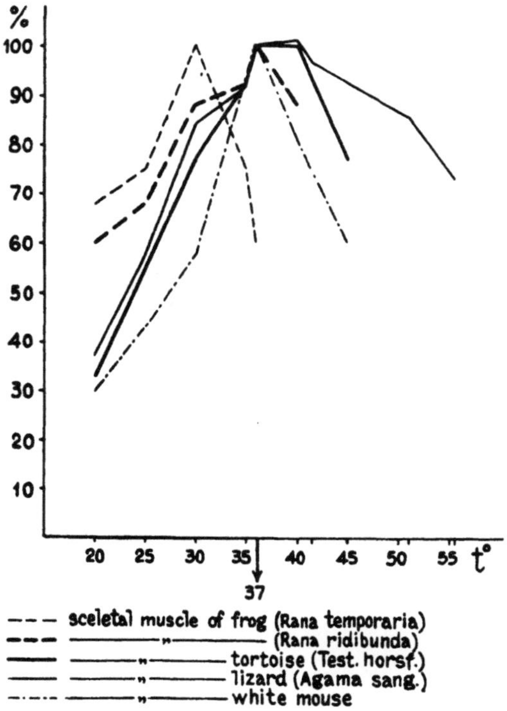

Fig. 1. Temperature optimum of creatine kinase reaction.

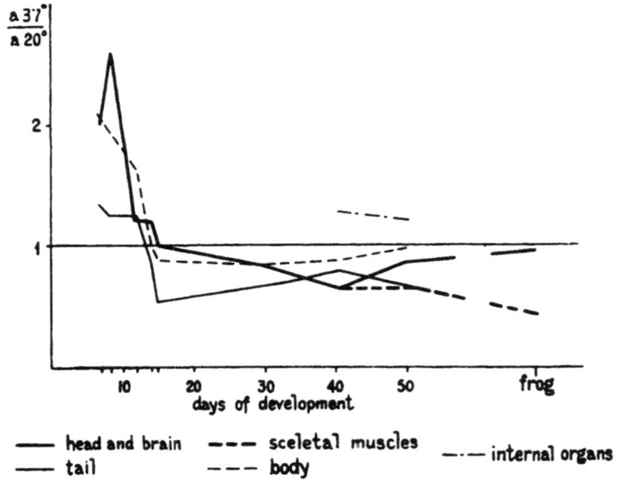

Fig. 2. Correlation of creatine kinase activities at two temperatures (enzyme from the tissues of Rana temporaria developing tadpoles).

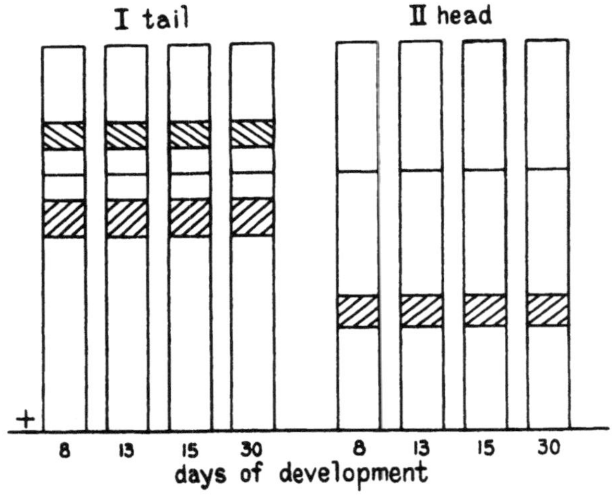

Fig. 3. Creatine kinase isoenzyme variations in the tissues of Rana temporaria developing tadpoles.

Bach Institute of Biochemistry USSR Academy of Sciences, Moscow

K. B. SEREBROVSKAYA

THE BEHAVIOR OF TRANSFERASES IN HETEROGENIC LIPID-WATER SYSTEMS

In early 1930-ies A. I. OPARIN and his coworkers published in the "Biochemische Zeitschrift" several papers concerning inactivation of a number of enzymes during their adsorption on colloidal precipitates. The enzymes could be reactivated through their desorption by other proteins (1).
At that time (2) it was already known that the protoplasm is a gel-like structure. On the basis of their experiments with regards to a reversible inactivation of enzymes A. I. OPARIN and his followers developed a hypothesis on the mechanism of reversible desorption in the regulation of enzyme activity in a living cell (3). They believed that cellular lipids played an important role in the process. Today these concepts are evidely develop by P. SIEKEWITZ and coworkers (4).
The second importand stage in the study of this problem was the work of D. L. TALMUD and his collaborators in early 1940-ies. The investigated the role of hydrophobic forces in the protein - protein and protein-lipid interactions (5). Their ideas are widely used in up-to-day studies of protein conformation transitions (6).
The third stage in the study of the problem was the investigations carried out in this country in 1950-ies with respect to a reversible action of enzymes in a living cell. It has been shown both by Soviet biochemists and scholars abroad that invertase appears to be transferase capable to transfer β-fructosyl of sucrose to different acceptors (water, alcohols and other sugars) (7).
The reaction develops according to MYRBACK' s scheme:

$$E + S \rightleftharpoons ES \rightleftharpoons E - P_2 + P_1 \longrightarrow P_2 + E$$

This scheme has been confirmed by JOUNG (8) and is applicable, in his opinion, to most transferases.
Bearing in mind the above concepts we investigated activity of transferases

(ribonuclease, hexokinase and ATPase of myosin, in model lipid-water systems. The lecithin fraction produced from the brain tissue of cattle or from the egg yolk served as lipid. The lipid-water system was prepared according to the BUNGENBERG, de JONG method (9) used in the production of lipid coacervates and examined with the JEM-7A J. KUCHARA (Wrozlav Institute of Agricultural Research) and W. I. STEIN-MARGOLINA (Bach Institute of Biochemistry Moskow) (10).

Fig. 1, represents structures of the lipid-water system multiplied by 300.000 times. These structures prepared by condensing the molecular-dispersed lipid (9) are completely identical to BENGHAM's liposomes (2) prepared through dispergation of condensed lipids.

Fig. 2, shows serum albumin treated liposomes wich are entirely identical to liposomes obtained with the aid of subtilisin treatment.

The liposome-protein interaction according to Bengham's terminology, results in protein-lipid penetration. It seems probable that the penetration is followed by a chemical protein-lipid interaction.

Fig. 3 presents data on the influence of liposomes upon RNase activity (12) Curves 1, 2, 3, 4 represent control measurements enzyme activity in a homogeneous solution as a function of time (the enzyme concentration varied from $2.5 . 10^{-3}$ mg/ml to $2 . 10^{-2}$ mg/ml).

Curves 5, 6, 7, 8 indicate measurements of enzyme activity in the presence of liposomes prepared from lecithin stored at -78^{0}. As seem from the data, a decrease of the enzyme concentration in the reation mixture brings about a gradual inhibition of its hydrolytic activity of the enzyme occurs its.

The presence of liposomes in the mixture results in the development of a synthetic process whose rate is the greater, the less the concentration of the enzyme. By assaying the iodine number, it was shown in parallel experiments that lecithin can be dehydrated in the cold, on the other hand, a study of extra-weak luminescence in lipid-water systems indicated that the lipid-water interaction results in the disappearance of lipid peroxides wich are always present in an initial lecithin preparation.

These date a comparison of the effect of liposomes produced from different age lecithin preparations on the RNase activity (13).

Fig. 4 shows the date obtained with two enzyme concentrations of $2.5 . 10^{-3}$ and $2 . 10^{-2}$ mg/ml, the substrate concentration being the same (curves 1 and 2). Curves 3 and 4 represent the enzyme activity in the presence of liposomes

prepared from two weaks old lecithinand curves 5 and 6 - in the presence of liposomes from three months old lecithin.
These data demonstrate that at high enzyme concentrations "young" liposomes inhibit its but slightly. A decrease of the enzyme concentration induces synthetic a activity during the first minutes of incubation which seems to be the ligher the "older" are the liposomes. In the course of incubation with the substrate the hydrolytic activity of the enzyme re-appears.
Simular date were obtained in regard to the effect of liposomes on activity of ATPase of myosin (14, 15, 16) and yeast hexokinase (17, 18). In these experiments exclusively a fresh lecithin fraction of brain was used for the preparation of liposomes and the influence of the lipid-water system of different age upon enzyme activity was studied. It was established that one-day liposomes greatly increase the rate of the direct reaction (Fig. 5, a and b, curves 2), whereas 3-5 day liposomes reverse the action of the tested enzymes (curves 3).
The difference of action of "young" and "aged" liposomes may be ascribed to variations in their donor-acceptor capacity: "young" liposomes may act as proton donors while the "aged" - as acceptors.
To verify this assumption specific experiments with respect to the influence of "aged" liposomes on the photooxidation of myosin ATP-ase in the presence of chlorophyll were carried out (19).
As known sensitized oxidation of myosin results in an increased viscosity of its solution due to polymerization of the protein oxidized molecules. In our experiments an introduction of liposomes sharply increased the rate of the polymerization (Fig. 6).
This may favour the idea that the lipid acts as a catalyst in this reaction.
The process may be represented by the followed scheme:

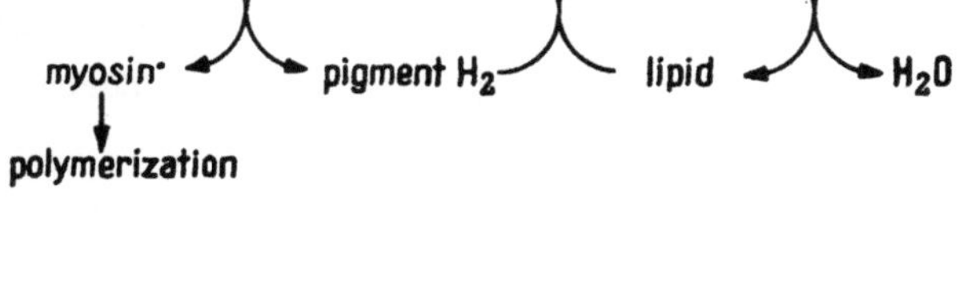

According to SHERAGA (20) the mechanism of protein polymerization of the myosin type involves an internal redox reaction with the participation of histidine and tyrosine. In a monomolecular protein these sites are closed intramolecularly. Polymerization follows the disruption of intramolecular bonds and the formation of intermolecular ones, i. e. in status nascendi.

In our opinion, it is the redox mechanism wich underlies the reversal of the transferase reaction. If it is assumed (21) that the dimer is an active form of RNase in the hydrolysis reaction and that the monomer is responsible for its synthetic activity, then the lipid may be regarded as a substance wich shifts the dimer $\rightleftharpoons$ monomer equilibrium towards monomerization of the enzyme by removing a proton from the active center, (e. g. from histidine). A study of the relationships between the specific synthetic activity and the concentration of the enzyme carried out in accordance with the method suggested by KURGANOV (22) has shown that in the lipid presence the activity increases along with a decline of the concentration.

These events testify in favour of the monomer synthetic activity.

In the case of hexokinase and ATPase "young" liposomes may also serve as proton donors, accelerating the rate of the direct reaction, whereas "aged" ones act as proton acceptors, contributing to the rate of the back reaction as with RNase (possibly at the expense monomerization on the lipid structure). According to SOLS (23) both enzymes work in this event as the some synthetic ATPase . It should be noted that a subunit ot yeast hexokinase with a molecular weight of 27. 5000 is fully identical to that of mitochondrial ATPase (6).

The presence of substance in the reaction mixture which may act as group acceptors in the transferase reaction may be regarded as a second indispensable condition for a reversal. In the case of RNase the product of the reaction first stage (cyclophosphate) may be transferred to acid-soluble oligonucleotides making them acid-nonsoluble. In the experiments with the reversal of ATPase hexokinase action a decrease of mineral phosphate in the first case and an increase of labile phosphate in the second one may be explained by the transfer of phosphorous of the phosphorylated enzyme to the medium ADP in the lipid presence. This transfer may take place only in status nascendi, i. e. during the disruption of phosphorous - enzyme bonds.

In conclusion I wish to express my deep feelings of gratitude to my teacher Academician A. I. OPARIN for his constant interest advice and valuable remarks.

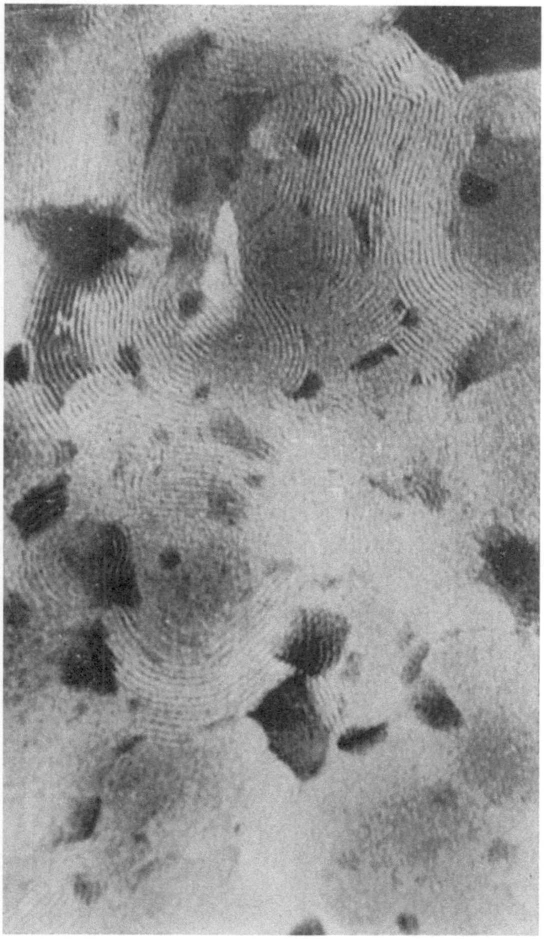

Fig. 1. Phosphatide-water system liposomes × 300000.

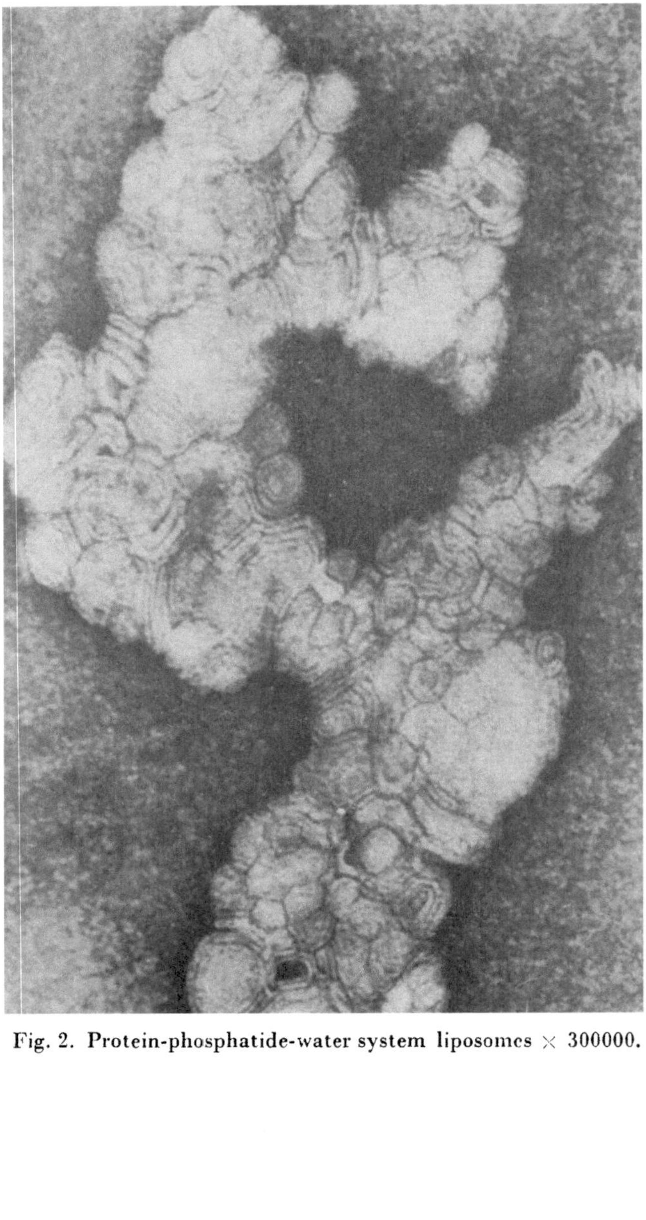

Fig. 2. Protein-phosphatide-water system liposomes × 300000.

REFERENCES

(1) OPARIN, A.I., A.I. KURSANOV, Biochem. Z. 239, (1931), 1

(2) LOEB. J., The Dynamics of living matter

(3) OPARIN, A.I., Isw. AN USSR, Biol. Ser. 6, (1937)

(4) SIKEVITS, P., In "The molecular control of cellular activity", N.Y. McGrow-Hill Book Co. 1962

(5) TALMUD, D.L., Journ. phys. chem. (USSR) 15, (1941), 532

(6) KLOTZ, J.M.. N.R. LANGERMAN, D.W. DARNALL, Annual Rev. Biochem., 39, (1970), 25

(7) OPARIN, A.I., La nature et le mechanisme d' action de l' invertase de la levure. Communications et rapports au III Congres International de Biochemie Bruxelles, 1-6 aout 1955

(8) YOUNG, Jui.H., Ann.Rev. Biochem. 38, (1969), 913

(9) BUNGENBERG de JONG, H.I., R.F. WASTERKAMP, Biochem. Z. 234, (1931), 347

(10) SEREBROVSKAYA, K.B., Ja. KUCHARA, W.J. STEIN-MARGOLINA, Dokl. AN USSR (in press)

(11) RACKER, E., Membranes of mitochondria and chloroplasts 1970

(12) OPARIN, A.I., K.B. SEREBROVSKAYA, N.W. VASILIEVA, W.M. SAMSONOVA, Dokl. AN USSR 179, (1968), 976

(13) SEREBROVSKAYA, K.B., Coacervates and protoplasm, Nauka, 1971

(14) SEREBROVSKAYA, K.B., W.M. SAMSONOVA, N.S. MENAJEDIBOVA, Dokl. AN USSR 180, (1968), 1250

(15) SEREBROVSKAYA, K.B., T.O. BALAEVSKAYA, T.I. OSIPOVA, Dokl. AN USSR 187, (1969), 939

(16) OPARIN, A.I., K.B. SEREBROVSKAYA, Dokl. AN USSR 185, (1969), 707

(17) OPSTIN, A.I., K.B. SEREBROVSKAYA, L.S. GOGILASCHWILI, Dokl. AN USSR, 185, (1969), 952

(18) SEREBROVSKAYA, K.B., L.S. GOGILASCHVILI, L.J. ANDRIADZE, Journ. evol. biochem. and physiol. (USSR), 5, (1969), 583

(19) SEREBROVSKAYA, K.B., T.K. LUZIK, T.I. OSIPOVA, Biochimia, USSR (in press)

(20) SHERAGA, H.A., In: The Proteins, vol. 1, H. Neurath (Ed.) N.Y.-London, 1963

(21) SCHAPOT, V.S., Nucleases. M. Medicina, 1968

(22) KURGANOV, B.I., Mol. biol. (USSR), 2, (1968), 430

(23) DELAFUENTE, G., In: FEBS Symposium 19, (1969), 249

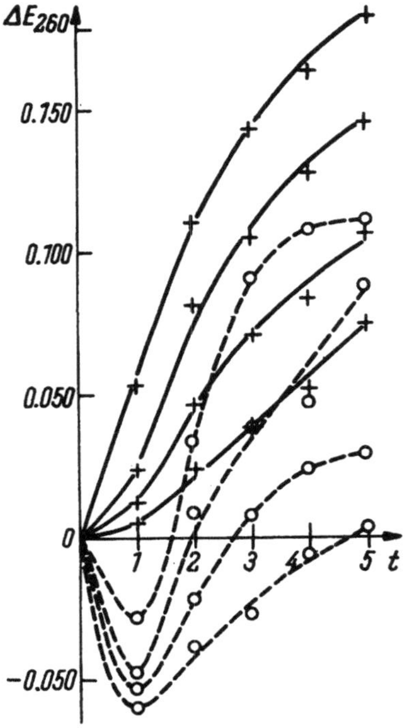

Fig. 3. Effect of liposomes on the ribonuclease activity by the different enzyme concentration.
1, 2, 3, 4 - in the absence of liposomes
5, 6, 7, 8 - in the presence of liposomes.

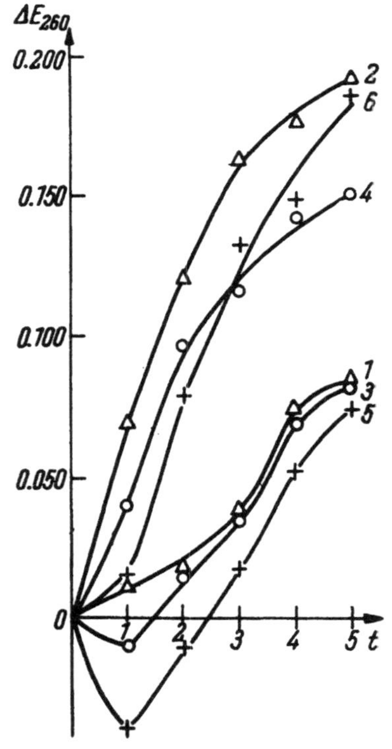

Fig. 4. Effect of liposomes on the ribonuclease activity by the different enzyme concentration.
1, 2 - in the absence of liposomes
3, 4 - in the presence of "young" liposomes
5, 6 - in the presence of "old" liposomes.

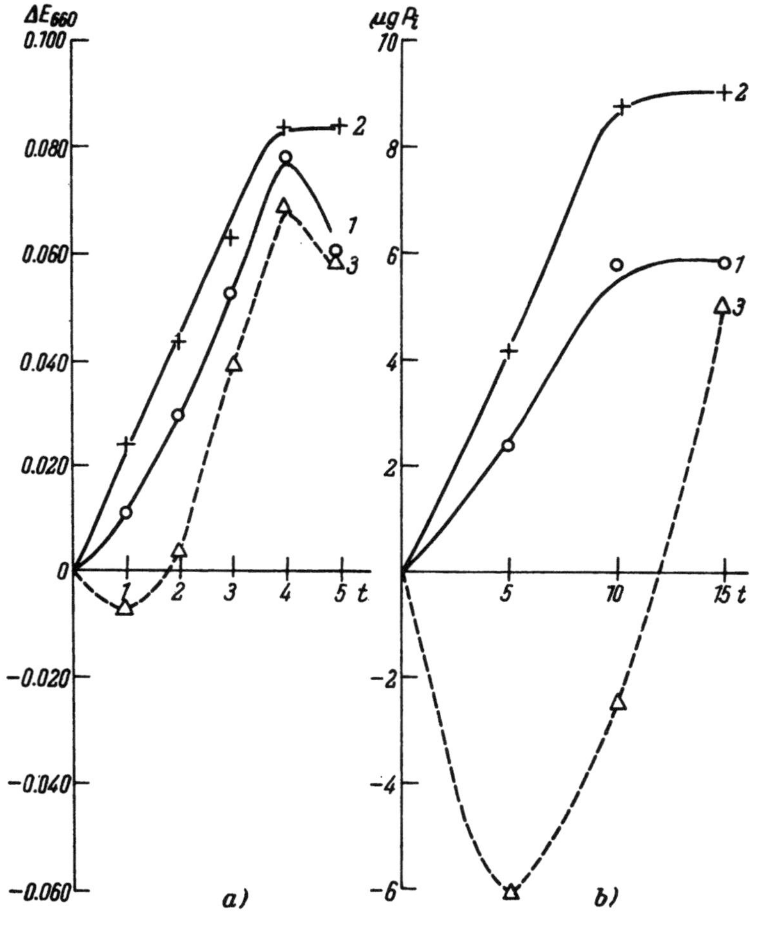

Fig. 5. Effect of liposomes on the ATPase and hexokinase activity.
a - ATPase, b - hexokinase
1 - in the absence of liposomes
2 - in the presence of "young" liposomes
3 - in the presence of "old" liposomes.

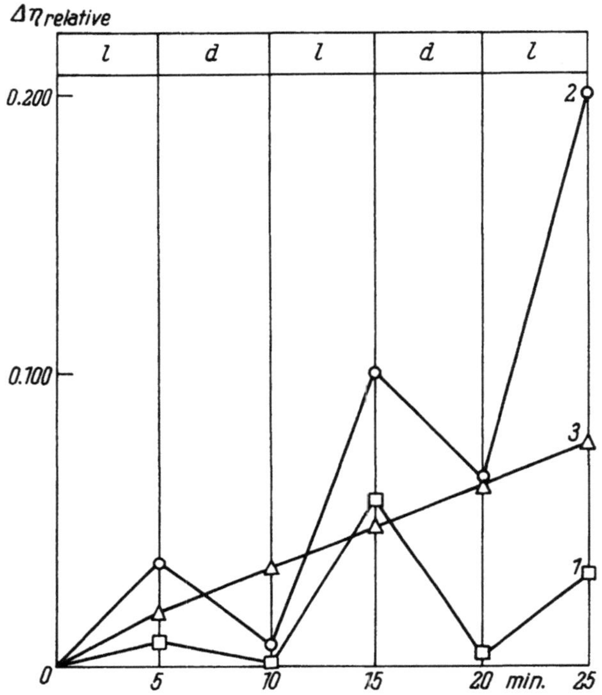

Fig. 6. Effect of "old" liposomes on the polymerization rate of the myosin ATPase in the light - dark change.
1 - in the absence of liposomes
2 - in the presence of liposomes ($8. 10^{-3}$ mg/ml lecithin)
3 - in the presence of liposomes ($22. 10^{-3}$ mg/ml).

Forschungszentrum für Molekularbiologie und Medizin der Deutschen Akademie der Wissenschaften zu Berlin, Berlin-Buch

J. G. REICH, G. WANGERMANN, MARGRIT FALCK

A GENERAL COMPUTER STRATEGY FOR THE ANALYSIS OF KINETIC AND BINDING CURVES

The kinetic analysis of an enzyme is usually the first step of a more detailed investigation either of its molecular mechanism or of its importance within a metabolic network. A series of experiments is carried out with varied ligand concentrations and a suitably recorded output vector, such as binding values or initial velocities. These data are then explored by graphical analysis, a system of plots and secondary plots, which serves to establish the numerical values of the kinetic parameters in question, and to discriminate among rival mechanisms.

But this graphical analysis becomes very tedious and often rather unsystematic, when more complicated mechanisms, in particular allosteric models, are to be considered (1). We have therefore made an attempt to develop a general strategy of parameter estimation from initial velocity or binding data which could allow for a more systematic approach to the problem of kinetic analysis of enzyme mechanisms.

The kinetic model in its most general form is

$$y_k = f(p, X_k),$$

where

$p = (p_1, p_2, p_3 \dots p_n)$ (set of kinetic constants)

$X_k = (lig_1, lig_2 \dots lig_d)$ (ligand concentrations)

$k = 1\,(1)\,m$ (number of experiments).

$y = (y_1, y_2 \dots y_m)$ is the fit to the experimental data $v = (v_1, v_2 \dots v_m)$. Since the experiment is complete at the time when the analysis is performed, X_k, the matrix of ligand concentrations, can be assumed to be constant so that y is a function of p, the values of the kinetic constants, alone.

The object of the strategy may then be stated as to find numerical values of these constants so that a good fit is obtained. A fit is intuitively considered as good, if, in a certain sense, its distance from the actually observed data is small.
Thus the goodness of fit is introduced as a distance within the space of experimental variables (the sample space). This concept may be stated more rigorously as follows:

$$\Phi(p) = \sum_{k=1}^{m} \left\{ w_k(v_k - y_k(p)) \right\}^2$$

where w_k is the weight factor of the k-th sample.
Thus Φ , the goodness of fit criterion, is introduced as the Euclidean distance between predicted fit and weighted data. Again, this criterion is a function of the parameter values, but also of the scale factor w_k the choice of which defines the statistical relevance of the fit criterion (for instance, weighted least squares, maximum likelihood, minimum χ^2, or minimum error).
The estimation of the parameters may now be formulated as a series of distance explorations in the sample space. The leading principle is that a fit is acceptable if the parameter values lead to a small distance between fit and data:

$$\Phi(p_{acc}) \leqslant \Phi_{eim}$$

One point p_{min} of this family of acceptable points p_{acc} is the minimum distance point:

$$\Phi(p) = \text{Min !}$$

the estimation of which is carried out by the following iterative algorithm:

$$p_{i+1} := p_i + dp \tag{1}$$

beginning with a guessed starting point and succesively applying a correction vector dp, until the minimum is found:

$$dp := - M \cdot g \tag{2}$$

The correction is taken in the direction of the negative gradient

$$g = \left(\frac{\partial d}{\partial p_i} \right) \qquad i = 1(1)n$$

of the Φ -criterion in the parameter speace. g is estimated by a 1:1000-perturbation difference formula.

M is the metric applied to this space (an n^2-matrix). Its choice determines the convergence properties of the algorithm. If simply

$$M = I \quad \text{(identity matrix)}$$

the method of steepest descents results, which has usually safe but extremely slow convergence.
If, on the other hand

$$M = G^{-1}$$

where

$$G = \left(\frac{\partial^2 \Phi}{\partial \rho_i \, \partial p_i} \right) \qquad i, j = 1(1)n \qquad \text{(information matrix)}$$

a modified Gauss-Newton search is obtained.
G is estimated, to accuracy of order Φ, by

$$G \approx g^T \cdot g$$

so that the iteration scheme with G^{-1} as matrix will usually not converge as long as Φ is high. Its eventual convergence when Φ is low (i. e. in the neighbourhood of p_{min}) is however very fast.
To combine the advantages of I and of G^{-1} as metric, we have followed an idea by MARQUARDT (2) and applied a linear combination of both:

$$M = G^{-1} + \lambda \cdot I \qquad \lambda \geq 0 \tag{3}$$

which contains the former ones as limiting cases. During the iteration scheme, λ is chosen as small as permissible to achieve stable convergence.
The algorithm (1), (2), and (3) was programmed in FORTRAN and Algol 60 and showed excellent convergence and numerical behaviour on a number of digital machines. It was applied to data from Phosphorylase b (paper by H. WILL, this symposium), Phosphofructokinase of yeast (presented in a discussion by E. HOFMANN and his group), and of Pyrophosphatase (discussion by T. RAPOPORT).
The input is prepared by the biochemist and contains:

1. the specifications of the models proposed (as addressed from a model catalogue or by an algebraic or topological code),
2. a starting point p_0 (stating the order of magnitude of p_{min}. It is guessed by inspection of the data and need not, as in other programs, be very precise),
3. the data, i. e. ligand concentrations and binding or velocity values.

Then the search for p_{min} is performed, and usually after 5 to 10 iterations the output from the computer is:

1. a statement whether or not the model and the data did fit together,
2. if so, the best parameter values, followed by
3. a region of statistical variation of each parameter as obtained by a linear expansion of Φ around p_{min},
4. graphs, tables and supplementary information on the best fit obtained.

If no convergence was achieved, informal statements on the reason of failure are given, which may include

1. Outliers in the data,
2. Incompatible model, which can be imitated by an extremely wrong starting point,
3. Nonunique model, imitable by a bad experimental design,
4. Statistical fluctuations in the data.

These cases are discussed in more detail in ref. (1). The next goal of our group is to develop a formalized diagnostic logic for the case that model and data do not fit together. This case is, of course, not rare, and its elucidation can help to discriminate among rival models or to design better experimental conditions.

REFERENCES

(1) REICH, J. G., FEBS Letters 9, (1970), 245

(2) MARQUARDT, D. W., J. Soc. Industr. Appl. Math. 11; (1963), 432

Physiologisch-chemisches Institut der Friedrich-Schiller-Universität Jena

A. HORN, G. CUMME und H. BÖRNIG

KINETIC INVESTIGATIONS OF ENZYMES WITH SUBSTRATES FORMING METAL COMPLEXES

Sigmoid kinetics of an enzyme may not only be produced by homotropic cooperativity (as e.g. is assumed in the allosteric models), but also by chemical interactions between the effectors, which cause interdepences of their concentrations.

A computer programme is described, by which from given total ligand and metal concentrations the free and complex concentrations may be calculated and thereupon inserted into any given velocity equation of the enzymatic model under study. This programme has been used to investigate the kinetic behaviour of inorganic pyrophosphatase, assuming four different non-allosteric models with Mg^{2+} and $P_2O_7^{4-}$ acting as effectors and the $MgP_2O_7^{2-}$ complex as substrate. The results demonstrate, that sigmoid kinetics (which have been established experimentally for this enzyme) may well be stem from the substrate level. Moreover, by aid of the programme experimental conditions may be determined, under which cooperativity at the substrate level may be discerned from cooperativity at the enzyme level. In the special case under study, this experimental condition consists of keeping constant the free Mg concentrations by which way the dependences of effector concentrations as well as enzymatic velocities on substrate concentration are greatly simplified. At constant $[Mg^{++}]$, the kinetic response is not cooperative (Hill coefficient ≤ 1) for the non-allosteric models, whereas an allosteric enzyme would exhibit $n > 1$ also at buffered $[Mg^{++}]$. Thus, a clear experimental distinction between given allosteric and non-allosteric models is possible.

Institut für Physiologische und Biologische Chemie der Humboldt-Universität Berlin
Physikochemisches Zentrum der Deutschen Akademie der Wissenschaften zu Berlin, Berlin-Buch
Physiologisch-Chemisches Institut der Karl-Marx-Universität Leipzig

G. GERBER, I. RAPOPORT, R. HINTSCHE, S. RAPOPORT, H. -J. BÖHME

INTERACTION OF IMIDAZOLE WITH MULTIVALENT ANIONS AS MODEL FOR BINDING OF PHOSPHOCOMPOUNDS TO PROTEINS

From investigations on binding of multivalent anions to haemoglobin we deduced a participation of imidazole groups (1, 2, 3). Fig. 1 shows a steep decline in the binding of ATP to oxygenated and deoxygenated human haemoglobin by changing the pH from 6 to 8, 5. The binding goes down to zero in the case of oxygenated haemoglobin. If binding of phosphocompounds to haemoglobin would be accomplished by electrostatic forces only other positively charged groups besides the imidazole should also be implicated and binding ought to be expected above pH 8, 5.
In Fig. 2 the influence of temperature on binding of ATP to methaemoglobin is shown on a Scatchard plot. Elevation of temperature decreases the binding. The thermodynamic parameters were reckoned from these and other data (3) not given here and they are summarized in the Table 1. The enthalpy of binding (presented in the column on the right) amounts to about -6, 8 kcal/mole. This value corresponds well with -6 to -8 kcal/mole given by WYMAN for the ionization enthalpy of the imidazole group (4).
To check wether the binding is restricted to phosphocompounds or wether anions in general will be bound methaemoglobin was titrated with several acids (Fig. 3). The curve on the right represents titration of haemoglobin with hydrochloric acid between pH 9 and 5. With the other acids quite different curves were obtained. That indicates an interaction of the anions added during titration with the haemoglobin. Binding of anions to proteins is accompanied by proton binding. The greater the charge of the anion added in this titrations the more the curves are shifted towards the alkaline range. The most pronounced effect has the fourfold negatively charged ferrocyanide. There is a maximal shift of the titration curves between pH 6 and 6, 5. pH-static experiments have been done, too, and they confirm this

Table 1

THERMODYNAMIC DATA OF THE ATP BINDING TO METHAEMOGLOBIN

pH	T	n	10^2xK	G	S	H	H[1]
	°C		M^{-1}	kcal/mole	cal x $mole^{-1}$ x $deg.^{-1}$	kcal/mole	
6.2	5	10	2.10	-2.94	-13.7	-5.10	-6.75
	10		1.78	-2.89	-13.6	-7.88	
	14		1.47	-2.84	-13.6	-7.35	
	21		1.08	-2.71	-13.7		
7.2	5	4	2.70	-3.08	-13.6	-6.85	-6.85
	17		1.59	-2.67	-14.4		

1) Mean value

finding (5). Binding to haemoglobin is, therefore, not restricted to phospho-compounds, but it is worth for anions commonly.

In view of these evidences on the role of imidazole residues in binding of multivalent anions to haemoglobin we performed model studies with imidazole, histidine and ferrocyanide. Since ferrocyanide is completely dissociated above pH 5, 5 no difficulties owing to changes in its protonation arise in interpretation of experiments with imidazole between pH 6 and 9.

Fig. 4 shows the influence of ferrocyanide on the titration of the amino and imidazole group of histidine. Histidine was 10 mM; ferrocyanide 10 mM. In order to check the influence of ionic strength titrations were also carried out with 100 mM sodium chloride. The shift of pK' estimated according to the DEBYE-HÜCKEL approximation amounts to 0, 11. The change of pK' by sodium chloride amounts to 0, 15 for the amino group and 0, 10 for the imidazole group and it fits well with the theory within experimental uncertainties.

Ferrocyanide influences the titration of the amino group in the manner expected from its effect on the ionic strength. But there occurs an additional shift in pK' of the imidazole moiety which is significant. It indicates a somewhat specific interaction between imidazolium and ferrocyanide. The association constant was calculated from this additional shift according to TANFORD (6) and amounts 44 M^{-1}. For this computation a 1:1 interaction was assumed. With imidazole alone quite similar results were obtained and the association constant was calculated to 51 M^{-1}.

Interaction of ferrocyanide and imidazole causes diminution of the number of osmotically active particles in solution. It ought to be possible, therefore, to draw the number of interacting molecules from the difference of the osmotic activity of an imidazole-ferrocyanide solution and an appropriate solution of identical concentrations.

The results of such a study are presented in Fig. 5. The values are plotted like Scatchard's method for binding of small molecules to macromolecules (7). The intercept on the abscisse indicates formation of an 1:1 complex of imidazole and ferrocyanide. From the slope of the curve an association constant of 31 M^{-1} was determined. That corresponds fairly well with the value found by the titrimetric assay.

We tried to investigate this interaction by calorimetric measurements and to calculate the association constant with the method given by BJURULF et al. (8). The heat of dilution of ferrocyanide is very great compared to the heat of association of imidazole with ferrocyanide. One can, however, roughly estimate an association constant below 200 M^{-1}.

The difference spectrum for imidazole-ferrocyanide was investigated with the double-beam Cary-14 recording spectrophotometer using slit compartment cells (Fig. 6). A broad absorption band with a maximum at 236 nm was observed. There could not be detected such a difference spectrum with potassium chloride of identical ionic strength.

SHINITZKI et al. have pointed out (9, 10) that the protonated imidazole may act as electron acceptor in electron donoracceptor complexes. Studies with iodide and indole revealed an absorption band at about 230 and 240 nm respectively for the complexes of imidazolium with these electron donors. The difference spectrum for imidazolium and a multivalent anion presented here would formally well correspond with SHINITZKI's observations.

Nuclear magnetic resonance studies were performed to aquire more information on this type of interaction. Concentrations of imidazole and ferrocyanide were varied between 20 and 100 mM. In the case of sodium chloride as well as in that of ferrocyanide there was a shift of the C(2)-proton by 13 to 14 cycles per second and of the C(4)- and C(5)-proton by 6 to 7 cycles per second. This shift of the imidazole protons identically for chloride and ferrocyanide seems to be primarely different from the results of the other investigations. The descrepancies remain to be further clarified.

The interaction of imidazolium with multivalent anions presented here seems to be of biological importance. The binding of phosphocompounds to the class of low-affinity binding sites of haemoglobin has an association constant of about 10^2 M^{-1} (3) which is close to the association constant of the interaction of imidazole and ferrocyanide reported here. An association constant about 100 times higher was found in preliminary measurements on interaction of imidazole and ferrocyanide in dimethyl-sulphoxide. This may to some extent simulate imidazole groups in a hydrophobic surrounding of proteins.

In quite a lot of enzymes including phototransferases imidazole seems to play an important role in the binding of the substrates and effectors and in the catalytic reactions. Many of these substrates, cofactors and effectors are anions, especially phosphocompounds and so the interaction with the imidazolium presented might be of significance, especially in a hydrophobic surrounding.

REFERENCES

(1) GARBY, L., G. GERBER, C.-H. de VERDIER, European J. Biochem. 10, (1969), 110

(2) JUNG, F., VIth International Symposium on Structure and Function of Erythrocytes, Berlin 1970

(3) JÄNIG, G.-R., G. GERBER, K. RUCKPAUL, S. RAPOPORT, F. JUNG, European J. Biochem. 17, (1970), 441

(4) WYMAN, J., J. Biol. Chem. 127, (1939), 1

(5) RAPOPORT, S., G. GERBER, K. RUCKPAUL, G.-R. JÄNIG, F. JUNG, H. FRUNDER, ALFRED-BENZON-Symposium IV: Oxygen Affinity of Hemoglobin and Red Cell Acid-Base Status, Copenhagen 1971

(6) TANFORD, Ch., Adv. Protein Chem. 17, (1962), 127

(7) SCATCHARD, G., Ann. N. Y. Acad. Sci. 51, (1949), 660

(8) BJURULF, C., J. LAYNEZ, I. WADSÖ, European J. Biochem. 14, (1970), 47

(9) SHINITZKY, M., E. KATCHALSKI, V. GRISARO, N. SHARON, Arch. Biochem. Biophys. 116, (1966), 332

(10) MAUCHAMP, J., M. SHINITZKY, Biochemistry 8, (1969), 1554

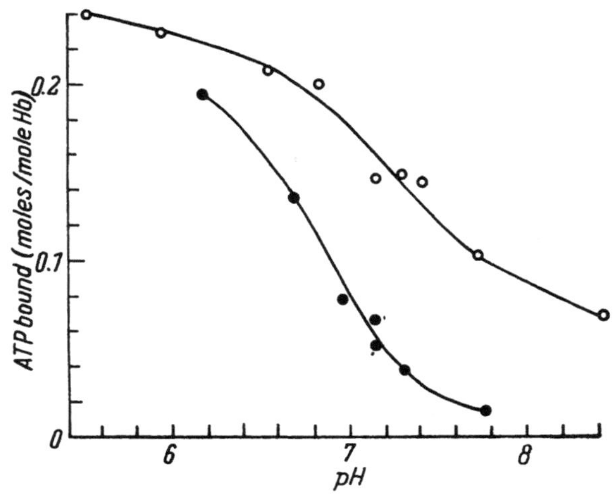

Fig. 1. Binding of ATP to deoxygenated (o) and oxygenated (o) haemoglobin as a function of pH.
Hb: 3, 1 mM; ATP: 1, 0 mM; Mg: 3, 0 mM; KCl: 110 mM; NaCl: 20 mM; inorganic phosphate: 1 mM.

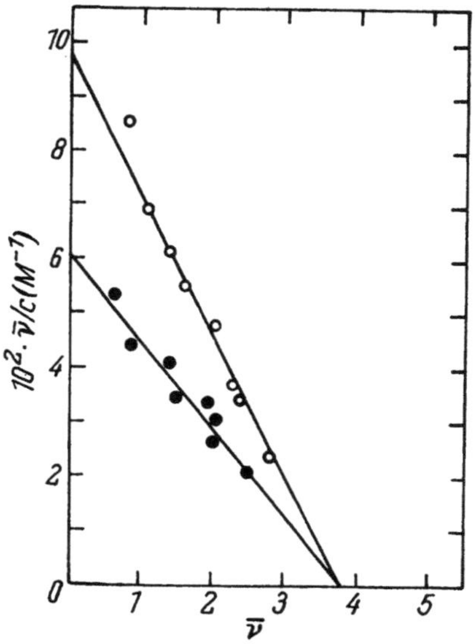

Fig. 2. Binding of ATP to methaemoglobin at pH 7, 2 (Scatchardplot). ATP concentration was varied at different temperatures.
Hb: 1, 55 mM; KCl: 0, 1 M; temperature: 5°, 17°.

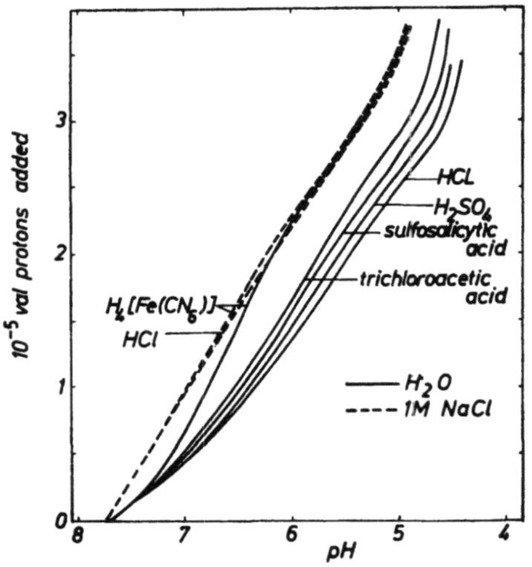

Fig. 3. Titration of methaemoglobin with acids.
Hb: 1, 34 mM; Vol. : 15 ml, temperature: 20 ^{0}C.

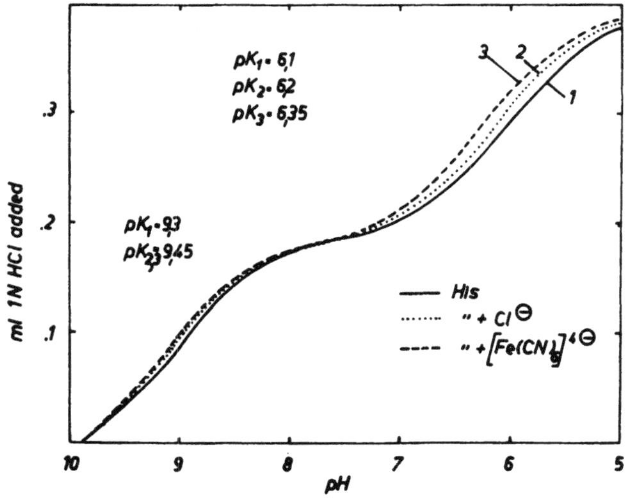

Fig. 4. Effect of anions on the titration of histidine.
histidine 10 mM; potassium ferrocyanide 10 mM; sodium chloride 100 mM; volume: 20 ml; temperature 25 ^{0}C.

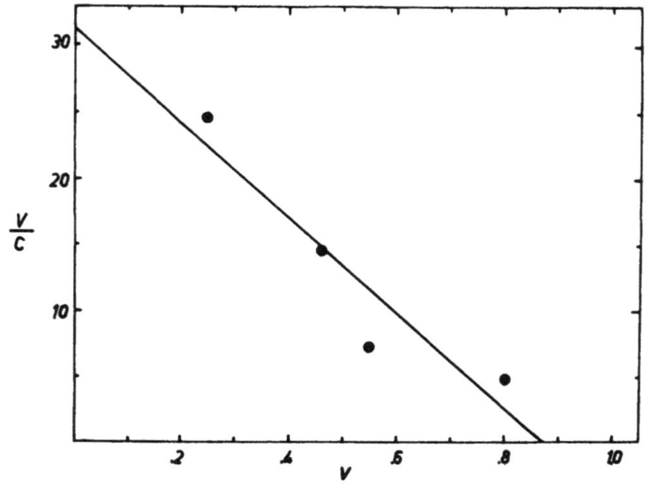

Fig. 5. Osmometric measurement of the imidazolium-ferrocyanide interaction. 0, 1 ml sample contained: 4 μ moles K_4 $Fe(CN)_6$, 4 to 40 μ moles imidazole, 2 to 20 μmoles HCl (pH 6, 8). 0, 1 ml reference solution contained 1 to 10 μ moles NaCl (pH 9, 2) instead of HCl.

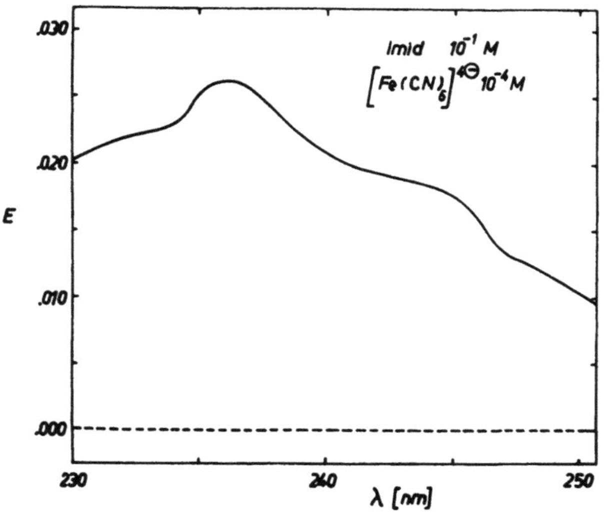

Fig. 6. Difference spectrum of the imidazolium-ferrocyanide interaction. Imidazole: 0, 1 M; potassium ferrocyanide 10^{-4} M. Cary-14 spectrophotometer, 10 mm slit-compartment cells. temperature: 20 °C.

Institute of Biochemistry, Hungarian Academy of Sciences, Budapest

L. POLGAR, P. HALASZ

ENHANCED REACTIVITY OF THE SH GROUP OF THIOL-SUBTILISIN

Thiol-enzymes, like papain or glyceraldehyde-3-phosphate dehydrogenase, have an SH group at their active site which is much more reactive than simple thiol-compounds. The origin of this enhanced reactivity has been studied in many laboratories, still we do not know the exact mechanism which would account for the reactivity of the SH group of thiol enzymes.

In the present talk, I am going to deal with the reactivity of an artificial cysteine enzyme, thiol subtilisin, which can be prepared from the serine protease, Carlsberg type subtilisin, by replacing the OH-group of its reactive serine residue with an SH-group (1). The reason for choosing thiol-subtilisin for such a study resides in the fact that we know the mechanism of action of the parent serine protease in considerable detail.

Fig. 1 shows the basic features of the mechanism of action of serine proteases. Serine proteases catalyze the hydrolysis of ester or amide substrates through the formation of an intermediate acyl-enzyme. A particular serine residue as a nucleophile and one histidine residue as a general base, general acid catalyst participate in the formation of the acyl-enzyme. The hydrolysis of the acyl-enzyme proceeds through the reverse mechanism. In the hydrolysis, the hydroxyl group of a water molecule is the nucleophile rather than the hydroxyl group of the serine residue.

Thiol-subtilisin is an ineffective enzyme towards the specific substrates of subtilisin, nevertheless it catalyzes the hydrolysis of some non-specific substrates through the acyl-enzyme intermediate as the original serine enzyme. Our previous studies have indicated that in the hydrolysis of p-nitrophenyl acetate catalyzed by thiol-subtilisin, the histidine residue in the proximity of the SH-group participates both in the formation and decomposition of the acetylthiol-enzyme intermediate (2). On the other hand, NEET et al. have suggested that the reactivity of the SH-group of thiol-subtilisin towards iodoacetamide is similar to that of a simple thiol-compound (3). Of course, the chemical mechanism of acyl-enzyme formation, which proceeds through the formation of a tetrahedral intermediate, is significantly different from that of alkylation, which is a simple nucleophilic subtitution on the sp^3 carbon atom. However, if the histidine residue contributes to the reactivity of the SH-group of thiol-subtilisin, this participation of the imidazole ring should manifest itself not only in acylation but also in alkylation. Therefore, we have studied the reaction of thiol-subtilisin with iodoacetamide in a wide pH-interval where the participation of the histidine residue in alkylation may be demonstrated.

Fig. 2 shows the pH-dependence of the second order rate constants of alkylation of thiol-subtilisin and glutathione with iodoacetamide in a semilogarithmic plot. The rate constants were measured in the presence of 1 M potassium chloride under second-order conditions or with excess iodoacetamide under pseudo first order conditions. The reaction followed the given reaction order for about 2-3 half lives. The solid lines in the figure are theoretical curves corresponding to the dissociation of one ionizable group. The pK of this ionizable group is 10 in the case of the enzyme and 9 in the case of glutathione which can be regarded as a model compound of a protein containing a simple thiol group.

It is seen from the figure that the experimental points in the alkaline pH-range fit well to the theoretical curve indicating that the reation of the enzyme, similarly to the reaction of the glutathione, depends on the dissociation of the SH-group. The pK of the enzyme is one unit higher than that of glutathione, and the second-order rate constants of the enzyme containing the fully dissociated SH-group are in the same order of magnitude. This is a fairly good agreement when an enzyme and a model compound are compared.

On the other hand, around neutral pH the experimental points do not follow the theoretical curve calculated for the ionizable group with the pK of 10. The rate constants are independent of pH in the range of 6 to 8. As the pH decreases, the rate of alkylation of the enzyme becomes increasingly higher relative to that of glutathione. Thus, at pH 7 there is a 300-fold and at pH 6 a 3000 fold difference in the rate constants. At even lower pH values this differenceis still higher but it is difficult to obtain exact values for the rate constants because of the denaturation of the protein.

The considerable rate-enhancement of the reaction at lower pH values should be due to the participation of the histidine residue. One of the possible forms of this participation would be a general base catalysis by the imidazole ring. However, the pH-dependence of alkylation does not support this possibility.

Namely, a pK of about 7 characteristic of the imidazole base is not seen in the pH-dependence curve, although it would be expected if a simple general base catalysis were to operate, as it was found in the case of the parent serine protease.

Another reasonable explanation which can account for the pH-dependence of alkylation is the formation of a hydrogen bond between the imidazole and the SH-groups. In this case, the apparent pK of the histidine residue should be shifted to lower values and that of the SH-group to higher values with a few pH units. In contrast to acylation by p-nitrophenyl acetate, alkylation by iodoacetamide clearly shows the pK of the SH-group of thiol-subtilisin, and this is only slightly higher than the pK of an ordinary SH-group. Accordingly, only a weak hydrogen bond, if any, can be formed between the imidazole and the SH-groups.

It appeared that the nature of the interaction between histidine and cysteine residues could be determined by the D_20 effect on the rate constants. A D_20 effect of about 3 would occur in the pH-range of 6 to 8 if a general base catalysis were to operate by the histidine residue (4). This value may be lower if a hydrogen bond is formed between the imidazole and the SH-groups. According to our measurements the ratio of the second-order rate constants of alkylation in water and in D_20 is 0.95 ± 0.15 rather than 3. This rules out the possibility of general base catalysis, and does not indicate that a significant hydrogen bond between the imidazole and the SH-groups would exist.

In order to explain the enhancement of the reactivity of the thiol-enzyme, we have to introduce the idea of cage pair or intimate pair formation in enzymes as a possible and important factor of catalytic function. This idea is applied in physical chemistry when molecules separated by the solvent and intimate pairs inside a common solvent shell are distinguished. Of course, intimate pairs of single molecules are very unstable in solution, since the two members of the pair react with each other or separate instantaneously. In a protein, however, the tertiary structure may be sufficiently rigid to hold two functional groups in the cage in which the separation by solvent is not possible.

Now we would interpret the reaction between thiol-subtilisin and iodoacetamide as follows. The reactivity of the SH-group above pH 9 corresponds to that of an ordinary thiol-compound. In the pH range of 6 to 8, the participation of a histidine residue accounts for the enhancement of the reactivity. The imidazole and the thiol groups form an intimate pair in which the proton is covalently bound to the nitrogen atom of the ring. Thus the intimate pair consists of a positive imidazolium ion and a negative mercaptide ion. If a weak hydrogen bond existed between these two groups, the donor atom should be the nitrogen and the acceptor should be the sulfur atom.

The lower reactivity of the mercaptide ion in the cage pair compared to the free mercaptide ion may be due to the effect of the positive charge of the imidazolium ion.

Finally, we would like to note that similar mechanism may account for the enhanced reactivity of the thiol group of creatine phosphotransferase and perhaps other thiol enzymes, too. In addition, intimate pair formation is not restricted to thiol and imidazole groups. This may be a general phenomenon which can significantly contribute to the catalytic power of enzymes.

REFERENCES

(1) POLGAR, L., Acta Biochim. et. Biophys. Acad. Sci. Hung. 3, (1969), 397

(2) POLGAR, L., M. L. BENDER, Biochemistry 6, (1967), 610

(3) NEET, K. E., A. NANCI, D. E. KOSHLAND Jr., J. Biol. Chem. 243, (1968), 6392

(4) BENDER, M. L., F. J. KEZDY, Ann. Rev. Biochem. 34, (1965), 49

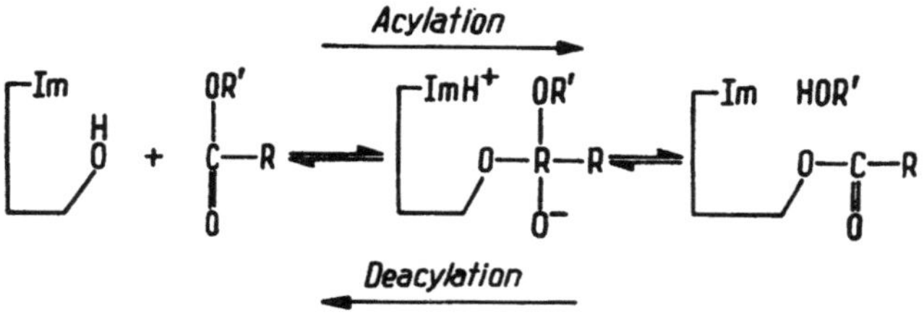

Fig. 1. Mechanisms of action of serine proteases.

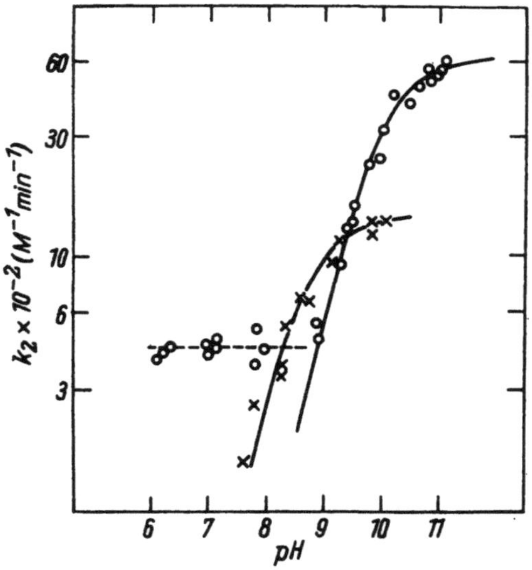

Fig. 2. pH-dependence of the second-order rate constants of alkylation with iodoacetamide. o-o: thiol-subtilisin, x-x: glutathione.

Institut für medizinische Chemie der Universität Wien

E. GRÜNDIG, R. F. L. MARUNA

URSPRUNG UND EIGENSCHAFTEN VON ISOZYMEN DER PHOSPHOMONOESTERASEN

In der medizinischen Literatur werden 2 Gruppen von Phosphomonoesterasen des Blutserums bzw. Plasmas beschrieben, die als saure und alkalische Phosphatasen bezeichnet werden. Diese Enzyme kommen ubiquitär vor. Im Hinblick auf die Beurteilung des Aussagewertes von Phosphatasebestimmungen im klinischen Laboratorium wäre es interessant zu wissen, welchen Ursprung diese Plasmaenzyme haben, falls sie im Serum in erhöhter Aktivität vorliegen.
Zur Charakterisierung dieser Enzyme haben wir folgende Bestimmungen eingeführt: Messung der Substratspezifität gegen Phenylphosphat, Nitrophenylphosphat, - und ß-Glycerophosphat, Bestimmung der pH-Optima, Beeinflussung der Aktivität durch Cu^{++}, Formaldehyd, F', ÄDTA, L-Tartrat und Cystein und elektrophoretische Auftrennung im Dünnschicht-Stärkegel. Als Puffer diente zur Aktivitätsbestimmung grundsätzlich Azetat- oder Veronalpuffer, zur Elektrophorese 0,02 M Trisazetat oder Carbonatpuffer. Als Substrat wurde - wenn nicht anders angegeben - Na-Phenylphosphat verwendet.
Nach der Elektrophorese im Stärkegelstreifen (20 x 3 cm) von 2 mm Dicke werden diese mit einem mit Substrat-Puffer-Gemisch getränkten, sehr saugfähigen Filterpapierstreifen bedeckt - die Kapazität des Puffers ist so groß, daß die Kombination Gel-Papier den gewünschten pH-Wert 4,8 oder 9,5 erreicht. Nach etwa 1 stündiger Inkubation bei 37 ^{o}C werden die Papierstreifen mit Phosphorwolframsäure versetzt. Das freigesetzte Phenol bewirkt die Umsetzung zu Wolfraumblau, das mit Ammoniakdämpfen in Form blauer Zonen sichtbar gemacht wird. Gegenfärbung der Gelstreifen zeigt, daß darin kein freies Phenol enthalten ist. Die Streifen werden densitometrisch ausgewertet, die Färbung ist im gewählten Bereich der Enzymaktivität proportional.
Zuerst sollen Versuche an Rattengeweben besprochen werden, an denen die Brauchbarkeit der Analysenmethodik und der Aussagewert der Ergebnisse getestet wurden. Die Extraktion der Organe erfolgte immer durch Homogenisieren

im Pottergerät mit Teflonstab mit 0,9 %iger Kochsalzlösung bei + 4° und anschließendem Zentrifugieren bei 7000 x g. Es war dies das schonendste Vorgehen, wir erhielten die höchstmöglichen Gesamtaktivitäten; letztere konnten durch Beschallung nicht erhöht werden.

Abb. 1 gibt nun schematisch die Lokalisierung und Bezeichnung der Isozymfraktionen auf den Pherogrammen in bezug auf Serumproteine wieder. Die Streckenlänge bezeichnet die Streuungsbreite der Bandenmittelpunkte bei den durchgeführten Bestimmungen. Die an der Auftragsstelle verbleibende Fraktion nennen wir 0, die bei pH 8,6 anodisch wandernden A_1-A_8, die kathodisch wandernden K_1-K_2. Der Nullfraktion entsprechen im Serum die γ_M Globuline (IgM), der Fraktion A_6 die Albumine, kathodisch wandernde Proteinfraktionen sind im Serum unter den gewählten Bedingungen die γ_G Globuline (IgG).

Abb. 2 zeigt am Beispiel von Rattennierenextrakten, daß wir bis zu 10 Fraktionen von sauren Phosphatasen erhalten haben. Aus dem Pherogramm geht hervor, daß die Gewebsextrakte nicht aufbewahrt werden dürfen. Die Beweglichkeit der Enzymproteine ändert sich. Es erfolgt eine Anreicherung an der Auftragsstelle und die mit den Präalbuminen wandernden Fraktionen A_7 und A_8 akkumulieren in der Albuminbande.

Aus Abb. 3, der Isozymverteilung von Rattenmuskelextrakten, ist zu entnehmen, daß sich auch beim Einfrieren und Wiederauftauen Veränderungen in der Verteilung der Fraktionen ergeben. Diese sind bei den 8-9 Fraktionen der sauren Phosphatasen nicht sehr stark - am auffallendsten ist der Befund, daß die Fraktion A_3 verschwindet. Bei den alkalischen Phosphatasen verschwinden die kathodisch wandernden und die mittleren Fraktionen A_1-A_4, es erfolgt Anreicherung an der Auftragsstelle und im Albuminbereich.

Des weiteren interessierte uns, ob Proteinkonzentration und Phosphataseaktivität der einzelnen Fraktionen parallel sind.

Abb. 4 zeigt das Ergebnis der Untersuchungen am Beispiel der Rattenmilz. Kathodisch wandernde Proteine lagen in so niedriger Konzentration vor, daß sie mit den üblichen Färbemethoden (Amidoschwarz) nicht mehr nachweisbar waren, Phosphataseaktivitäten waren jedoch einwandfrei vorhanden. Während die Phosphataseaktivitäten an der Auftragsstelle am höchsten waren, lagen am Ort der höchsten Proteinkonzentration mit der Wanderungsgeschwindigkeit der Serumalbumine (A_6) nur geringe Aktivitäten vor.

Die nächste Abb. Nr. 5 zeigt am Beispiel des Oberschenkelknochens der Ratte, wie sehr die Isozymmuster innerhalb eines Gewebes verschieden und wie empfindlich die Phosphatasen gegenüber äußeren Einflüssen sind. Wird der ganze Knochen schnell homogenisiert und zentrifugiert, findet man kathodisch wandernde Fraktionen. Schon durch den Zeitaufwand der weiteren Aufbereitung des Knochens - Zerteilen, Ausspülen des Markes etc. treten die ersten Veränderungen ein - die kathodisch wandernden Fraktionen sind nicht mehr nachweisbar. Vergleicht man Epiphyse und Diaphyse, ergibt sich, daß bei ersterer bei den sauren Phosphatasen die Aktivität der Fraktion 0 niedriger ist als bei den letzteren, während in der Fraktion A_2 die Aktivität in der Epiphyse höher ist als in der Diaphyse. Bei den alkalischen Phosphatasen sind die relativen Differenzen nicht so groß. Vom Knochenmark haben wir 2 Aufbereitungen gemacht: bei einer wurde das Mark lediglich aus dem Knochen herausgespült und zentrifugiert (Eluat), bei der anderen der Rückstand des "Eluates" noch in der üblichen Weise homogenisiert und dann zentrifugiert ("Homogenat"). Bei den sauren Phosphatasen sieht man hervorstechende Unterschiede in der Isozymverteilung: Im "Eluat" sind die Fraktionen A_3, A_4 und A_5 höher, im Homogenat die Fraktionen A_2 und A_7. Bei den alkalischen Phosphatasen ist im "Eluat" die Aktivität der Fraktion 0 mit 53 % die weitaus höchste, im "Homogenat" überwiegen die Fraktionen A_2 und A_3. Diese Unterschiede sind ein deutlicher Hinweis auf die verschiedene Lokalisation der einzelnen Isozymfraktionen innerhalb der Zelle.

Table 1

PROTEIN CONTENT AND TOTAL ACTIVITY OF ACID AND ALKALINE PHOSPHATASES IN TISSUE EXTRACTS OF HIGH-BONE (RAT) APPLIED FOR INVESTIGATIONS

	content of proteins	Activity of acid phosphatases	Activity of alkaline phosphatases
Total bone	0,42 %	190 mg/Phenol/g/h	1940 mg Phenol/g/h
Epiphysis	0,32 %	146 "	145 "
Diaphysis	0,04 %	12 "	28 "
"Eluate"	0,42 %	650 "	3940 "
Marrow			
"Homogenate"	0,42 %	560 "	5950 "

Die Tabelle 1 gibt Proteingehalt und Gesamtaktivität der sauren und alkalischen Phosphatasen in den zur Isozymtrennung verwendeten Knochenextrakten an. Mit Ausnahme des Diaphysenextraktes waren Proteingehalt und Aktivität der sauren Phosphatasen in der gleichen Größenordnung. Bei den alkalischen Phosphatasen fällt auf, daß praktisch die gesamte Aktivität im Knochenmark enthalten war. Die Tabelle 2 gibt abschließend einen vergleichenden Überblick über Verteilung der Phosphataseisozyme in einigen Gewebsarten der Ratte. Angegeben sind die Zahl der aufgefundenen Fraktionen sowie die Fraktionen mit der höchsten relativen Aktivität. Im Muskelextrakt fanden wir je 9 Fraktionen, im Milzextrakt je 8 der sauren und alkalischen Phosphatasen mit der höchsten Aktivität jeweils an der

Table 2

SUMMARY: DISTRIBUTION OF PHOSPHATASE ISOZYMES IN SOME RATTISSUES

	Acid phosphatases			Alkaline phosphatases		
	Number of fractions	Highest activity		Number of fractions	Highest activity	
		Name of the fractions	rel %		Name of the fractions	rel %
Muscle	9	0	23,4	9	0	26,4
Spleen	8	0	51,0	8	0	26,5
Kidney	10	0	20,6	10	0	26,6
		A 7/8	22,6		A_1	21,0
Bone total	12	A_3	14,0	10	A_3	37,6
		K 2/3	17,6		0	21,9
"Eluate"	8	A_3	23,5	7	0	53,2
Marrow						
"Homogenate"	9	A 6/7	38,5	7	0	34,5
					A_2	21,9
Diaphysis	9	0	24,4	7	0	28,0
		A 6/7	21,6		A_2	10,2

Auftragsstelle. Im Nierenextrakt lagen je 10 Fraktionen vor mit hoher Aktivität an der Auftragsstelle. Bei den sauren Phosphatasen wandern etwa 23 % mit den Präalbuminen ($A_{7, 8}$), bei den alkalischen bleiben weitere 21 % im Bereich der -Globuline (A_1). Im ganzen Knochen finden wir 12 bzw. 10 Fraktionen, besonders hohe Aktivitäten im Bereich der A_3-Fraktion: bei den sauren Phosphatasen finden wir zusätzlich hohe Aktivitäten, die kathodisch wandern, bei den alkalischen an der Auftragsstelle. Der Aussagewert dieses Befundes ist nur gering, wenn man sich an die Aktivitätsverteilung (Tab. 1) erinnert. Besonders soll nochmals auf das unterschiedliche Isozymmuster in unseren Markpräparaten hingewiesen werden: Bei den sauren Phosphatasen im "Eluat" höchste Aktivität in der A_3-Fraktion, im "Homogenat" in den Fraktionen A_6 und A_7, während bei den alkalischen Phosphatasen in beiden Präparationen die Hauptmenge der Aktivität an der Auftragsstelle liegen bleibt.
An zwei Beispielen soll noch kurz die Wirkung von Effektoren auf unsere Gewebeextrakte aufgezeigt werden.

Table 3

THE INFLUENCE OF EFFECTORS ON THE ACTIVITY OF ACID PHOSPHATASES IN RAT-KIDNEY MEASURED AT THE 3 pH-OPTIMA (SUBSTRATE: PHENYLPHOSPHATE)

	pH 4, 0	pH 5, 0	pH 6, 4
10^{-4} m $CuSO_4$	- 38 %	- 60 %	- 84 %
10^{-4} m NaF	- 28 %	- 33 %	- 19 %
0, 5 % HCHO	- 19 %	- 52 %	-
10^{-2} m L-Tartrate	- 92 %	- 33 %	- 44 %
$2.5.10^{-3}$ m EDTA	+ 83 %	+ 116 %	+ 260 %
$2.5.10^{-5}$ m EDTA	+ 20 %	+ 20 %	+ 100 %

Tabelle 3 illustriert den Einfluß auf die Aktivität von sauren Phosphatasen aus Rattennieren bei den 3 pH-Optima, bei pH 4, 0, pH 5, 0 und pH 6, 4, wenn als Substrat Phenylphosphat eingesetzt wird. Cu^{++}, F' und Formaldehyd wirken als Inhibitoren. Besonders ist darauf hinzuweisen, daß L-Tartrat speziell bei nie-

drigen pH-Werten stark hemmt. Dieser Befund ist im Hinblick auf die Annahme vieler Kliniker äußerst wichtig, daß man im Serum Prostataphosphatasen selektiv durch ihre Hemmbarkeit durch L-Tartrat nachweisen könne. Durch EDTA sind saure Nierenphosphatasen stark aktivierbar.
Auch saure Milzphosphatasen haben 3 pH-Optima (pH 4, 0, pH 5, 0 und pH 5, 5) (Tabelle 4).

Table 4

THE INFLUENCE OF EFFECTORS ON THE ACTIVITY OF ACID PHOSPHATASES IN RAT-SPLEEN (pH-OPTIMA: 4, 0 - 5, 0 - 5, 5 - SUBSTRATE: PHENYLPHOSPHAT)

	pH 4, 0	pH 5, 2
10^{-4} m $CuSO_4$	- 46 %	- 71 %
10^{-4} m NaF	- 58 %	- 43 %
0, 5 % HCHO	- 34 %	- 53 %
10^{-2} m L-Tartrate	- 18 %	- 43 %
2. 5. 10^{-3} m EDTA	+ 36 %	- 19 %

Gegen Cu^{++}, F' und Formaldehyd verhalten sie sich ähnlich wie Nierenphosphatasen, sind aber durch EDTA weniger aktivierbar und sehr viel weniger tartratempfindlich. Diese Beispiele zeigen, daß auch die Empfindlichkeit gegen Effektoren als Unterscheidungsmerkmal dienen kann.
Im folgenden soll nun kurz über die Anwendung der geschilderten Erkenntnisse auf Fragestellungen der Humanmedizin eingegangen werden. Zunächst einige Befunde bei M. Gaucher. Dabei handelt es sich um eine in Mitteleuropa selten auftretende Erbkrankheit, u.a. gekennzeichnet durch Speicherung des Pipoids Kerasin im rediculoendothelialen System; Speicherzellen sind vor allem in Knochenmarks- und Organpunktat - oft, aber nicht immer, tritt ein Milztumor auf. Der Nachweis der Speicherzellen ist schwierig, so daß eine weitere diagnostische Möglichkeit gesucht wurde. Da die Aktivitäten der sauren Serumphosphatasen stets erhöht waren, wurde ihre Charakterisierung versucht.
Abb. 6 zeigt, daß die Isozymverteilung im Plasma stark von der Norm abweicht.

Sie entspricht aber praktisch der Verteilung im Milzhomogenat, das viele Speicherzellen enthielt. Diese Untersuchungen wurden noch mit der Stärkeblockelektrophorese nach Smithies durchgeführt, die quantitative Aussagen nicht ohne weiteres zuläßt und weniger Fraktionen ergibt. Die Plasmaphosphatasen sind, wie aus der Tabelle 5 hervorgeht, bei den beiden pH-Optima 3, 85 und 5, 1 durch Cu^{++} und Formaldehyd wenig, durch F' stark hemmbar, werden durch EDTA kaum beeinflußt und wie die Phosphatasen aus Milzhomogenat durch Cystein aktiviert.

Table 5

THE INFLUENCE OF EFFECTORS ON THE ACTIVITY OF ACID PHOSPHATASES IN PLASMA AND SPLEEN-HOMOGENATE OF PATIENTS SUFFERING FROM M. GAUCHER (SUBSTRATE: PHENYLPHOSPHATE, FOR ESTIMATIONS WITH CYSTEINE: β-GLYCEROPHOSPHATE)

	pH 3, 85	pH 5, 1
Plasma: 10^{-4} m $CuSO_4$	- 4 %	- 37 %
10^{-4} m NaF	- 77 %	- 70 %
0, 5 % HCHO	-	- 33 %
$2.5.10^{-3}$ m EDTA	+ 6 %	+ 10 %
10^{-4} m Cystein	+ 42 %	+ 70 %
Spleen-homogenate:		
10^{-4} m Cystein	+ 40 %	+ 25 %

Durch die Summe dieser Merkmale unterscheiden sie sich deutlich von Phosphotasen, die bei anderen Erkrankungen im Plasma - meist mit erhöhter Aktivität - auftreten. Einschlägige Untersuchungen wurden bereits mit Erfolg differentialdiagnostisch eingesetzt.

Als weiteres Beispiel möchten wir die wichtigsten Befunde bei primärem Hyperparathyreoidismus mitteilen, der zu einer generalisierten Erkrankung des Knochengerüstes führt. Die Serumphosphatasen sind erhöht.

Tabelle 6 gibt einen Überblick über die Verteilung der Phosphataseisozyme in Epithelkörperchen (Autopsiematerial) und exstirpierten Parathyreoideaadenomen. Elektrophoretisch treten je 9 Fraktionen von sauren und je 8 von alkalischen

Table 6

SUMMARY: DISTRIBUTION OF PHOSPHATASE ISOZYMES IN PARATHYREOIDEA AND PARATHYREOIDEA-ADENOMA

	Acid phosphatases			Alkaline phosphatases		
	Number of fractions	Highest activity		Number of fractions	Highest activity	
		Name of the fraction	rel %		Name of the fraction	rel %
Parathyreoidea	9	0 A 2/3	32,2 21,8	8	0 A 2/3	44,0 24,0
Adenoma	9	0 A 2/3	35,2 21,5	8	K 1/2 0 A 2/3	17,0 18,0

Phosphatasen auf. Die Isozymverteilung ist bei den sauren Phosphatasen in beiden Präparaten praktisch identisch, unterscheidet sich aber stark bei den alkalischen Phosphatasen. Hier liegt die Hauptaktivität bei den Epithelkörperchen im Bereich der Auftragsstelle sowie in den Fraktionen A_2 und A_3, bei den Adenomen ist außerdem ein hoher Prozentsatz kathodisch wandernder Anteile zu sehen.
Es wurden nun bei 5 Patienten (Tabelle 7) die Phosphataseaktivitäten im Plasma vor der Operation und 2 Tage nach Entfernung der Adenome untersucht.
Die Aktivitäten der sauren Phosphatasen waren vor der Operation erhöht, hatten aber sofort nach der Operation praktisch Normalwerte erreicht.
Die erhöhten Aktivitäten der alkalischen Phosphatasen blieben hoch; nach 12 Tagen erreichten sie bei einem Patienten (geringe Knochenaffektionen) sehr niedrige Werte. Daher vermuten wir, daß die im Plasma in erhöhter Aktivität auftretenden sauren Phosphatasen aus den Adenomen stammen könnten. Die Vermutung wird durch den in Abb. 14 dargestellten Einfluß von Effektoren auf die sauren Phosphatasen in Adenomextrakt, Epitelkörperchen und Patientenplasma untermauert.
Soweit die Messungen durchgeführt werden konnten, stimmen die Ergebnisse überein: Hemmbarkeit durch Cu^{++}, F', Formaldehyd und Tartrat, Aktivierbarkeit durch EDTA. Diese Phosphatasen sind also zum Unterschied von den bei M. Gaucher auftretenden durch EDTA aktivierbar. Die durch Elektrophorese

Table 7

ACTIVITIES OF PHOSPHATASE IN PLASMA OF PATIENTS SUFFERING FROM PRIMARY HYPERPARATHYREOIDISM BEFORE RESECTION (A.O.) AND ON THE 2nd DAY AFTERWARDS (P.O.) IN mU/ml (SUBSTRAT: NITROPHENYLPHOSPHATE)

Patient Nr.	Acid phosphatases		Alkaline phosphatases	
	A.O.	P.O.	A.O.	P.O.
1	26,6	13,3	400	384
2	6,0	5,0	83	60
3[1]	22,4	5,0	23	24
4	28,3	15,0	116	110
5	17,1	10,9	139	130
Standard	till 11 mU/ml		20 - 48 mU/ml	

1 12 days after resection: alkaline phosphatases 3,0 mU/ml

Table 8

THE INFLUENCE OF EFFECTORS ON THE ACTIVITY OF ACID PHOSPHATASES IN PARATHYREOIDEA ADENOMA AND IN PLASMA OF PATIENTS SUFFERING FROM PRIMARY HYPERPARATHYREOIDISM COMPARED WITH NORMAL PARATHYREOIDEA

	Adenoma-extract		Plasma		Parathyreoidea-extract
	pH 4,4	pH 5,2	pH 4,4	pH 5,1	pH 4,4
10^{-4} m $CuSO_4$	- 75 %	- 39 %	-	- 7 %	-
10^{-4} m NaF	- 25 %	- 61 %	- 21 %	- 9 %	-
0,5 % HCHO	- 90 %	- 58 %	- 10 %	- 17 %	-
10^{-2} m L-Tartrate	- 25 %	- 100 %	- 9 %	- 33 %	- 63 %
$2.5.10^{-3}$ m EDTA	+ 100 %	+ 43 %	+ 42 %	+ 70 %	+ 32 %

bewirkte Auftrennung der alkalischen Phosphatasen hingegen entspricht der bei anderen Knochenerkrankungen z. B. M. PAGET (Osteitis deformans) gefundenen, und ist ähnlich der des unfraktionierten Knochens und des "Homogenates" aus Knochenmark.

Zum Schluß danken wir Dr. CZITOBER, unserem ärztlichen Mitarbeiter von der 1. Medizinischen Universitätsklinik Wien, der uns das Material für die Untersuchungen an Humanpräparaten zur Verfügung stellte.

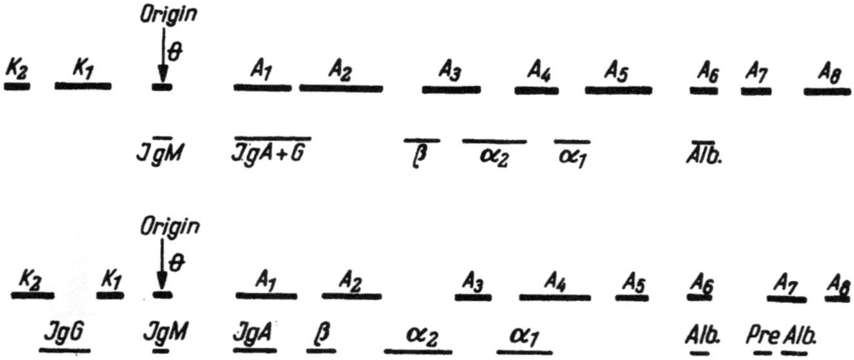

Fig. 1. The mobility of the phosphatase fractions in relation to the serum protein fractions. Top: alkaline phosphatases (carbonate buffer); Bottom: acid phosphatases (tris buffer).

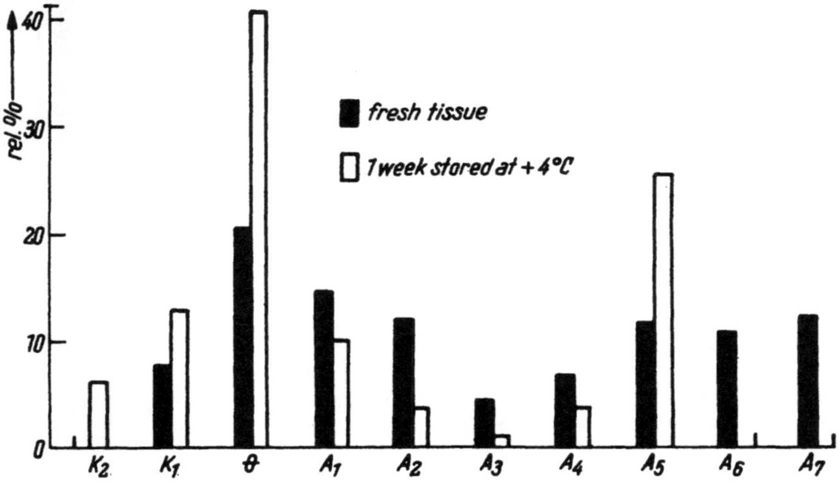

Fig. 2. The isozyme pattern of the acid phosphatases of rat-kidney.

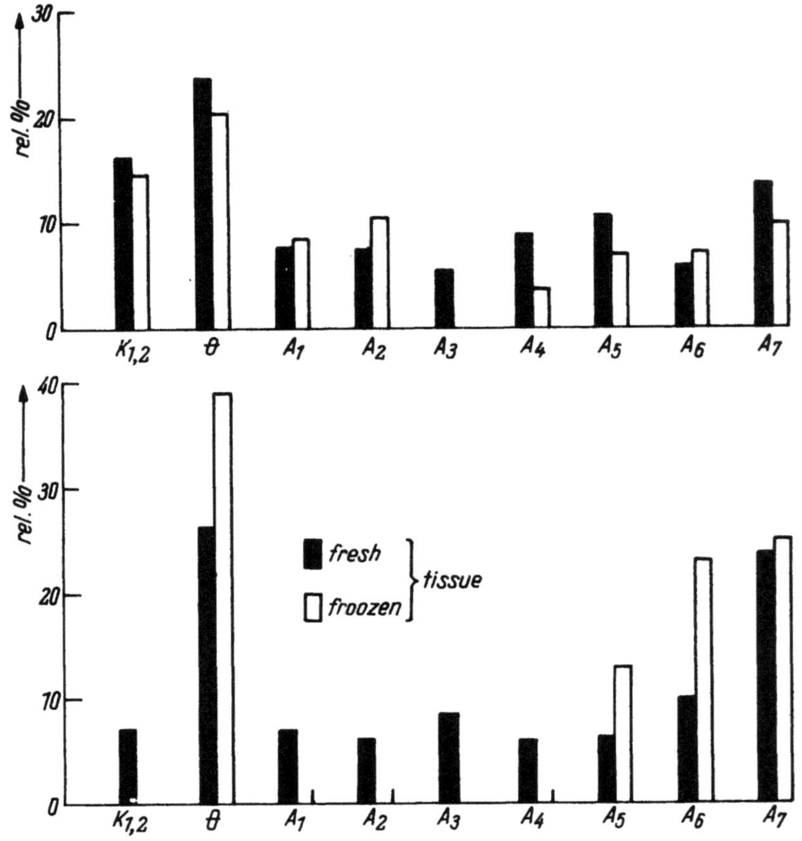

Fig. 3. The isozyme pattern of rat-muscle phophatases. Top: acid phophatases. Bottom: alkaline phosphatases.

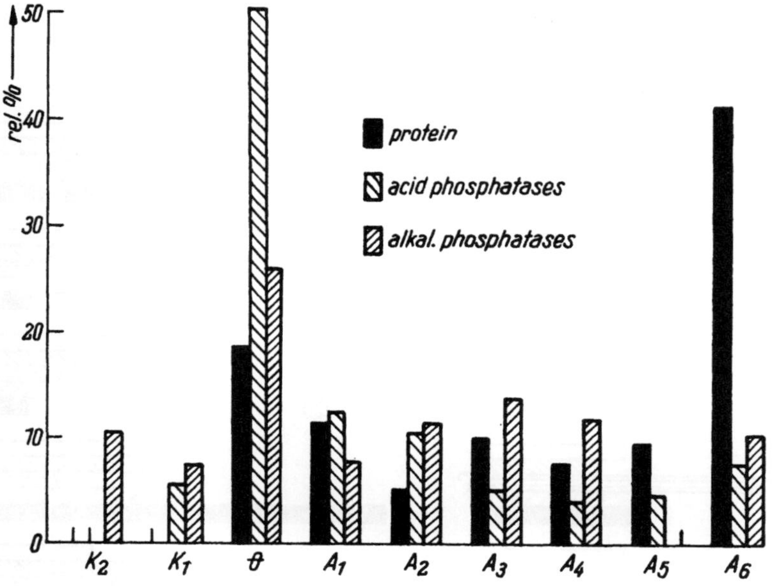

Fig. 4. The protein and phosphatase-isozyme pattern of rat-spleen homogenates.

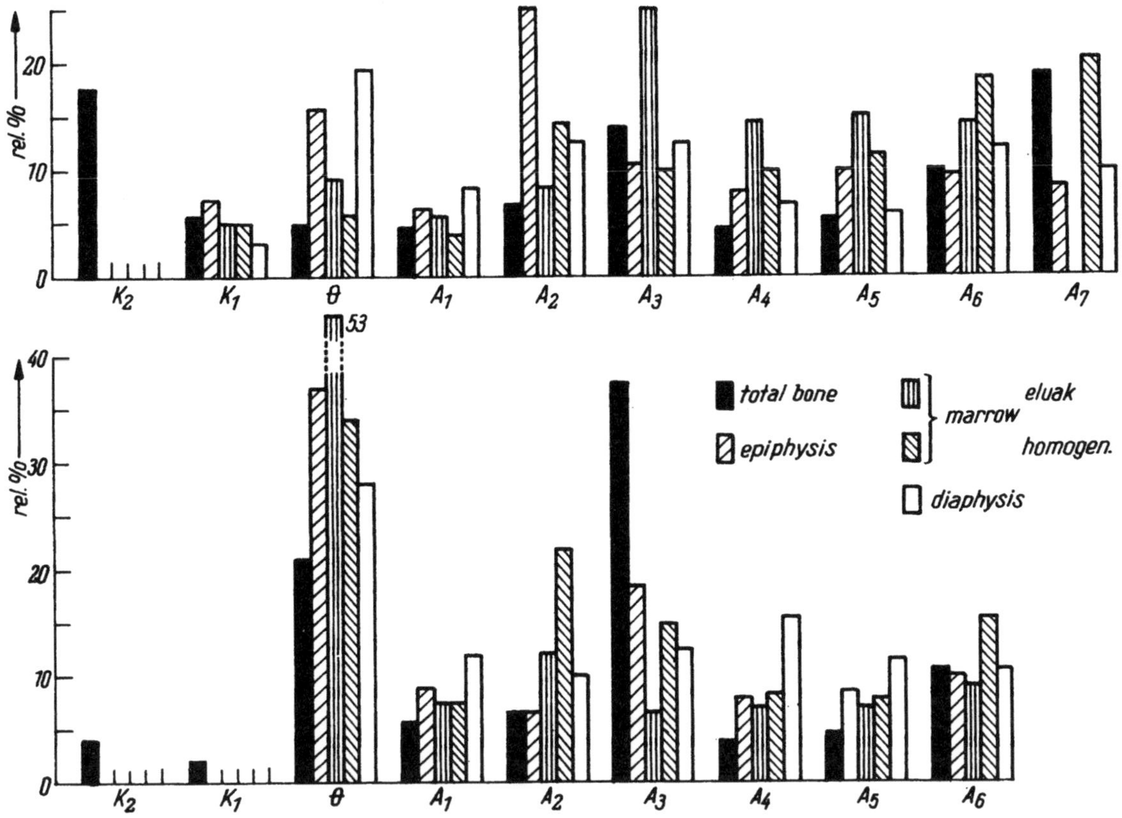

Fig. 5. The isozyme pattern of the thigh-bone of the rat. Top: acid phosphatases. Bottom: alkaline phosphatases.

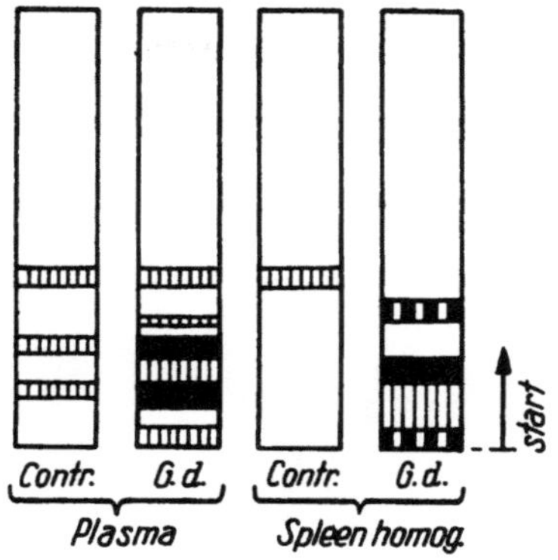

Fig. 6. The isozyme pattern of acid plasma phosphatases and the related enzymes of spleen homogenate in Gaucher' s disease.

Institut für Pharmakologie und Toxikologie der Ernst-Moritz-Arndt-Universität Greifswald

J. BEHLKE, J. LAMPE, W. SCHELER

LIGANDENINDUZIERTE ÄNDERUNGEN VON CD-SPEKTREN UND VOM ASSOZIATIONS-DISSOZIATIONS-GLEICHGEWICHT DES NEUNAUGEN-HÄMOGLOBINS

Proteine und speziell Enzyme stellen vielfach assoziierende Systeme dar (1). Unter einem Assoziat verstehen wir ein aus mindestens zwei Untereinheiten bestehendes Makromolekül, in dem die Monomeren durch nichtkovalente Bindungen zusammengehalten werden. Durch Variation der Milieubedingungen (pH, Ionenstärke usw.), aber auch durch Substrat- oder Ligandenbindung, durch Modifizierung der prosthetischen Gruppe oder des Apoproteins können die sog. Zwischenkettenbeziehungen im Makromolekül beeinflußt und konzentrationsabhängige Veränderungen der Molekulargewichte nachgewiesen werden. Als Modell für unsere Untersuchungen wählten wir verschiedene Hb. Sie bestehen meistens aus zwei Arten von Hämpolypeptidketten gemäß der Formel $\alpha_2\beta_2$ für die Vertebraten-Hb. Die Dissoziation der tetrameren Hb erfolgt symmetrisch nach folgendem Schema

(1) $\alpha_2\beta_2 \rightleftharpoons 2\,\alpha\beta \rightleftharpoons \alpha_2 + \beta_2$ (2, 3) oder allgemein

(2) $Hb_4 \rightleftharpoons 2\,Hb_2 \rightleftharpoons 4\,Hb.$

Dabei werden zunächst Halb- und dann Viertelmoleküle erhalten. Eine stufenweise Dissoziation oder umgekehrt eine Assoziation scheint bei dem aus nur einer Polypeptidkettenart bestehenden Hb der Flußneunaugen abzulaufen. Dabei kann unter bestimmten Bedingungen folgende Gleichgewichtsbeziehung existieren:

(3) $i\ Hb_4 \rightleftharpoons j\ Hb_3 \rightleftharpoons k\ Hb_2 \rightleftharpoons 1\ Hb$

Wieweit die Assoziation abläuft, hängt jedoch vom gewählten Milieu, dem Oxydationszustand des Eisens und der Ligandenbindung ab.

Die Molekulargewichtsbestimmungen am Neunaugen-Hb wurden mit einer analytischen Ultrazentrifuge durchgeführt. Für die CD-Untersuchungen verwendeten wir einen Dichrographen der Fa. Jouan (Typ CD 185).

Die Untersuchungen führten zu folgenden Ergebnissen:

1. Molekulargewichtsstudien

1.1. MetHb:

Das Flußneunaugen-MetHb mit dem dreiwertigen Eisen als Zentralatom sedimentiert verhältnismäßig langsam. Die Sedimentationskoeffizienten variieren nur geringfügig mit der Konzentration (4, 5) (Abb. 1).
Es liegt eine schwach negative Konzentrationsabhängigkeit der $s_{20,W}$-Werte vor, $s_{20,W} = 1,94 - 0,015 \cdot c\,[S]$ (c = g/100 ml). Mit Hilfe der empirischen Gleichung von ATASSI und GANDHI (6):

(4) $M = (s_o - s_i) / k$ s_o = Sedimentationskonstante; s_i = 0,68; $k = 7,416 \cdot 10^{-5}$

läßt sich eine Molmasse von 17 000 bestimmen.
Auf Grund der Aminosäureanalyse konnte für das monomere Molekül ein Molekulargewicht von 16 450 ermittelt werden (7). Durch Variation des pH-Wertes zwischen 4,5 und 10 wurden keine Veränderungen im Sedimentationsverhalten und somit im Molekulargewicht beobachtet (Abb. 2).
Die Ionisation des am Eisen gebundenen H_2O-Moleküls mit einem pK' = 8,15 hat somit keinen Einfluß auf die hydrodynamischen Eigenschaften des Neunaugen-MetHb.

1.2. MetHb-L-Komplexe

In Gegenwart der Liganden N_3^-, F^-, CN^-, OCN^- und SCN^- bilden sich die entsprechenden MetHb-L-Komplexe aus. Dabei variieren auf Grund der unterschiedlichen Affinitäten auch die dazu notwendigen Ligandenkonzentrationen. Die Komplexbildung macht sich nicht im alkalischen pH-Bereich, aber im schwach sauren Milieu im Sedimentationsverhalten bemerkbar (Abb. 2). Die Sedimentationskoeffizienten steigen dabei von 1,9 [S] auf 3,1 [S] an. Bereits in geringen Konzentrationen des MetHb·N_3 beobachten wir die entsprechende Zunahme der $s_{20,W}$-Werte (Abb. 1) und eine Zunahme des Molekulargewichtes auf das Doppelte (Mg = 33 000). Die Reversibilität der Dimerisierung kann entweder durch Verdünnung oder Alkalisieren der Proben mit nachfolgender Dissoziation nachgewiesen werden. Die pH-Abhängigkeit der Sedimentationskoeffizienten gehorcht einem Dimer-Monomer-Übergang unter Mitwirkung eines Protons pro Monomer

(5) $2\,P + 2\,H \rightleftharpoons (PH)_2$

$$(6)\quad K_a = \frac{[P_2]}{[P]^2\ [H]^2} = \frac{1}{2 \cdot c \cdot [H]^2} \cdot \frac{(1-\alpha)}{\alpha^2}$$

Daraus ergibt sich für den Dissoziationsgrad

$$(7)\quad \alpha = -\frac{1}{4\ c \cdot K_a\ [H]^2} \pm \sqrt{\left(\frac{1}{4\ c \cdot K_a \cdot [H]^2}\right)^2 + \frac{1}{2\ c \cdot K_a \cdot [H]^2}}$$

(Abb. 3).

Die Kurven (Abb. 3) für die Dissoziation und Assoziation zeigen in Abhängigkeit vom pH-Wert die entsprechende Asymmetrie (siehe auch Abb. 2).

Der Ablauf der Dimerisierungsreaktion muß dementsprechend wie folgt formuliert werden (5)

$$(8)\quad 2\ \text{MetHb} + 2\ H^+ + 2\ L \rightleftharpoons (H^+ \cdot \text{MetHb} \cdot L)_2$$

Diesem Gesamtgleichgewicht liegen drei Partialgleichgewichte zugrunde

$$(9)\quad K_a = [H^+ \cdot \text{MetHb} \cdot L] \ / \ [\text{MetHb} \cdot L]\ [H^+]$$

$$(10)\quad K_b = [H^+ \cdot \text{MetHb} \cdot L] \ / \ [H^+ \cdot \text{MetHb}]\ [L]$$

$$(11)\quad K_c = [(H^+ \cdot \text{MetHb} \cdot L)_2] \ / \ [H^+ \cdot \text{MetHb} \cdot L]^2$$

Zur Berechnung der Assoziationskonstanten K_c müssen die Komplexbildungskonstante K_b und die Ionisationskonstante K_a zuvor auf direktem bzw. indirektem Wege bestimmt werden. Werden z. B. die Untersuchungen in Gegenwart ausreichend hoher Ligandenkonzentrationen durchgeführt, so kann die Gleichung (10) vernachlässigt und die Assoziationskonstante K_c unter Berücksichtigung der vorgeschalteten Ionisationsgleichgewichte (Gleichung 9) ermittelt werden

$$(12)\quad K_c = \frac{(1-\alpha)\ (1 + K_a \cdot [H^+])^2}{\alpha^2 \cdot c \cdot 2\ K_a^{\ 2}\ [H^+]^{\ 2}}$$

Dabei gelten für die Gesamtkonzentration (c) und den Dissoziationsgrad (α) folgende Beziehungen

$$(13)\quad c = 2\ [(H^+ \cdot \text{MetHb} \cdot L)_2] + [H^+ \cdot \text{MetHb} \cdot L] + [\text{MetHb} \cdot L]$$

$$(14)\quad \alpha = ([H^+ \cdot \text{MetHb} \cdot L] + [\text{MetHb} \cdot L]) \ / \ c$$

Der Dissoziationsgrad α läßt sich aus den experimentell bestimmten Sedimentationskoeffizienten s_x bzw. Molekulargewichten M_x bei Kenntnis jener Werte für die monomere und dimere Form entsprechend Gleichung (15) errechnen

$$(15)\quad \alpha = \frac{s_2 - s_x}{s_2 - s_1} = \frac{M_2 - M_x}{M_2 - M_1}$$

In Tab. 1 wurden die Assoziationskonstanten K_c für einige MetHb-L-Komplexe zusammengestellt. Hierbei ist keine Korrelation zwischen der Komplexbildungskonstante K_b und K_c zu erkennen (4. 5).

Tabelle 1

ASSOZIATIONSKONSTANTEN K_c EINIGER MetHb· L-KOMPLEXE

Ligand	$K_a [M^{-1}]$	$K_c [M^{-1}]$
N_3^-	$5,60 \cdot 10^6$	$9,48 \cdot 10^4$
F^-	$8,92 \cdot 10^5$	$3,81 \cdot 10^4$
OCN^-	$5,62 \cdot 10^5$	$1,40 \cdot 10^4$
CN^-	$5,00 \cdot 10^5$	$1,03 \cdot 10^4$
SCN^-	$4,17 \cdot 10^5$	$9,63 \cdot 10^3$

Wird in Gegenwart einer genügend hohen $[H^+]$-Konzentration gearbeitet, so kann Gleichung (9) vernachlässigt werden. K_c ist dann nur noch unter Berücksichtigung des Komplexbildungsgleichgewichtes (Gleichung (10)) nach Gleichung (16) zu bestimmen

$$(16)\quad K_c = \frac{(1-\alpha)\;(1 + K_b[L])^2}{\alpha^2 \cdot c \cdot 2\, K_b^{\,2}\; [L]^{\;2}}$$

Ein Beispiel für die Abhängigkeit der Dimerisierungsreaktion des sauren Neunaugen-MetHb von der Ligandenkonzentration ist in der Abb. 4 dargestellt. Der Anstieg der Sedimentationskoeffizienten mit zunehmender N_3^--Konzentration und somit Komplexbildung verläuft auf Grund der Gleichgewichtsbeziehung nicht linear.

1. 3. HbCO:

Im schwach sauren Milieu steigen die Sedimentationskoeffizienten des Neunaugen-HbCO sowie auch des HbO_2 und DesoxyHb mit zunehmender Konzentration über den Wert des dimeren Moleküls an (8, 9). Die $s_{20,W}$-Werte betragen z. B. im Konzentrationsbereich um 1, 5 - 2, 0 [g/dl] 3, 8 ± 0, 2 [S] (Abb. 5).

Da auch das scheinbare Molekulargewicht M_{Wapp} mit steigender Konzentration von 17 000 auf Werte um 40 000 zunimmt (Abb. 6), kann unter den entsprechenden Milieubedingungen mit einem Monomer-Dimer-Trimer-Gleichgewicht gerechnet werden.

Die Berechnung der Assoziationskonstanten K_n erfolgt nach ADAMS u. FILMER (10) entsprechend der allgemeinen Beziehung

(17) $n\,P_1 \rightleftharpoons P_n$ (n = 2, 3..., P_1 = monomeres Protein, P_n = assoziiertes Protein)

durch Bestimmung der jeweiligen Konzentrationsverhältnisse

$$(18) \qquad K_n = \frac{c_n}{c_1^{\,n}}$$

Da es sich hier um nichtideale Systeme handelt, entspricht das ermittelte scheinbare Molekulargewicht M_{Wapp} nicht dem wahren Molekulargewicht $M_{W(c)}$; sondern beinhaltet noch den Nichtidealitätsterm BM_1, gemäß der Gleichung (19)

$$(19) \qquad \frac{M_1}{M_{Wapp}} = \frac{M_1}{M_{W(c)}} + BM_1 \cdot c$$

Die Größe des Nichtidealitätsterms sowie die Konzentrationen der Einzelkomponenten kann aus der Konzentrationsabhängigkeit des scheinbaren Molekulargewichtes bzw. des Quotienten M_1 / M_{Wapp} nach ADAMS u. FILMER (10) ermittelt werden.

Für die angenäherte (gestrichelte) Kurve in Abb. 6 gelten die folgenden Parameter: BM_1 = 0, 01, $\overline{K}_2$ = 2, 36 [g/l] bzw. 2, 00 · 10^4 [Mol/l] und $\overline{K}_3$ = 17, 5 [g/l] bzw. 1, 68 · 10^9 [Mol/l]. (Abb. 7).

Abb. 7 beinhaltet die aus diesen Konstanten errechneten Partialkonzentrationen c_i in Abhängigkeit von der Gesamtkonzentration c.

Dementsprechend werden im Assoziationsprozeß des Neunaugen-HbCO immer Dimere gebildet. Ihr Gehalt liegt relativ konstant zwischen 15 und 20 %. Mit steigender Konzentration nimmt der Anteil der Trimeren auf Kosten der Monomeren stärker zu.

2. CD-Spektren:

Der Beeinflussung der Dissoziations-Assoziations-Gleichgewichte liegen zwar Veränderungen der Elektronendichteverteilung an der Moleküloberfläche bzw. im Zwischenkettenbereich zugrunde. Der auslösende Vorgang, z.B. die Ligandenbindung findet jedoch an der prosthetischen Gruppe, dem Häm(in)eisen bzw. in den Aminosäuregruppen des distalen Proteinbereiches statt. Die sich im Ligandenrezeptor vollziehenden elektronischen Veränderungen mit einer Modifizierung der Asymmetrie im Hb-Chromophor sollten weiterhin Aufschluß über den möglichen Zusammenhang zwischen Assoziations-Dissoziations-Verhalten und dichroistischer Absorption des Neunaugen-Hb bzw. seiner Komplexe erbringen (Abb. 8).

In Abb. 8 sind einige der CD-Absorptionsspektren aufgezeigt bzw. die Abhängigkeit der Elliptizität von der dichroistischen Absorptionsbande bzw. der relativen Assoziationstendenz des Neunaugen-Hb bzw. seiner Komplexe zusammengestellt. Obwohl hier nur erste Befunde vorliegen und diskutiert werden können, sei folgende Schlußbemerkung erlaubt. Der Zirkulardichroismus gestattet lediglich Aussagen zur Symmetriebänderung chromophober Gruppen bei den lokalen Konformationsänderungen. Eine Übertragung jener Strukturmodifikationen auf entferntere oder oberflächlich gelegene Gruppen, die die Voraussetzung für die Beeinflussung der Assoziations-Dissoziations-Gleichgewichte darstellen, kann nur in grober Weise oder gar nicht mit dieser Methode erfaßt werden.

LITERATUR

(1) REITHEL, F.J., Advances Protein Chemistry 18, (1963), 123

(2) GUIDOTTI, G., L.C. CRAIG, Proc. natn. Acad. Sci. U.S.A. 50, (1963), 47

(3) GUIDOTTI, G., W. KONIGSBERG, Proc. natn. Acad. Sci. U.S.A. 50, (1963), 774

(4) BEHLKE, J., W. SCHELER, Acta biol. med. germ. 21, (1968), 739

(5) BEHLKE, J., W. SCHELER, Europ. J. Biochem. 15, (1970), 245

(6) ATASSI, M.Z., S.K. GANDHI, Naturwissenschaften 52, (1965), 259

(7) BRAUNITZER, G., H. FUJIKI, Naturwissenschaften 56, (1969), 322

(8) BRIEHL, R.W., J. biol. Chemistry 238, (1963), 2361

(9) BEHLKE, J., W. SCHELER, FEBS Letters 7, (1970), 177

(10) ADAMS, E.T., D.L. FILMER, Biochemistry 5, (1966), 2971

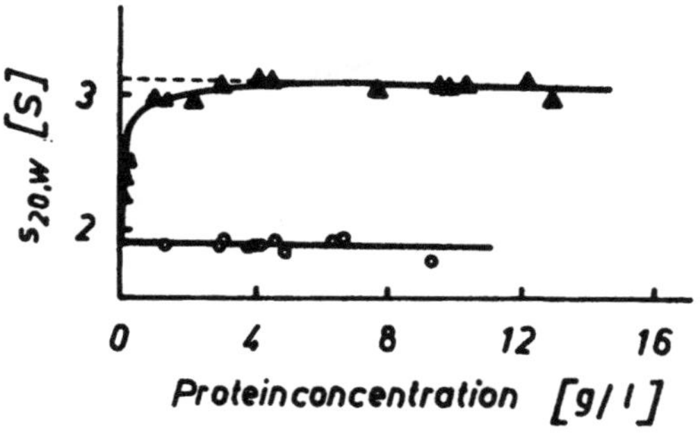

Fig. 1. Sedimentationskoeffizienten des Neunaugen-MetHb (o) bzw. -MetHb· N_3 (▲) in Abhängigkeit von der MetHb-Konzentration. NaN_3-Konzentration 1-5 mM, pH 4,9-5,7. Puffer: 0,1 M Acetat (5).

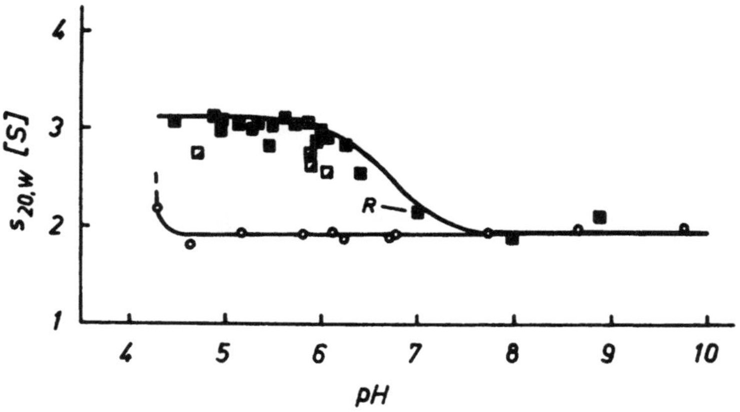

Fig. 2. pH-Abhängigkeit der Sedimentationskoeffizienten vom Neunaugen-MetHb (o) und -MetHb· N_3 (■). MetHb-Konzentration 2-12 g/l . Partiell gefüllte Karos: Unvollständige Assoziation. Die ausgezogene obere Kurve wurde rechnerisch entsprechend Gleichung (7) unter Berücksichtigung eines pK-Wertes = 6,75 (4) ermittelt. R: Reneutralisierte Probe eines MetHb· N_3-Ansatzes von pH 5,68. Puffer: 0,1 M Acetat bzw. 0,13 M Borax-Phosphat (5).

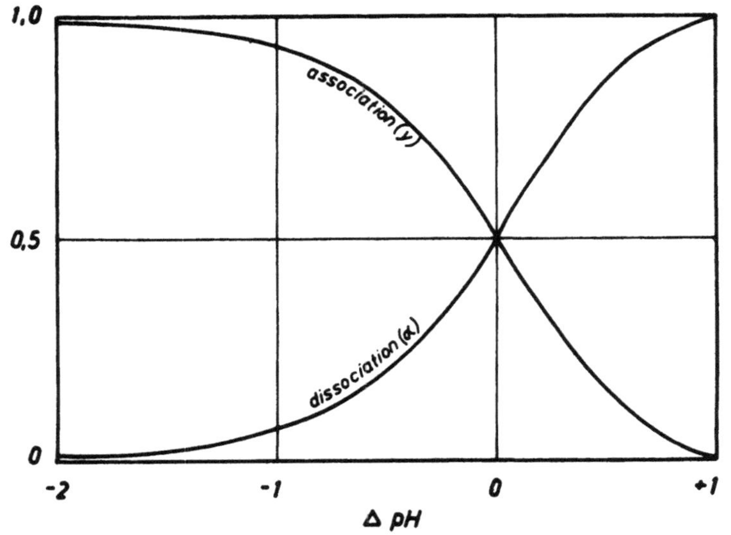

Fig. 3. Theoretische Kurvenverläufe für einen Monomer-Dimer-Übergang unter Einbeziehung eines Protons/monomere Einheit nach Gleichung (5) - (7) für c = 1, 0 und K_a = 1, 0.

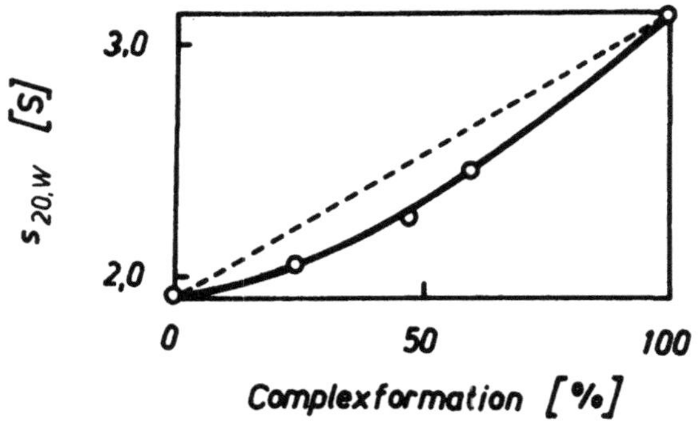

Fig. 4. Sedimentationskoeffizienten eines Neunaugen-MetHb/MetHb·N_3-Gemisches in Abhängigkeit vom Grad der Komplexbildung (5). MetHb-Konzentration 6, 6 [g/l], pH 5, 5, 0, 1 M Acetat.

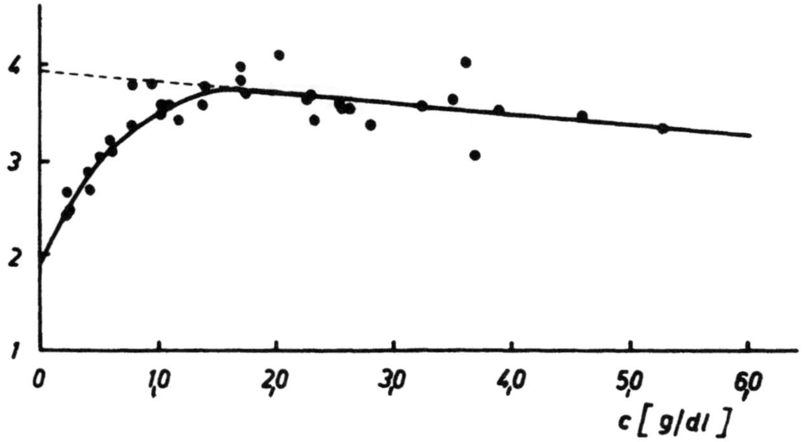

Fig. 5. Sedimentationskoeffizienten des Neunaugen-HbCO in Abhängigkeit von der Konzentration. Puffer: 0, 1 M Acetat, pH 5, 5 ± 0, 2.

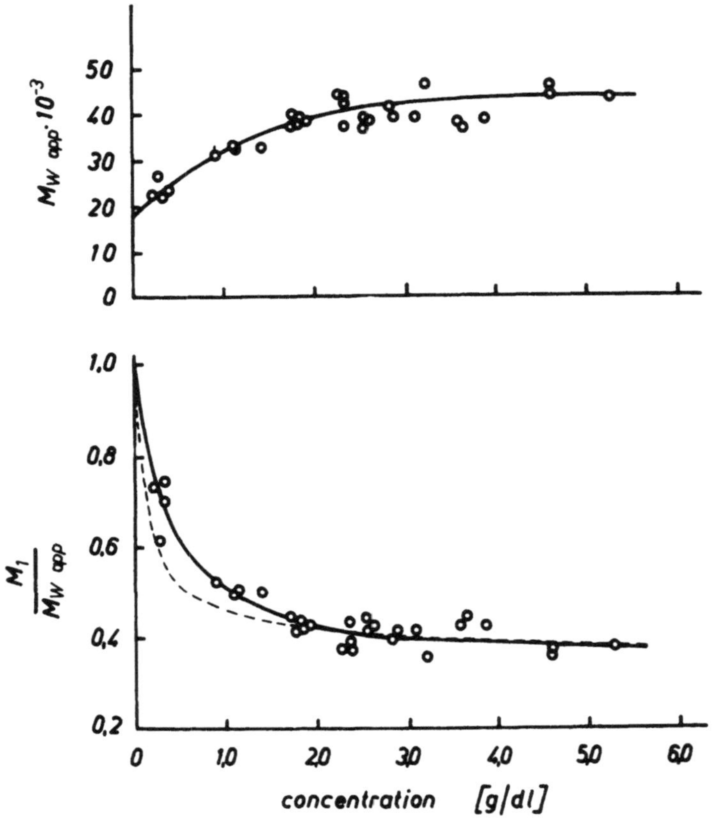

Fig. 6. Scheinbares Molekulargewicht (obere Kurve) bzw. Quotient monomeres/scheinbares Molekulargewicht des Neunaugen-HbCO in Abhängigkeit von der Konzentration. Bedingungen wie in Abb. 5. Gestrichelte Kurve: Theoretischer Verlauf für die Parameter: $BM_1 = 0,01$, $K_2 = 2,36$ [g/l] und $K_3 = 17,5$ [g/l].

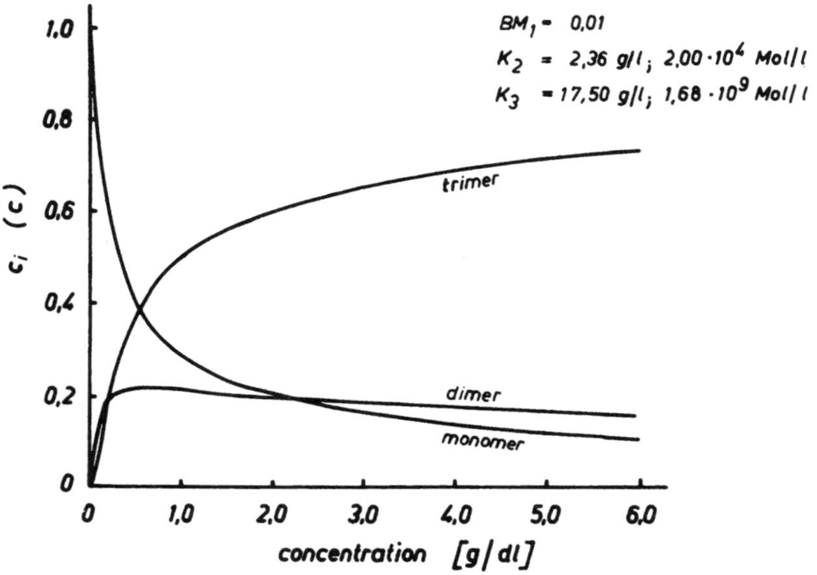

Fig. 7. Molfraktionen des monomeren, dimeren und trimeren Neunaugen-HbCO entsprechend Abb. 6.

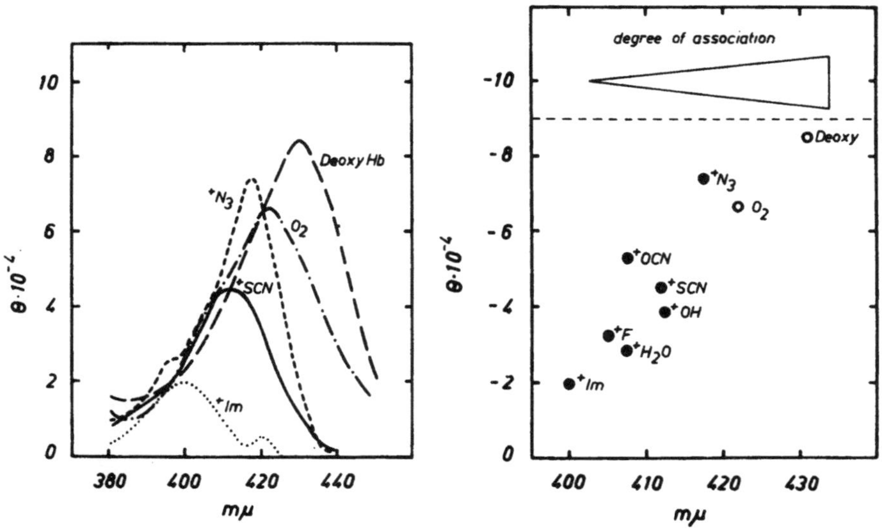

Fig. 8. Dichroistische Absorptionsbanden des Neunaugen-Hb und einiger Komplexe (links) bzw. Abhängigkeit der Elliptizität von der dichroistischen Absorptionsbande und dem Grad der Assoziation.
Hb-Konzentrationen: 2,1 - 5,0 [g/l], Puffer: 50 mM-NaCl-Phosphat, pH 5,5 - 8,7.

Physiologisch-chemisches Institut der Martin-Luther-Universität Halle/S.

R. KLEINE, J. SCHUBERT

PROPERTIES OF A HIGH MOLECULAR ARYLAMIDASE-PHOSPHATASE-COMPLEX FROM RAT KIDNEY MICROSOMES AND ATTEMPTS FOR THE SEPARATION OF BOTH ENZYME ACTIVITIES

The present paper deals with the characterization of an high molecular weight enzyme complex derived from microsomal and endothelial membrane materials. It consists of two different enzyme species, an alkaline phosphatase and a proteolytic enzyme. The latter enzyme belongs to a group of aminopeptidases, named arylamidases whose existence was first shown in 1962. Arylamidases can differentiated from the other peptidases by means of their ability to liberate aromatic amines, e. g. ß-naphthylamide, from amino acid arylamides and by means of their behaviour against metal ions.

Extremely rich in arylamidase activity proved to be the kidney (1). One of the best examined arylamidases, arylamidase N, is localized predominantly in microsomes (2) which also contain most of the alkaline phosphatase (3) and carboxylic esterases (4).

The enzyme complex was detected in rat kidney microsomes during purification of carboxylic esterases (5). For this reason the microsomes were carefully treated with the non ionic detergent Triton-X-100 and the extract was subsequently given on a Sepharose 4B-column. The results of such a separating experiment are shown in Fig. 1.

As it can be seen a sharp protein peak emerges with the void volume of the column, suggesting a high-molecular behaviour. Detailed studies about his catalytic properties revealed as complete absence of esterolytic and proteolytic activities with exception of arylamidase and alkaline phosphatase. Furthermore these two enzyme activities could' nt detect in the following fractions. Besides its behaviour on Sepharose column, its high molecular weight character also manifests in the inability to immigrate agarose and low cross-linked polyacrylamide gels. Since the molecular weights of the up to date known isolated kidney arylamidases and phosphatases lies between 136. 000 (6) and 280. 000 (7) and

150. 000 to 170. 000 (8) daltons respectively an association of the two enzyme moieties to inert membrane protein material was conceivable. In this direction only one case of an associate occurence of arylamidase and alkaline phosphatase due to be to the existence of sialic acid was reported in the literature for chicken serum (9).

Furthermore, since glycoproteids besides lipoproteids are essential constituents of membraneous proteins (10) and most of the as yet characterized phosphatases represents sialo proteins (11) experiments for the detection of carbohydrates in the present enzyme complex were undertaken.

The results are summarized in the Fig. 1. The main constituent are the protein-bound hexoses followed by uronic acids. The complete absence of sialic acid was the most remarkable result.

During the isolation procedure the arylamidase was enriched nearly 50 fold, whereas the purification factor for phosphatase was to be found in the range of 10.

Substrate specificity and kinetical parameters

Fig. 2 represents the substrate specificity of the arylamidase against 16 different amino acid-ß-naphthylamides.

In accordance with the substrate spectrum of the arylamidase solubilized by an autolysic step (12) L-alanyl-β-naphthyl-amide (ANA) was the optimal substrate, whereas L-arginyl-and L-aspartyl-β-naphthylamide, the typical substrates of

Table 1

KINETIC PARAMETERS OF THE ENZYME COMPLEX AT 25 °C

	ANA	L-pNA	p-NPP
pH-Opt.	7, 5	7, 5	9, 5
K_m	0, 14x10^{-4}m	7, 2x10^{-4}m	2, 8x10^{-4}m
V_{max}	1, 0 μMxmin^{-1}	7, 1 μMxmin^{-1}	2, 7 μMxmin^{-1}
V_{max}/K_m	1, 5xmin^{-1}	0, 98xmin^{-1}	0, 96xmin^{-1}

arylamidase B and A respectively, proved to be poor substrates.. Furthermore the present arylamidase possesses a significant activity against alanyl-, leucyldi- and -tripeptides. On the other hand endopeptidases and amino acid esterase activities were not detectable.

Phosphatase activity was determined with the optimal substrate p-nitrophenyl phosphate (p-NPP). For the detection of phosphatase action in agarose and polyacrylamide gels served the chromogenic substrate α-naphthyl phosphate. Some kinetical parameters of the complex against the arylamidase substrates ANA and L-Leucine-p-nitroanilide (L-pNA) and the phosphatase substrate p-NPP such as K_m and V'_{max} values are given in Table 1.

Influence of effectors

To clarify and differentiate both activities of the complex the influence of typical activator and inhibitor substances was tested. An excerpt of our results is given in Fig. 3.

Some of the typical phosphatase inhibitors used display the expected effect, such as EDTA, ortho phosphate an Cu^{++}- ions. Also, the slight activation effect of Mg^{++}-and Zn^{++}-ions is a characteristic feature for phosphatases. On the other hand the conspicuous puromycin sensitivity of this enzyme was unexpected and refer to an arylamidase-like property. With the exception of Zn^{++}- and Ni^{++}-ions (inhibit arylamidase only) all the other metal ions and chelating agents tested displayed an equal effect on both enzyme activities.

A different action showed N-ethylmaleimide (NEM); in contrast to arylamidase which is a SH-independent enzyme (2;7) phosphatase was enhanced by NEM maximal 3-fold. The estimation of the activator constant by means of the DIXON-WEBB-plot (13) reveals a relative low value of $1 \cdot 10^{-6}$ Mole/l (Fig. 4).

Reciprocal influence of the substrates on both enzyme activities

Furthermore the possible influence of an arylamidase substrate on phosphatase action and vice versa was studied. Unlike the arylamidase whose kinetical parameters were uneffected by α-naphthyl phosphate the p-nitrophenyl phosphate hydrolysis endures in the presence of L-leucine-β-naphthyl-amide (LNA) a relevant retardation of their initial phase. In mixtures with small LNA additions this lag phase was overcome after reaching the terminal phase of the phosphatase timeturnover-curve. A possible reason for the transient nature of this inhibitor action might be the hydrolysis of LNA because neither leucine nor β-naphthylamine proved to be phosphatase inhibitors. There exists a direct relation between LNA concentration and degree of phosphatase inhibition. The type of inhibition proved to be competitive as it is shown by the methods of LINEWEAVER and BURK (14) and HUNTER and DOWNS (15) (compare in this contect Fig. 5).

Separation of both enzyme activities

Finally attempts for the careful dissociation and subsequent separation of both enzyme activities were undertaken.

As mentioned above the high molecular weight character of the complex is deduceable from his attachment to inert glycoprotein material, which is sparely broken down by the use of the non-ionic detergent Triton-X-100 (19).

As it is known since the investigations performed mainly by BINKLEY (16), MATHESON (17) and PFLEIDERER (18) solubilization of the particulate aminopeptidases is reached by means of proteolytic digestion. A disadvantage of these auto- or heterolytic procedures consists in their unspecificity as other sensitive bonds in these crude extracts were hydrolyzed too. As the enzyme complex in question is delivered from other hydrolytic impurities incubation with several enzymes were performed to reach a dissociation of the two enzyme activities of the complex. Also, the same aim was tried by incubation the complex at different H^+ concentrations without any enzyme addition. The criterion for a successful degradation was the rate of migration into 5 per cent polyacrylamide gel during electrophoresis. The results are summarized in Fig. 6.

Trypsin, chymotrypsin, pronase, and with a certain degree lipase cause a selective solubilization of two protein fractions derived from the complex with different electrophoretic mobility. Whereas the lower migrating zone represents phosphatase activity the faster migrating one shows arylamidase activity only.

Incubation of the complex with papain, neuraminidase or hyaluronidase at weak acid conditions also yields in the appearance of two protein bonds. In contrast to the above mentioned experiments arylamidase activity was detected also in the phosphatase containing zone. After lipase-treatment a faint positive arylamidase reaction was observed in the phosphatase band too.

Since the same effect of the emergence of two arylamidase bands was reached in the experiments without enzyme addition in the pH-range of 6, 0 to 4, 5 it is conceivable that the above described results with the three acid hydrolases might be due to a spontane dissociation phenomenon of the enzyme complex in acid media. The optimal pH of the nonenzymatic release of the two enzymes is pH 5.0.

The determination of the molecular weights of the two liberated enzymes after the method of HEDRICK and SMITH (20) resulted in a approximated value of 170.000 and 180.000 for arylamidase and alkaline phosphatase respectively (Fig. 7).

These values correspond with the molecular weights communicated for both enzymes, isolated by auto- or heterolysis and butanol treatment respectively.

To study the interactions between these two enzymes in more detail their complete separation is an indispensable condition. To reach this task we submit the tryptic digest of the complex several gel filtration experiments. Apart from a complete separation of the undegraded complex the two liberated smaller enzymes were not separated from each other. They were eluted with an approximated molecular weight of 200.000.

Preliminary experiments suggest however that a sufficient separation of both solubilized enzymes seems possible by means of preparative polyacrylämide gel electrophoresis.

REFERENCES

(1) NAGEL, W., F. WILLIG, Z. Vit.-Horm.-u. Ferm. Forsch. 14, (1965), 89

(2) MAHADEVAN, S., A. L. TAPPEL, J. biol. Chem. 242, (1967), 2369

(3) GRIFFIN, M. J., R. P. COX, Nature 204, (1964), 476

(4) DE DUVE, C., R. WATTIAUX, E. A. BARNARD, Adv. Enzymol. 24, (1962), 291

(5) KLEINE, R., Habilitationsschrift, Halle/S., 1968

(6) KENNY, A. G., S. G. GEORGE, S. G. R. APARICIO, Biochem. J. 115, (1969), 18

(7) WACHSMUTH, E. D., I. FRITZE, G. PFLEIDERER, Biochemistry 5, (1966), 169

(8) BUTTERWORTH, P. J., Biochem. J. 107, (1968), 467

(9) LAW, G. R., Science 156, (1967), 1106

(10) COOK, G. M. W., Biol. Rev. 43, (1968), 363

(11) SAINI, P. K., J. DONE, Febs Letters 7, (1970), 86

(12) HANSON, H., H.-J. HÜTTER, K. MANNSFELDT, C. KRETSCHMER, C. SOHR, Z. physiol. Chem. 348, (1967), 680

(13) DIXON, M., E.C. WEBB, Enzymes, Longmans, Green C o, Ltd, London 1964

(14) LINEWEAVER, H., D. BURK, J. Amer. Chem. Soc. 56, (1934), 658

(15) HUNTER, A., E.C. DOWNS, J. biol. Chem. 157, (1945), 427

(16) BINKLEY, F., J. Amer. Chem. Soc. 82, (1960), 987

(17) MATHESON, A., S. BJERRE, C.S. HANES, Canad, J. Biochem. 41, (1963), 1741

(18) PFLEIDERER, G., P.G. CELLIERS, Biochem. Z. 339, (1963), 186

(19) SINGH, J., A.R. WASSERMANN, Biochim. Biophys. Acta 221, (1970), 379

(20) HEDRICK, J.L., A.J. SMITH, Arch. Biochem. Biophys. 126, (1968), 155

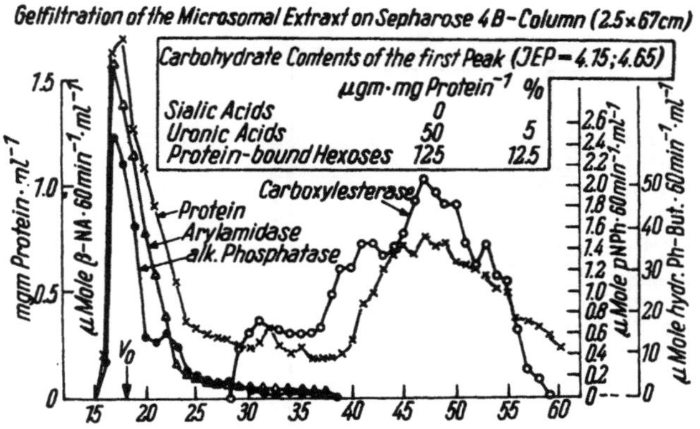

Fig. 1. Gelfiltration of the microsomal extract on Sepharose 4 B-column (2, 5 x 67 cm). 80 mg protein was added to the column. Protein was estimated by the method of LOWRY et al. (J. biol. Chem. 193, (1951) 265). Arylamidase was determined with L-leucyl-β-naphthylamide (LNA), alkaline phosphatase with p-nitrophenyl phosphate (pNPh), and carboxylic esterase with phenyl butyrate (Ph-But) as substrate respectively. The isoelectric points (= IEP) were determined by means of isoelectric focusing with LKB Ampholine in the pH range 3. 0 to 10. 0.

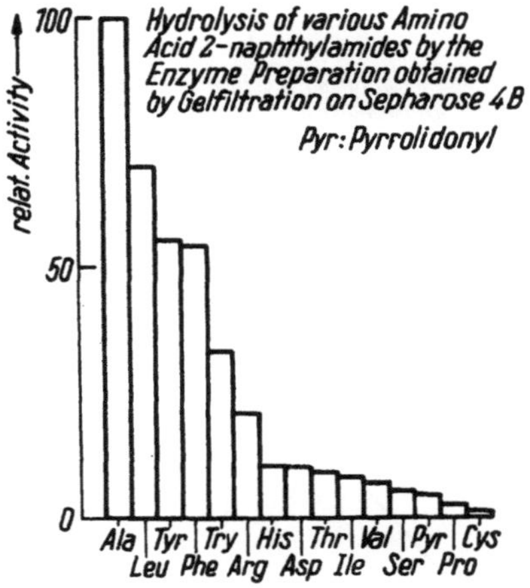

Fig. 2. Hydrolysis of various amino acid-β-naphtylamides by the enzyme complex obtained by gelfiltration on Sepharose 4 B (cf. Fig. 1); L-ala-β-NA hydrolysis is taken as 100.

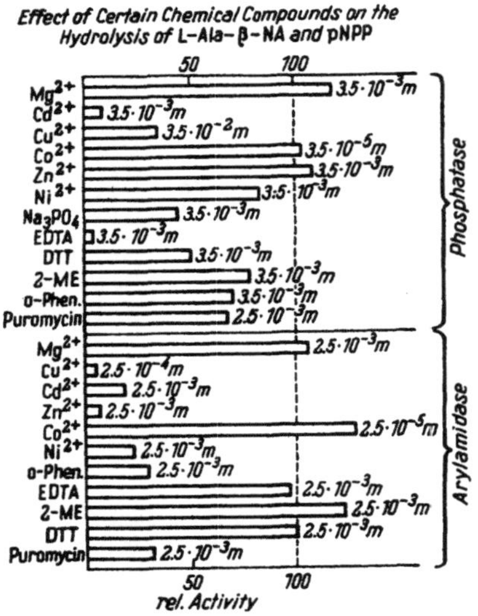

Fig. 3. Effect of various effectors on the hydrolysis of p-nitrophenyl phosphate and L-ala-β-NA. The enzyme preparation was preincubated with the effector 15 min at room temperature and after addition of substrate the activity was measured. Values represent relative activity (hydrolysis in absence of effector = 100).

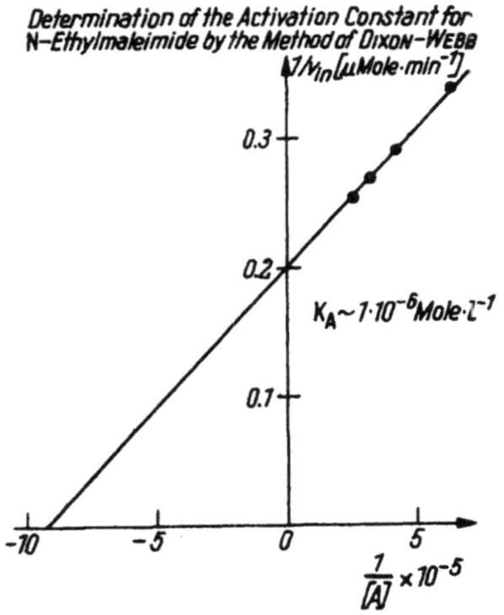

Fig. 4. Determination of the activator constant for N-ethylmaleimide TNEM) by the method of DIXON and WEBB (13); Substrate: p-NPP.

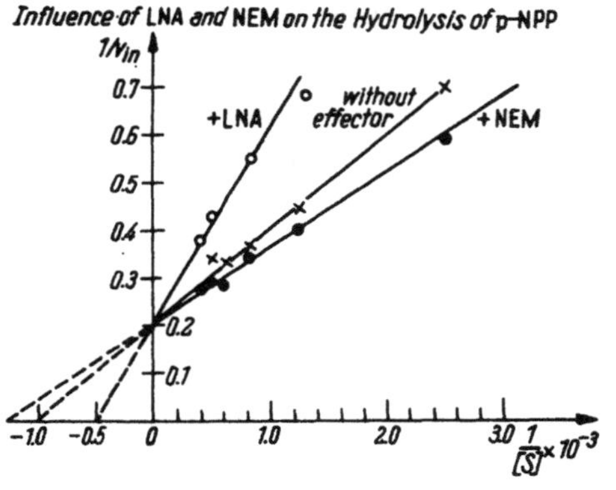

Fig. 5. LINEWEAVER and BURK-plot (14) of the influence of N- ethylmaleimide and L-leucyl-β-naphthylamide on the hydrolysis of p-nitrophenyl phosphate at pH 10, 2.

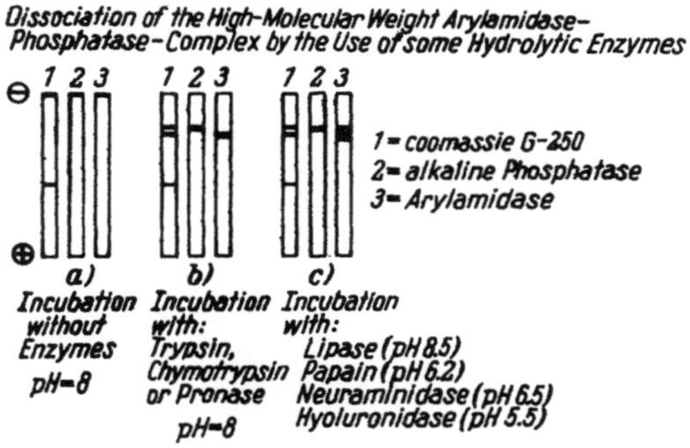

Fig. 6. Dissociation of the high molecular weight arylamidase-phosphatase complex by enzymatic cleavage in presence of various hydrolytic enzymes. The incubation of the enzyme complex with the different enzymes was performed by application of the pH-stat-technique.

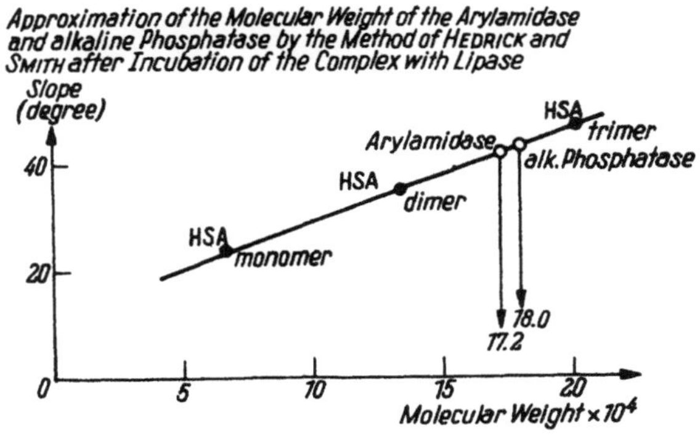

Fig. 7. Approximation of the molecular weight of the arylamidase and alkaline phosphatase respectively by the method of HEDRICK and SMITH (20) after incubation of the complex with lipase.

ACTIVE MEMBERS OF THE SYMPOSIUM

AFANASIEVA, T. P., Bach Institute of Biochemistry of the USSR Academy of Sciences, Moscow, USSR

AFTING, E. G., Biochemisches Institut der Universität Freiburg im Breisgau, BRD

AVAEVA, S. M., Laboratory of Bioorganic Chemistry, Moscow State University, USSR

BAUTZ, E. K. F., Institut für Molekulare Genetik der Universität Heidelberg, BRD

BEHLKE, J., Institut für Pharmakologie und Toxikologie der Ernst-Moritz-Arndt-Universität Greifswald, DDR

BEZBORODOVA, S. I., Institute of Biochemistry and Physiology of Microorganisms; Academy of Sciences of the USSR, Puschino on the Oka, USSR

BÖHME, H. -J., Physiologisch-Chemisches Institut der Karl-Marx-Universität Leipzig, DDR

BÖRNIG, H., Physiologisch-Chemisches Institut der Friedrich-Schiller-Universität Jena, DDR

BOLDYREV, A. A., Department of Animal Biochemistry, Moscow State University, USSR

BONTING, S. L., Department of Biochemistry, University of Nijmegen, The Netherlands

BRAND, K., Max-Planck-Institut für Ernährungsphysiologie Dortmund, BRD

CHETVERIKOVA, E. P., Institute of Biophysics, Puschino on Oka, USSR

CUMME, G., Physiologisch-Chemisches Institut der Friedrich-Schiller-Universität Jena, DDR

DESSEREVSKI, V. I., Institute of Biological Physics; Academy of Sciences of USSR, Puschino on Oka, USSR

DETTMER, D., Physiologisch-Chemisches Institut der Karl-Marx-Universität Leipzig, DDR

DOMAGK, G., Unite de Biochimie Universite de Louvain, Leuven, Belgium

EIFLER, R., Institut für Physiologische und Biologische Chemie der Humboldt-Universität Berlin, DDR

FRUNDER, H., Physiologisch-Chemisches Institut der Friedrich-Schiller-Universität Jena, DDR

GANCEDO, C., Instituto de Enzimologia, Centro de Investigaciones Biologicas, Madrid, Spain

GAUER, J., Physiologisch-Chemisches Institut der Karl-Marx-Universität Leipzig, DDR

GERBER, G., Institut für Physiologische und Biologische Chemie der Humboldt-Universität Berlin, DDR

GRÜNDIG, E., Institut für medizinische Chemie der Universität Wien, Österreich

HAHN, V., Institut für Physiologische und Biologische Chemie der Humboldt-Universität Berlin, DDR

HANSEN, G., Institut für Physiologische und Biologische Chemie der Humboldt-Universität Berlin, DDR

HANSON, H., Physiologisch-Chemisches Institut der Martin-Luther-Universität Halle, DDR

HEITMANN, P., Institut für Physiologische und Biologische Chemie der Humboldt-Universität Berlin, DDR

HELMREICH, E., Institut für Physiologische Chemie der Universität Würzburg, BRD

HESS, B., Max-Planck-Institut für Ernährungsphysiologie Dortmund, BRD

HINTSCHE, P., Forschungszentrum für Molekularbiologie und Medizin der Deutschen Akademie der Wissenschaften zu Berlin, Berlin-Buch, DDR

HOFER, H.W., Fachbereich Biologie der Universität Konstanz, BRD

HOFMANN, E., Physiologisch-Chemisches Institut der Karl-Marx-Universität Leipzig, DDR

HOLY, A., Institute of Organic Chemistry and Biochemistry; Czecholovak Academy of Science, Prague, Czechoslovakia

HOLZER, H., Biochemisches Institut der Universität Freiburg im Breisgau, BRD

HORN, A., Physiologisch-Chemisches Institut der Friedrich-Schiller-Universität Jena, DDR

HORSKA, K., Institute of Organic Chemistry and Biochemistry; Czechoslovak Academy of Sciences, Prague, Czechoslovakia

HULE, V., Department of Clinical Biochemistry, Hospital of Medical Faculty, University J. E. Purkyne, Brno, Czechoslovakia

JENSEN, J., Institute of Physiology, University of Aarhus, Denmark

JUNGNICKEL, F., Sektion Biologie - Pflanzenphysiologie - der Friedrich-Schiller-Universität Jena, DDR

KELETI, T., Institute of Biochemistry, Hungarian Academy of Sciences, Budapest, Hungary

KLEINE, R., Physiologisch-Chemisches Institut der Martin-Luther-Universität Halle, DDR

KONOSHENKO, G.I., Moscow State University, Moscow, USSR

KOPPERSCHLÄGER, G., Physiologisch-Chemisches Institut der Karl-Marx-Universität Leipzig, DDR

KRASAVINA, M. S., Timiryazev Institute of Plant Physiology, USSR Academy of Sciences, Moscow, USSR

KRASHENINNIKOV, I. A., Faculty of Biology and Soil Sciences, Moscow Lomonossov State University, USSR

KRAUSE, E. G., Forschungszentrum für Molekularbiologie und Medizin der Deutschen Akademie der Wissenschaften zu Berlin, Berlin-Buch, DDR

KRITSKY, M. A., Bach Institute of Biochemistry, Academy of Sciences of the USSR, Moscow, USSR

KÜHN, B., Institut für Physiologische und Biologische Chemie der Humboldt-Universität Berlin, DDR

KULAEV, I. S., Faculty of Biology and Soil Sciences, Moscow Lomonossov State University, USSR

LEJKIN, Ju. N., Department of Animal Biochemistry, State University of Moscow, USSR

LIEBE, St., Physiologisch-Chemisches Institut der Karl-Marx-Universität Leipzig, DDR

LIVANOVA, N. V., Institute of Biochemistry, Academy of Sciences of the USSR, Moscov, USSR

LOHMANN, K., Forschungszentrum für Molekularbiologie und Medizin der Deutschen Akademie der Wissenschaften zu Berlin, Berlin-Buch, DDR

LYSLOVA, S. N., Department of Biochemistry, State University, Leningrad, USSR

MANSUROVA, S. E., Faculty of Biology and Soil Sciences, Moscow State University, USSR

MARETZKI, D., Institut für Physiologische und Biologische Chemie der Humboldt-Universität Berlin, DDR

MESHKOVA, N. P., Department of Animal Biochemistry, Moscov State University, USSR

MÜLLER, F., Physiologisch-Chemisches Institut der Karl-Marx-Universität Leipzig, DDR

NESMEJANOVA, M. A., Institute of Biochemistry and Physiology of Microorganisms, USSR Academy of Sciences, Puschino on Oka, USSR

NORBY, J., Institute of Physiology, University of Aarhus, Denmark

OVADI, J., Institute of Biochemistry, Hungarian Academy of Sciences, Budapest, Hungary

PALM, D., Physiologisch-Chemisches Institut der Universität Würzburg, BRD

PANTELEJEVA, N. S., Department of Biochemistry, State University Leningrad, USSR

POLGAR, L., Institute of Biochemistry, Hungarian Academy of Sciences, Budapest, Hungary

RAPOPORT, I., Institut für Physiologische und Biologische Chemie der Humboldt-Universität Berlin, DDR

RAPOPORT, S. M., Institut für Physiologische und Biologische Chemie der Humboldt-Universität Berlin, DDR

RAPOPORT, T., Institut für Physiologische und Biologische Chemie der Humboldt-Universität Berlin, DDR

REICH, J. G., Forschungszentrum für Molekularbiologie und Medizin der Deutschen Akademie der Wissenschaften zu Berlin, Berlin-Buch, DDR

REMKE, H., Physiologisch-Chemisches Institut der Karl-Marx-Universität Leipzig, DDR

REPKE, K. R. H., Forschungszentrum für Molekularbiologie und Medizin der Deutschen Akademie der Wissenschaften zu Berlin, Berlin-Buch, DDR

RICARD, J., Centre Universitaire de Marseille-Luminy, Department de Biologie Biochimie Vegetale, Marseille, France

SEBESTA, K., Institute of Organic Chemistry and Biochemistry, Czechoslovak Academy of Sciences, Prague, Czechoslovakia

SEREBROVSKAJA, K. B., Bach Institute of Biochemistry, Academy of Sciences of the USSR, Moscow, USSR

SEVERIN, S. E., Department of Animal Biochemistry, State University of Moscow, USSR

SIMUTH, J., Biological Institute, Slovak Academy of Sciences, Bratislava, Czechoslovakia

SOKOLOVA, S. V., Timiryazev Institute of Plants Physiology, USSR Academy of Sciences, Moscow, USSR

SOLS, A., Instituto de Enzimologia, Centro de Investigaciones Biologicas, Madrid, Spain

VINOGRADOV, A. D., Department of Animal Biochemistry, Moscow State University, USSR

WENZEL, K. W., Physiologisch-chemisches Institut der Karl-Marx-Universität Leipzig, DDR

WIEKER, H.-J., Max-Planck-Institut für Ernährungsphysiologie Dortmund, BRD

WILL, H., Forschungszentrum für Molekularbiologie und Medizin der Deutschen Akademie der Wissenschaften zu Berlin, Berlin-Buch, DDR

WITZEL, H., Biochemisches Institut der Universität Münster, BRD

WOHLHUETER, R., Biochemisches Institut der Universität Freiburg im Breisgau, BRD

ZELINKOVA, E., Biological Institute of Slovak Academy of Sciences, Bratislava, Czechoslovakia

ZILLIG, W., Max-Planck-Institut für Biochemie, München, BRD

ZIMMERMANN, G., Physiologisch-Chemisches Institut der Karl-Marx-Universität Leipzig, DDR

Bei Fragen zur Produktsicherheit wenden Sie sich bitte an:
If you have any questions regarding product safety,
please contact:

Walter de Gruyter GmbH
Genthiner Straße 13
10785 Berlin
productsafety@degruyterbrill.com